6th International Conference on Physics Opportunities at an Electron-Ion Collider

EPJ Web of Conferences Volume 112 (2016)

Palaiseau, France
7 – 11 September 2015

Editors:

C. Marquet
B. Pire
F. Sabatie

ISBN: 978-1-5108-2206-1

Printed from e-media with permission by:

Curran Associates, Inc.
57 Morehouse Lane
Red Hook, NY 12571

Some format issues inherent in the e-media version may also appear in this print version.

Printed by Curran Associates, Inc. (2016)

For additional information, please contact EDP Sciences – Web of Conferences at the address below.

EDP Sciences – Web of Conferences
17, Avenue du Hoggar
Parc d'Activité de Courtabœuf
BP 112
F-91944 Les Ulis Cedex A
France

Phone: +33 (0) 1 69 18 75 75
Fax: +33 (0) 1 69 28 84 91

contact-edps@webofconferences.org

Additional copies of this publication are available from:

Curran Associates, Inc.
57 Morehouse Lane
Red Hook, NY 12571 USA
Phone: 845-758-0400
Fax: 845-758-2633
Email: curran@proceedings.com
Web: www.proceedings.com

TABLE OF CONTENTS

POETIC VI

September 7th-11th 2015
Palaiseau, France

6th International Conference on Physics Opportunities at an Electron-Ion Collider

Sixth International Conference on Physics Opportunities at an Electron-Ion Collider

September 7-11, 2015

Ecole Polytechnique, Palaiseau, France

http://poetic6.sciencesconf.org

The International Advisory Committee:

Nestor Armesto	(Santiago de Compostela, Spain),
Elke Aschenauer	(Upton, United States),
Daniel Boer	(Groningen, Netherlands),
Marco Contalbrigo	(Ferrara, Italy),
Markus Diehl	(Hamburg, Germany),
Rolf Ent	(Newport News, United States),
Max Klein	(Liverpool, United Kingdom),
Andrzej Sandacz	(Warsaw, Poland),
Marco Stratmann	(Tuebingen, Germany),
Lech Szymanowski	(Warsaw, Poland),
Tony Thomas	(Adelaide, Australie),
Thomas Ullrich	(New Haven, United States),
Raju Venugopalan	(Upton, United States).

Sponsors:

List of Participants

Last Name	First Name	Country	Affiliation
Agadjanov	Andria	Germany	University of Bonn
Albacete	Javier	Spain	Universidad de Granada
Altinoluk	Tolga	Spain	Universidade de Santiago de Compostela
Anikin	Igor	Russia	JINR
Anselmino	Mauro	Italy	INFN Torino
Arleo	François	France	LLR - Ecole polytechnique
Armesto	Nestor	Spain	Universidade de Santiago de Compostela
Aschenauer	Elke-Caroline	USA	Brookhaven National Laboratory
Avakian	Harut	USA	Jefferson Lab
Balitsky	Ian	USA	Jlab / OPDU
Ball	Jacques	France	SPhN - CEA Saclay
Benali	Meriem	France	LPC Clermont Ferrand
Boer	Daniel	The Netherlands	University of Groningen
Boer	Marie	FRANCE	IPN - Orsay
Boussarie	Renaud	France	LPT - Orsay
Braun	Vladimir	Germany	University of Regensburg
Bressan	Andrea	Italy	INFN Trieste
Camsonne	Alexandre	USA	Jefferson Lab
Chhabra	Rahul	India	NIT Jalandhar
Chirilli	Giovanni Antonio	USA	The Ohio State University
Contalbrigo	Marco	Italy	INFN Ferrara
Cosyn	Wim	Belgium	Ghent University
d'Hose	Nicole	France	SPhN - CEA Saclay
De Gruttola	Daniele	France	Centro Fermi
Defurne	Maxime	FRANCE	SPhN - CEA Saclay
Deshpande	Abhay	USA	Stony Brook University
Desnault	Camille	France	IPN - Orsay
Diehl	Markus	Germany	DESY
Ducloue	Bertrand	Finland	University of Jyvaskyla
Dumitru	Adrian	USA	Baruch College (CUNY)
Dupré	Raphaël	France	IPN - Orsay
Engelhardt	Michael	USA	New Mexico State University
Ent	Rolf	USA	Jefferson Lab
Fuchey	Eric	France	SPhN - CEA Saclay
Garçon	Michel	France	SPhN - CEA Saclay
Gay Ducati	Maria Beatriz	Brasil	UFRGS
Gelis	Francois	France	IPhT- CEA Saclay
Goldstein	Gary	USA	Tufts University
Golec-Biernat	Krzysztof	Poland	IFJ Krakow
Goncalves	Victor	Brasil	Universidade Federal de Pelotas
Gorshteyn	Mikhail	Germany	Johannes Gutenberg Universität

Goutte	Hèloïse	France	SPhN - CEA Saclay
Grabovsky	Andrey	Russia	Novosibirsk State University
Guerrero Rodriguez	Pablo	Spain	Universidad de Granada
Guidal	Michel	France	IPN - Orsay
Gwenlan	Claire	UK	Oxford
Hafidi	Kawtar	USA	Argonne National Lab
Hatta	Yoshitaka	Japan	Yukawa Institute for Theoretical Physics
Hyde	Charles	USA	Old Dominion University
Iancu	Edmond	FRANCE	IPhT - CEA Saclay
Jalilian-Marian	Jamal	USA	Baruch College (CUNY)
Joo	Kyungseon	USA	University of Connecticut
Kiselev	Alexander	USA	Brookhaven National Laboratory
Kivel	Nikolay	Germany	Helmholtz-Institut Mainz
Klein	Max	United Kingdom	University of Liverpool
Kopeliovich	Boris	Chile	Universidad Tecnica Federico Santa Maria
Kovchegov	Yuri	USA	The Ohio State University
Kovner	Alex	USA	University of Connecticut
Kumano	Shunzo	Japan	KEK
Kumar	Krishna	USA	Stony Brook University
Kumericki	Kresimir	Croatia	University of Zagreb
Kusina	Aleksander	France	LPSC Grenoble
Levillain	Maxime	France	SPhN - CEA Saclay
Litvinenko	Vladimir	USA	Stony Brook University
Liuti	Simonetta	United States	University of Virginia
Lorce	Cedric	Belgique	Université de Liège
Lyonnet	Florian	USA	SMU
Madrigal	José Daniel	France	IPhT - CEA Saclay
Marquet	Cyrille	France	CPHT - Ecole Polytechnique
Matas	Marek	Czech Republic	Technical University Prague
McNulty	Ronan	Ireland	University College Dublin
Mezrag	Cédric	France	SPhN - CEA Saclay
Milhano	José Guilherme	Portugal	IST Lisboa
Mocsy	Agnes	USA	Pratt Institute
Moutarde	Hervé	France	SPhN - CEA Saclay
Mueller	Alfred	United States	Columbia University
Mueller	Dieter	Germany	University of Cape Town
Mukherjee	Maitreyee	India	Variable Energy Cyclotron Centre
Mulders	Piet	Netherlands	Nikhef
Munier	Stéphane	France	CPHT - Ecole Polytechnique
Munoz Camacho	Carlos	France	IPN - Orsay
Navarra	Fernando	Brazil	University of Sao Paulo
Newman	Paul	United Kingdom	University of Birmingham
Nielsen	Marina	Brazil	University of Sao Paulo
Ollitrault	Jean-Yves	France	IPhT - CEA Saclay

Olness	Fredrick	United States	SMU
Park	Kijun	USA	Old Dominion University
Pasquini	Barbara	Italy	University of Pavia
Peigné	Stéphane	France	Subatech
Peschanski	Robi	france	IPhT - CEA Saclay
Petreska	Elena	France	CPHT - Ecole Polytechnique
Petti	Richard	USA	Brookhaven National Laboratory
Pham	Tri Nang	France	CPHT - Ecole Polytechnique
Pilat	Fulvia	USA	Jefferson Lab
Pire	Bernard	France	CPHT - Ecole Polytechnique
Platchkov	Stephane	France	SPhN - CEA Saclay
Rai	Gulshan	USA	U.S. Department of Energy
Rezaeian	Amir	Chile	Universidad Tecnica Federico Santa Maria
Sabatie	Franck	France	SPhN - CEA Saclay
Sapeta	Sebastian	Switzerland	CERN
Schienbein	Ingo	France	LPSC Grenoble
Scimemi	Ignazio	Spain	Universidad Complutense Madrid
Semenov-Tyan-Shansky	Kirill	France	CPHT - Ecole Polytechnique
Sivers	Dennis	USA	Portland Physics Institute
Soffer	Jacques	France	Temple University
Soto Ontoso	Alba	Spain	Universidade de Granada
Soyez	Gregory	France	IPhT - CEA Saclay
Spiesberger	Hubert	Germany	Johannes-Gutenberg-Universität
Steffens	Fernanda	Germany	DESY - Zeuthen
Stolarski	Marcin	Portugal	LIP - Lisboa
Strikman	Mark	USA	Penn State University
Szymanowski	Lech	Poland	NCBJ Warsaw
Taliotis	Anastasios	Belgium	Vrije Univeriteit Brussel
Tomasi-Gustafsson	Egle	France	SPhN - CEA Saclay
Ullrich	Thomas	USA	Brookhaven National Laboratory
Van Hulse	Charlotte	Spain	University of the Basque Country
van Leeuwen	Marco	Netherlands	Nikhef
Vandenbroucke	Maxence	France	SPhN - CEA Saclay
Vega	Alfredo	Chile	Universidad de Valparaiso
Venugopalan	Raju	Germany	Brookhaven National Laboratory
Wagner	Jakub	Poland	NCBJ Warsaw
Wallon	Samuel	France	LPT Orsay
Wang	Yu-Ming	Germany	Technische Universität München
Weigert	Heribert	Germany	University of Cape Town
Weiss	Christian	USA	Jefferson Lab
Wilfert	Malte	Germany	Universität Mainz
Wu	Bin	France	IPhT - CEA Saclay
Zhou	Jian	Netherlands	Nikhef

6th International conference on Physics Opportunities at Electron-Ion collider
POETIC VI
7-11 September 2015
École Polytechnique, Palaiseau, France

EPJ Web of Conferences **112**, 00001 (2016)
DOI: 10.1051/epjconf/201611200001

POETIC VI

September 7th-11th 2015
Palaiseau, France

6th International Conference on Physics Opportunities at an Electron-Ion Collider

Sixth International Conference on Physics Opportunities at an Electron-Ion Collider

September 7-11, 2015

Ecole Polytechnique, Palaiseau, France

http://poetic6.sciencesconf.org

Following earlier workshops at Stellenbosch, Bloomington, Valparaiso, Jyvaskyla and Yale, the sixth edition of the International Conference on Physics Opportunities at an Electron-Ion Collider (POETIC) took place in École Polytechnique (Palaiseau, France) from September 7th to 11th 2015. Organized by the Centre de Physique Théorique (CPHT) of École Polytechnique and the Service de Physique Nucléaire (SPhN) of the Commissariat à l'Énergie Atomique (Saclay), and sponsored by the CNRS, the P2IO labex, Brookhaven Lab (New York) and Jefferson Lab (Virginia), the conference gathered 125 scientists from all over the world.

An electron-ion collider (EIC) can be thought of as a microscope to study gluons, those particles that bind quarks together, ensuring the cohesion of protons and neutrons, as well as that of atomic nuclei. Gluons are responsible for 98% of the visible mass of the universe, yet the mechanisms at play, described by quantum chromodynamics (QCD), remain to be understood. This is why the US nuclear physics community announced that they had unaminously identified an EIC as the number one priority for future construction.

This was made public during the conference opening presentation, and by now has been further endorsed by the National Science Advisory Committee in their October 2015 recommendations to the United States' DOE and NSF. The future EIC project will open a unique and decisive window on the dynamics of gluons, and in particular on confinement, one of QCD's most mysterious feature.

The conference went on with five days of presentations by internationally renowned experts, covering the recent theoretical, phenomenological and experimental developments that advanced the field of EIC physics. Several other relevant topics related to the physics in the CEBAF, RHIC, and LHC experimental programs were also discussed. These presentations, often followed by vivid and passionate discussions, are reported in these conference proceedings. The next edition of POETIC will take place in Temple University (Philadelphia) in the fall 2016.

Cyrille Marquet, Bernard Pire, Franck Sabatié

edp sciences

EPJ Web of Conferences

Statement of Peer review

In submitting conference proceedings to EPJ Web of Conferences, I certify to the Publisher that I adhere to the **Policy on Publishing Integrity** of the journal in order to safeguard good scientific practice in publishing.

1. All articles have been subjected to peer review administered by the proceedings editors.
2. Reviews have been conducted by expert referees, who have been requested to provide unbiased and constructive comments aimed, whenever possible, at improving the work.
3. Proceedings editors have taken all reasonable steps to ensure the quality of the materials they publish and their decision to accept or reject a paper for publication has been based only on the merits of the work and the relevance to the journal

Title, date and place of the conference

6th International Conference on Physics Opportunities at an Electron-Ion Collider
7-11 September 2015
École Polytechnique, Palaiseau, France

Proceedings editor(s):

Cyrille MARQUET
Bernard PIRE
Franck SABATIE

Date and editor's signature

05 - 02 - 2016

17, avenue du Hoggar - PA de Courtabœuf – BP 112 - 91944 Les Ulis Cedex A (France)
Tél. 33 (0)1 69 18 75 75 - Fax 33 (0)1 69 07 45 17 – www.edpsciences.org

EPJ Web of Conferences **112**, 01001 (2016)
DOI: 10.1051/epjconf/201611201001

Resonance matrix elements on the lattice

Andria Agadjanov[1,a], Véronique Bernard[2,b], Ulf-G. Meißner[1,3,c], and Akaki Rusetsky[1,d]

[1]*Helmholtz-Institut für Strahlen- und Kernphysik (Theorie), Bethe Center for Theoretical Physics, Universität Bonn, D-53115 Bonn, Germany*
[2]*Institut de Physique Nucléaire, UMR 8608, CNRS, Univ. Paris-Sud, Université Paris-Saclay, F-91406 Orsay Cedex, France*
[3]*Institute for Advanced Simulation (IAS-4), Institut für Kernphysik (IKP-3) and Jülich Center for Hadron Physics, Forschungszentrum Jülich, D-52425 Jülich, Germany*

Abstract. The notion of resonance matrix elements is discussed. The analogue of the Lellouch-Lüscher formula for the pion photo(electro)production amplitude in the $\Delta(1232)$ resonance region is given. A prescription for the pole extraction of the $\Delta N\gamma^*$ transition form factors is provided. Theoretical prospects on the lattice calculation of $B \to K^*(892)l^+l^-$ form factors are briefly summarized.

1 Introduction

Recently, the $\Delta N\gamma^*$ transition form factors have been calculated on the lattice [1, 2]. The knowledge on these form factors, in particular, is important to calculate the quark transition charge density, which could be visualized [3]. However, due to large values of the quark masses in these simulations, the Δ is treated as a stable particle. Consequently, the standard well-known techniques of lattice QCD are applicable. This does not hold anymore for such values of the pion mass, where the Δ resonance eventually decays.

The resolution of the problem is given by the Lellouch-Lüscher method [4], which relates the electroweak decay amplitude to the finite-volume matrix element of the weak Hamiltonian. The method properly takes into account the different normalizations of the matrix elements in the infinite and finite volume.

Moreover, a meaningful *definition* should be given to the matrix elements involving resonances, such as the Δ or ρ. It is well known that resonances manifest themselves as complex poles of the scattering T-matrix on unphysical Riemann sheets. It appears natural to introduce the resonance matrix elements in a similar manner [5, 6]. Such a definition involves an analytic continuation of the matrix element to the resonance pole position. Thus, the corresponding form factor could be only defined at the resonance pole. Other possible definitions would be model- or process-dependent.

In the present write-up, we review the results of Ref. [6] on the photo(electro)production $\gamma^*N \to \pi N$ near the $\Delta(1232)$ resonance. We apply systematically a non-relativistic effective field theory in a

[a]e-mail: aagadjanov@hiskp.uni-bonn.de
[b]e-mail: bernard@ipno.in2p3.fr
[c]e-mail: meissner@hiskp.uni-bonn.de
[d]e-mail: rusetsky@hiskp.uni-bonn.de

finite volume in the covariant version [12, 13]. It is shown, along the lines of the Lellouch-Lüscher method, how the corresponding multipole amplitudes can be found from the lattice data. We discuss also the extraction of the $\Delta N\gamma^*$ transition form factors at the resonance pole. At the end, we give a brief outlook.

2 The $\Delta N\gamma^*$ transition on the lattice

2.1 Infinite volume

The $\Delta N\gamma^*$ transition is governed by the three-point function of the form

$$F^{\mu\rho}(P,Q) = i^2 \int d^4X d^4Y e^{iPX-iQY} \langle 0|T O^\mu(X) J^\rho(0)\bar{\psi}(Y)|0\rangle, \tag{1}$$

where $O^\mu(X)$ and $\psi(Y)$ denote the Δ and nucleon local field operators, respectively, and J^ρ is the electromagnetic current. By separating the contribution to this three-point function near a double pole, we can *define* the current matrix element involving the resonance as

$$\langle P,\lambda|J^\rho(0)|Q,\varepsilon\rangle = \lim_{P^2\to s_R,\, Q^2\to m_N^2} \frac{Z_R^{-1/2}}{2E_R} \frac{Z_N^{-1/2}}{2m_N} (s_R - P^2)(m_N^2 - Q^2)\, \bar{u}_\mu(P,\lambda) F^{\mu\rho}(P,Q) u(Q,\varepsilon)\,. \tag{2}$$

Here, Z_R and Z_N denote the renormalization constants of the Δ and nucleon fields, respectively, s_R is the *complex* Δ resonance pole, and m_N is the nucleon mass. $u_\mu(P,\lambda)$ describes the free spin-3/2 particle with complex mass E_R and spin projections $\lambda = -3/2, -1/2, 1/2, 3/2$. Similarly, $u(Q,\varepsilon)$ is a free Dirac spinor with $\varepsilon = -1/2, 1/2$. Although the resonance can not be associated with any single-particle state in the Fock space of the theory, nevertheless, Eq. (2) gives a meaning to the corresponding resonance matrix element. To the best of our knowledge, this is the only field-theoretical definition possible. Clearly, the definition Eq. (2) is process-independent.

On the other hand, the presence of the resonance is usually associated with the change of the scattering phase shift by the $\pi/2$ degree value. Then, in the narrow width approximation, the imaginary part of the amplitude $\mathcal{A}$ can be related to the corresponding current matrix element F (see Refs. [9, 10]):

$$|\mathrm{Im}\,\mathcal{A}| = \sqrt{\frac{8\pi}{p_A\Gamma}}|F^A|, \tag{3}$$

where all quantities are taken at the *real* Breit-Wigner pole $p = p_A$ and Γ is the resonance width. However, since the non-resonant background is unknown, the extracted form factors will be process-dependent.

In principle, both definitions converge for infinitely narrow resonances [6]. In practice, the result of Ref. [13] which extracts the residues of the photoproduction amplitude at the complex pole positions corresponding to the different baryon resonances from the experimental data, suggests that for some resonances the discrepancies in the results could be sizable.

Below, we provide formulae, which can be used on the real axis, as well as in the complex energy plane. Still, we emphasize again, that only the definition, which implies the analytic continuation, leads to the process-independent values of the resonance form factors.

2.2 Finite volume

The Lorentz decomposition of the $\Delta N\gamma^*$ current matrix element requires three scalar form factors $G_M(t)$, $G_E(t)$, $G_C(t)$, where $t = Q^2$ is a photon four-momenta squared. In order to determine them separately, we choose the following kinematics: we consider the Δ at rest $\mathbf{P} = 0$ and put the 3-momenta of the nucleon $\mathbf{Q}$ along the third axis. Then, after rotation to the Euclidean space, we pick up the following matrix elements

$$\langle 1/2|J^3(0)|1/2\rangle = \frac{E_\Delta - Q^0}{E_\Delta} A\, G_C(t),$$

$$\langle 1/2|J^+(0)|-1/2\rangle = \sqrt{\frac{1}{2}} A(G_M(t) - 3G_E(t)),$$

$$\langle 3/2|J^+(0)|1/2\rangle = \sqrt{\frac{3}{2}} A(G_M(t) + G_E(t)), \tag{4}$$

where $t = (E_\Delta - E_N)^2 - \mathbf{Q}^2$ and

$$J^+(0) = \frac{J^1(0) + iJ^2(0)}{\sqrt{2}}, \qquad A = \frac{E_\Delta + m_N}{2m_N}\sqrt{2E_\Delta(Q^0 - m_N)}. \tag{5}$$

In the following, we denote them as F_i, i=1,2,3.

In order to extract these matrix elements on the lattice, we define the following local field operators,

$$O_{3/2}(X) = \frac{1}{2}(1 + \Sigma_3)\frac{1}{2}(1 + \gamma_4)\frac{1}{\sqrt{2}}(O^1(X) - i\Sigma_3 O^2(X)),$$

$$O_{1/2}(X) = \frac{1}{2}(1 - \Sigma_3)\frac{1}{2}(1 + \gamma_4)\frac{1}{\sqrt{2}}(O^1(X) + i\Sigma_3 O^2(X)),$$

$$\tilde{O}_{1/2}(X) = \frac{1}{2}(1 + \Sigma_3)\frac{1}{2}(1 + \gamma_4)O^3(X),$$

$$\bar{\psi}_{\pm 1/2}(Y) = \bar{\psi}(Y)\frac{1}{2}(1 \pm \Sigma_3)\frac{1}{2}(1 + \gamma_4), \quad \Sigma_3 = \mathrm{diag}(\sigma_3, \sigma_3), \tag{6}$$

which produce the Δ and nucleon particles with a given spin projection, denoted by the lower index on the l.h.s of Eq. (6). We refer the reader to Ref. [6] for more details

The quantities F_i can be measured on the lattice, however, one gets the matrix elements of the current between a given eigenstate of the finite-volume Hamiltonian and a one-nucleon state. The dependence of the energy E_Δ on the volume is not suppressed exponentially (unlike the case of a stable Δ), but only as inverse power of the volume [14]. The same statement holds for the matrix elements F_i.

3 Multipoles and form factors

The volume dependence of the quantities F_i arises, due to the initial-state pion-nucleon interaction. The essence of the Lellouch-Lüscher method is to exactly cancel this dependence on the volume.

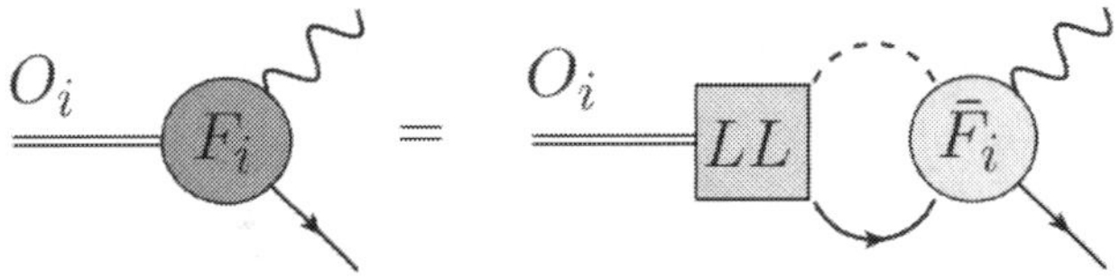

Figure 1. The $\Delta N\gamma^*$ vertex in a finite volume

The diagrammatic representation of the $\Delta N\gamma^*$ vertex within the non-relativistic effective field theory in a finite volume [1] is shown in Fig. 1. The grey box contains all two-particle reducible diagrams, which are summed up to the well-known Lellouch-Lüscher factor. This factor contains all dependence on the volume. The amplitudes $\bar{F}_i$ are volume-independent (up to the contributions exponentially suppressed in volume). Moreover, they can be related to particular $\gamma^* N \to \pi N$ helicity amplitudes, to which the Δ resonance might contribute (see Ref. [6]).

Finally, we summarize our results on the $\Delta N\gamma^*$ transition form factors:

1. The matrix elements $F_i = F_i(p, |\mathbf{Q}|)$ are functions of the relative momenta in the πN system p and $|\mathbf{Q}|$. In order to extract the form factors at the resonance pole (see below), the quantities F_i must be measured at different values of p in the Δ-region, keeping $|\mathbf{Q}|$ fixed. This can be achieved, e.g., by applying asymmetric boxes or the twisting in the nucleon line. The scattering phase shift $\delta(p)$ should be calculated through the Lüscher equation at the same values of p: $\delta(p) + \phi(q) = n\pi$, $n = 0, 1, \ldots$, where $q = pL/2\pi$ and $\phi(q)$ is related to the Lüscher zeta-function.

2. The helicity amplitudes $\mathcal{A}_i$ for pion photo(electro)production in the elastic region at a given discrete momenta p are given by

$$\mathcal{A}_i(p, |\mathbf{Q}|) = e^{i\delta(p)}\, V^{1/2}\Big(\frac{p^2}{2\pi|d\delta(p)/dp + d\phi(q)/dp|}\Big)^{-1/2} |F_i(p, |\mathbf{Q}|)|\,, \tag{7}$$

   where V is a volume. The relation between the helicity and multipole amplitudes can be found in Refs. [6, 9].

3. The $\Delta N\gamma^*$ matrix elements $F_i^A(p_A, |\mathbf{Q}|)$, defined at real energies, are given by Eq. (3), with additional index $i = 1, 2, 3$ attached.

4. The extraction of the $\Delta N\gamma^*$ matrix elements at the resonance pole proceeds as follows. One first calculates the irreducible amplitudes $\bar{F}_i(p, |\mathbf{Q}|)$ from the equation

$$\bar{F}_i(p, |\mathbf{Q}|) = V^{1/2}\Big(\frac{\cos^2\delta(p)}{|d\delta(p)/dp + d\phi(q)/dp|}\frac{p^2}{2\pi}\Big)^{-1/2} F_i(p, |\mathbf{Q}|). \tag{8}$$

5. Since the functions $p^3 \cot\delta(p)\, \bar{F}_i(p, |\mathbf{Q}|)$ are low-energy polynomials in p^2 (see Ref. [6]), they can be fitted by the effective-range expansion to lattice data:

$$p^3 \cot\delta(p)\, \bar{F}_i(p, |\mathbf{Q}|) = A_i(|\mathbf{Q}|) + p^2 B_i(|\mathbf{Q}|) + \cdots . \tag{9}$$

6. The resonance matrix elements at the pole are evaluated by substitution

$$F_i^R(p_R, |\mathbf{Q}|) = i\, p_R^{-3}\, Z_R^{1/2}(A_i(|\mathbf{Q}|) + p_R^2 B_i(|\mathbf{Q}|) + \cdots)\,, \tag{10}$$

[1] For alternative, but equivalent approaches, see, e.g., Refs. [4, 15, 16].

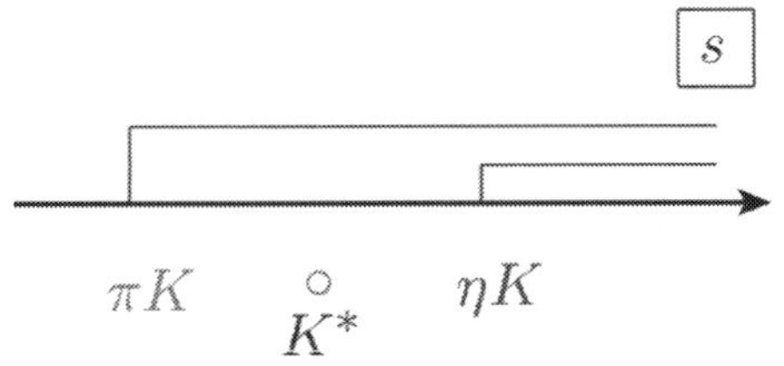

Figure 2. Thresholds near $K^*(892)$ resonance

where p_R is the Δ resonance complex pole position, and Z_R is given by

$$Z_R = \left(\frac{p}{8\pi E}\right)^2 \left(\frac{16\pi p^3 E^3}{w_N w_\pi (2p\, d(p^3 \cot\delta(p))/dp + 3ip^2)}\right)\Bigg|_{p=p_R} . \tag{11}$$

Here, the quantities w_N, w_π denote the on-shell energies of the nucleon and pion, respectively. The pole position p_R must be found separately, e.g., by applying the effective-range expansion:

$$-\frac{1}{a} + \frac{1}{2} r p_R^2 + \cdots = -ip_R^3 , \tag{12}$$

where the constants a, r should be fitted.

7. Finally, the form factors can be derived from Eq. (4).

4 Summary and Outlook

This write-up summarizes our findings on the theoretical framework for the extraction of the $\Delta N\gamma^*$ transition form factors on the lattice. The framework is based on the non-relativistic effective field theory in a finite volume. We have derived the analogue of the Lellouch-Lüscher formula for the photo(electro)production amplitude in the elastic region. The prescription for the extraction of the form factors at the resonance pole has been also given.

The result might be improved in one particular way. It would be important to consider the Δ resonance in the moving frames. Then, one can carry out a similar procedure as given above.

The rare $B \to K^*(892)l^+l^-$ decay is another example of the phenomenologically interesting process. It can be studied along the same lines. However, this problem is more complicated, due to the presence of the second $K\eta$ channel (see Fig. 2). The multichannel generalization of the Lellouch-Lüscher method has been derived in Refs. [17, 18]. On the contrary, the extraction of the form factors at the resonance pole in the multichannel case has not been studied before. Recently, we have filled this gap and considered the $\pi K - \eta K$ coupled-channel system in a finite volume [19]. A similar theoretical framework might be useful for the extraction of the electromagnetic form factors of the $\Lambda(1405)$ resonance on the lattice [20].

Acknowledgments

We acknowledge the support by the DFG (CRC 16, "Subnuclear Structure of Matter") and by the Bonn-Cologne Graduate School of Physics and Astronomy. This research is supported in part by Volkswagenstiftung under contract no. 86260.

References

[1] C. Alexandrou, G. Koutsou, J. W. Negele, Y. Proestos and A. Tsapalis, Phys. Rev. D **83** (2011) 014501
[2] C. Alexandrou, G. Koutsou, H. Neff, J. W. Negele, W. Schroers and A. Tsapalis, Phys. Rev. D **77** (2008) 085012
[3] C. E. Carlson and M. Vanderhaeghen, Phys. Rev. Lett. **100** (2008) 032004
[4] L. Lellouch and M. Lüscher, Commun. Math. Phys. **219** (2001) 31
[5] V. Bernard, D. Hoja, U.-G. Meißner and A. Rusetsky, JHEP **1209** (2012) 023
[6] A. Agadjanov, V. Bernard, U.-G. Meißner and A. Rusetsky, Nucl. Phys. B **886** (2014) 1199
[7] M. T. Hansen and S. R. Sharpe, Phys. Rev. D **86** (2012) 016007.
[8] R. A. Briceno, M. T. Hansen and A. Walker-Loud, arXiv:1406.5965 [hep-lat] (2014).
[9] I. G. Aznauryan, V. D. Burkert and T. -S. H. Lee, arXiv:0810.0997 [nucl-th].
[10] D. Drechsel, O. Hanstein, S. S. Kamalov and L. Tiator, Nucl. Phys. A **645** (1999) 145
[11] R. L. Workman, L. Tiator and A. Sarantsev, Phys. Rev. C **87** (2013) 6, 068201
[12] G. Colangelo, J. Gasser, B. Kubis and A. Rusetsky, Phys. Lett. B **638** (2006) 187
[13] J. Gasser, B. Kubis and A. Rusetsky, Nucl. Phys. B **850** (2011) 96
[14] M. Lüscher, Nucl. Phys. B **354** (1991) 531
[15] C. H. Kim, C. T. Sachrajda and S. R. Sharpe, Nucl. Phys. B **727** (2005) 218
[16] N. H. Christ, C. Kim and T. Yamazaki, Phys. Rev. D **72** (2005) 114506
[17] M. T. Hansen and S. R. Sharpe, Phys. Rev. D **86** (2012) 016007
[18] R. A. Briceño, M. T. Hansen and A. Walker-Loud, Phys. Rev. D **91** (2015) 3, 034501
[19] A. Agadjanov, V. Bernard, U.-G. Meißner and A. Rusetsky, in preparation.
[20] B. J. Menadue, W. Kamleh, D. B. Leinweber, M. Selim Mahbub and B. J. Owen, PoS LATTICE **2013** (2014) 280

EPJ Web of Conferences **112**, 01002 (2016)
DOI: 10.1051/epjconf/201611201002

Factorization theorem, gluon poles and new contributions in semi-inclusive processes

I. V. Anikin[1,a] and O. V. Teryaev[1,b]

[1] *Bogoliubov Laboratory of Theoretical Physics, JINR, 141980 Dubna, Russia*

Abstract. We discuss the role of gluon poles and the gauge invariance for the hadron tensors of Drell-Yan and direct photon production processes with the transversely polarized hadron.

1 Introduction

We discuss the direct photon production in two hadron collision where one hadron is transversely polarized. We present the hadron tensor for this process and study the effects which lead to the soft breaking of factorization (or the universality breaking) through the QED and QCD gauge invariance [1]. As in [2], the special role is played by the contour gauge for gluon fields. We demonstrate that the prescriptions for the gluonic poles in the twist 3 correlators are dictated by the prescriptions in the corresponding hard parts. We argue that the prescriptions in the gluonic pole contributions differ from each other depending upon the initial or final state interactions of the diagrams under consideration. The different prescriptions are needed to ensure the QCD gauge invariance. We treat this situation as a breaking of the universality condition resulting in factorization soft breaking. The extra diagram contributions, which naively do not have an imaginary phase, is discussed in detail. We also show that the new ("non-standard") terms do contribute to the hadron tensor exactly as the "standard" terms known previously. This is exactly similar to the case of Drell-Yan process studied in [2].

We also present the results of the detailed analysis of hadron tensor in the Feynman gauge with the particular emphasis on the QED gauge invariance [3]. We find that the QED gauge invariance can be maintained only by taking into account the non-standard diagram. Moreover, the results in the Feynman and contour gauges coincide if the gluon poles in the correlators $\langle \bar\psi \gamma_\perp A^+ \psi \rangle$ are absent. This is in agreement with the relation between gluon poles and the Sivers function which corresponds to the "leading twist" Dirac matrix γ^+. We confirm this important property by comparing the light-cone dynamics for different correlators. As a result, we obtain the QED gauge invariant hadron tensor which completely coincides with the expression obtained within the light-cone contour gauge for gluons, see [2].

[a] e-mail: anikin@theor.jinr.ru
[b] e-mail: teryaev@theor.jinr.ru

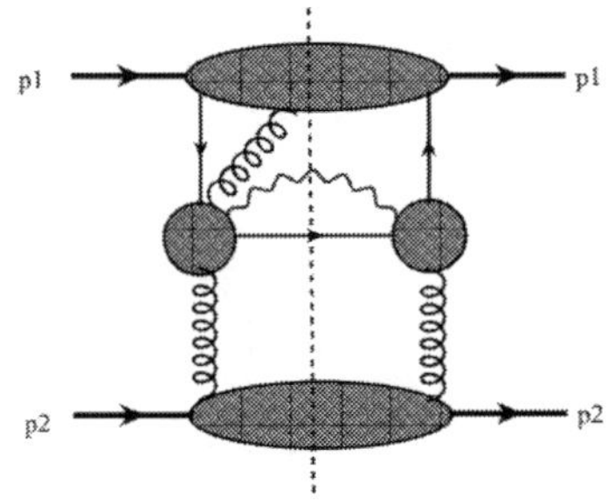

Figure 1. The Feynman diagram describing the hadron tensor of the direct photon production.

2 Direct Photon Production

We study the semi-inclusive process where the hadron with the transverse polarization collides with the other unpolarized hadron to produce the direct photon in the final state in:

$$N^{(\uparrow\downarrow)}(p_1) + N(p_2) \to \gamma(q) + q(k) + X(P_X). \tag{1}$$

For (1) (also for the Drell-Yan process), the gluonic poles manifest. We perform our calculations within a *collinear* factorization with kinematics defined as in [1]. So, we have the hadron tensor related to the corresponding asymmetry:

$$d\sigma^{\uparrow} - d\sigma^{\downarrow} \sim \mathcal{W} = \sum_{i=1}^{2} \sum_{j=1}^{8} \mathcal{A}_i^{\rm LO} * \mathcal{B}_j^{\rm NLO}. \tag{2}$$

Here, we will mainly discuss the hadron tensor rather than the asymmetry itself. So, the hadron tensor as an interference between the LO and NLO diagrams, $\mathcal{A}_i^{\rm LO} * \mathcal{B}_j^{\rm NLO}$, can be presented by Fig. 1.

As a result of factorization, we should reach the factorized form for the considered subject:

$$\text{Hadron tensor} = \{\text{Hard part (pQCD)}\} \otimes \{\text{Soft part (npQCD)}\}. \tag{3}$$

Usually, both the hard and soft parts, see (3), are independent of each other, UV- and IR-renormalizable and, finally, various parton distributions, parametrizing the soft part, have to manifest the universality property. However, the hard and soft parts of the DPP hadron tensor are not fully independent each other [2]. Actually, the DY hadron tensor has formally factorized with the mathematical convolution and the twist-3 function $B^V(x_1, x_2)$ satisfies still the universality condition. In contrast to the DY-process, the DPP tensor will include the functions $B^V(x_1, x_2)$ that will not manifest the universality.

We now dwell on the QCD gauge invariance of the hadron tensor for the direct photon production (DPP). First of all, let us remind that having used the *contour gauge conception* [1, 4], one can check that the representation

$$B_+^V(x_1, x_2) = \frac{T(x_1, x_2)}{x_1 - x_2 + i\epsilon} \tag{4}$$

belongs to the gauge defined by $[x, -\infty] = 1$, while the representation

$$B_-^V(x_1, x_2) = \frac{T(x_1, x_2)}{x_1 - x_2 - i\epsilon} \tag{5}$$

corresponds to the gauge that defined by $[+\infty, x] = 1$. In both (4) and (5), the function $T(x_1, x_2)$ related to the following prametrization:

$$\langle p_1, S^T|\bar{\psi}(\lambda_1\tilde{n})\,\gamma^+\,\tilde{n}_\nu G_T^{\nu\alpha}(\lambda_2\tilde{n})\,\psi(0)|S^T, p_1\rangle = \varepsilon^{\alpha+S^T-}(p_1p_2)\int dx_1dx_2\, e^{ix_1\lambda_1+i(x_2-x_1)\lambda_2}\,T(x_1, x_2)\,. \quad (6)$$

Roughly speaking, it resembles the case where two different vectors have the same projection on the certain direction. In this sense, the usual axil gauge $A^+ = 0$ can be understood as a "projection" which corresponds to two different "vectors" represented by two different contour gauges.

Further, to check the gauge invariance, we have to consider all typical diagrams, depicted in Fig.1, that correspond to the certain ξ-process (see, [5]). Notice that, for the QCD gauge invariance, we have to assume that all charged particles are on its mass-shells. That is, we will deal with only the physical gluons.

As a result of our calculation, the QCD gauge invariance (or the Ward identity) takes place provided only the presence of the different complex prescriptions in gluonic poles dictated by the final or initial state interactions:

$$\left.\begin{array}{l}\text{FSI} \;\Rightarrow\; \frac{1}{\ell^+ + i\epsilon} \Rightarrow \text{gauge } [+\infty^-, z^-] = 1 \Rightarrow \frac{T(x_1,x_2)}{x_1-x_2-i\epsilon} \\ \text{ISI} \;\Rightarrow\; \frac{1}{-\ell^+ + i\epsilon} \Rightarrow \text{gauge } [z^-, -\infty^-] = 1 \Rightarrow \frac{T(x_1,x_2)}{x_1-x_2+i\epsilon}\end{array}\right\} \Rightarrow \text{QCD Gauge Invariance}\,, \quad (7)$$

where FSI and ISI imply the final and initial state interactions, respectively. We emphasize the principle differences between the considered case and the proof of the QCD gauge invariance for the perturbative Compton scattering amplitude with the physical gluons in the initial and final states. The latter does not need any external condition, like the presence of gluon poles.

We now calculate the full expression for the hadron tensor which involves both the standard and new contributions to the gluon pole terms. After computing the corresponding traces and performing simple algebra within the frame we are choosing, it turns out that the only nonzero contributions to the hadron tensor come from the following terms

$$\begin{aligned} d\mathcal{W}(\text{H1}) &= \frac{d^3\vec{q}}{(2\pi)^3 2E}\int\frac{d^3\vec{k}}{(2\pi)^3 2\varepsilon}\delta^{(2)}(\vec{\mathbf{k}}_\perp + \vec{\mathbf{q}}_\perp)\,\mathrm{C}_2\int dx_1dy\delta(x_1 - x_B)\,\delta(y - y_B)\mathcal{F}^g(y)\,\times \\ &\int dx_2\,\frac{2S^2\,x_1\,y^2}{[x_2yS + i\epsilon][x_1yS + i\epsilon]^2}\,\frac{\varepsilon^{q_\perp+S_\perp-}}{p_1^+}\,B_-^V(x_1, x_2)\,, \end{aligned} \quad (8)$$

$$\begin{aligned} d\mathcal{W}(\text{H7}) &= \frac{d^3\vec{q}}{(2\pi)^3 2E}\int\frac{d^3\vec{k}}{(2\pi)^3 2\varepsilon}\delta^{(2)}(\vec{\mathbf{k}}_\perp + \vec{\mathbf{q}}_\perp)\,\mathrm{C}_1\int dx_1dy\delta(x_1 - x_B)\,\delta(y - y_B)\mathcal{F}^g(y)\,\times \\ &\int dx_2\,\frac{(-2)S\,T\,x_1\,(y - 3y_B)}{[x_2T + i\epsilon][x_1T + i\epsilon]^2}\,\frac{\varepsilon^{q_\perp+S_\perp-}}{p_1^+}\,B_+^V(x_1, x_2)\,, \end{aligned} \quad (9)$$

$$\begin{aligned} d\mathcal{W}(\text{D4}) &= \frac{d^3\vec{q}}{(2\pi)^3 2E}\int\frac{d^3\vec{k}}{(2\pi)^3 2\varepsilon}\delta^{(1)}(\vec{\mathbf{k}}_\perp + \vec{\mathbf{q}}_\perp)\,\mathrm{C}_1\int dx_1dy\delta(x_1 - x_B)\,\delta(y - y_B)\,\frac{2}{S}\mathcal{F}^g(y)\,\times \\ &\frac{2S^2\,x_1\,(y - 2y_B)}{[x_1T + i\epsilon]^2}\,\frac{\varepsilon^{q_\perp+S_\perp-}}{2x_1p_1^+ + i\epsilon}\int dx_2\,B_+^V(x_1, x_2)\,, \end{aligned} \quad (10)$$

and

$$\begin{aligned} d\mathcal{W}(\text{H10}) &= \frac{d^3\vec{q}}{(2\pi)^3 2E}\int\frac{d^3\vec{k}}{(2\pi)^3 2\varepsilon}\delta^{(2)}(\vec{\mathbf{k}}_\perp + \vec{\mathbf{q}}_\perp)\,\mathrm{C}_3\int dx_1dy\delta(x_1 - x_B)\,\delta(y - y_B)\mathcal{F}^g(y)\,\times \\ &\int dx_2\,\frac{2T(x_1 - x_2)(2T + Sy)}{[x_1T + i\epsilon][x_2T + i\epsilon][(x_1 - x_2)yS + i\epsilon]}\,\frac{\varepsilon^{q_\perp+S_\perp-}}{p_1^+}\,B_+^V(x_1, x_2)\,. \end{aligned} \quad (11)$$

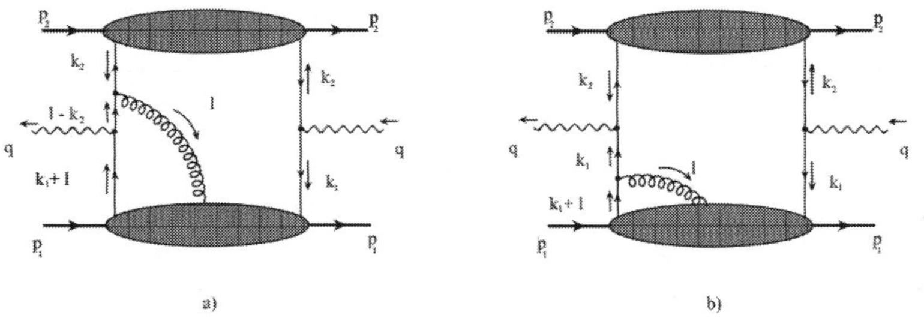

Figure 2. The Feynman diagrams which contribute to the polarized Drell-Yan hadron tensor.

Here, $C_1 = C_F^2 N_c$, $C_2 = -C_F/2$, $C_3 = C_F N_c C_A/2$. The other diagram contributions disappear owing to the following reasons: (i) the γ-algebra gives $(\gamma^-)^2 = 0$; (ii) the common pre-factor $T + yS$ goes to zero, (iii) cancelation between different diagrams.

Analysing the results for the terms H1, H7, D4 and H10 (see Eqns. (8)–(11)), we can see that

$$d\mathcal{W}(\text{H1}) + d\mathcal{W}(\text{H7}) + d\mathcal{W}(\text{D4}) = d\mathcal{W}(\text{H10})\,. \tag{12}$$

In other words, as similar to the Drell-Yan process, the new ("non-standard") contributions generated by the terms H1, H7 and D4 result again in the factor of 2 compared to the "standard" term H10 contribution to the corresponding hadron tensor.

3 Drell-Yan process

We now discuss the hadron tensor which contributes to the single spin (left-right) asymmetry measured in the Drell-Yan process with the transversely polarized nucleon (see Fig. 2):

$$N^{(\uparrow\downarrow)}(p_1) + N(p_2) \to \gamma^*(q) + X(P_X) \to \ell(l_1) + \bar{\ell}(l_2) + X(P_X). \tag{13}$$

Here, the virtual photon producing the lepton pair ($l_1 + l_2 = q$) has a large mass squared ($q^2 = Q^2$) while the transverse momenta are small and integrated out. The left-right asymmetry means that the transverse momenta of the leptons are correlated with the direction $\mathbf{S} \times \mathbf{e}_z$ where S_μ implies the transverse polarization vector of the nucleon while $\mathbf{e}_z$ is a beam direction [6]. We perform our calculations within a *collinear* factorization with kinematics as in [2].

We work within the Feynman gauge for gluons. After calculation of all relevant traces in the factorized hadron tensor and after some algebra, we arrive at the following contributions for the unintegrated hadron tensor (which involves all relevant contributions except the mirror ones): the standard diagram depicted in Fig. 2(a) gives us

$$\overline{\mathcal{W}}^{(\text{Stand.})}_{\mu\nu} + \overline{\mathcal{W}}^{(\text{Stand.},\partial_\perp)}_{\mu\nu} = \bar{q}(y)\Big\{ -\frac{p_{1\mu}}{y}\varepsilon_{\nu S^T - p_2}\int dx_2 \frac{x_1 - x_2}{x_1 - x_2 + i\epsilon} B^{(1)}(x_1, x_2) - \tag{14}$$
$$\Big[\frac{p_{2\nu}}{x_1}\varepsilon_{\mu S^T - p_2} + \frac{p_{2\mu}}{x_1}\varepsilon_{\nu S^T - p_2}\Big] x_1 \int dx_2 \frac{B^{(2)}(x_1, x_2)}{x_1 - x_2 + i\epsilon} + \frac{p_{1\mu}}{y}\varepsilon_{\nu S^T - p_2}\int dx_2 \frac{B^{(\perp)}(x_1, x_2)}{x_1 - x_2 + i\epsilon}\Big\},$$

while the non-standard diagram presented in Fig. 2(b) contributes as

$$\overline{\mathcal{W}}^{(\text{Non-stand.})}_{\mu\nu} = \bar{q}(y)\frac{p_{2\mu}}{x_1}\varepsilon_{\nu S^T - p_2}\int dx_2 \Big\{ B^{(1)}(x_1, x_2) + B^{(2)}(x_1, x_2)\Big\}. \tag{15}$$

In Eqns. (14) and (15), all indices have to be treated as contravariant ones independently of the real position in the formulae. We also introduce the shorthand notation: $\varepsilon_{ABCD} = \varepsilon_{\mu_1\mu_2\mu_3\mu_4} A_{\mu_1} B_{\mu_2} C_{\mu_3} D_{\mu_4}$. Moreover, the parametrizing functions are associated with the following correlators:

$$B^{(1)}(x_1, x_2) = \frac{T(x_1, x_2)}{x_1 - x_2 + i\varepsilon} \Longleftarrow \mathcal{F}_2\Big[\langle p_1, S^T|\bar{\psi}(\eta_1)\gamma^+ A^T(z)\psi(0)|S^T, p_1\rangle\Big], \tag{16}$$

$$B^{(2)}(x_1, x_2) \Longleftarrow \mathcal{F}_2\Big[\langle p_1, S^T|\bar{\psi}(\eta_1)\gamma^\perp A^+(z)\psi(0)|S^T, p_1\rangle\Big], \tag{17}$$

$$B^{(\perp)}(x_1, x_2) \Longleftarrow \mathcal{F}_2\Big[\langle p_1, S^T|\bar{\psi}(\eta_1)\gamma^+ (\partial^\perp A^+(z))\psi(0)|S^T, p_1\rangle\Big], \tag{18}$$

where $\mathcal{F}_2$ denotes the corresponding Fourier transformation. Summing up all contributions from the standard and non-standard diagrams, we finally obtain the expression for the unitegrated hadron tensor. We have

$$\overline{\mathcal{W}}_{\mu\nu} = \overline{\mathcal{W}}^{(\text{Stand.})}_{\mu\nu} + \overline{\mathcal{W}}^{(\text{Stand.}, \partial_\perp)}_{\mu\nu} + \overline{\mathcal{W}}^{(\text{Non-stand.})}_{\mu\nu} = \tag{19}$$

$$\bar{q}(y)\Bigg\{\Big[\frac{p_{2\mu}}{x_1} - \frac{p_{1\mu}}{y}\Big]\varepsilon_{\nu S^T - p_2}\int dx_2 B^{(1)}(x_1, x_2) + \frac{p_{2\mu}}{x_1}\varepsilon_{\nu S^T - p_2}\int dx_2 B^{(2)}(x_1, x_2) -$$

$$\Big[\frac{p_{2\nu}}{x_1}\varepsilon_{\mu S^T - p_2} + \frac{p_{2\mu}}{x_1}\varepsilon_{\nu S^T - p_2}\Big]x_1\int dx_2 \frac{B^{(2)}(x_1, x_2)}{x_1 - x_2 + i\epsilon} + \frac{p_{1\mu}}{y}\varepsilon_{\nu S^T - p_2}\int dx_2 \frac{B^{(\perp)}(x_1, x_2)}{x_1 - x_2 + i\epsilon}\Bigg\},$$

Notice that the first term in Eqn. (19) coincides with the hadron tensor calculated within the light-cone gauge $A^+ = 0$.

Let us now discuss the QED gauge invariance of the hadron tensor. From Eqn. (19), we can see that the QED gauge invariant combination is

$$\mathcal{T}_{\mu\nu} = \big[p_{2\mu}/x_1 - p_{1\mu}/y\big]\varepsilon_{\nu S^T - p_2}, \quad q_\mu \mathcal{T}_{\mu\nu} = q_\nu \mathcal{T}_{\mu\nu} = 0. \tag{20}$$

At the same time, there is a single term with $p_{2\nu}$ which does not have a counterpart to construct the gauge-invariant combination $p_{2\mu}/x_1 - p_{1\mu}/y$. Therefore, the second term in Eqn. (14) should be equal to zero. This also leads to nullification of the second term in Eqn. (15). Now, to get the QED gauge invariant combination (see (20)) one has no the other way rather than to combine the first terms in Eqns. (14) and (15) which justifies the treatment of gluon pole in $B^{(1)}(x_1, x_2)$ using the complex prescription. In addition, we have to conclude that the only remaining third term in (14) should not contribute to SSA.

Moreover, we can conclude that, in the case with the substantial transverse component of the momentum, there are no any sources for the gluon pole at $x_1 = x_2$. As a result, the function $B^{(2)}(x_1, x_2)$ has no gluon poles and, due to T-invariance, $B^{(2)}(x_1, x_2) = -B^{(2)}(x_2, x_1)$, obeys $B^{(2)}(x, x) = 0$. On the other hand, if we have γ^+ in the correlator, see (16), the transverse components of gluon momentum are not substantial and can be neglected. That ensures the existence of the gluon poles for the function $B^{(1)}(x_1, x_2)$. This corresponds to the fact that the Sivers function, being related to gluon poles, contains the "leading twist" projector γ^+. So, we may conclude that in the Feynman gauge the structure $\gamma^+(\partial^\perp A^+)$ does not produce the imaginary part and SSA as well.

Working within the Feynman gauge, we derive the QED gauge invariant (unitegrated) hadron tensor for the polarized DY process:

$$\overline{\mathcal{W}}^{\text{GI}}_{\mu\nu} = \bar{q}(y)\big[p_{2\mu}/x_1 - p_{1\mu}/y\big]\varepsilon_{\nu S^T - p_2}\int dx_2 B^{(1)}(x_1, x_2). \tag{21}$$

Further, after calculation the imaginary part (addition of the mirror contributions), and after integration over x_1 and y, the QED gauge invariant hadron tensor takes the form

$$W^{\text{GI}}_{\mu\nu} = \bar{q}(y_B)\big[p_{2\mu}/x_B - p_{1\mu}/y_B\big]\varepsilon_{\nu S^T - p_2} T(x_B, x_B). \tag{22}$$

This expression fully coincides with the hadron tensor which has been derived within the light-cone gauge for gluons. Moreover, the factor of 2 in the hadron tensor, we found within the axial-type gauge [2], is still present in the frame of the Feynman gauge.

4 Conclusions

We explore the QCD gauge invariance of the hadron tensor for the direct photon production in two hadron collision where one of hadrons is transversely polarized. We argue the effects which lead to the soft breaking of factorization by inspection of the QCD gauge invariance. We demonstrate that the initial or final state interactions in diagrams define the different prescriptions in the gluonic poles. Moreover, the different prescriptions are needed to ensure the QCD gauge invariance. This situation can be treated as a soft breaking of the universality condition resulting in factorization breaking. We find that the "non-standard" new terms, which exist in the case of the complex twist-3 B^V-function with the corresponding prescriptions, do contribute to the hadron tensor exactly as the "standard" term known previously. This is another important result of our work. We also observe that this is exactly similar to the case of Drell-Yan process studied in [2].

We argue the absence of gluon poles in the correlators $\langle\bar{\psi}\gamma_\perp A^+\psi\rangle$ based on the light-cone dynamics. At the same time, we also show that the Lorentz and QED gauge invariance of the hadron tensor calculated within the Feynman gauge requires the function $B^{(2)}(x_1, x_2)$ to be without any gluon poles. This property seems to be natural from the point of view of gluon poles relation [7] to Sivers functions as the latter is related to the projection γ^+. As for the function $B^{(\perp)}(x_1, x_2)$, the transverse derivative of Sivers function resulting from taking its moments, may act on both integrand and boundary value. Our result suggests that only the action on the boundary value, related to $B^{(1)}(x_1, x_2)$ should produce SSA. It is not unnatural bearing in mind that the integrand differentiation is present even for simple straight-line contours which are not producing SSA.

References

[1] I. V. Anikin and O. V. Teryaev, Eur. Phys. J. C **75**, 184 (2015) [arXiv:1501.04536 [hep-ph]].

[2] I. V. Anikin and O. V. Teryaev, Phys. Lett. B **690**, 519 (2010) [arXiv:1003.1482 [hep-ph]]; I. V. Anikin and O. V. Teryaev, J. Phys. Conf. Ser. **295**, 012057 (2011) [arXiv:1011.6203 [hep-ph]].

[3] I. V. Anikin and O. V. Teryaev, Phys. Lett. B **751**, 495 (2015) [arXiv:1508.00227 [hep-ph]].

[4] S. V. Ivanov, G. P. Korchemsky and A. V. Radyushkin, Yad. Fiz. **44**, 230 (1986) [Sov. J. Nucl. Phys. **44**, 145 (1986)]; S. V. Ivanov and G. P. Korchemsky, Phys. Lett. B **154**, 197 (1985); S. V. Ivanov, Fiz. Elem. Chast. Atom. Yadra **21**, 75 (1990).

[5] N. N. Bogolyubov and D. V. Shirkov, "Introduction To The Theory Of Quantized Fields," Intersci. Monogr. Phys. Astron. **3**, 1 (1959).

[6] V. Barone, A. Drago and P. G. Ratcliffe, Phys. Rept. **359**, 1 (2002) [arXiv:hep-ph/0104283].

[7] D. Boer, P. J. Mulders and F. Pijlman, Nucl. Phys. B **667**, 201 (2003) [hep-ph/0303034]. X. Ji, J. W. Qiu, W. Vogelsang and F. Yuan, Phys. Rev. Lett. **97**, 082002 (2006) [hep-ph/0602239]. P. G. Ratcliffe and O. V. Teryaev, hep-ph/0703293; Mod. Phys. Lett. A **24**, 2984 (2009) [arXiv:0910.5348 [hep-ph]].

EPJ Web of Conferences **112**, 01003 (2016)
DOI: 10.1051/epjconf/201611201003

Studies of the nucleon structure in back-to-back SIDIS

H. Avakian[1,a]

[1] *Thomas Jefferson National Accelerator Facility, 12000 Jefferson Ave. Suite 5,Newport News, VA 23606*

Abstract. The Deep Inelastic Scattering (DIS) proved to be a great tool in testing of the theory of strong interactions, which was a major focus in last decades. Semi-Inclusive DIS (SIDIS), with detection of an additional hadron allowed first studies of 3D structure of the nucleon, moving the main focus from testing the QCD to understanding of strong interactions and quark gluon dynamics to address a number of puzzles accumulated in recent years. Detection of two hadrons in SIDIS, which is even more complicated, provides access to details of quark gluon interactions inaccessible in single-hadron SIDIS, providing a new avenue to study the complex nucleon structure. Large acceptance of the Electron Ion Collider, allowing detection of two hadrons, produced back-to-back in the current and target fragmentation regions, combined with clear separation of two regions, would provide a unique possibility to study the nucleon structure in target fragmentation region, and correlations of target and current fragmentation regions.

1 Introduction

The quark-gluon dynamics manifests itself in a set of non-perturbative functions describing all possible spin-spin and spin-orbit correlations. Many experiments worldwide are currently trying to pin down various effects related to the nucleon structure through semi-inclusive deep-inelastic scattering (HERMES at DESY [1–4], COMPASS at CERN [5], Jefferson Lab [6–9]) polarized proton-proton collisions (PHENIX, STAR and BRAHMS at RHIC) [10–12], and electron-positron annihilation (Belle and BaBar) [13–16].

Azimuthal distributions of final state particles in in hard scattering processes, in particular, are sensitive to the orbital motion of quarks and play an important role in the study of transverse momentum distributions (TMDs) of quarks in the nucleon. Transverse momentum dependent distributions of partons, encoded in TMDs, and transverse space distributions of partons, encoded in Generalized Parton Distributions (GPDs), have been widely recognized as key objectives of the JLab 12 GeV upgrade and a driving force behind construction of the Electron Ion Collider (EIC). Correlations of the spin of the target or/and the momentum and the spin of quarks, combined with final state interactions define the azimuthal distributions of produced particles.

During the last few years, the first results on transverse single-spin asymmetries (SSAs) have become available. HERMES and COMPASS [3, 5, 17–21] measurements for the first time directly indicated significant azimuthal moments generated both by the Collins and the Sivers effects. Measurements of SSAs at JLab, performed with longitudinally polarized NH_3 [22] and transversely polarized ^{3}He

[a] e-mail: avakian@jlab.org

[23–26] indicate that spin orbit correlations may be significant for certain combinations of spins of quarks and nucleons and transverse momentum of scattered quarks. Large spin-azimuthal asymmetries have been observed at JLab for longitudinally polarized beam [6] and transversely polarized target [27], consistent with corresponding measurements at HERMES [28] and COMPASS [29], which have been interpreted in terms of higher twist contributions, related to quark-gluon correlations. Recent measurements of multiplicities and double spin asymmetries as a function of the final transverse momentum of pions in SIDIS at JLab [8, 22] suggest that transverse momentum distributions may depend on the polarization of quarks and possibly also on their flavor. Calculations of transverse momentum dependence of TMDs in different models [30–33] and on lattice [34, 35] indicate that the dependence of the transverse momentum distributions on the quark polarization and flavor may be very significant.

Combination of measurements in a large Q^2 range from HERMES, COMPASS and JLab, extended to EIC would allow studies of evolution effects and control possible higher twist contributions in the measurements of TMD observables in general, and of the Sivers asymmetry in particular. Wide acceptance of CLAS12 detector at Jefferson Lab (JLab) and EIC would allow also checks of Sivers effect in the target fragmentation region, where it is expected to change the sign [36]. Much higher Q^2 range accessible at JLab12 with CLAS12 and EIC would allow for studies of Q^2-dependence of different higher twist spin-azimuthal asymmetries, which, apart from providing important information on quark-gluon correlations are needed for understanding of possible corrections from higher twists to leading twist observables. An important process which can provide independent information on twist-3 TMDs is the di-hadron production in SIDIS described by interference functions [37–43]. In fact, the measurement of SSAs with longitudinally polarized target or beam is sensitive in particular to the twist-3 chiral-odd distribution functions e and h_L, in combination with the chiral-odd interference fragmentation function $H_1^{\sphericalangle}$ [41]. This effect survives after integration over quark transverse momenta and can be analyzed in the framework of collinear factorization. The dihadron production, thus, becomes a unique tool to study the higher twist effects appearing as $\sin\phi$ modulations in target or beam spin dependent azimuthal moments of the SIDIS cross section [44, 45].

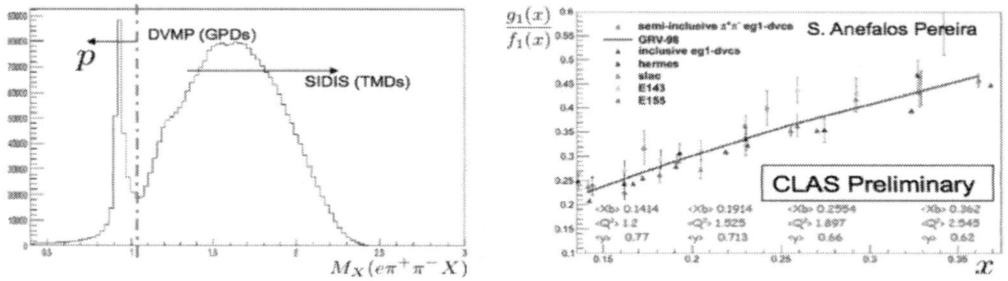

Figure 1. Missing mass of the dihadron system showing the separation of exclusive process (left) and dihadron double spin asymmetry measured at 5.7 GeV in dihadron production, $e\pi^+\pi^-X$ (red triangles), compared with DIS in different experiments.

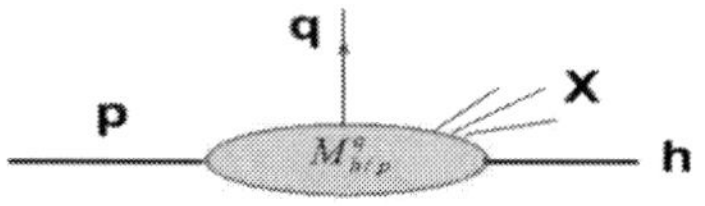

Table 1. TABLE of Leading Twist FFs

	U	L	T
U	$\hat{u}_1$	$\hat{l}_1^{\perp h}$	$\hat{t}_1^h, \hat{t}_1^{\perp}$
L	$\hat{u}_{1L}^{\perp h}$	$\hat{l}_{1L}$	$\hat{t}_{1L}^h, \hat{t}_{1L}^{\perp}$
T	$\hat{u}_{1T}^h, \hat{u}_{1T}^{\perp}$	$\hat{l}_{1T}^h, \hat{l}_{1T}^{\perp}$	$\hat{t}_{1T}, \hat{t}_{1T}^{hh}, \hat{t}_{1T}^{\perp h}, \hat{t}_{1T}^{\perp\perp}$

Figure 2. The fracture function defining the conditional probability to produce the hadron h when a quark q is struck in a proton target(left) and the full list of leading twist FFs, where U,L,T stand for unpolarized, longitudinally polarized and transversely polarized nucleons (rows) and quarks (columns).

The interference fragmentation function $H_1^{\sphericalangle}$ has been used to obtain information on the transversity parton distribution function [46]. Although, dihadron production in SIDIS requires higher energies and Q^2, than single hadron SIDIS, measurements of double-spin asymmetries at CLAS are already at 5.7GeV compatible with simple leading twist predictions for equality of double spin asymmetries in eX,$e\pi^+\pi^-X$, and $e\pi^0X$, assuming the sea quark contributions are negligible at large x_B and fragmentation functions sum of charged pions are flavor independent. CLAS measurement of the double spin asymmetry from inclusive DIS (also from HERMES and SLAC) are consistent with CLAS measurements of double spin asymmetry in charged pion pair production (see Fig. 1).

Although, the Target Fragmentation Region (TFR) of DIS, when the hadrons are created from the target remnant, carries important information about the spin and flavor structure of the nucleon, it has not been studied systematically in experiments due to lack of theory fundamentals. The main physical question in the TFR is how the diquark-like remnant system after the DIS process dresses itself up to become a full-fledged hadron, i.e., by which mechanism the quark-antiquark pairs restoring color neutrality are produced, and how this process is correlated with the spin of the target or/and the produced particles. In the valence-quark region ($x > 0.1$) accessible at JLab with CLAS detector at 5.7 GeV the polarization transfer from the beam is expected to be significant. High luminosity and high polarization of the electron beam makes CLAS an ideal place for studies of correlations between target and current fragmentation regions.

Recently the leading twist formalism for spin and transverse-momentum dependent fracture functions (FFs) has been developed [47]. The production of two hadrons in polarized SIDIS, with one spinless hadron produced in the current fragmentation region (CFR) and another in the TFR, would provide access to the full set of leading twist FFs [48]. In case of double hadron production process $l(\ell) + N(P) \to l(\ell') + h_1(P_1) + h_2(P_2) + X$, at leading order, the cross-section includes all fracture functions, describing conditional probabilities to produce a hadron of certain type in a target fragmentation for a given flavor of struck quark [47, 48]:

$$\frac{\mathrm{d}\sigma^{l(\ell,\lambda)+N(P,S)\to l(\ell')+h_1(P_1)+h_2(P_2)+X}}{\mathrm{d}x\,\mathrm{d}y\,\mathrm{d}z_1\,\mathrm{d}\zeta_2\,\mathrm{d}^2\vec{P}_{T1}\,\mathrm{d}^2\vec{P}_{T2}\,\mathrm{d}\phi_S} =$$

$$\frac{\alpha^2 x_B}{Q^4 y}\left[1+(1-y)^2\right]\left(\sigma_{UU} + S_{\|}\,\sigma_{UL} + S_{\perp}\,\sigma_{UT} + \lambda\, D_{ll}\,\sigma_{LU} + \lambda\, S_{\|} D_{ll}\,\sigma_{LL} + \lambda\, S_{\perp} D_{ll}\,\sigma_{LT}\right),$$

where $D_{ll}(y) = y(2-y)/1 + (1-y)^2$. The subscripts in the structure functions $\sigma_{UT,UL,LT}$, specify the beam (first index) and target (second index) polarization (U, L, T for unpolarized, longitudinally and transversely polarized targets, and U, L for unpolarized and longitudinally polarized beam).

For an unpolarized target expressions there are two contributions σ_{UU} and σ_{LU}, involving convolutions of unpolarized, longitudinally and transversely polarized quark FFs (see Fig. 2) and fragmentation functions of unpolarized, D_1, and transversely polarized quarks, $H_1^{\perp}$, [47]:

$$\sigma_{UU} = F_0^{\hat{u}\cdot D_1} - D_{nn}\left[\frac{P_{T1}^2}{m_1 m_N} F_{kp1}^{\hat{l}^{\perp}\cdot H_1^{\perp}} \cos(2\phi_1) + \frac{P_{T1}P_{T2}}{m_1 m_2} F_{p1}^{\hat{t}^{h}\cdot H_1^{\perp}} \cos(\phi_1+\phi_2) + \left(\frac{P_{T2}^2}{m_1 m_N} F_{kp2}^{\hat{l}^{\perp}\cdot H_1^{\perp}} + \frac{P_{T2}^2}{m_1 m_2} F_{p2}^{\hat{t}^{h}\cdot H_1^{\perp}}\right)\cos(2\phi_2)\right]. \tag{1}$$

$$\sigma_{LU} = -\frac{P_{T1}P_{T2}}{m_2 m_N} F_{k1}^{\hat{l}^{\perp h}\cdot D_1} \sin(\phi_1-\phi_2), \tag{2}$$

The structure functions $F_{...}^{...}$ are specific convolutions [48] of fracture and fragmentation functions depending on $x, z_1, \zeta_2, P_{1\perp}^2, P_{2\perp}^2, \vec{P}_{1\perp}\cdot\vec{P}_{2\perp}$. The hadron 1 produced in the CFR ($x_{F1} > 0$) is described by standard scaled variable $z_1 \simeq P{\cdot}P_1/P{\cdot}q$, and its transverse momentum $\vec{P}_{T1}$ (with magnitude P_{T1} and azimuthal angle ϕ_1) and the hadron 2, h_2 in the TFR ($x_{F2} < 0$) is described by similar variables, $\zeta_2 \simeq E_2/E$ and $\vec{P}_{T2}$ (P_{T2} and ϕ_2), where x_F is Feynman variable defining the fraction of the longitudinal momentum of the hadron in the virtual photon-proton center of mass (CM) frame.
In electroproduction, the polarized lepton emits a virtual photon with non–zero longitudinal polarization, which in turn selects preferentially one polarization state of the struck quark. The opposite polarization of a remnant $s\bar{s}$ pair can again be transferred to the final–state Λ polarization, with the efficiency extracted from eN collisions. After removing a polarized scattered quark from an unpolarized nucleon, the remnant diquark may combine with an s quark, which could originate from the nucleon sea or from a color string between the diquark and the scattered quark to form a Λ hyperon. Significant polarization effects (~ 15-20%) have been predicted in Intrinsic Strangeness Model (ISM) for Λ production in the TFR in deep-inelastic scattering [49]. Longitudinal Λ polarization transfer coefficient as a function of x_F has been already measured with 5.5 GeV electron beam at CLAS (e1f data set). In Fig. 3, the measured asymmetry is shown with the projected results for the future measurements using the CLAS12 detector and 11 GeV beam [50], compared with the ISM predictions [49]. The spin transfer, is proportional to the beam polarization, P_B, and the depolarization factor, $D(y)$. The meson cloud model [51], due to the pseudoscalar nature of the $NK\Lambda$ coupling, predicts the polarization of final-state Λ hyperons to be strongly anticorrelated to that of the nucleon, vanishing for an unpolarized target. The large acceptance of the EIC (see Fig. 4) would provide a unique possibility to study the nucleon structure in the target fragmentation region.

2 SSA in back-to-back hadron production

With hadrons detected in the TFR, the beam SSAs appear already at leading order. With two hadrons detected in the final state, structure functions may depend also on the relative azimuthal angle of the two hadrons, generating a long range correlation between hadrons produced in CFR and TFR. The kinematic plane for back-to-back hadron production is shown on Fig. 5. First measurements of single spin asymmetry defined by Eq.2 in semi-inclusive production of protons and charged pions in coincidence with the scattered electron in hard scattering kinematics ($Q^2 > 1$ GeV^2, $W^2 > 4$ GeV^2) have been performed by the CLAS collaboration using 5.5 GeV and 5.7 GeV longitudinally polarized electron beams scattering off a 5-cm-long liquid-hydrogen target (CLAS e1f and e16 experiments). Target and current fragmentation regions were selected by cuts on the x_F variables of protons ($x_F < 0$) and pions ($x_F > 0$), in addition to standard data quality, vertex, acceptance,and fiducial cuts. Based on previous studies we have chosen $0.7 > z > 0.4$ as the canonical cut on the final state π^+ to exclude contamination from exclusive events and decay pions from baryon resonances produced in the target fragmentation region. With a cut on the missing mass of the $ep\pi^+X$ system, $M_X > 1.4$

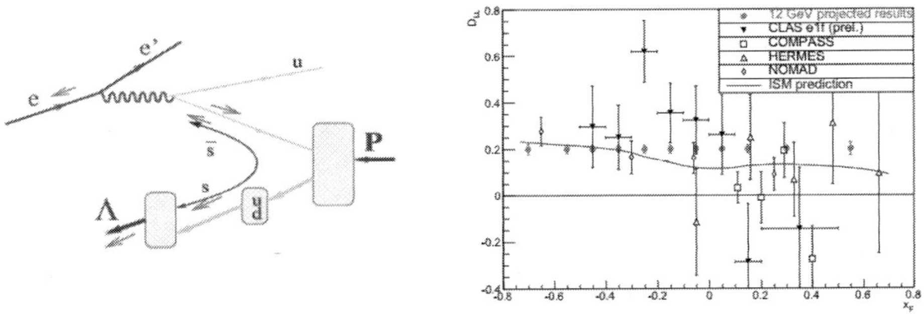

Figure 3. Dominant diagram for Λ production in the target fragmentation region due to scattering on a valence u quark (left) and Longitudinal Λ polarization transfer coefficient as a function of x_F from CLAS 5.5 GeV e1f data set (right).

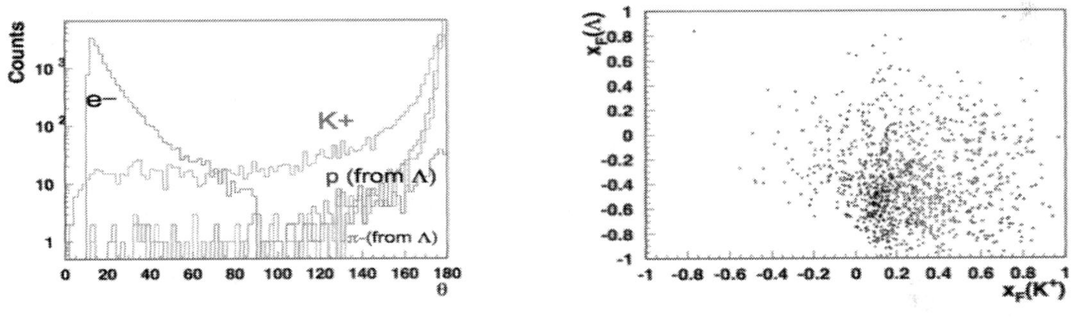

Figure 4. Angular distributions of $e'\Lambda K^+X$ events for EIC configuration with 5 GeV electrons and 50 GeV protons in the Lab frame (left) and x_F distributions for Λs and Kaons for different z fractions of Kaons (from left to wright the lower cut on z increasing from $z > 0.1$ to $z > 0.5$).

we have almost no data at $z > 0.7$, but we use this cut as it was determined in Hall C that strong deviations for the quark-parton model occur at high z [52]. Additional cuts were imposed on the final state π^+ ($0.7 > z > 0.4$) to exclude contamination from exclusive events and decay pions from baryon resonances produced in the target fragmentation region. The contamination from target fragmentation, higher twist, or other effects are important for $z < 0.3$.

Choosing as independent azimuthal angles $\Delta\phi = \phi_2 - \phi_1$ and ϕ_2, the beam spin asymmetry could be defined as

$$A_{LU}(x, z_1, \zeta_2, P_{T1}^2, P_{T2}^2, \Delta\phi) = \frac{\int d\phi_2\, \sigma_{LU}}{\int d\phi_2\, \sigma_{UU}} = \frac{-\frac{P_{T1}P_{T2}}{m_2 m_N} F_{k1}^{\hat{f}_{1,k}\cdot D_1} \sin(\Delta\phi)}{F_0^{\hat{u}\cdot D_1}}. \tag{3}$$

The beam spin asymmetry, $A_{LU}^{\sin\Delta\phi}$, has been calculated as a sinusoidal modulation of the difference of azimuthal angles of proton and π^+ with respect to the lepton scattering plane, for different electron helicity states. The modulation was extracted for different bins in x, z of the pion and the product of transverse momenta of final state proton and pion with respect to the virtual photon in the CM frame.

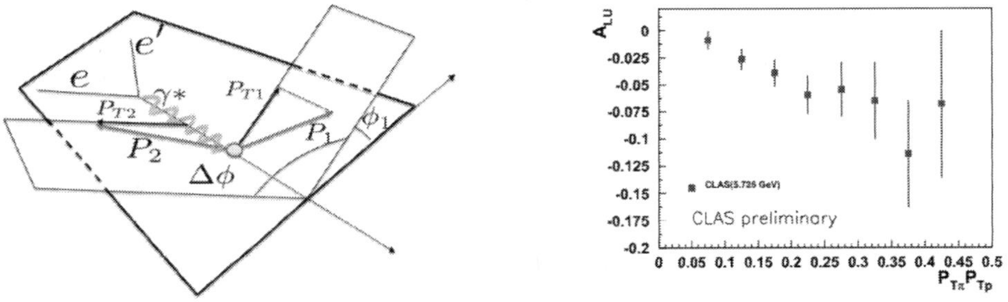

Figure 5. Kinematic plane for back-to-back (b2b) hadron production in SIDIS (left) and measured SSA in back-to-back proton and π^+ production as a function of the product of their transverse momenta $P_{T1}P_{T2}$.

The P_T-dependence of the SSA shows a trend for asymmetry to increase with increasing transverse momenta of pion and proton P_{T1},P_{T2}, (see Fig.5), consistent with expectations from theory. The x-dependence of the the $A_{LU}^{\sin\phi}$ is consistent with asymmetry being large in the large-x_B region, were the valence quark presence is very significant. Similar behavior has been observed also for events with neutral and negative pions in the final state, detected in coincidence with scattered electron and proton in the target fragmentation region.

For quantitative comparison with ongoing measurements as well as projections for future measurements of different b2b processes using the CLAS12 and EIC one will need modeling of FFs, which can be modeled using different partonic models used to predict polarization of Λ hyperons in the target fragmentation region of DIS, such as the meson cloud model [51] or intrinsic strangeness model for Λ production in the target fragmentation region in deep-inelastic scattering [53, 54].

3 Summary

In conclusion, kinematic dependences of single spin asymmetries are measured in a wide kinematic range at CLAS with polarized beam and unpolarized hydrogen target. Significant single-spin asymmetries have been observed in back-to-back pion and proton electroproduction for the first time. Measurements of single-spin asymmetries indicate that spin-orbit correlations may play an important role in description of the structure of nucleon in terms of elementary quarks and gluons going beyond the simple collinear partonic representation.

Within the canonical ranges of x, Q^2, z, P_T, ϕ, M_X, a LO pQCD model of SIDIS, observed asymmetries may be interpreted in a framework using FFs to describe the conditional probabilities of finding partons and hadronization functions describing the fragmentation of the parton to final state hadrons in the current fragmentation. A non-zero A_{LU} in b2b SIDIS, measured for the first time, indicate that spin-orbit correlations between hadrons may be very significant opening a new avenue for studies of the complex nucleon structure in terms of quark and gluon degrees of freedom.

The JLab 12-GeV upgrade will provide the unique combination of wide kinematic coverage, high beam intensity (luminosity), high energy, high polarization, and advanced detection capabilities necessary to study the transverse momentum and spin correlations in di-hadron production in double-polarized semi-inclusive processes both in the target and current fragmentation regions. The large acceptance of the EIC combined with clear separation of target and current fragmentation regions would provide a qualitatively new tool to study the nucleon structure beyond the traditional current fragmentation.

References

[1] A. Airapetian et al. (HERMES), Phys. Rev. Lett. **84**, 4047 (2000), hep-ex/9910062
[2] A. Airapetian et al. (HERMES), Phys. Rev. **D64**, 097101 (2001), hep-ex/0104005
[3] A. Airapetian et al. (HERMES), Phys. Rev. Lett. **94**, 012002 (2005), hep-ex/0408013
[4] A. Airapetian et al. (HERMES), Phys. Lett. **B648**, 164 (2007), hep-ex/0612059
[5] V.Y. Alexakhin et al. (COMPASS), Phys. Rev. Lett. **94**, 202002 (2005), hep-ex/0503002
[6] H. Avakian et al. (CLAS), Phys. Rev. **D69**, 112004 (2004), hep-ex/0301005
[7] H. Avakian, P. Bosted, V. Burkert, L. Elouadrhiri (CLAS), AIP Conf. Proc. **792**, 945 (2005), nucl-ex/0509032
[8] H. Mkrtchyan et al., Phys. Lett. **B665**, 20 (2008), hep-ph/0709.3020
[9] M. Osipenko et al. (CLAS), Phys. Rev. **D80**, 032004 (2009), hep-ex/0809.1153
[10] J. Adams et al. (STAR), Phys. Rev. Lett. **92**, 171801 (2004), hep-ex/0310058
[11] M. Chiu (PHENIX), AIP Conf. Proc. **915**, 539 (2007), nucl-ex/0701031
[12] I. Arsene et al. (BRAHMS), Phys. Rev. Lett. **101**, 042001 (2008), nucl-ex/0801.1078
[13] K. Abe et al. (Belle), Phys. Rev. Lett. **96**, 232002 (2006), hep-ex/0507063
[14] A. Vossen et al. (BELLE Collaboration), Phys.Rev.Lett. (2011), 1104.2425
[15] I. Garzia et al., Nuovo Cimento **33C**, 269 (2010)
[16] I. Garzia (BaBar), Nuovo Cim. **C036**, 181 (2013)
[17] E.S. Ageev et al. (COMPASS), Nucl. Phys. **B765**, 31 (2007), hep-ex/0610068
[18] A. Airapetian et al. (HERMES), Phys. Lett. **B693**, 11 (2010), 1006.4221
[19] M.G. Alekseev et al. (The COMPASS), Phys. Lett. **B692**, 240 (2010), 1005.5609
[20] M.G. Alekseev et al., Eur. Phys. J. **C70**, 39 (2010), 1007.1562
[21] C. Adolph et al. (COMPASS Collaboration), Phys.Lett. **B713**, 10 (2012), 1202.6150
[22] H. Avakian et al. (CLAS), Phys. Rev. Lett. **105**, 262002 (2010), hep-ex/1003.4549
[23] X. Qian et al. (Jefferson Lab Hall A), Phys. Rev. Lett. **107**, 072003 (2011), 1106.0363
[24] J. Huang et al. (Jefferson Lab Hall A), Phys. Rev. Lett. **108**, 052001 (2012), 1108.0489
[25] Y.X. Zhao et al. (Jefferson Lab Hall A), Phys. Rev. **C90**, 055201 (2014), 1404.7204
[26] Y. Zhang et al. (Jefferson Lab Hall A), Phys. Rev. **C90**, 055209 (2014), 1312.3047
[27] Y.X. Zhao et al. (Jefferson Lab Hall A) (2015), 1502.01394
[28] A. Airapetian et al. (HERMES), Phys. Lett. **B622**, 14 (2005), hep-ex/0505042
[29] C. Adolph et al. (COMPASS), Nucl. Phys. **B886**, 1046 (2014), 1401.6284
[30] Z. Lu, B.Q. Ma, Nucl. Phys. **A741**, 200 (2004), hep-ph/0406171
[31] M. Anselmino, A. Efremov, A. Kotzinian, B. Parsamyan, Phys. Rev. **D74**, 074015 (2006), hep-ph/0608048
[32] B. Pasquini, S. Cazzaniga, S. Boffi, Phys. Rev. **D78**, 034025 (2008), hep-ph/0806.2298
[33] C. Bourrely, F. Buccella, J. Soffer, Phys.Rev. **D83**, 074008 (2011), 1008.5322
[34] P. Hagler, B.U. Musch, J.W. Negele, A. Schafer, Europhys. Lett. **88**, 61001 (2009), hep-lat/0908.1283
[35] B.U. Musch, P. Hagler, J.W. Negele, A. Schäafer, Phys.Rev. **D83**, 094507 (2011), 1011.1213
[36] H.H. Matevosyan, A. Kotzinian, E.C. Aschenauer, H. Avakian, A.W. Thomas, Phys. Rev. **D92**, 054028 (2015), 1502.02669
[37] R.L. Jaffe, X.m. Jin, J. Tang, Phys. Rev. Lett. **80**, 1166 (1998), hep-ph/9709322
[38] A. Bianconi, S. Boffi, R. Jakob, M. Radici, Phys. Rev. **D62**, 034008 (2000), hep-ph/9907475
[39] M. Radici, R. Jakob, A. Bianconi, Phys. Rev. **D65**, 074031 (2002), hep-ph/0110252

[40] A. Bacchetta, M. Radici, Phys. Rev. **D67**, 094002 (2003), hep-ph/0212300

[41] A. Bacchetta, M. Radici, Phys. Rev. **D69**, 074026 (2004), hep-ph/0311173

[42] A. Bacchetta, M. Radici, Phys. Rev. **D74**, 114007 (2006), hep-ph/0608037

[43] F.A. Ceccopieri, M. Radici, A. Bacchetta, Phys. Lett. **B650**, 81 (2007), hep-ph/0703265

[44] H. Avakian et al. (Jefferson Lab Hall B), PAC38 Proposal (2011)

[45] S. Pisano et al., JLab Experiment E12-06-112B/E12-09-008B (2014), http://www.jlab.org/exp_prog/proposals/14/E12-06-112B_E12-09-008B.pdf

[46] A. Bacchetta, A. Courtoy, M. Radici (2011), 1104.3855

[47] M. Anselmino, V. Barone, A. Kotzinian, Phys. Lett. **B699**, 108 (2011), 1102.4214

[48] M. Anselmino, V. Barone, A. Kotzinian, Phys. Lett. **B706**, 46 (2011), 1109.1132

[49] J.R. Ellis, D. Kharzeev, A. Kotzinian, Z. Phys. **C69**, 467 (1996), hep-ph/9506280

[50] M. Mirazita et al., JLab Experiment E12-06-112A/E12-09-008A (2014), http://www.jlab.org/exp_prog/proposals/14/E12-06-112A_E12-09-008A.pdf

[51] W. Melnitchouk, A.W. Thomas, Z. Phys. **A353**, 311 (1995), hep-ph/9508205

[52] T. Navasardyan et al., Phys. Rev. Lett. **98**, 022001 (2007), hep-ph/0608214

[53] M.A. Alberg, J.R. Ellis, D. Kharzeev, Phys. Lett. **B356**, 113 (1995), hep-ph/9503333

[54] J.R. Ellis, M. Karliner, D.E. Kharzeev, M.G. Sapozhnikov, Phys. Lett. **B353**, 319 (1995), hep-ph/9412334

EPJ Web of Conferences **112**, 01004 (2016)
DOI: 10.1051/epjconf/201611201004

Deeply Virtual Compton Scattering on the Neutron: JLab Experiment E08-025

Meriem BENALI[1,2,3,a], Malek MAZOUZ[2,3], Helene FONVIEILLE[1]

[1] *Laboratoire de Physique Corpusculaire Clermont-Ferrand, IN2P3, 63177 Aubière Cedex, France*
[2] *Faculté des Sciences de Monastir, 5000, Tunisia*
[3] *For the Hall A DVCS collaboration*

Abstract.
This paper gives the preliminary results of the experimental cross section for deeply virtual Compton scattering on the neutron ($en \to en\gamma$). The E08-025 experiment was performed at Jefferson Lab Hall A. We measured the ($D(e, eX - H(e, e'\gamma)X)$ unpolarized cross section and we extracted, for the first time, a non-zero contribution of (neutron-DVCS + coherent-deuteron-DVCS) at $Q^2 = 1.75\ GeV^2$ and $x_B = 0.36$.

1 Introduction

The electromagnetic probe is a very important tool to investigate the properties of the nucleon. This technique has been used to measure the nucleon form factors by elastic scattering [1] and the partons distribution functions through inclusive deep inelastic scattering [2]. While elastic scattering allowed access to spatial distributions of charge and current in the nucleon, the deep inelastic scattering provides information about parton's longitudinal momentum and helicity distributions in the nucleon. Unfortunately, these two separate informations do not allow multi-dimensional description of the nucleon, any spatial information remains inaccessible in inclusive DIS. In the mid-90's, a new theoretical framework has been developed, namely the generalized parton distributions (GPDs) [3–5] , which relates the elastic form factors and the ordinary parton distributions. In addition, these observables give information about the angular momentum of partons which is an important element to solve the "spin crisis" problem [4, 6]. Indeed, the longitudinal projection of the nucleon spin (= 1/2) is written according to Jaffe an Manohar's decomposition as [7]:

$$1/2 = \frac{\Delta\Sigma}{2} + \Delta G + L_q + L_g$$

where $\Delta\Sigma$ is the intrinsic spin contribution of quarks which is measured ($\Delta\Sigma \approx 0.34$) [8, 9], ΔG is the gluons spin contribution and is also determined [9, 10] and L_q, L_g are respectively the orbital momentum of quarks and gluons wich are unknown so far. However, Ji [4] demonstrated that the GPDs give access to the quarks (and gluons) angular momentum contribution to the nucleon's spin (Ji's sum rule):

[a]e-mail: benali@jlab.org

$$J_q = \frac{1}{2}\Delta\Sigma_q + L_q = \frac{1}{2}\int_{-1}^{+1} xdx[H_q(x,\xi,t=0) + E_q(x,\xi,t=0)].$$

At leading order, the GPDs are four universal functions $H_q(x,\xi,t)$, $E_q(x,\xi,t)$, $\tilde{H}_q(x,\xi,t)$ and $\tilde{E}_q(x,\xi,t)$ defined for each quark flavor q. They depend on the momentum fraction (x), the DVCS scaling variable(ξ) and the momentum transfer (t). The GPDs can be accessed by the hard exclusive processes. The simplest one is the "Deeply Virtual Compton Scattering". At the Bjorken limit, defined by a large virtuality $Q^2 \to \infty$, high energy of the virtual photon $\nu \to \infty$ and fixed Bjorken variable $x_B = \frac{Q^2}{2M\nu}$, where M is the nucleon mass, this process can be factorized into two parts: a known perturbative part ($\gamma^* q \to \gamma q$) and non-perturbative part which is parametrized by GPDs as shown in Figure 1.

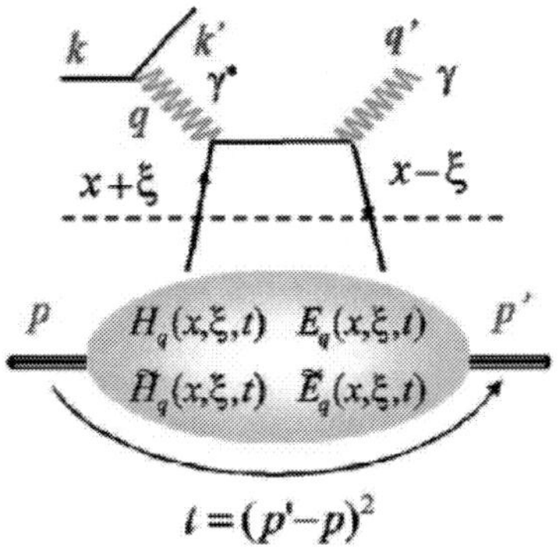

Figure 1. Handbag diagram for the DVCS process.

Motivated by the importance of these observables, different DVCS experiments were performed. The first Beam spin asymmetries were measured by HERMES [11] and CLAS [12]. The H1 [13] and ZEUS [14] collaborations measured the DVCS cross section at low x_B. The first measurements of DVCS cross section in the valence quark region were performed at JLAB Hall A [15]. The first extraction of DVCS on the neutron and the first experimental constraint on the GPD E was obtained at JLab Hall A [16]. DVCS beam-spin asymmetries and target spin asymetries were determined in JLab Hall B using CLAS [17, 18]. The E08-025 experiment was performed at JLab Hall A in 2010 in order to measure the n-DVCS cross section, which is sensitive to the GPD E (the less constrained GPD). The data were taken at fixed $Q^2 = 1.75\ GeV^2$ and $x_B = 0.36$ but at two beam energies $E_{beam} = 4.45\ GeV$ and $E_{beam} = 5.54\ GeV$. In this experiment, 15 cm liquid deuterium (D_2) and hydrogen (H_2) targets are used. The photon electroproduction on a deuterium target can be decomposed into: coherent $d(e,e'\gamma)d$ and incoherent $p(e,e'\gamma)p$ and $n(e,e'\gamma)n$ contributions:

$$D(e,e'\gamma)X = p(e,e'\gamma)p + n(e,e'\gamma)n + d(e,e'\gamma)d + ...$$

As shown in figure 2, each contribution is the sum of the DVCS process and the Bethe-Heitler (BH) process where the real photon is emitted by the incoming or the scattered electron.

The BH amplitude is fully calculable using the form factors so measuring the experimental cross section of the $n(e,e'\gamma)n$ process leads to the determination of the $|DVCS|^2$ plus the interference term. Finally, two measurements at fixed x_B and Q^2 but at two different beam energies allow the separation of the $|DVCS|^2$ and the interference terms.

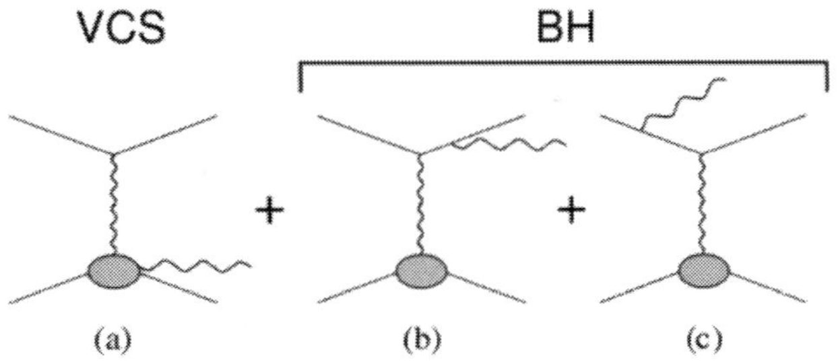

Figure 2. (a): The DVCS process, (b) and (c) Bethe-Heitler process diagrams

The DVCS on the neutron is deduced by comparing the data taken on hydrogen and deuterium targets. The scattered electron is detected in a High Resolution Spectrometer (Left-HRS) in coincidence with the DVCS photon detected in an electromagnetic calorimeter. The recoil nucleon is identified with the missing mass technique. The electromagnetic calorimeter is a matrix of 13×16 PbF_2 blocks, placed at $1.1m$ from the target, centered around the virtual photon direction which leads to a momentum transfer t ranging roughly from -0.1 to -0.5 GeV^2. Each block is connected to (PM + electronic base + ARS) which are equivalent to a numeric oscilloscope recording signals on $128ns$. The main goal of using ARS is to deal with pile up signals in our high luminosity experiment. Due to radiation damage the optical properties of the PbF2-blocks deteriorate during the experiment leading to an overall decline of blocks gain. For this reason, the first step in our analysis is the calorimeter energy calibration.

2 Calorimeter energy calibration

Two elastic calibrations, based on $ep \to e'p'$ elastic reaction, are performed during our experiment but cannot provide a daily calibration [20]. So we performed a second calibration where we used $H(e, e'\pi^0)X$ and $D(e, e'\pi^0)X$ data which have been taken simultaneously with DVCS data. The calibration method is based on the comparison between the measured energy of a detected π^0 (using the energy deposit in calorimeter) and its expected energy calculated with its scattering angle (also given by the calorimeter) [19]. The invariant mass of $\gamma\gamma$ events detected in the calorimeter provides a good test of this calibration. Figure 3 shows the invariant mass peak position versus data group number. After the calibration, the points representing the new invariant mass peak position are aligned with the exact π^0 mass. To check our calibration, we compared the missing mass squared peak position (of the same data), corresponding to $H(e, e'\pi^0)p$ exclusive events, before and after this calibration for each group of data. After the calibration we reproduce, as shown in Figure 4, the nucleon mass squared (0.88 GeV^2) for each group number.

Thanks to this π^0 calibration method and to the daily switching between hydrogen and deuterium targets during data taking, we obtained a daily calibration of the calorimeter energy. This allows to reduce the systematic error on the results coming from the energy calibration of the calorimeter.

3 Selection of exclusive photon electroproduction events

The DVCS events are selected with a cut on the missing mass squared $M_X^2 = (k + p - k' - q')^2 < (M + M_{\pi^0})^2 \approx 1.15\ GeV^2$. Below this cut, we have only $H(e, e'\gamma)p$ (or $D(e, e'\gamma)pn$) events (called DVCS events), but these events are contaminated by:

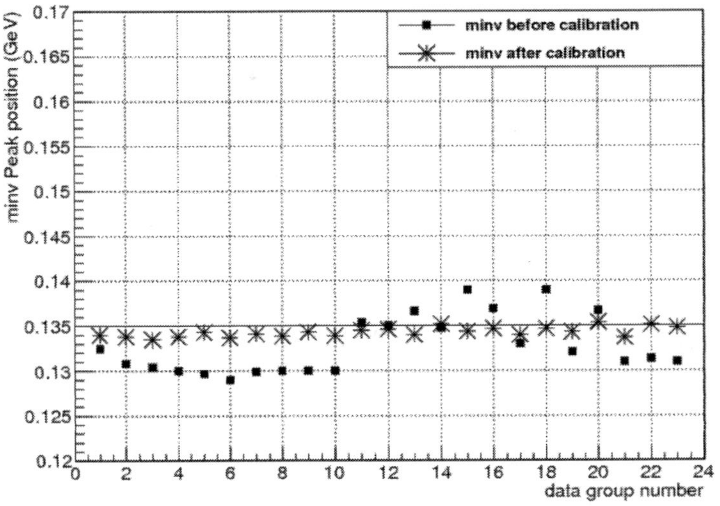

Figure 3. Invariant mass peak position as a function of data group number before and after π^0 calibration. The solid horizontal line represents the π^0 mass.

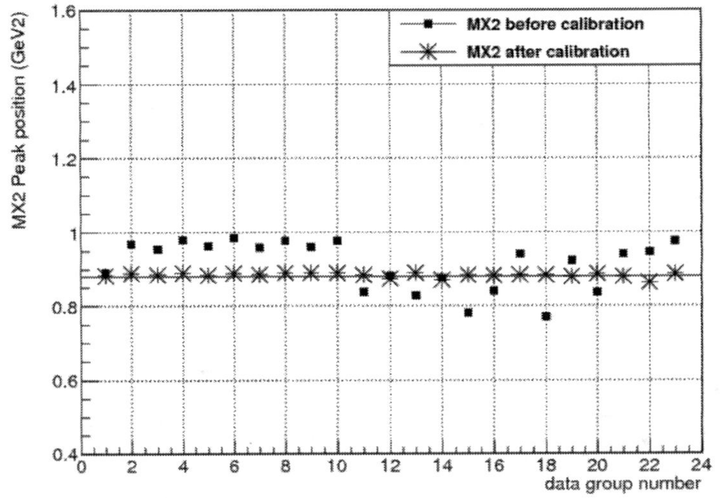

Figure 4. Missing mass squared peak position as a function of data group number before and after π^0 calibration. The solid horizontal line represents the nucleon mass squared.

- Accidental coincidences : when the detected photon in the calorimeter is in fortuitous coincidence with the scattered electron.
- π^0 contamination : when the detected photon in the calorimeter comes from the asymmetric decay of π^0, this photon resembles kinematically to a DVCS photon. We substract the π^0 contamination by using a clean sample of detected π^0 from $H(e, e'\pi^0)X$ or $D(e, e'\pi^0)X$. For these events, a simulation of all possible $\pi^0 \to \gamma\gamma$ decays tells us which fraction of asymmetric decays is seen as DVCS events [20].

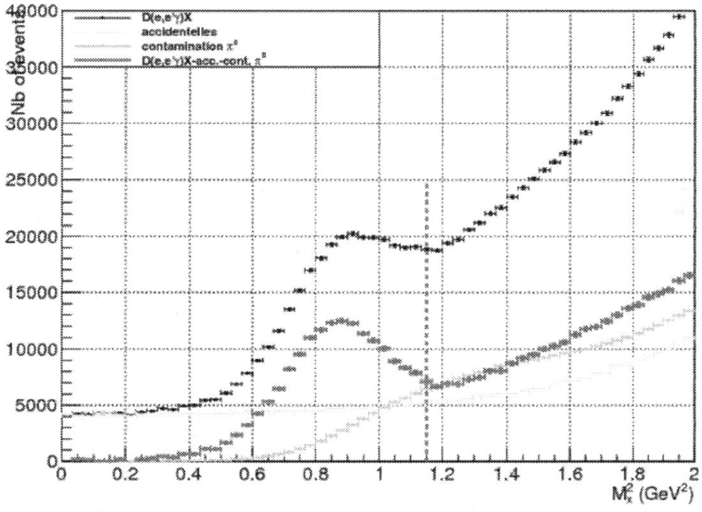

Figure 5. Missing mass squared spectrum of raw data (black), accidental data (yellow), π^0 contamination (green), and after all substractions (blue spectrum) for $D(e, e'\gamma)X$, "Kin2 low" kinematics.

Figure 5 shows the spectrum of the missing mass squared M_X^2 for $D(e, e'\gamma)X$ raw data, the accidental coincidences and the π^0 contamination events for one kinematic (called Kin2Low: $E_{Beam} = 4.45 GeV$). Before subtracting the $H(e, e'\gamma)p$ events from $D(e, e'\gamma)pn)$ events, we added Fermi momentum [21] to H2 data and normalized H2 and D2 data to the same luminosity. The $D(e, e'\gamma)X - H(e, e'\gamma)X$ data are represented by the purple spectrum in figure 6. This spectrum contains the DVCS on the neutron events (n-DVCS) and also the DVCS on the coherent-deuteron events (d-DVCS). The vertical line represents the missing mass squared cut at the pion production threshold. These experimental data are used to extract the (n-DVCS + d-DVCS) cross section.

4 Smearing of the simulation data

The extraction of the cross section is done using the experimental data and the simulation data. First, it is necessary to have a good matching between the experimental data and simulated data (having the same exclusive peak position and having the same resolution). Our method is based on smearing the photon energy for each event in the simulation, because the experimental resolution is dominated by the energy resolution of the calorimeter. Since the extraction of the cross section is performed by bins in t and in ϕ (the angle between the hadronic and leptonic planes), our smearing method was applied for each of these bins independently. Figure 7 shows a spectrum of the missing mass squared of the $H(e, e'\gamma)p$ data (blue) and the proton simulated data after smearing (red), a good agreement between the two spectra is clear below the pion production threshold.

5 Extraction of the cross section

The experimental data ($D(e, e'\gamma)X - H(e, e'\gamma)X$) contain a neutron contribution (n-DVCS) and a coherent deuteron contribution (d-DVCS). In this case, the unpolarized photon electroproduction cross section can be written as [22]:

$$\frac{d^4\sigma}{dQ^2 dx_B dt d\phi} = \sigma(BH_n) + \sigma(BH_d) + \sum_{in}(\Gamma_{in}(Q^2, x_B, t, \phi)X_{in}) + \sum_{id}(\Gamma_{id}(Q^2, x_B, t, \phi)X_{id}), \quad (1)$$

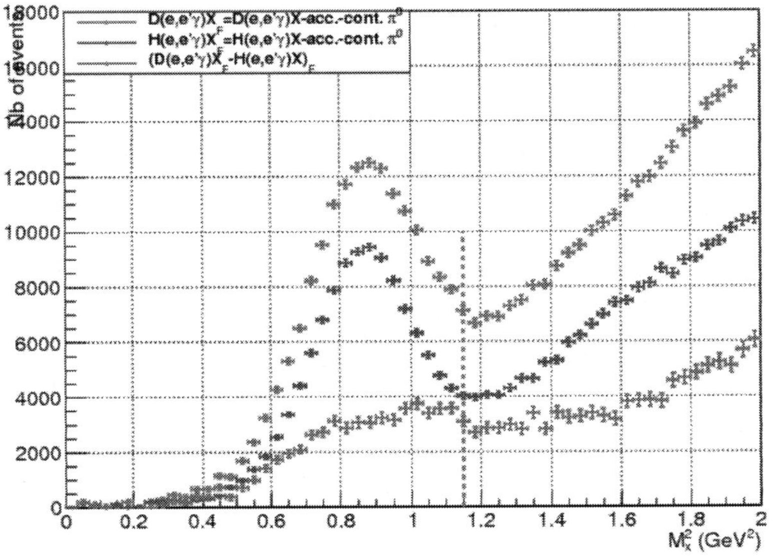

Figure 6. Missing mass squared spectrum of ($D(e, eX$- accidental - π^0contamination) events (red), ($H(e, e'\gamma)X$- accidental - π^0 contamination) events (blue) and ($D(e, e'\gamma)X - H(e, e'\gamma)X$). The purple spectrum represents the (n-DVCS+deuteron-coherent-DVCS) events at the left of the vertical line which is the pion production threshold.

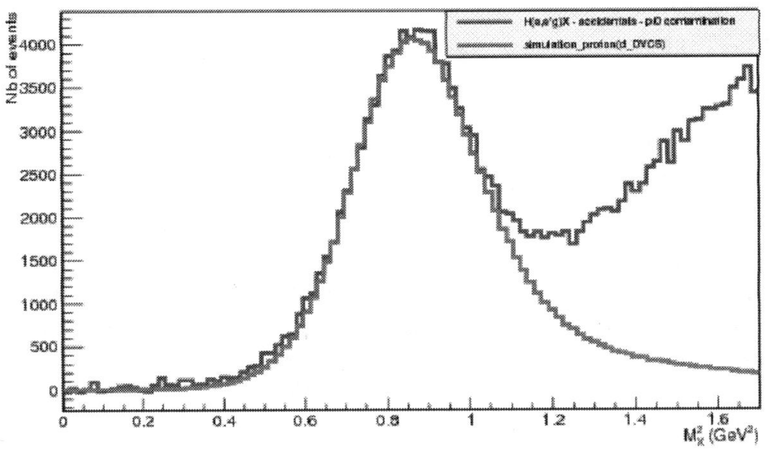

Figure 7. Missing mass squared spectrum of the $H(e, e'\gamma)p$ data (blue specrum) and the proton simulation data (red spectrum).

where the indices (n/d) signify respectively (neutron/deuteron-coherent), BH is the Bethe-Heitler contribution and $\Gamma_i(Q^2, x_B, t, \phi)$ is a calculable kinematical factor which depends on the kinematics variables. (X_{in}, X_{id}) are the unknown harmonics that we want to extract; these observables are linear combinations of Compton Form Factors (CFFs) which contain the GPDs observables. The dependence on ϕ in the kinematical factors allows us to separate the contribution of the different harmonics (for the neutron or the coherent-deuteron) and a binning in M_X^2 is important to separate the contributions of the neutron from the ones of coherent-deuteron.

In order to extract the observables X_{in} and X_{id} given by equation 1, we performed a χ^2 minimization between the experimental number of events, N^{exp}, and the simulated number of events, N^{sim}.

$$\chi^2 = \sum_{k=1}^{Nbin} \left(\frac{N_k^{sim} - N_k^{exp}}{\Delta N_k^{exp}}\right)^2. \tag{2}$$

Note the summation over all bins at the vertex k (5 bins in t $\times$ 20 bins in ϕ= 100 bins). ΔN_k^{exp} is the experimental statistical error in one bin k.

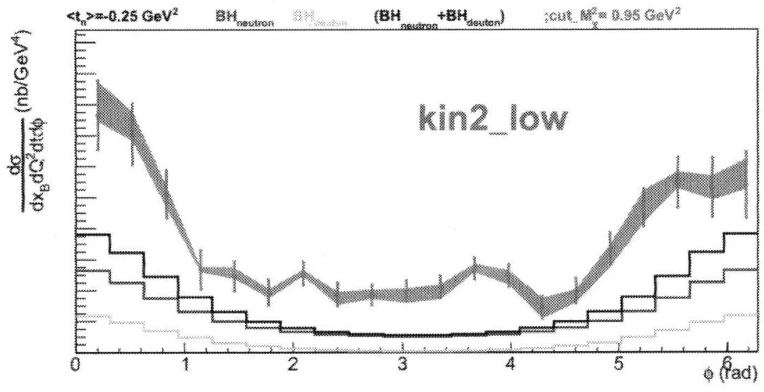

Figure 8. Preliminary results of $(n(e,e'\gamma)n + d(e,e'\gamma)d)$ experimental cross section (black points), neutron BH contribution (blue histogram), coherent-deuteron BH contribution (green histogram) and BH-sum (black histogram) for one bin at $< t >= -0.25\ GeV^2$ and the kinematic Kin2Low. Error bars are statistical and the shaded area is the systematic error.

Figure 8 shows the preliminary results of the experimental cross section corresponding to the sum $(n(e,e'\gamma)n + d(e,e'\gamma)d)$, for one bin at $< t >= 0.25\ GeV^2$ for one kinematic Kin2Low. These results show for the first time the existence of a positive contribution of n-DVCS (+ d-DVCS) which is observed at both beam energies. These results are obtained after testing their stability as a function of the upper M_X^2 cut.

5.1 Conclusion

In summary, our analysis of this experiment provides preliminary results of the experimental cross section $(n(e,e'\gamma)n + d(e,e'\gamma)d)$. For the first time, we have found a positive contribution of n-DVCS (+ d-DVCS) at both beam energies. We plan to make more stability and correlation studies to estimate separately the contributions n-DVCS and d-DVCS and a global fit will be performed using data at both energies to extract CFFs.

References

[1] V. Punjabi et al., Eur.Phys.J. A51, 79 (2015) and references therein.

[2] H. Abramowicz et al, Eur.Phys.J. C75 (2015) 12, 580 and references therein.

[3] D. Muller et al., Fortschr. Phys. 42, 2101 (1994).
[4] X. Ji, Phys. Rev. Lett. 78, 610 (1997).
[5] A.V. Radyushkin, Phys. Lett. B 380, 417 (1996).
[6] J. Ashman t al, Nucl. Phys. B 328, 1 (1989).
[7] R.L. Jaffe and A. Manohar, Nucl. Phys. B 337 (1990) 509.
[8] D. Hash, Nucl. Phys. B (Proc. Suppl.) 191 (2009) 79-87.
[9] V.Y. Alexakhin et al., Phys. Lett. B 647 (2007) 8.
[10] E.S. Ageev et al., Phys. Lett. B 633 (2006) 25.
[11] A. Airapetian et al. (HERMES), Phys. Rev. Lett. 87, 182001 (2001).
[12] S. Stepanyan et al. (CLAS), Phys. Rev. Lett. 87, 182002(2001).
[13] C. Adloff et al. (H1), Phys. Lett. B 517, 47 (2001).
[14] S. Chekanov et al. (ZEUS), Phys. Lett. B573, 46 (2003).
[15] C. Munoz Camacho et al., Phys. Rev. Lett. 97, 262002 (2006).
[16] M. Mazouz et al. Phys.Rev.Lett. 99, 242501 (2007).
[17] F.X Girod et al. Phys.Rev.Lett. 100, 162002 (2008).
[18] S. Pisano et al. Phys.Rev. D91, 052014 (2015).
[19] M. Mazouz, DVCS internal note, (2011).
[20] C. Munoz Camacho, PhD thesis, Université Paris VI,(2005).
[21] M. Garcon and J.W. Van Orden. Adv. Nucl. Phys. 26, 293 (2001).
[22] A.V. Belitsky et al, Phys.Rev. D82, 074010 (2010).

EPJ Web of Conferences **112**, 01005 (2016)
DOI: 10.1051/epjconf/201611201005

Timelike Compton Scattering off the nucleon: observables and experimental perspectives for JLab at 12 GeV

Marie Boër[1,2,a]

[1] *Thomas Jefferson National Accelerator Facility, Newport News, Virginia 23606*
[2] *Institut de Physique Nucléaire and Laboratoire de Physique Théorique, CNRS-IN2P3, Université Paris-Sud, Université Paris-Saclay, 91406 Orsay Cedex, France.*

Abstract.
Hard exclusive processes such as photoproduction or electroproduction of photon or meson off the nucleon provide access to the Generalized Parton Distributions (GPDs), in the regime where the scattering amplitude is factorized into a hard and a soft part. GPDs contain the correlation between the longitudinal momentum fraction and the transverse spatial densities of quarks and gluons in the nucleon. Timelike Compton Scattering (TCS) correspond to the reaction $\gamma N \to \gamma^* N \to e^+e^- N$, where the photon is scattered off a quark. It is measured through its interference with the associated Bethe-Heitler process, which has the same final state. TCS allows to access the GPDs and test their universality by comparison to the results obtained with the DVCS process ($eN \to e\gamma N$). Also, results obtained with TCS provide additional independent constrains to the GPDs parameterization.
We will present the physical motivations for TCS, with our theoretical predictions for TCS observables and their dependencies. We calculated for JLab 12 GeV energies all the single and double beam and/or target polarization observables off the proton and off the neutron. We will also present the experimental perspectives for the next years at JLab. Two proposals were already accepted at JLab: in Hall B, with the CLAS12 spectrometer, in order to measure the unpolarized cross section and in Hall A, with the SoLID spectrometer, in order to measure the unpolarized cross section and the beam spin asymmetry at high intensity. A Letter Of Intent was also submitted in order to measure the transverse target spin asymmetries in Hall C. We will discuss the merits of this different experiments and present some of the expected results.

1 Introduction

Timelike Compton Scattering (TCS) correspond to the scattering of a high energy photon off a quark. A timelike virtual photon is emitted and decays into a lepton pair (Fig. 1, left panel). We will focus on TCS off the nucleon and on the case where the virtual photon decays into an electron-positron pair of invariant mass between 2 and 3 GeV. This condition ensures to avoid the region of vector mesons resonances. It also ensures a high enough virtuality (typically $Q'^2 > 2$ GeV^2) to allow for factorization of TCS amplitude between a hard perturbative part, which is calculable, and a non perturbative

[a] e-mail: mboer@jlab.org

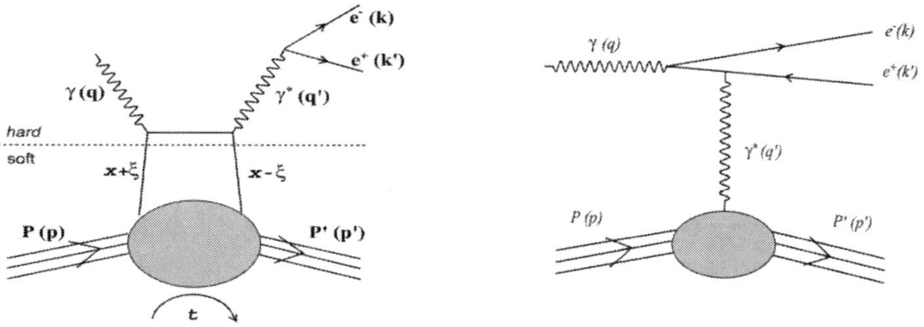

Figure 1. Left panel: diagram for TCS off the nucleon, right panel: Bethe-Heitler process. Crossed diagrams are not represented.

part corresponding to low energy interactions in the nucleon. The later could be parameterized by Generalized Parton Distributions (GPDs) [1–5]. The GPDs are Fourier transform into momentum space of the QCD bilocal operators involved in the soft part of the amplitude. Without taking acount evolution effects (running in Q'^2), the GPDs depend on 3 independent variables, x, ξ and t, which represent respectively the longitudinal momentum fraction carried by the struck quark, the longitudinal momentum transfer to the quark (with $\xi \simeq \frac{Q'^2}{2s-Q'^2}$ for TCS on a fixed nucleon target) and the 4-momentum transfer squared $t = (p' - p)^2$. GPDs contain the correlation between the longituninal momentum fraction x of the struck quark and its transverse spatial distribution, implicitely contained in the t (momentum transfer squared) dependence. For instance, it can be shown in a model independent way [6] that the Fourier transform of GPD$(x, \xi = 0, t)$ gives the probability to find a quark with momentum fraction x of the nucleon at a transverse distance $b_\perp$ of the center of the nucleon, where $b_\perp$ is the conjugate variable of t. At leading twist, the amplitude for massless quarks involves 4 chiral-even GPDs (without quark helicity flip) that are associated to 4 γ matrix structures: vector (GPD H), tensor (GPD E), axial (GPD $\tilde{H}$) and pseudo-scalar (GPD $\tilde{E}$).

TCS interfers with Bethe-Heitler (BH) process (Fig. 1, right panel), where the lepton pair comes from splitting of the initial photon and one lepton of the pair interacts with the nucleon. BH non perturbative part is parameterized by Form Factors.

TCS off the proton was originally theoretically investigated in [7, 8]. The TCS and BH unpolarized, circularly and linearly beam polarized cross sections were derived. In our study, we calculated the BH and TCS amplitudes, where we waived some of the t/Q'^2 and M^2/Q'^2 (where M is the proton mass) approximations, compared to the previous theoretical works. We also included correction terms to restaure gauge invariance to the twist 2 amplitudes, which are not gauge invariant. We studied the impact of these corrections on our results and found few percent deviations for the TCS cross sections and $< 1\%$ deviations for the spin asymmetries, in the phase space of concern. We then calculated all the unpolarized and beam and/or target polarized cross sections, with circularly or linearly polarized beam and with longitudinally or transversally polarized proton or neutron [9, 10]. In this proceeding, we briefly summarize the formalism that we used and we present some of these results. We will discuss the interest of the measurement of some of these observables, for TCS off the proton and off the neutron, and we will present the experimental project at for JLab at 12 GeV.

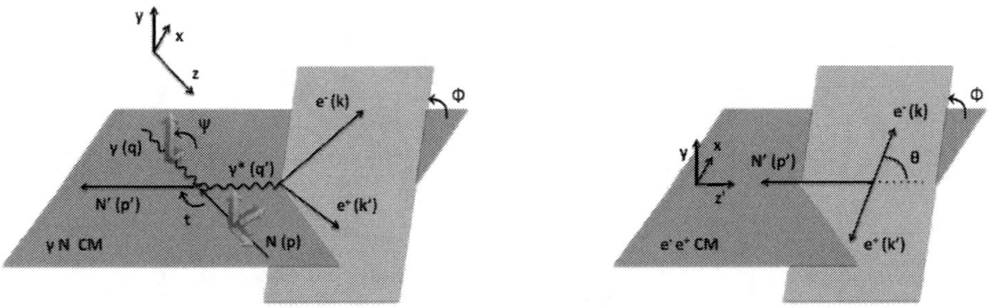

Figure 2. Left panel: scheme of the TCS reaction in the γP C.M.. The red arrows represent the 3 polarization states for the proton (along the x, y or z axis, where the xz plane is the reaction plane) and 2 polarization states for a linearly polarized photon (along the $x-$ and $y-$axis). The angle between the beam polarization vector and the reaction plane is noted Ψ. Right panel: we boost the system and we define the 2 decay angles of the electron in the e^+e^- C.M., where θ and ϕ are respectively the polar and azimuthal angles with respect to the γP plane and the direction of the γ^* axis.

2 Observables

2.1 Unpolarized cross section and formalism

The unpolarized $\gamma P \to e^+e^-P'$ (where P stands for a proton or a neutron) cross section depends on 5 independent kinematic observables. We decided to derive it at fixed beam energy or at fixed ξ, where it can be expressed as a function of the kinematic variables Q'^2 and t and as a function of the solid angle of the electron in the γ^* rest frame, $\Omega = (\theta, \phi)$ (Fig. 2). At fixed beam energy the unpolarized cross section reads

$$\frac{d^4\sigma}{dQ'^2 dt d\Omega} = \frac{1}{64\ (2\pi)^4\ (2ME_\gamma)^2}\ |T^{BH} + T^{TCS}|^2, \tag{1}$$

where T^{BH} and T^{TCS} are respectivelly the BH and the TCS amplitudes. The $|T^{BH} + T^{TCS}|^2$ term is averaged over the proton and photon polarization and summed over the outgoing proton helicities.

We use a frame were the average photon momenta $\bar{q} = \frac{1}{2}(q + q')$ and nucleon momenta $P = \frac{1}{2}(P + P')$ are collinear along the $z-$axis and in opposite directions. We define the lightlike vectors along the positive and negative z directions as:

$$\begin{aligned} \tilde{p}^\mu &= P^+/\sqrt{2}(1,0,0,1), \\ n^\mu &= 1/P^+ \cdot 1/\sqrt{2}(1,0,0,-1), \end{aligned} \tag{2}$$

and we define the light-cone components $a^\pm$ by $a^\pm \equiv (a^0 \pm a^3)/\sqrt{2}$. The light cone momentum fractions x and ξ are defined respectively by $k^+ = xP^+$ and by $\Delta^+ = -2\xi P^+$ where $\Delta = (p' - p) = (q - q')$. In the asymptotic limit (neglecting terms in t/Q'^2), the variable ξ can be expressed as a function of Q'^2 and s, the squared center of mass energy, as $\xi = \frac{Q'^2}{2s - Q'^2}$. The results that we present in this proceeding are calculated for the asymptotic limit. Within this formalism, the BH amplitude reads

$$T^{BH} = -\frac{e^3}{\Delta^2}\ \bar{u}(p')\Gamma^\nu u(p)\ \epsilon^\mu(q)\ \bar{u}(k)\left(\gamma_\mu \frac{\not{k} - \not{q}}{(k-q)^2}\gamma_\nu + \gamma_\nu \frac{\not{q} - \not{k}}{(q-k')^2}\gamma_\mu\right) v(k'), \tag{3}$$

with the virtual photon-proton electromagnetic vertex matrix

$$\Gamma^{\nu} = \gamma^{\nu} F_1(t) + \frac{i\sigma^{\nu\rho}\Delta_{\rho}}{2M} F_2(t). \tag{4}$$

In this work, we used the Dirac and Pauli form factors $F_1(t)$ and $F_2(t)$ parameterization of [11, 12]. The TCS amplitude reads (using Ji convention for the GPDs [5]):

$$T^{TCS} = -\frac{e^3}{q'^2}\, \bar{u}(k)\gamma^{\nu} v(k')\, \epsilon^{\mu}(q) \Bigg[\tag{5}$$

$$\frac{1}{2}(-g_{\mu\nu})_{\perp} \int_{-1}^{1} dx \left(\frac{1}{x-\xi-i\epsilon} + \frac{1}{x+\xi+i\epsilon} \right) \left(H^p(x,\xi,t)\bar{u}(p')\, \not{n} u(p) + E^p(x,\xi,t)\bar{u}(p') i\sigma^{\alpha\beta} n_{\alpha} \frac{\Delta_{\beta}}{2M} u(p) \right)$$

$$-i(\epsilon_{\nu\mu})_{\perp} \int_{-1}^{1} dx \left(\frac{1}{x-\xi-i\epsilon} - \frac{1}{x+\xi+i\epsilon} \right) \left(\tilde{H}^p(x,\xi,t)\bar{u}(p')\, \not{n}\gamma_5 u(p) + \tilde{E}^p(x,\xi,t)\bar{u}(p')\gamma_5 \frac{\Delta.n}{2M} u(p) \right) \Bigg].$$

We have used the metrics

$$\begin{aligned} (-g_{\mu\nu})_{\perp} &= -g_{\mu\nu} + \tilde{p}_{\mu} n_{\nu} + \tilde{p}_{\nu} n_{\mu}, \\ (\epsilon_{\nu\mu})_{\perp} &= \epsilon_{\nu\mu\alpha\beta}\, n^{\alpha} \tilde{p}^{\beta}, \end{aligned} \tag{6}$$

and we used the VGG model for the GPDs parameterization [13, 14]. More details about the formalism may be found in [9].

2.2 Single spin asymmetries

Single spin asymmetries with a polarized target or a polarized photons beam are defined as

$$A_{\odot U}\ (A_{LU},\ A_{Ui}) = \frac{\sigma^+ - \sigma^-}{\sigma^+ + \sigma^-}, \tag{7}$$

where the notations $A_{\odot U}$, A_{LU} and A_{Ui} correspond respectivelly to the circularly polarized beam spin asymmetry, the longitudinally polarized beam spin asymmetry and the target spin asymmetries with a nucleon polarized along the i−axis, where $i = x, y$ or z. The $\sigma^{\pm}$ terms stand for the 4-differential polarized cross sections, with a polarization vector in the direction of the corresponding axis (+ sign) or in the opposite direction (− sign). The $A_{\odot U}$ and A_{Ui} asymmetries are proportionnal to the imaginary part of the amplitudes and are sensitive to the interference between BH and TCS. They would cancel if they were only BH in the reaction. This fact simplify the extraction of physics information about the imaginary part of TCS amplitude by fitting the asymmetries with a phenomenological model. We display on Fig. 3 the $A_{\odot U}$, A_{Ux} and A_{Uz} asymmetries as a function of ϕ. The different curves correspond to different scenarii of GPD parametrization. These results show that these observables are mostly sensitive to the GPD H, and present non negligible sensitivities to the GPDs $\tilde{H}$ and E.

Remark: whereas GPD are purely real, they appear in TCS amplitude through integral over x (Eq. 5). We access them indirectly through complex functions called Compton Form Factors (CFF). Indeed, previous asymmetries access $\Im(\mathcal{H})$ (CFF associated to GPD H), which is proportionnal to $H(\pm\xi, \xi, t)$ (it is the same for the other GPDs).

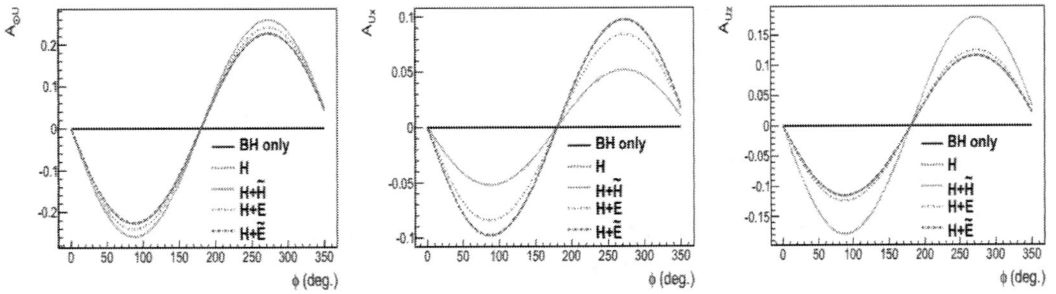

Figure 3. Single spin asymmetries as a function of ϕ for BH and for BH+TCS, with different scenarii of GPD parametrization. Left panel: $A_{\odot U}$, central panel: A_{Ux}, left panel: A_{Uz}. Calculation have been done for $\xi = 0.2$ ($E_\gamma \simeq 11.2$ GeV), $t = -0.4$ GeV^2, $Q'^2 = 7$ GeV^2 and θ integrated over $[45°, 135°]$.

2.3 Proton and neutron comparison: circularly polarized beam spin asymmetry

Nucleon GPDs decompose into quark GPDs as

$$H^p = \frac{4}{9}H^u + \frac{1}{9}H^d; \quad H^n = \frac{4}{9}H^d + \frac{1}{9}H^u, \tag{8}$$

where $H^{p,n}$ are proton or neutron GPDs and $H^{u,d}$ are GPDs for quarks u or d. The knowledge of both proton and neutron GPDs leads to quark flavor decomposition. An other interest of studying both proton and neutron TCS comes from that some factors in the amplitudes may enhance some GPD dependencies for one of the nucleon more than for the other one. Indeed proton and neutron TCS observables may present different sensitivities to GPDs and different kinematic distributions. Fig. 4 presents the circularly polarized beam spin asymmetries and its quark angular momenta J_u and J_d dependencies, entering GPD E parametrization. We notice a $\sim$ 3 times stronger sensitivity (for this kinematic) to quarks angular momenta for the neutron, accompagnied by a sign change of the asymmetry. More details about comparison of TCS off the proton and off the neutron may be found in [10].

3 Experimental perspectives for TCS at JLab at 12 GeV

Several programs aiming at measuring TCS off the proton at JLab at 12 GeV have been developped or are still in development. Two proposals [15, 16] have been already approved. Both of them aim at measuring the unpolarized BH+TCS cross section and the circularly polarized beam spin asymmetry. The proposal [15] plans to run for 100+20 days at a luminosity of 10^{35} cm^{-2}s^{-1} with the JLab Hall B large acceptance CLAS12 spectrometer. An analysis using the CLAS spectrometer and the 6 GeV electron beam has been performed [17]. It shows the experimental feasibility of such measurement in JLab Hall B. The proposal [16] plans to measure the same observables with the SoLID large acceptance spectrometer in Hall A, with a luminosity of 10^{37} cm^{-2}s^{-1} for 50 days. The luminosity improvement of Hall A will provide a better precision for the study of the Q'^2 and t evolution of the TCS cross section. These two experiments plan to measure TCS using quasi-real photons coming from the electron beam radiations, providing a photon beam energy in a range from 4 GeV to 10

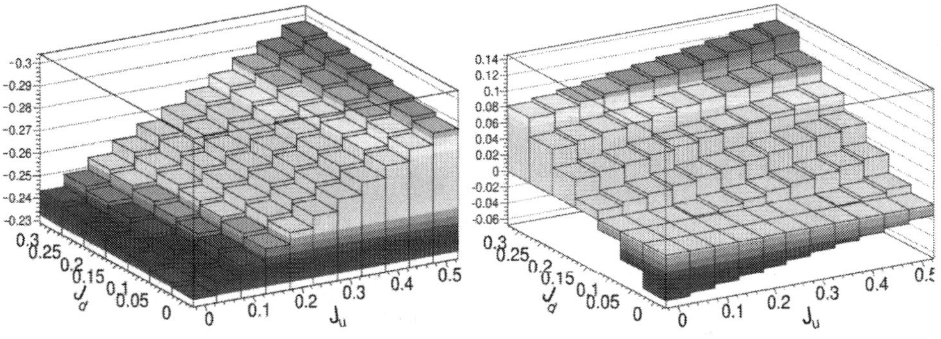

Figure 4. Evolution of the circularly polarized beam spin asymmetries for TCS+BH off the proton (left) or off the neutron (right) as a function of the quark angular momenta J_u and J_d. Calculation have been done for $\xi = 0.2$ ($E_\gamma \simeq 11.2$ GeV), $t = -0.4$ GeV^2, $Q'^2 = 7$ GeV^2, $\phi = 90°$ and θ integrated over [45°, 135°]. Four nucleon GPD enter TCS parametrization.

GeV. They are complementary. Another program is in development for the JLab Hall C with the NPS spectrometer. It aims at measuring the single target spin asymmetries using a transversally polarized NH_3 target [18]. The main interests of these measurement are the sensitivity of the transverse target spin asymmetries to the GPDs $\tilde{H}$ and E and the fact that these observables have never been measured and will provide new independent constrains to CFF fits.

The measurement of observables related to TCS off the neutron would be complementary to TCS off the proton and present lot of interests (see previous section). However, it is experimentaly more challenging to measure TCS off the neutron due to a lower cross section compared to proton and due to experimental issues such as neutron detection... No experimental programs have been developped yet for such measurement, but some studies may be done in a next future.

4 Conclusion

We presented a sample of our results for TCS and BH off the proton and off the neutron. We shown some of the single and double beam and/or target polarized spin asymmetries. We then discussed the interest of their experimental measurement in a purpose of extracting Compton Form Factors in order to parametrize the GPDs. We also presented some of the experimental perspectives for studying TCS at JLab at 12 GeV.

References

[1] K. Goeke, M. V. Polyakov and M. Vanderhaeghen, Prog. Part. Nucl. Phys. **47**, 401 (2001).
[2] M. Diehl, Phys. Rept. **388**, 41 (2003).
[3] A.V. Belitsky, A.V. Radyushkin, Phys. Rept. **418**, 1 (2005).
[4] M. Guidal, H. Moutarde and M. Vanderhaeghen, Rept. Prog. Phys. **76**, 066202 (2013).
[5] X. Ji, Phys.Rev.Lett. **78**, 610 (1997); Phys.Rev.D **55**, 7114 (1997).
[6] M. Burkardt, Int. J. Mod. Phys. A **18** (2003) 173

[7] E. R. Berger, M. Diehl and B. Pire, Eur. Phys. J. C **23**, 675 (2002).

[8] A. T. Goritschnig, B. Pire and J. Wagner, Phys. Rev. D **89** (2014) 9, 094031

[9] M. Boër, M. Guidal and M. Vanderhaeghen, Eur. Phys. J. A **51** (2015) 8, 103.

[10] M. Boër, M. Guidal and M. Vanderhaeghen, arXiv:1510.02880 [hep-ph].

[11] O. Gayou *et al.* [Jefferson Lab Hall A Collaboration], Phys. Rev. Lett. **88** (2002) 092301

[12] E. J. Brash, A. Kozlov, S. Li and G. M. Huber, Phys. Rev. C **65** (2002) 051001

[13] M. Vanderhaeghen, P.A.M. Guichon and M. Guidal, Phys. Rev. Lett. **80** 5064 (1998)

[14] M. Vanderhaeghen, P. A. M. Guichon and M. Guidal, Phys. Rev. D **60** 094017 (1999).

[15] M. Guidal, T. Horn, P. Nadel-Turonski, R. Paremuzyan, S. Stepanyan *et al.* and the CLAS Collaboration, *Jefferson Lab. PAC 39 proposal E12-12-001,* (2012).

[16] M. Boër, P. Nadel-Turonski, J. Zhang, Z. Zhao *et al.* and the SoLID coll., *Jefferson Lab. PAC 43 proposal E12-12-006A* (2015).

[17] R. Paremuzyan, PhD thesis, Yerevan Institute, *Time-like Compton Scattering,* (2010).

[18] M. Boër, A. Mkrtchyan, P. Nadel-Turonski, V. Tadevosyan *et al.* and the NPS coll., *Jefferson Lab. PAC 43 Letter Of Intent LOI12-15-007* (2015).

EPJ Web of Conferences **112**, 01006 (2016)
DOI: 10.1051/epjconf/201611201006

Revealing transversity GPDs through the photoproduction of a photon and a ρ meson

R. Boussarie[1,a], B. Pire[2,b], L. Szymanowski[3,c], and S. Wallon[1,4,d]

[1] *Laboratoire de Physique Théorique, UMR 8627, CNRS, Univ. Paris Sud, Université Paris-Saclay, 91405 Orsay, France*
[2] *Centre de physique théorique, École Polytechnique, CNRS, Université Paris-Saclay, 91128 Palaiseau, France*
[3] *National Centre for Nuclear Research (NCBJ), Warsaw, Poland*
[4] *UPMC Univ. Paris 06, Faculté de Physique, 75252 Paris, France*

Abstract. Photoproduction of a pair of particles with large invariant mass is a natural extension of collinear QCD factorization theorems which have been much studied for deeply virtual Compton scattering and deeply virtual meson production. We discuss the production of a photon and a meson, where the wide angle Compton scattering on a meson subprocess factorizes from generalized parton distribution. We calculate at dominant twist and leading order in α_s, the production cross-section of a transversely polarized ρ meson which is sensitive to chiral-odd GPDs, and show that it may be measurable in near future JLab experiments.

1 Introduction

Near forward photoproduction of a pair of particles with large invariant mass is a natural extension of collinear QCD factorization theorems which have been much studied for near forward deeply virtual Compton scattering (DVCS) and deeply virtual meson production [1]. We study here the reaction

$$\gamma^{(*)}(q) + N(p_1) \rightarrow \gamma(k) + \rho^0(p_\rho, \epsilon_\rho) + N'(p_2), \tag{1}$$

where a wide angle Compton scattering subprocess $\gamma(q\bar{q}) \rightarrow \gamma\rho$ characterized by the large scale $M^2_{\gamma\rho} = (p_\rho + k)^2$ (the final state $\gamma\rho$ invariant squared mass) factorizes from GPDs. This large scale $M_{\gamma\rho}$ is related to the large transverse momenta transmitted to the final photon and to the final meson, the pair having an overall small transverse momentum (since $\Delta^2 = (p_2 - p_1)^2$ is small). The study of such reactions was initiated in Ref. [2], where the process under study was the high energy diffractive photo- (or electro-) production of two vector mesons, the hard probe being the virtual pomeron exchange (and the hard scale being the virtuality of this pomeron), in analogy with the virtual photon exchange

[a] e-mail: Renaud.Boussarie@th.u-psud.fr
[b] e-mail: bernard.pire@polytechnique.edu
[c] e-mail: Lech.Szymanowski@ncbj.gov.pl
[d] e-mail: Samuel.Wallon@th.u-psud.fr

occuring in the deep inelastic electroproduction of a meson. The process we study here follows the same way of thinking as the one studied in Ref. [3], there with a π instead of a γ.

Finally, we note that a similar strategy has also been advocated in Ref. [4] to enlarge the number of processes which could be used to extract information on chiral-even GPDs.

The analogy to the timelike Compton scattering process [5] :

$$\gamma^{(*)}N \to \gamma^* N' \to \mu^+\mu^- N' , \tag{2}$$

where the lepton pair has a large squared invariant mass Q^2, is instructive. Although the photon-meson pair in the process (1) has a more complex momentum flow, one may draw on this analogy to ascribe the role of the hard scale to the photon-meson pair invariant mass.

The experimental study of such processes should not present major difficulties to modern detectors such as those developed for the 12 GeV upgrade of Jlab or for the Compass experiment at CERN. Probing various processes, thus verifying the universality of the GPDs, will increase our confidence in a meaningful extraction of dominant (i.e. chiral-even) GPDs from near future experiments.

A noteworthy feature of the process (1) is that, at dominant twist, separating the transverse (respectively longitudinal) polarization of the ρ meson allows one to get access to chiral-odd (respectively chiral-even) GPDs. This is much welcome since access to the chiral-odd transversity GPDs [6], noted H_T, E_T, $\tilde{H}_T$, $\tilde{E}_T$, which decouple from deeply virtual Compton scattering and deeply virtual meson production at leading order, has turned out to be most challenging [7]. Quark mass effects [8] or production of a meson described by a twist 3 distribution amplitude [9] are two other ways which have been proposed to evade this difficulty.

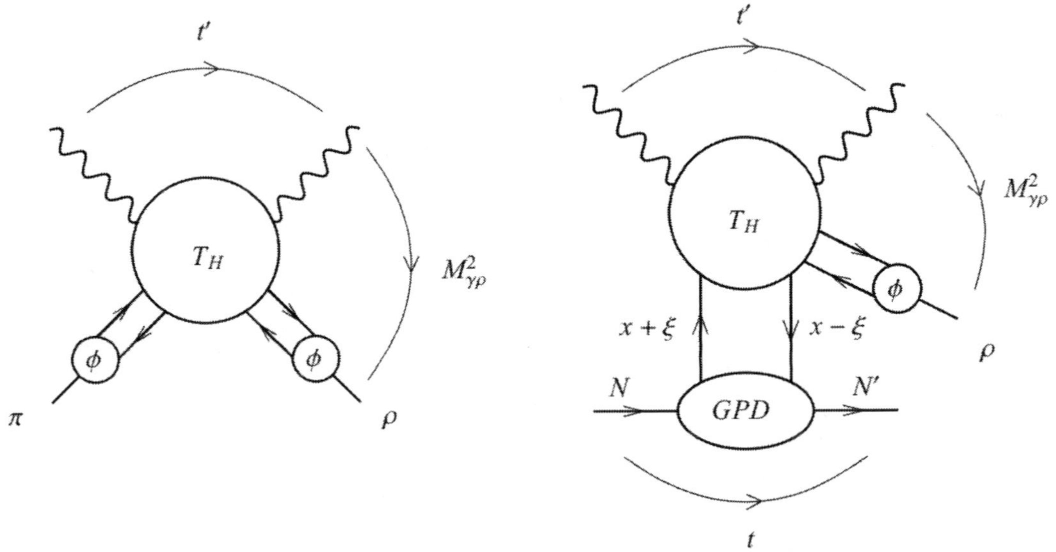

Figure 1. The wide angle Compton scattering process (left) and its generalization to the photoproduction of a $\gamma\rho$ pair (right).

The calculation of the amplitude of process (1) relies on the use of the now classical proof of the factorization of exclusive scattering at fixed angle and large energy [10]. The amplitude for the wide angle Compton scattering process $\gamma + \pi \to \gamma + \rho$ is written [11] as the convolution of mesonic distribution amplitudes (DAs) and a hard scattering subprocess amplitude $\gamma + (q + \bar{q}) \to \gamma + (q + \bar{q})$ with the final meson states replaced by a collinear quark-antiquark pair. In a second stage, we extract from the factorization procedure of the deeply virtual Compton scattering amplitude near the forward region the right to replace one entering meson DA by a $N \to N'$ GPD, and thus get Fig. 1 (right panel). The required skewness parameter ξ is written in terms of the final photon - meson squared invariant mass $M^2_{\gamma\rho}$ as

$$\xi = \frac{\tau}{2-\tau} \quad , \quad \tau = \frac{M^2_{\gamma\rho} - t}{S_{\gamma N} - M^2} . \tag{3}$$

Indeed the same collinear factorization property bases the validity of the leading twist approximation which either replaces the meson wave function by its DA or the $N \to N'$ transition by nucleon GPDs. A slight difference is that light cone fractions $(z, 1-z)$ leaving the DA are positive, while the corresponding fractions $(x+\xi, \xi - x)$ may be positive or negative in the case of the GPD. Our Born order calculation shows that this difference does not ruin the factorization property.

In order for the leading twist factorization of a partonic amplitude to be valid, one should avoid the dangerous kinematical regions where a small momentum transfer is exchanged in the upper blob, namely small $t' = (k-q)^2$ or small $u' = (p_\rho - q)^2$, and the resonance regions for each of the invariant squared masses $(p_\rho + p_{N'})^2$, $(k + p_\rho)^2$.

Although we here limit ourselves to an initial on-shell photon, we note that our present discussion applies as well to the case of electroproduction where a moderate virtuality of the initial photon may help access the perturbative domain with a lower value of the hard scale $M_{\gamma\rho}$.

2 Kinematics

We use the following conventions. We decompose all momenta on a Sudakov basis as $v^\mu = a\,n^\mu + b\,p^\mu + v^\mu_\perp$, with p and n the light-cone vectors $p^\mu = \frac{\sqrt{s}}{2}(1,0,0,1)$, $n^\mu = \frac{\sqrt{s}}{2}(1,0,0,-1)$, $v^\mu_\perp = (0, v^x, v^y, 0)$ and $v^2_\perp = -\vec{v}^2_t$. The particle momenta read

$$p^\mu_1 = (1+\xi)\,p^\mu + \frac{M^2}{s(1+\xi)}\,n^\mu\,, \quad p^\mu_2 = (1-\xi)\,p^\mu + \frac{M^2 + \vec{\Delta}^2_t}{s(1-\xi)} n^\mu + \Delta^\mu_\perp\,, \quad q^\mu = n^\mu\,, \tag{4}$$

$$k^\mu = \alpha\,n^\mu + \frac{(\vec{p}_t - \vec{\Delta}_t/2)^2}{\alpha s}\,p^\mu + p^\mu_\perp - \frac{\Delta^\mu_\perp}{2}, \quad p^\mu_\rho = \alpha_\rho\,n^\mu + \frac{(\vec{p}_t + \vec{\Delta}_t/2)^2 + m^2_\rho}{\alpha_\rho s}\,p^\mu - p^\mu_\perp - \frac{\Delta^\mu_\perp}{2},$$

with $\bar{\alpha} = 1 - \alpha$ and M, m_ρ are the masses of the nucleon and the ρ meson. The total center-of-mass energy squared of the γ-N system is thus

$$S_{\gamma N} = (q + p_1)^2 = (1+\xi)s + M^2 . \tag{5}$$

From these kinematical relations it follows that :

$$2\,\xi = \frac{(\vec{p}_t - \frac{1}{2}\vec{\Delta}_t)^2}{s\,\alpha} + \frac{(\vec{p}_t + \frac{1}{2}\vec{\Delta}_t)^2 + m^2_\rho}{s\,\alpha_\rho}\,, \tag{6}$$

and

$$1 - \alpha - \alpha_\rho = \frac{2\,\xi\,M^2}{s\,(1-\xi^2)} + \frac{\vec{\Delta}^2_t}{s\,(1-\xi)} . \tag{7}$$

On the nucleon side, the transferred squared momentum is

$$t = (p_2 - p_1)^2 = -\frac{1+\xi}{1-\xi}\vec{\Delta}^2_t - \frac{4\xi^2 M^2}{1-\xi^2} . \tag{8}$$

The other various Mandelstam invariants read

$$s' = (p_\gamma + p_\rho)^2 = M^2_{\gamma\rho} = 2\xi\, s\left(1 - \frac{2\xi M^2}{s(1-\xi^2)}\right) - \vec{\Delta}_t^2 \frac{1+\xi}{1-\xi}, \tag{9}$$

$$-t' = -(p_\gamma - q)^2 = \frac{(\vec{p}_t - \vec{\Delta}_t/2)^2}{\alpha}, \tag{10}$$

$$-u' = -(p_\rho - q)^2 = \frac{(\vec{p}_t + \vec{\Delta}_t/2)^2 + (1-\alpha_\rho)\, m_\rho^2}{\alpha_\rho}. \tag{11}$$

Let us remind that we are interested in the "large angle" kinematical domain where every invariant $s', -t', -u'$ is large (as compared to Λ^2_{QCD}), with the constraints $0 < \alpha, \alpha_\rho < 1$.

3 The scattering amplitude

In the spirit of the above described factorizations, the scattering amplitude of the process (1) is written as

$$\mathcal{A}(t, M^2_{\gamma\rho}, u') = \sum_{q,i} \int_{-1}^{1} dx \int_0^1 dz\; T_i^q(x, v, z)\, H_i^q(x, \xi, t)\Phi_{\rho_{L,T}}(z), \tag{12}$$

where T_i^q is the hard part of the amplitude and H_i^q the corresponding (chiral-even and chiral-odd) GPDs of a parton q in the nucleon target, and $\Phi_{\rho_{L,T}}(z)$ the leading twist chiral-even (resp. chiral-odd) distribution amplitude of the ρ_L (resp. ρ_T) meson.

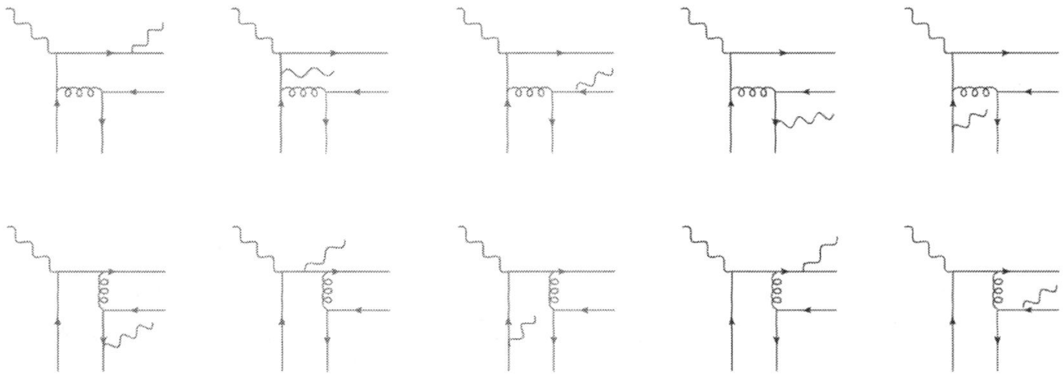

Figure 2. The Feynman diagrams describing the subprocess at leading order; in Feynman gauge only the 4 diagrams on the right contribute to the ρ_T case

The scattering sub-process is described by 20 Feynman diagrams, but an interesting (quark-antiquark interchange) symmetry allows to deduce the contribution of half of the diagrams from the 10 diagrams shown on Fig. 2 through a $(x \leftrightarrow -x\,;\, z \leftrightarrow 1-z)$ interchange. Moreover, in Feynman gauge, only the 4 diagrams on the right of Fig. 2 contribute to the chiral-odd case.

The scattering amplitudes get both real and imaginary parts. Focusing on the chiral-odd amplitude (since accessing transversity GPDs was the first motivation of our study), we get the following results. The z and x dependence of this amplitude can be factorized as

$$T_i^q = e_q^2\, \alpha_{em}\, \alpha_s\, \mathcal{N}(z, x)\, \mathcal{T}^i \tag{13}$$

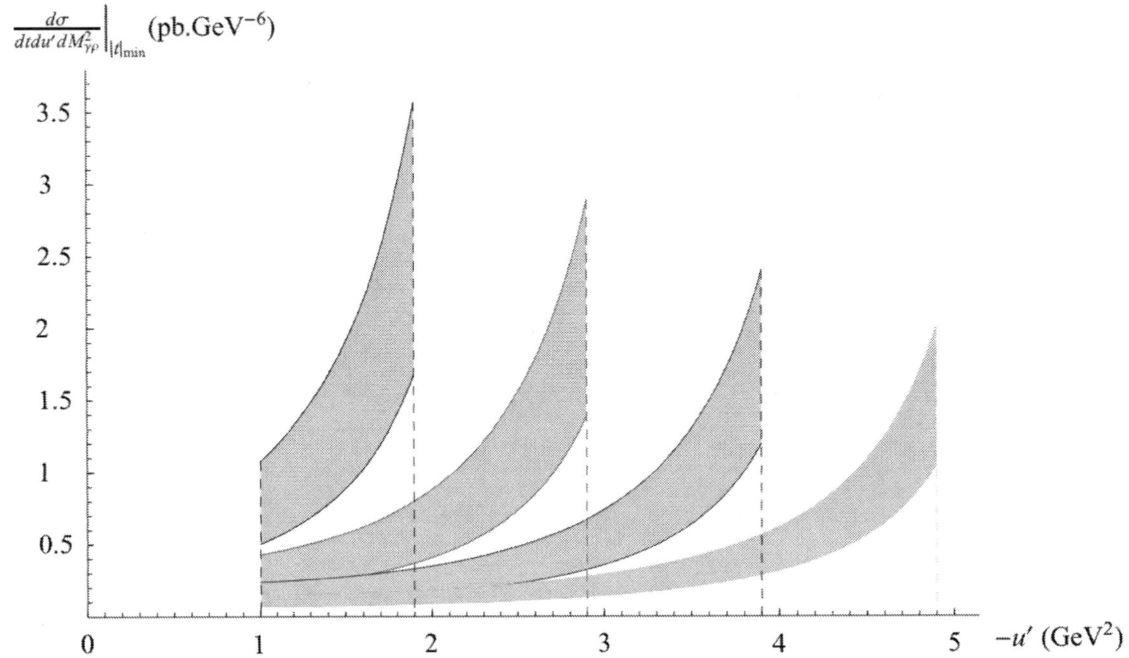

Figure 3. The differential cross section (15) for the production of $\gamma\rho_T$ involving chiral-odd GPDs, as a function of $-u'$ at $S_{\gamma N} = 20$ GeV^2 for (from left to right, resp. in black, red, blue, green) $M^2_{\gamma\rho} = 3, 4, 5, 6$ GeV^2. The gray bands indicate the theoretical uncertainties.

with (in the gauge $p.\epsilon_k = 0$):

$$\begin{aligned} \mathcal{T}^i &= (1-\alpha)\Big[\big(\epsilon_{q\perp}.p_\perp\big)\big(\epsilon_{k\perp}.\epsilon_{\rho\perp}\big) - \big(\epsilon_{k\perp}.p_\perp\big)\big(\epsilon_{q\perp}.\epsilon_{\rho\perp}\big)\Big]\,p^i_\perp \\ &- (1+\alpha)\big(\epsilon_{\rho\perp}.p_\perp\big)\big(\epsilon_{k\perp}.\epsilon_{q\perp}\big)\,p^i_\perp + \alpha\big(\alpha^2-1\big)\xi s\big(\epsilon_{q\perp}.\epsilon_{k\perp}\big)\,\epsilon^i_\rho \\ &- \alpha\big(\alpha^2-1\big)\xi s\Big[\big(\epsilon_{q\perp}.\epsilon_{\rho\perp}\big)\epsilon^i_{k\perp} - \big(\epsilon_{k\perp}.\epsilon_{\rho\perp}\big)\epsilon^i_{q\perp}\Big]\,. \end{aligned} \tag{14}$$

Using as a first estimate the asymptotic form of the ρ−meson distribution amplitude, we perform analytically the integration over z. Inserting a model for the transversity GPDs [3], we use numerical methods for the integration over x.

Starting with the expression of the scattering amplitude (12), the differential cross-section as a function of t, $M^2_{\gamma\rho}$, $-u'$ reads

$$\left.\frac{d\sigma}{dt\,du'\,dM^2_{\gamma\rho}}\right|_{|t|=|t|_{min}} = \frac{|\mathcal{M}|^2}{32 S^2_{\gamma N} M^2_{\gamma\rho} (2\pi)^3}\,. \tag{15}$$

We show in Fig. 3 this cross section (15) as a function of $-u'$ at $S_{\gamma N} = 20$ GeV^2 for various fixed values of $M^2_{\gamma\rho}$ (namely 3, 4, 5, 6 GeV^2). There are rather huge uncertainties for these cross-sections, mainly of course because of our poor knowledge of chiral-odd GPDs. One of the main uncertainties is due to the fact that the transversity PDFs [12], on which we rely to build our model for transversity GPDs, are themselves not very constrained by existing data on semi-inclusive deep inelastic scattering.

The cross section grows with $(-u')$ but its normalization is rather small. We expect a larger cross-section for the longitudinal ρ case where chiral-even GPDs contribute; this will not help disentangling the transverse ρ cross section. A complete analysis of the angular distribution of the emerging $\pi^+\pi^-$ pair allows in principle to access the chiral-odd sensitive contribution at the amplitude level. If such an analysis is performed, the high luminosity expected at JLab@12GeV might be sufficient for an experimental evidence of a needed contribution of transversity GPDs at twist 2, in the case of a ρ_T in the final state.

Acknowledgments

L. Sz was partially supported by grant of National Science Center, Poland, No. 2015/17/B/ST2/01838. This work is partly supported by the French grant ANR PARTONS (Grant No.ANR-12-MONU-0008-01).

References

[1] M. Diehl, Phys. Rept. **388**, 41 (2003); A. V. Belitsky and A. V. Radyushkin, Phys. Rept. **418**, 1 (2005).
[2] D. Yu. Ivanov *et al.*, Phys. Lett. B **550**, 65 (2002); R. Enberg *et al.*, Eur. Phys. J. C **47**, 87 (2006).
[3] M. El Beiyad *et al.*, Phys. Lett. B **688**, 154 (2010); cross-sections presented here should be corrected due to the omission of numerical prefactors.
[4] S. Kumano *et al.*, Phys. Rev. D **80**, 074003 (2009).
[5] D. Mueller, D. Robaschik, B. Geyer, F. M. Dittes and J. Horejsi, Fortsch. Phys. **42** (1994) 101; E. R. Berger, M. Diehl and B. Pire, Eur. Phys. J. C **23** (2002) 675; B. Pire, L. Szymanowski and J. Wagner, Phys. Rev. D **83**, 034009 (2011).
[6] M. Diehl, Eur. Phys. J. C **19**, 485 (2001).
[7] M. Diehl *et al.*, Phys. Rev. D **59**, 034023 (1999); J. C. Collins and M. Diehl, Phys. Rev. D **61**, 114015 (2000).
[8] B. Pire and L. Szymanowski, Phys. Rev. Lett. **115** (2015) 9, 092001 and arXiv:1510.01869 [hep-ph].
[9] S. Ahmad, G. R. Goldstein and S. Liuti, Phys. Rev. D **79**, 054014 (2009); S. V. Goloskokov and P. Kroll, Eur. Phys. J. C **65**, 137 (2010).
[10] A. V. Efremov and A. V. Radyushkin, Theor. Math. Phys. **44** (1980) 573 [Teor. Mat. Fiz. **44** (1980) 17]; *ibid* Theor. Math. Phys. **44** (1981) 664 [Teor. Mat. Fiz. **44** (1980) 157]; G. P. Lepage and S. J. Brodsky, Phys. Rev. D **22** (1980) 2157.
[11] or its crossed process $\gamma\gamma \to M_1 M_2$, which is known at next to leading order, B. Nizic, Phys. Rev. D **35**, 80 (1987); G. Duplancic and B. Nizic, Phys. Rev. Lett. **97** (2006) 142003.
[12] M. Anselmino, M. Boglione, U. D'Alesio, S. Melis, F. Murgia and A. Prokudin, Phys. Rev. D **87** (2013) 094019.

EPJ Web of Conferences **112**, 01007 (2016)
DOI: 10.1051/epjconf/201611201007

The low Q^2 chicane and Compton polarimeter at the JLab EIC

Alexandre Camsonne[1,a]

[1] *Jefferson Laboratory*
12000 Jefferson Avenue
Newport News, VA 23606
USA

Abstract. The JLAB EIC (JLEIC) design includes a chicane after the interaction point to detect electron associated with production of quasi-real photon at the interaction. This chicane layout can also be used for Compton polarimetry to measure the electron beam polarization. This proceeding will present the layout of the low Q^2 chicane and the implementation and current R&D of a Compton polarimeter which would be located in the middle of this chicane.

1 Introduction

The future Electron Ion Collider will feature availability of both polarized electron and ion beams coupled with a high luminosity from $10^{33}cm^{-2}s^{-1}$ up to a few $10^{34}cm^{-2}s^{-1}$. Energy of the electrons will be similar to the existing CEBAF from 3 to 10 GeV. The protons or ions will have an energy from 20 to 100 GeV [1].

Polarization measurement is important since the machine is designed to run at high luminosity, many measurements will be limited by systematic errors. In the case of inclusive electron measurement, electron polarization can be a major contribution. Polarization also enters the photon production cross-section which can be used to determine the luminosity needed for all the cross section measurements.

1.1 Layout

The Jefferson Laboratory EIC design (JLEIC) is a ring ring collider design in a figure of 8 [2] which makes it easier to conserve polarization of ions and electrons since the precession cancels at first order at the interaction point.

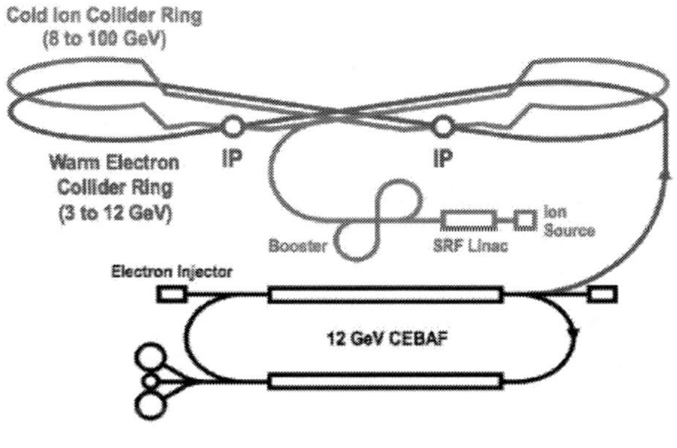

Figure 1. Figure of 8 layout of the JLEIC electron ring with the chicane after in the interaction point

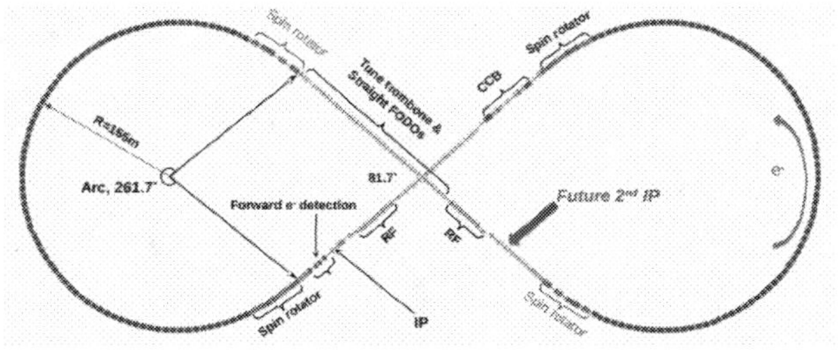

Figure 2. Figure of 8 layout of the JLEIC electron ring with the chicane after in the interaction point

Two interactions points are planned, the main interaction point is based on a solenoidal magnet giving an almost hermetic detection as show in Fig. 4 where on can see the detector with the magnets for the final focusing and the low Q^2 chicane.

1.2 Beam helicity structure

The electron beam will use CEBAF as full energy injector. The CEBAF injector is using a circularly polarized laser on a GaAS source, by switching the laser polarization one can switch the electron polarization to the opposite direction for a certain period of time. The direction of the spin of the polarized electron can be manipulated using Wien filters. Using this CEBAF feature, two large macrobunches of about 2.3 μs with opposite spin direction will be filled in the ring 6 composed of the small bunch of at 476 MHz. As shown in Fig. 5

[a]e-mail: camsonne@jlab.org

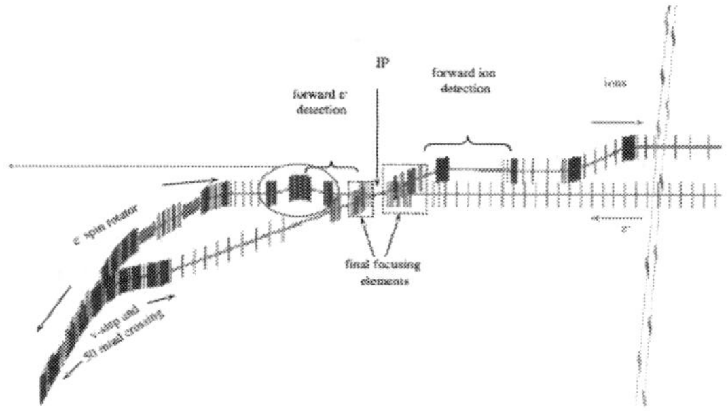

Figure 3. Detailed layout of the magnet lattice of the chicane at the interaction point with the forward electron detection chicane

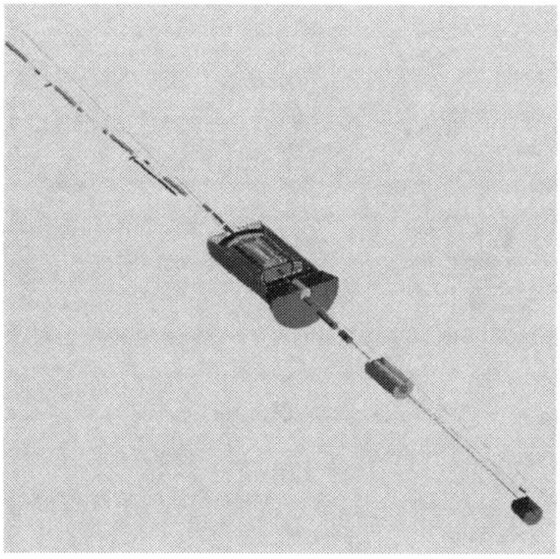

Figure 4. Overview of the interaction point main detector with the final focusing magnets and the low Q^2 chicane

2 The low Q^2 chicane

Quasi-real photons can be produced at the interaction point, since they are almost colinear with the beam the corresponding electrons will not be detected in the main detector. In order to detect those electrons a chicane was designed after the interaction point. It consists of 4 dipoles with the first dipole which takes out electrons corresponding to high photon energy which will be bent out and detected by a tracker detector. The designed momentum resolution is 10^{-3} on the electron energy assuming a 1 mm detector resolution. This will allow to study J/ψ, charm production, real and Time like Compton Scattering for example. A detector closer to the beam similar to roman pots or Compton electron detector could be added to improve the low energy photon energy acceptance if needed. The optimization of the chicane acceptance is being studied. All the magnets are modeled in the GEMC simulation. In Fig. 7 one can see electrons deflected in the detector at the first dipole.

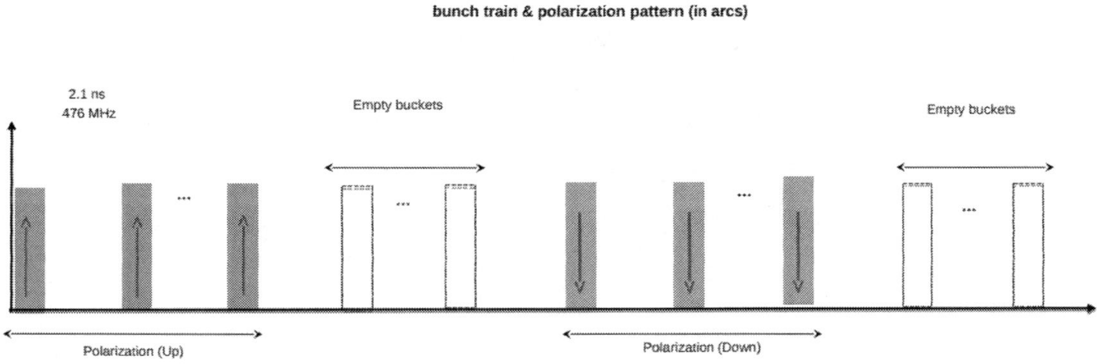

Figure 5. Beam polarization structure, the two 2.3 μs macrobunch of opposite

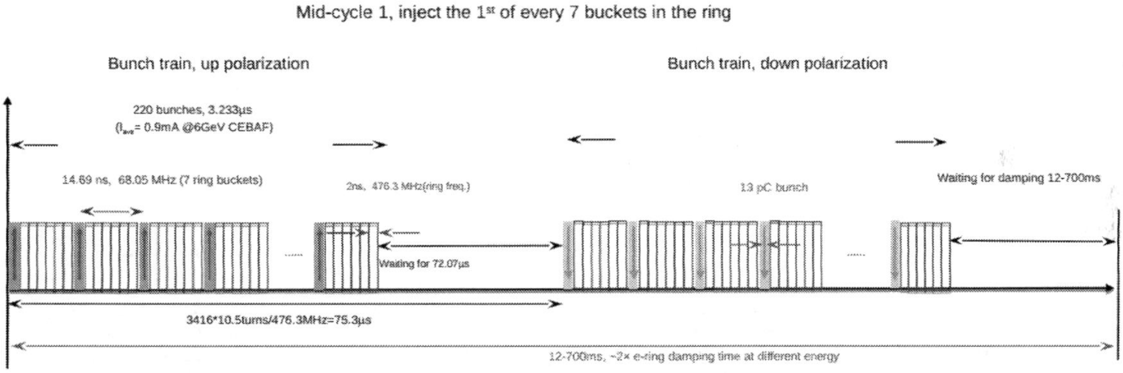

Figure 6. Injection scheme and microbunch helicity structure of the JLEIC beam

3 Performances of the Compton polarimeter at Jefferson Laboratory

3.1 Chicane layout

Compton polarimetry is commonly used to measure polarization for experiment where accuracy on the polarization beam is critical, the interaction cross-section of circularly polarized photons with longitudinally polarized electrons is different depending of the direction of the polarization of either the electron or photons. By varying one of the polarization one can measure the Compton asymmetry which is proportional to both photon and electron polarization. Compton polarimeters are considered non invasive meaning they can continuously monitor the beam polarization as long as the amount of Compton interaction is just a fraction of each electron bunch.

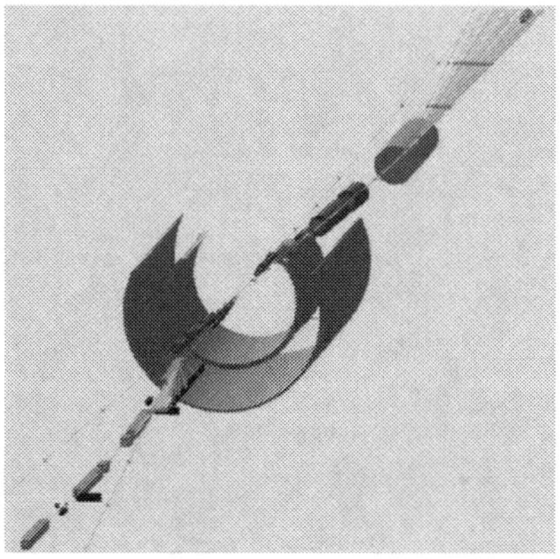

Figure 7. Display of a few events detected in the low Q^2 chicane

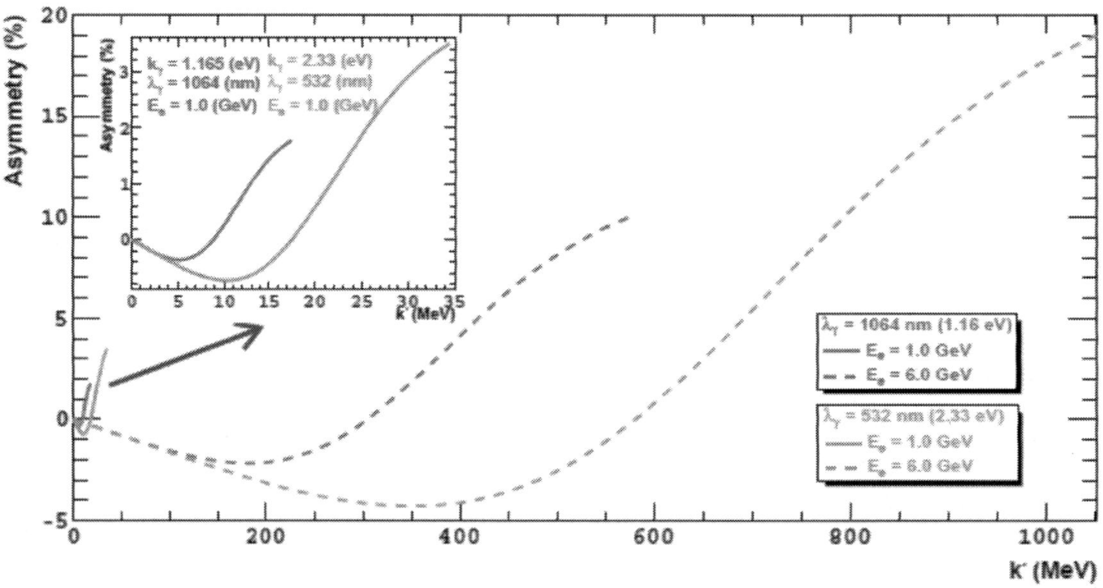

Figure 8. Compton asymmetry for different laser and beam energies

The Compton Scattering crossection can be accurately computed with QED, so one can extract the absolute beam polarization from the measured asymmetry. The Compton asymmetry is particularly well suited for measurement at higher energy where the Compton asymmetry is larger as shown in Fig. 8.

The photon polarization is known accurately to a few tenth of a percent level giving a measurement of the beam polarization. Sub-percent measurement of the polarization was achieved at Jefferson Laboratory in Hall A and Hall C.

The low Q^2 chicane is a good configuration for Compton polarimetry, it allows to separate the Compton electron from the Compton photon. For the current JLEIC design we will rely on a design similar to the one used at Jefferson Laboratory as show in the schematics Fig. 9.

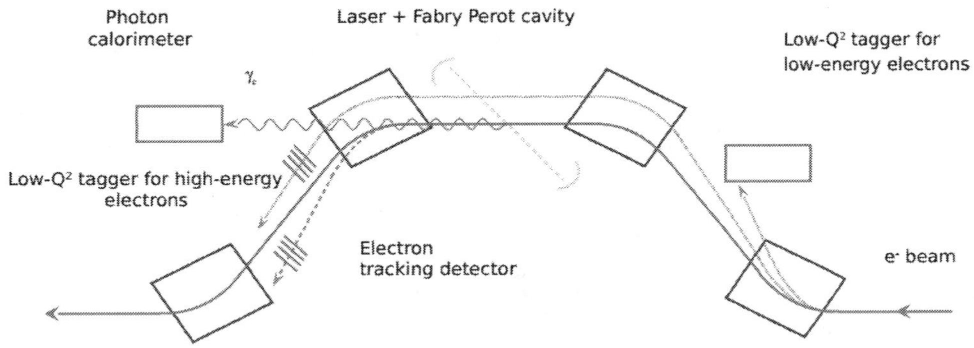

Figure 9. Schematic of the low Q^2 chicane layout including the Compton polarimeter

The photon source used at Jefferson Laboratory currently is a green CW laser amplified by a Perot Fabry cavity with a typical gain around 1500. It is placed between dipoles 2 and 3 with a small crossing angle with the electron beam.

Compton polarimetry can be done by detecting either the Compton Photon or Compton Electron, both methods have different systematic and are sensitive to different backgrounds. They can give two independent measurement of the polarization which is a good crosscheck.

3.2 Hall A HAPPEX experiment : photon detection using integrated method

A calorimeter is placed in front of the Compton interaction point behind the third dipole of the chicane. The Compton photons go straight in the calorimeter while the electron beam is bent to stay inside of the chicane. For the 12 GeV beam at Jefferson Laboratory, a lead tungstate calorimeter will be used while for energies lower than 6 GeV a GSO crystal was used. Two methods will be used both having different systematic. The one which was used for the HAPPEX experiment is a digital integration. A gate is open corresponding to one helicity window and all the samples are summed digitally [3], this allows to compute the integrated asymmetry without use of any thresholds as shown in Fig. 10.

For the HAPPEX experiment the total errors at 3 GeV are summarized in table 11

This yield a measurement accuracy of 0.93% at 3 GeV, which is the best photon Compton measurement at this energy up to date.

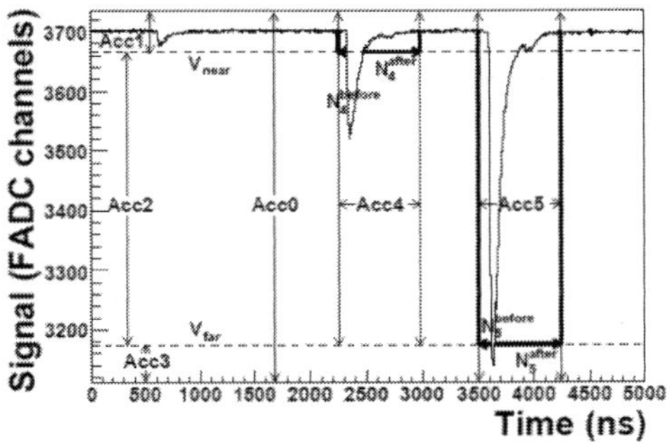

Figure 10. Digital FADC integration

Systematic Errors	
Laser Polarization	0.80%
Analyzing Power:	
Non-linearity	0.3%
Electron Energy Uncertainty	0.1%
Collimator Position	0.05%
MC Statistics	0.07%
Total on Analyzing Power	0.33%
Gain Shift:	
Background Uncertainty	0.31%
Pedestal Uncertainty	0.20%
Total on Gain Shift	0.37%
Total	**0.94%**

Figure 11. HAPPEX systematic errors summary

3.3 Hall C QWeak diamond electron detector

The Compton electrons having lost energy during the Compton interaction are bent more after the third dipole and are detected. A strip detector is placed in the dispersion plane then the electron energy is directly proportional to the position in the detector. In Hall C, a 4 planes with 96 strips per plane diamond detector was used as shown in Fig. 12 and installed as in Fig. 13.

Compton spectra are recorded for each beam helicity as shown in Fig. 14.

The Compton asymmetry can then be computed Fig. 15 strip by strip and by fitting the asymmetry one has access to the beam polarization.

Beam polarization was measured for the QWeak experiment at 0.6 % accuracy. [4]

4 EIC R&D and preliminary design

4.1 Photon source

For EIC, the current will be about 1000 higher than at Jefferson Laboratory, so one can envision the use of CW laser which have typical power up to 30 Watts. RF pulsed laser matched to the beam

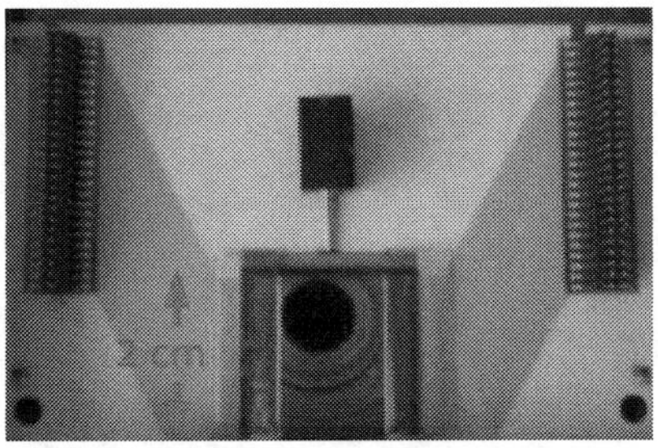

Figure 12. One of the Hall C Compton diamond detector plane

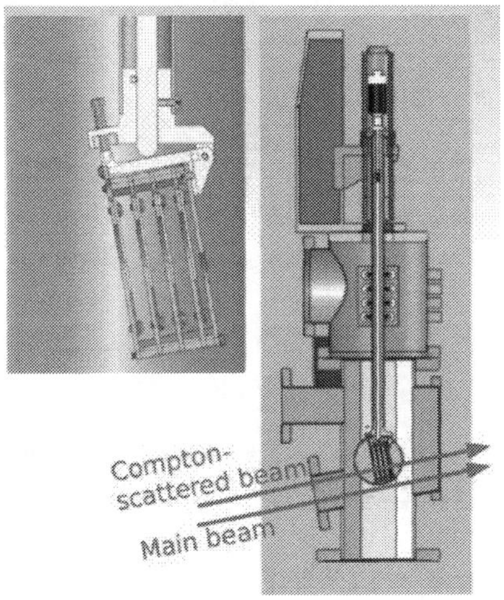

Figure 13. Layout of the Hall C Compton Electron detector

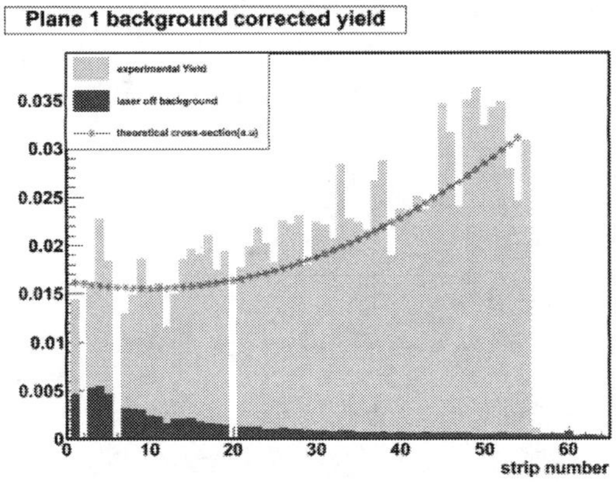

Figure 14. Typical Compton energy spectrum

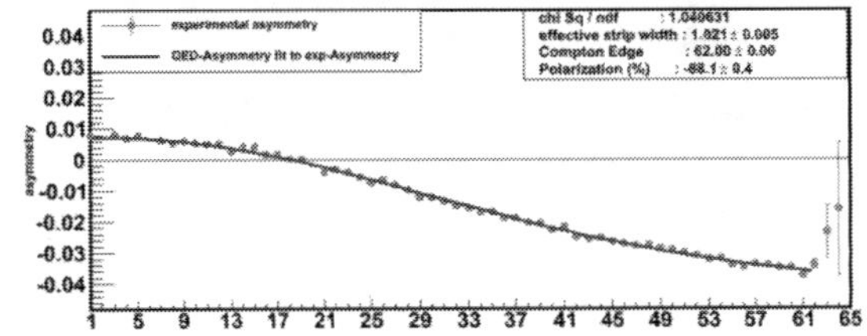

Figure 15. Compton electron detector asymmetry as a function of strip number (energy)

frequency gives a better luminosity with larger crossing angles by concentrating the energy in smaller bunches. This might prove to useful to fit the space constraints as shown in Fig. 16.

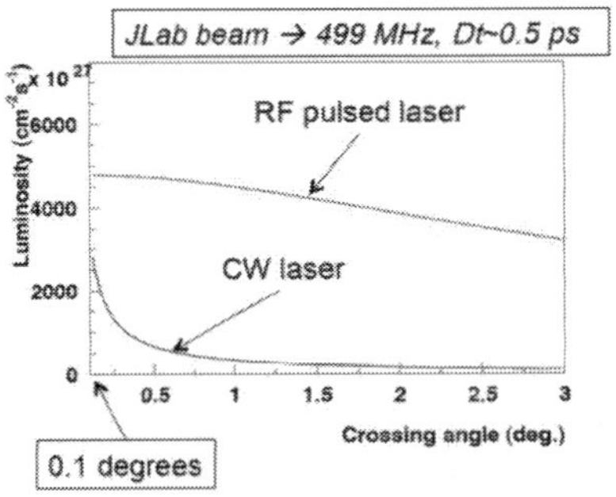

Figure 16. Luminosity as a function of crossing angle for CW and RF pulsed lasers

Having a single laser simplifies the design : there are less apertures which could interact with beam halo and one could also consider fast flipping of the laser polarization to be able to sample the polarization within a macrobunch. Though depending of the background, more powerful laser sources might be needed so optical cavities similar to the one used at Jefferson Laboratory could be chosen hence a major effort in simulation and beam tests to evaluate the sources of background. R&D for ILC is also ongoing for pulsed Perot Fabry cavity which would be the most powerful photon source available.

4.2 EIC electron detector design

Since the best measurement to date was the SLAC SLD experiment measurement at 21 GeV with a 0.52% measurement accuracy, we proposed to first focus on the Compton polarimetry detecting the Compton electrons for JLEIC. The Compton polarimetry at JLEIC is a bit more difficult than at SLAC because of a lower energy ranging from 3 to 11 GeV and dispersion, this implies having the detector fairly close from the beam making it very sensitive to background. Conditions at EIC are also very different to the current JLAB setup in terms of current, while 80 μA of current is available at JLAB, JLEIC will run high current going from 0.72A up to 3A. At such high currents, background from the electron beam can be significant and also shielding and cooling is needed to protect the detector from the large amount of RF power deposited from the beam. Such additional material can affect the shape of the detector response and introduce systematic errors on the polarization measurement.

Careful R&D and simulation is crucial to optimize the detector design, and make sure the detector reach the 1% accuracy while having sufficient radiation hardness to be able to run for at least a single run period.

With the successful experience of the Hall C detector the JLEIC baseline detector is diamond strip detector which has radiation hardness greater than 10 MRad.

In addition to the design and simulation effort a test stand is proposed at Jefferson Laboratory/ Since two sub-percent Compton polarimeters which can be crosschecked by Moller polarimeters are available at Jefferson Laboratory, it is an ideal place to cross check the simulation and make sure sub-percent polarization can be achieved at lower current in shielding conditions required for JLEIC allowing to control how well we can correct for the effect of additional material.

Another research point is optimizing the timing response of the detectors, silicon and diamond detectors usually require large amplification giving typical response time of the order of a microsecond with usual integration times. For example for eRHIC current design, several electron sources will be used switched at 10.8 MHz, so a detector response shorter than 100 ns is desirable to be able to measure the polarization of each individual source which usually have different polarization. A fast detector can also improve timing resolution which could be useful for example for the low Q^2 detector where an electron corresponding to a quasi-real photon will be recorded in coincidence with the main detector.

5 Conclusion

The JLAB EIC is a polarized electron and ion machine. It features a low Q^2 chicane allowing to look at quasi-real photons physics. Expected momentum resolution is 10^{-3}. This chicane will also host a Compton polarimeter. Compton polarization is a natural candidate for electron beam polarization measurements at EIC considering the high energy and high current available. Experience from Jefferson Laboratory has shown than sub-percent electron polarization measurements were feasible for both Compton photon and electron detection for an energy ranging 1 to 3 GeV. With the increase current additional shielding of the detector has to be done. An ongoing R&D effort (eRD15) focusing on the Compton electron detection has been funded from the EIC R&D fund. It will evaluate the effect of the shielding on the measurement will be evaluated through simulation. A test stand will be implemented at Jefferson Laboratory to validate the simulation, test radiation hardness of the detector and electronics and optimize the timing properties of the detector.

References

[1] F. Lin *et al.*, "PROGRESS ON THE DESIGN OF THE POLARIZED MEDIUM-ENERGY ELECTRON ION COLLIDER AT JLAB" Proceeding IPAC2015 TUYB3

[2] V. Morozov *et al.*, "Progress on the Interaction Region Design and Detector Integration at JLab's MEIC,"

[3] M. Friend *et al.*, "Upgraded photon calorimeter with integrating readout for Hall A Compton Polarimeter at Jefferson Lab," Nucl. Instrum. Meth. A **676**, 96 (2012) doi:10.1016/j.nima.2012.02.041 [arXiv:1108.3116 [physics.ins-det]].

[4] A. Narayan, Phd Thesis, Missisipi State University," Determination of electron beam polarization using electron detector in Compton polarimeter with less than 1% statistical and systematic uncertainty"

EPJ Web of Conferences **112**, 01008 (2016)
DOI: 10.1051/epjconf/201611201008

Lattice QCD calculations of transverse momentum-dependent parton distributions (TMDs)

M. Engelhardt[1,a], B. Musch[2], T. Bhattacharya[3], J. R. Green[4], R. Gupta[3], P. Hägler[2], S. Krieg[5,6], J. Negele[7], A. Pochinsky[7], A. Schäfer[2], S. Syritsyn[8], and B. Yoon[3]

[1] *Department of Physics, New Mexico State University, Las Cruces, NM 88003, USA*
[2] *Institut für Theoretische Physik, Universität Regensburg, 93040 Regensburg, Germany*
[3] *Theoretical Division, Los Alamos National Laboratory, Los Alamos, NM 87545, USA*
[4] *Institut für Kernphysik, Johannes Gutenberg-Universität Mainz, 55099 Mainz, Germany*
[5] *Bergische Universität Wuppertal, 42119 Wuppertal, Germany*
[6] *IAS, Jülich Supercomputing Centre, Forschungszentrum Jülich, 52425 Jülich, Germany*
[7] *Center for Theoretical Physics, Massachusetts Institute of Technology, Cambridge, MA 02139, USA*
[8] *Theory Center, Thomas Jefferson National Accelerator Facility, Newport News, VA 23606, USA*

Abstract. An ongoing program of evaluating TMD observables within Lattice QCD is reviewed, summarizing recent progress with respect to several challenges faced by such calculations. These lattice calculations are based on a definition of TMDs through hadronic matrix elements of quark bilocal operators containing staple-shaped gauge connections. A parametrization of the matrix elements in terms of invariant amplitudes serves to cast them in the Lorentz frame preferred for a lattice calculation. Data on the naively T-odd Sivers and Boer-Mulders effects as well as the transversity TMD are presented.

1 Introduction

In the description of hadron structure, transverse momentum-dependent parton distribution functions [1] (TMDs) play a role complementary to generalized parton distributions (GPDs). Whereas GPDs encode information about the transverse spatial distribution of partons (through Fourier transformation with respect to the momentum transfer), TMDs contain information about the transverse momentum distribution of partons. Cast in a Lorentz frame in which the hadron of mass m_h propagates with a large momentum in the 3-direction, $P^+ \equiv (P^0 + P^3)/\sqrt{2} \gg m_h$, the quark momentum components scale such that TMDs are principally functions $f(x, k_T)$ of the quark longitudinal momentum fraction $x = k^+/P^+$ and the quark transverse momentum vector k_T, with the dependence on the component $k^- \equiv (k^0 - k^3)/\sqrt{2} \ll m_h$ becoming ignorable in this limit. The function $f(x, k_T)$ will thus be regarded as having been integrated over k^-.

Experimentally, TMDs manifest themselves in angular asymmetries observed in processes such as semi-inclusive deep inelastic scattering (SIDIS) and the Drell-Yan (DY) process. Corresponding signatures have emerged at COMPASS, HERMES and JLab [2–4], and that has motivated targeting a significant part of the physics program at future experiments in this direction, e.g., at the upgraded

[a]e-mail: engel@nmsu.edu

JLab 12 GeV facility and at the proposed electron-ion collider (EIC). Relating the experimental signature to the hadron structure encoded in TMDs requires a suitable factorization framework, the one having been advanced in [5–8] being particularly well-suited for connecting phenomenology to Lattice QCD. Factorization in the TMD context is considerably more involved than standard collinear factorization, with the resulting TMDs in general being process-dependent, via initial and/or final state interactions between the struck quark and the hadron remnant.

2 Definition of TMD observables

The definition of TMD observables amenable to lattice evaluation has been laid out in detail in [9]. Summarizing briefly, the starting point is the fundamental correlator

$$\widetilde{\Phi}^{[\Gamma]}_{\text{unsubtr.}}(b, P, S, \ldots) \equiv \frac{1}{2}\langle P, S|\ \bar{q}(0)\ \Gamma\ \mathcal{U}[0, \eta v, \eta v + b, b]\ q(b)\ |P, S\rangle \tag{1}$$

where S denotes the spin of the hadron and Γ stands for an arbitrary γ-matrix structure. The staple-shaped gauge connection $\mathcal{U}$ follows straight-line paths connecting the positions given in its argument; the unit vector v thus specifies the direction of the staple, whereas η parametrizes its length. The presence of $\mathcal{U}$ introduces divergences in $\widetilde{\Phi}^{[\Gamma]}_{\text{unsubtr.}}$ additional to the wave function renormalizations of the quark operators; these divergences accordingly must ultimately be compensated by additional "soft factors", which are expected to be multiplicative and do not need to be specified in detail here, since only appropriate ratios in which they then presumably cancel will ultimately be considered. In order to regularize rapidity divergences, the staple direction v is taken slightly off the light cone into the space-like region [5, 6], with perturbative evolution equations governing the approach to the light cone [7]. A useful parameter characterizing how close v is to the light cone is the Collins-Soper evolution parameter $\hat{\zeta} = v \cdot P/(|v|\,|P|)$, in terms of which the light cone is approached for $\hat{\zeta} \to \infty$.

The correlator (1) can be decomposed in terms of invariant amplitudes $\widetilde{A}_{iB}$. The components relevant for the Sivers and Boer-Mulders effects, as well as the transversity, discussed below, are

$$\frac{1}{2P^+}\widetilde{\Phi}^{[\gamma^+]}_{\text{unsubtr.}} = \widetilde{A}_{2B} + im_h\epsilon_{ij}\, b_i\, S_j \widetilde{A}_{12B} \tag{2}$$

$$\frac{1}{2P^+}\widetilde{\Phi}^{[i\sigma^{i+}\gamma^5]}_{\text{unsubtr.}} = im_h\epsilon_{ij}\, b_j \widetilde{A}_{4B} - S_i \widetilde{A}_{9B} - im_h\Lambda b_i \widetilde{A}_{10B} + m_h[(b\cdot P)\Lambda - m_h(b_T\cdot S_T)]b_i \widetilde{A}_{11B} \tag{3}$$

where Λ denotes the hadron helicity (i.e., $S^+ = \Lambda P^+/m_h$, $S^- = -\Lambda m_h/2P^+$). These amplitudes are useful in that they can be evaluated in any desired Lorentz frame, including a frame that is particularly suited for the lattice calculation. Specializing, for present purposes, to TMDs integrated over momentum fraction x, by considering specifically[1] $b \cdot P = 0$, they serve to define the "generalized Sivers shift"

$$\langle k_y\rangle_{TU}(b_T^2, \ldots) = -m_h \widetilde{A}_{12B}(-b_T^2, \ldots)/\widetilde{A}_{2B}(-b_T^2, \ldots) = m_h \tilde{f}_{1T}^{\perp[1](1)}(b_T^2, \ldots)/\tilde{f}_1^{[1](0)}(b_T^2, \ldots) \tag{4}$$

where the right-hand expression introduces the notation in terms of Fourier-transformed TMD moments, for details, cf. [9, 10]; as discussed in [11], the use of Fourier-transformed TMDs is also advantageous in the analysis of experimental data. In the $b_T \to 0$ limit, (4) formally represents the average transverse momentum k_y of unpolarized ("U") quarks orthogonal to the transverse ("T") spin of the hadron, normalized to the corresponding number of valence quarks. Analogously, one can also extract the generalized Boer-Mulders shift

$$\langle k_y\rangle_{UT}(b_T^2, \ldots) = m_h \widetilde{A}_{4B}(-b_T^2, \ldots)/\widetilde{A}_{2B}(-b_T^2, \ldots) = m_h \tilde{h}_1^{\perp[1](1)}(b_T^2, \ldots)/\tilde{f}_1^{[1](0)}(b_T^2, \ldots) \tag{5}$$

[1] The treatment can be readily expanded to nonzero $b \cdot P$, providing access to the x-dependence of TMDs.

which, in the $b_T \rightarrow 0$ limit, formally represents the average transverse momentum k_y of quarks polarized in the transverse ("T") direction orthogonal to the k_y-direction in an unpolarized ("U") hadron, normalized to the corresponding number of valence quarks. Similarly, one can introduce a generalized tensor charge via the ratio of Fourier-transformed moments of the transversity and unpolarized TMDs,

$$\tilde{h}_1^{[1](0)}(b_T^2,\ldots)/\tilde{f}_1^{[1](0)}(b_T^2,\ldots) = -[\widetilde{A}_{9B}(-b_T^2,\ldots) - m_h^2 b^2 \widetilde{A}_{11B}(-b_T^2,\ldots)/2]/\widetilde{A}_{2B}(-b_T^2,\ldots)\,. \quad (6)$$

In the $b_T \rightarrow 0$ limit, this formally reduces to the tensor charge, normalized to the corresponding number of valence quarks. The ratios (4)-(6) are designed to cancel both multiplicative soft factors associated with the gauge connection $\mathcal{U}$ as well as wave function renormalizations attached to the quark operators in (1) at finite physical separation b. It should be emphasized that the construction presented up to this point has been developed within continuum QCD; transferring it verbatim to Lattice QCD constitutes a working assumption which ultimately requires further consideration. In particular, the breaking of space-time symmetries engendered by the lattice discretization may imply changes to the purely multiplicative nature of the soft factors and renormalization constants, and thus invalidate the cancellations invoked in the construction of the ratios (4)-(6). Pending a foundational investigation of these issues within the lattice formulation, empirical insight into possible renormalization effects can be obtained by studying the stability of TMD ratios such as (4)-(6) under substantial changes of the lattice discretization. This constitutes one of the topics addressed in the presentation of numerical results below.

3 Lattice evaluation and results

To access observables such as (4)-(6) within Lattice QCD, one calculates hadron matrix elements of the type (1) and then decomposes them into invariant amplitudes, as given in (2) and (3). For this to be possible, it is crucial to work in a scheme where the four-vectors b and v are generically space-like, for the following reason: By employing a Euclidean time coordinate to project out hadron ground states via Euclidean time evolution, Lattice QCD cannot straightforwardly accomodate operators containing Minkowski time separations. The operator of which one takes matrix elements thus has to be defined at a single time. Only if both b and v are space-like is there no obstacle to boosting the problem to a Lorentz frame in which b and v are purely spatial, and evaluating $\widetilde{\Phi}^{[\Gamma]}_{\text{unsubtr.}}$ in that frame. The results extracted for the invariant amplitudes $\widetilde{A}_{iB}$ are then immediately valid also in the original frame in which (1) was initially defined, thus completing the determination of quantities of the type (4)-(6).

Since, in a numerical lattice calculation, the staple extent η necessarily remains finite, two extrapolations must be performed from the generated data, namely, the one to infinite staple length, $\eta \rightarrow \infty$, and the extrapolation of the staple direction towards the light cone, $\hat{\zeta} \rightarrow \infty$. As discussed below, the former extrapolation is fairly straightforward. On the other hand, the extrapolation $\hat{\zeta} \rightarrow \infty$ constitutes a considerable challenge. Given that v is purely spatial in the Lorentz frame used for the lattice calculation, the accessible range of $\hat{\zeta}$ is determined by the available hadron spatial momenta P (in units of the hadron mass). The numerical signal achieved in lattice calculations rapidly deteriorates with rising hadron momentum, and thus only a very limited set of $\hat{\zeta}$ can be accessed. Data from a dedicated study of the Boer-Mulders shift in a pion [12] are discussed below to address this challenge; the lower mass of the pion compared to the nucleon facilitates reaching higher $\hat{\zeta}$ and makes a controlled extrapolation possible.

Finally, a general challenge faced by Lattice QCD calculations is reaching the physical pion mass. The computational expense of treating quark masses corresponding to the physical point has hitherto precluded the generation of lattice TMD data at that point. Instead, data are produced at artificially

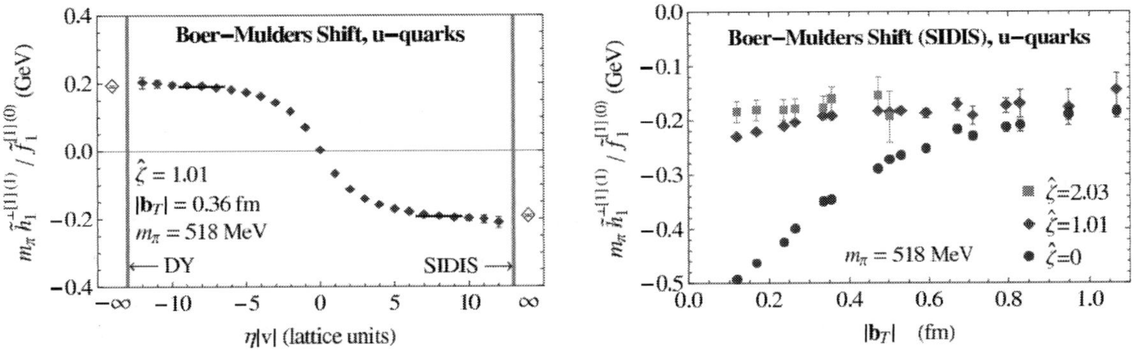

Figure 1. Generalized Boer-Mulders shift in a pion as a function of staple extent at fixed quark separation b_T and Collins-Soper evolution parameter $\hat{\zeta}$ (left); and as a function of b_T in the $\eta \to \infty$ SIDIS limit, at three different values of $\hat{\zeta}$ (right).

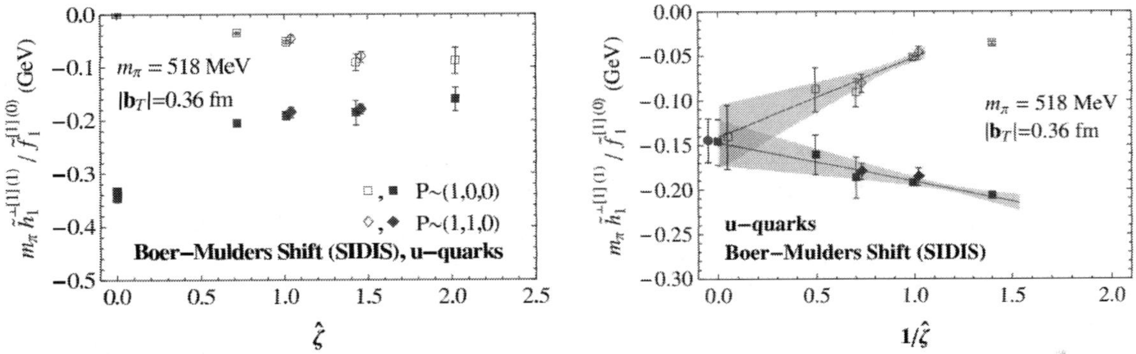

Figure 2. Generalized Boer-Mulders shift in a pion in the SIDIS limit as a function of Collins-Soper evolution parameter $\hat{\zeta}$ at fixed b_T (left); and as a function of $1/\hat{\zeta}$ together with fits using the form $c + d/\hat{\zeta}$ (right). Shown are both the full shift (filled symbols) as well as a partial contribution (open symbols), cf. comments in the main text. Fit curves are for separate fits to the two cases; the extrapolated data point shown with a circle symbol results from a simultaneous fit to both.

high pion masses. To also obtain preliminary insight into the pion mass dependence of TMD observables, data at differing pion masses are juxtaposed below.

Focusing first on the issue of $\hat{\zeta}$-extrapolation, Figs. 1 and 2 display data for the generalized Boer-Mulders shift (5) in a pion [12], obtained in a mixed action scheme employing domain wall valence quarks on a MILC 2+1-flavor asqtad quark ensemble with a lattice spacing of $a = 0.12$ fm and a pion mass of $m_\pi = 518$ MeV. In the pion case, the isovector quark combination vanishes; the shown data are for u-quarks alone, with the corresponding disconnected contributions omitted. Fig. 1 (left) exhibits results obtained at a given quark separation b_T and a given staple direction characterized by $\hat{\zeta}$, as a function of the staple extent. The T-odd behavior of the observable is evident, with $\eta \to \infty$ corresponding to the SIDIS limit, whereas $\eta \to -\infty$ yields the DY limit. The data level off to approach clearly identifiable plateaux as the staple length grows. The limiting SIDIS and DY values, represented by the open symbols, are extracted by imposing antisymmetry in η, allowing one to appropriately average the $\eta \to \pm\infty$ plateau values. Fig. 1 (right) summarizes the SIDIS limit values as a function of b_T for three different $\hat{\zeta}$. Remarkably, the b_T-dependence of the Boer-Mulders shift flattens as $\hat{\zeta}$ is increased, and the data for different $\hat{\zeta}$ approach each other at large b_T. It would be useful to understand

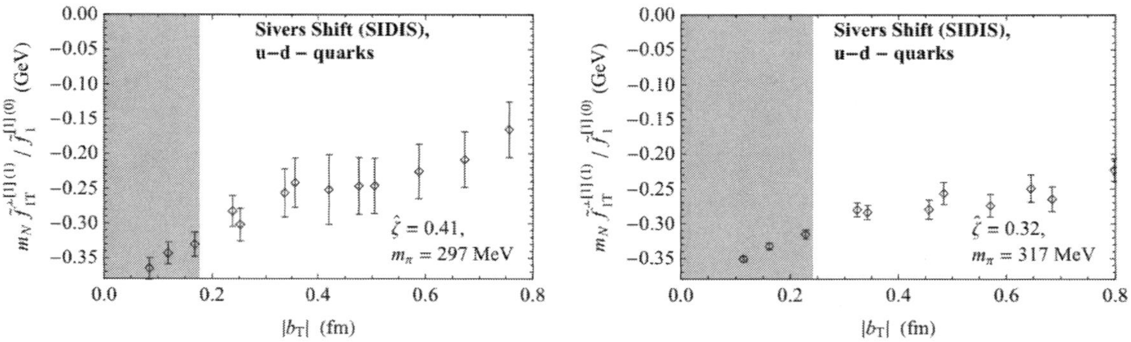

Figure 3. Generalized Sivers shift in a nucleon as a function of b_T in the $\eta \to \infty$ SIDIS limit, at a fixed $\hat{\zeta}$, in a fine lattice domain wall fermion calculation (left) and a coarse lattice clover fermion calculation (right), cf. main text for details.

this behavior in detail. Fig. 2 focuses on a particular value of b_T, displaying the $\hat{\zeta}$-dependence of both the full Boer-Mulders shift as well as a certain partial contribution which vanishes at $\hat{\zeta} = 0$, but dominates the quantity at large $\hat{\zeta}$; comparison of the full Boer-Mulders shift with the partial contribution thus can give an indication of convergence towards the large $\hat{\zeta}$ limit. For further details, cf. [9, 12]. The partial contribution already furnishes roughly one half of the full shift at the highest $\hat{\zeta}$ reached, signaling that the calculation has covered a significant part of the evolution to large $\hat{\zeta}$. The right-hand panel of Fig. 2 shows an extrapolation to the large $\hat{\zeta}$ limit using the functional form $c + d/\hat{\zeta}$ (analogous fits using the form $c + d/\hat{\zeta}^2$ are seen to be inferior [12]). Given that the range of $\hat{\zeta}$ accessed numerically does not clearly overlap with the regime in which perturbative evolution equations become applicable, this form should be viewed as no more than a physically motivated ad hoc ansatz. The fits to the full and partial data converge to compatible values, agreeing also with a combined fit to both data sets. This buttresses confidence in the extrapolations and demonstrates that lattice calculations can achieve a signal for the Boer-Mulders shift of sufficient quality such that it survives taking the $\hat{\zeta} \to \infty$ limit.

Turning to the issue of the dependence of TMD ratios on the lattice discretization, which provides an empirical test of whether renormalization effects indeed cancel in such observables, Fig. 3 shows representative results for the generalized Sivers shift (4) in the nucleon. Results for the isovector, $u-d$ quark combination are displayed; in this channel, couplings of the operator insertion to disconnected quark loops in the nucleon, which have not been evaluated, cancel. The figures juxtapose results obtained on a RBC/UKQCD 2+1-flavor domain wall fermion ensemble featuring a lattice spacing of $a = 0.084$ fm and a pion mass of $m_\pi = 297$ MeV with results obtained on a 2+1-flavor clover fermion ensemble provided by K. Orginos and collaborators in the Jefferson Lab lattice group; the latter has a lattice spacing of $a = 0.114$ fm and a pion mass of $m_\pi = 317$ MeV. The two ensembles thus have very similar pion masses, but differ substantially in the lattice discretization. The domain wall fermion ensemble not only features a considerably finer spacing, but also respects chiral symmetry to the largest extent possible, whereas the clover fermion ensemble strongly breaks the continuum symmetry. Fig. 3, analogously to Fig. 1 (right), shows the SIDIS limit results as a function of b_T at a given $\hat{\zeta}$, where the shaded area below $|b_T| = 2a$ indicates the region where the data may be significantly affected by finite lattice cutoff effects. The results for the Sivers shift obtained on the two ensembles in question are compatible, despite the considerable differences in terms of discretization scheme. This suggests that the renormalization effects embodied in soft factors and wave function renormalization constants indeed cancel in TMD ratios also in the lattice formulation.

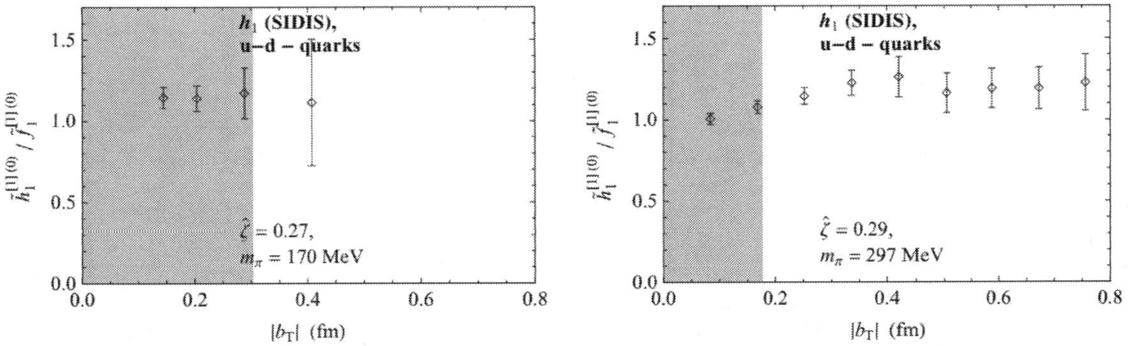

Figure 4. Generalized tensor charge as a function of b_T in the $\eta \to \infty$ SIDIS limit, at a fixed $\hat{\zeta}$, in domain wall fermion calculations at $m_\pi = 170$ MeV (left) and $m_\pi = 297$ MeV (right).

Finally, a preliminary indication of the pion mass dependence of TMD ratios in the chiral regime is given by the juxtaposition in Fig. 4. It compares data for the generalized tensor charge of the nucleon (6) obtained on two RBC/UKQCD 2+1-flavor domain wall fermion ensembles, one with a lattice spacing of $a = 0.144$ fm and a pion mass of $m_\pi = 170$ MeV, and the other with a lattice spacing of $a = 0.084$ fm and a pion mass of $m_\pi = 297$ MeV, cf. also Fig. 3 (left). The lighter pion mass data are affected by considerably stronger statistical fluctuations; the $m_\pi = 170$ MeV calculation employed 8 source-sink pairs on each of 310 lattices, i.e., 2480 samples, for each matrix element; the $m_\pi = 297$ MeV calculation 8 source-sink pairs on 533 lattices, i.e., 4264 samples. Fig. 4 displays, analogously to Fig. 3, the SIDIS limit data for the generalized tensor charge as a function of b_T at a given $\hat{\zeta}$. Keeping in mind that the sizeable statistical uncertainties at $m_\pi = 170$ MeV preclude a stringent comparison, the results at the two pion masses are compatible; no significant variation with m_π is apparent even this far into the chiral regime.

4 Summary and outlook

TMDs can be formally defined in terms of hadronic matrix elements of quark bilocal operators containing staple-shaped gauge connections, which incorporate final/initial state effects in SIDIS/DY processes. Evaluating such matrix elements within Lattice QCD, one encounters several challenges. For one, to cancel multiplicative soft factors and renormalization constants, appropriate ratios of Fourier-transformed TMDs, cf. (4)-(6), are constructed. However, the breaking of space-time symmetries engendered by the lattice discretization may imply changes to the purely multiplicative nature of the soft factors and renormalization constants contained in the continuum definition; this possibility was addressed empirically by studying the universality of TMD ratios under changes of the discretization scheme. Secondly, the gauge connection staples are generically taken off the light cone to regularize rapidity divergences, with the Collins-Soper parameter $\hat{\zeta}$ controlling the approach to the light cone. A dedicated study of pion TMD ratios demonstrated that the attendant extrapolation in $\hat{\zeta}$ is feasible within Lattice QCD calculations. Thirdly, TMD calculations must ultimately approach the physical pion mass; the pion mass dependence of TMD ratios has been explored, and no significant variations with pion mass have hitherto been observed, although more accurate, higher statistics data at light pion masses are needed to draw any substantive conclusions about the m_π-dependence of TMD ratios as the physical pion mass is approached. These studies buttress the ongoing program of calculating TMD observables within Lattice QCD presented here. Beyond the TMD observables discussed above, the developed methods are also being put to use to study generalized observables containing a nonzero momentum transfer in (1). These can be related to quark orbital angular momentum in the nucleon, since they provide mixed quark position and momentum information in the transverse plane. First exploratory data in this direction are becoming available.

Acknowledgements

Computations were performed using resources provided by the U.S. DOE Office of Science through the National Energy Research Scientific Computing Center (NERSC), a DOE Office of Science User Facility, under Contract No. DE-AC02-05CH11231, as well as through facilities of the USQCD Collaboration, employing the Chroma software suite [13]. The MILC and RBC/UKQCD collaborations are gratefully acknowledged for providing gauge ensembles analyzed in this work, as are K. Orginos (supported by DOE grant DE-FG02-04ER41302) and the Jefferson Lab lattice group (supported by DOE grant DE-AC05-06OR23177, under which Jefferson Science Associates, LLC, operates Jefferson Laboratory). Support by the Heisenberg-Fellowship program of the DFG (P.H.), SFB/TRR-55 (S.K.,A.S.), the PRISMA Cluster of Excellence at the University of Mainz (J.G.), and the RIKEN Foreign Postdoctoral Researcher Program (BNL) as well as the Nathan Isgur Fellowship (JLab) (S.S.) is acknowledged. This work was furthermore supported by the U.S. DOE and the Office of Nuclear Physics through grants DE-FG02-96ER40965 (M.E.), DE-SC0011090 (J.N.) and DE-FC02-06ER41444 (A.P.). R.G., T.B. and B.Y. are supported by DOE grant DE-KA-1401020 and the LDRD program at LANL.

References

[1] D. Boer, M. Diehl, R. Milner, R. Venugopalan, W. Vogelsang, *et al.*, `arXiv:1108.1713`.
[2] M. Alekseev, *et al.*, COMPASS Collaboration, Phys. Lett. **B673**, 127 (2009).
[3] A. Airapetian, *et al.*, HERMES Collaboration, Phys. Rev. Lett. **103**, 152002 (2009).
[4] H. Avakian, *et al.*, CLAS Collaboration, Phys. Rev. Lett. **105**, 262002 (2010).
[5] S. M. Aybat and T. C. Rogers, Phys. Rev. **D 83**, 114042 (2011).
[6] J. C. Collins, *Foundations of Perturbative QCD* (Cambridge University Press, Cambridge, 2011).
[7] S. M. Aybat, J. C. Collins, J.-W. Qiu and T. C. Rogers, Phys. Rev. **D 85**, 034043 (2012).
[8] J. C. Collins and T. C. Rogers, Phys. Rev. **D 87**, 034018 (2013).
[9] B. Musch, P. Hägler, M. Engelhardt, J. W. Negele and A. Schäfer, Phys. Rev. **D 85**, 094510 (2012).
[10] D. Boer, L. Gamberg, B. Musch and A. Prokudin, JHEP **1110**, 021 (2011).
[11] M. Aghasyan, *et al.*, JHEP **1503**, 039 (2015).
[12] M. Engelhardt, P. Hägler, B. Musch, J. W. Negele and A. Schäfer, `arXiv:1506.07826`.
[13] R. G. Edwards and B. Joó, SciDAC Collaboration, Nucl. Phys. Proc. Suppl. **140**, 832 (2005).

EPJ Web of Conferences **112**, 01009 (2016)
DOI: 10.1051/epjconf/201611201009

Probing Spin Dependent Quark and Gluon Distributions Through Azimuthal and Polarization Asymmetries

Gary R. Goldstein[1,a] and Simonetta Liuti[2,3,b]

[1] *Department of Physics and Astronomy, Tufts University, Medford, MA 02155 USA.*
[2] *Department of Physics, University of Virginia, Charlottesville, VA 22904, USA.*
[3] *Laboratori Nazionali di Frascati, INFN, Frascati, Italy*

Abstract. Spin and transverse momentum dependent parton distributions - Generalized Parton Distributions, Transverse Momentum Distributions and related distributions - are at the interface between the QCD structure of the hadrons and observable quantities. The distributions are contained as linear superpositions within helicity amplitudes that factorize into universal forms at leading or next to leading order. These amplitudes are probed in high energy leptoproduction processes through angular dependent cross sections and polarization asymmetries. The phenomenological extraction of the amplitudes and the distributions is a challenging task. We will present some models for the distributions in leptoproduction processes, as well as the observables that connect with the quark-nucleon helicity amplitudes for different Deeply Virtual production possibilities.

1 Introduction

The spin of the hadrons depends on the distributions of spin and orbital angular momenta (OAM) of the fundamental constituents, quarks and gluons. The quark and gluon field correlations in the nucleon are indirectly measurable through electroproduction processes. The angular momenta associated with the quark and gluon fields within QCD, are encoded in the transverse momentum distributions (TMDs) and the Generalized Parton Distributions (GPDs). These can provide a 3-dimensional picture of the angular momentum structure.

Of special interest among pdf's are the nucleon's *transversity* structure functions, $h_1(x)$ - the probability of finding a definite transversity quark inside a transversely polarized nucleon. They are chiral odd, and can be observed in Semi Inclusive Deep Inelastic Scattering (SIDIS), where they are convoluted with fragmentation functions, or in the Drell-Yan process in conjunction with another chiral-odd partner. They also contribute to exclusive electroproduction processes, particularly Deeply Virtual Meson Production (DVMP), through chiral odd GPDs. The *transversity* GPD $H_T^q(x,\xi,t)$ has the limiting form $H_T^q(x,0,0) = h_1^q(x)$. Hence, from the DVMP processes, transversity can be determined. This will be a main focus of this presentation.

In advance of the development of the formalism and the brief presentation of our model, we show some important results - the transversity $h_1(x)$ in Fig 1(a) and the predicted asymmetry $A_{UT}^{\sin(\phi-\phi_s)}(x_{Bj},t,Q^2)$, that is particularly sensitive to the transversity, in Fig. 1(b).

[a]e-mail: gary.goldstein@tufts.edu
[b]e-mail: sl4y@virginia.edu

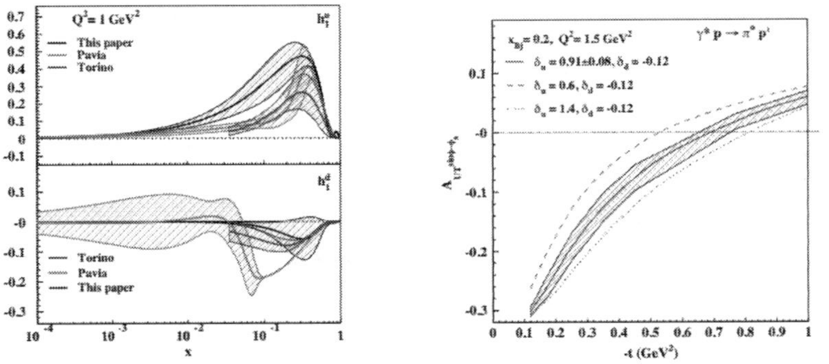

Figure 1. Left (a): transversity, h_1^q plotted vs. x at $Q^2 = 1$ GeV^2, for the u quarks (upper panel) and for the d quark (lower panel). Besides our analysis, the recent extractions from the Pavia group [1] obtained from dihadron production in a collinear framework, and from the Torino group [2] obtained combining data on polarized SIDIS single hadron production [3, 4], and dihadron production from e^+e^- annihilation [5]. Right (b): asymmetry $A_{UT}^{\sin(\phi-\phi_S)}$, plotted vs. $-t$, at $x_{Bj} = 0.2$, $Q^2 = 1.5$ GeV^2 for the $\gamma^* p \to \pi^0 p'$ reaction. The graph shows the sensitivity of the asymmetry to variations of the tensor charge, or the precision that is needed in measurements of this quantity in order to reduce the size of the errors. Adapted from Ref. [6].

2 Formalism

DVCS and DVMP can be described within QCD factorization, through the convolution of specific GPDs and hard scattering amplitudes. There are four chiral-even GPDs, $H, E, \widetilde{H}, \widetilde{E}$ [9] and four additional chiral-odd GPDs, known to exist by considering twist-two quark operators that flip quark helicity by one unit, $H_T, E_T, \widetilde{H}_T, \widetilde{E}_T$ [10, 11]. All GPDs depend on two additional kinematical invariants besides the parton's Light Cone (LC) momentum fraction, x, and the DVCS process' four-momentum transfer, Q^2, namely $t = \Delta^2$ where $\Delta = P - P'$ is the momentum transfer between the initial and final protons, and ξ, or the fraction of LC momentum transfer, $\xi = \Delta^+/(P^+ + P'^+)$. The observables containing the various GPDs are the Compton Form Factors (CFFs) - convolutions over x of GPDs with the struck quark propagator. The quark GPDs are defined (at leading twist) as the matrix elements of the following projection of the unintegrated quark-quark proton correlator (see Ref.[11] for a detailed overview),

$$W^{\Gamma}_{\Lambda',\Lambda}(x,\xi,t) \;=\; \frac{1}{2}\int \frac{dz^-}{2\pi} e^{ix\overline{P}^+ z^-} \langle P', \Lambda' \mid \overline{\psi}\left(-\frac{z}{2}\right)\Gamma\psi\left(\frac{z}{2}\right) \mid P, \Lambda\rangle\Big|_{z^+=0, z_T=0}, \qquad (1)$$

where $\Gamma = \gamma^+, \gamma^+\gamma_5, i\sigma^{i+}\gamma_5 (i = 1, 2)$, and the target's spins are Λ, Λ'.

For the chiral-odd case, $\Gamma = i\sigma^{i+}\gamma_5$, $W^{\Gamma}_{\Lambda',\Lambda}$ was parametrized as [11],

$$\begin{aligned} W^{[i\sigma^{i+}\gamma_5]}_{\Lambda',\Lambda}(x,\xi,t) \;&=\; \frac{1}{2\overline{P}^+}\overline{U}(P',\Lambda')\Bigg(i\sigma^{+i}H_T(x,\xi,t) + \frac{\gamma^+\Delta^i - \Delta^+\gamma^i}{2M}E_T(x,\xi,t) \\ &+\; \frac{P^+\Delta^i - \Delta^+P^i}{M^2}\widetilde{H}_T(x,\xi,t) + \frac{\gamma^+P^i - P^+\gamma^i}{2M}\widetilde{E}_T(x,\xi,t)\Bigg)U(P,\Lambda) \end{aligned} \qquad (2)$$

The spin structures of GPDs that are directly related to spin dependent observables are most effectively expressed in term of helicity dependent amplitudes, developed extensively for the covariant description of two body scattering processes (see also Ref.[11]).

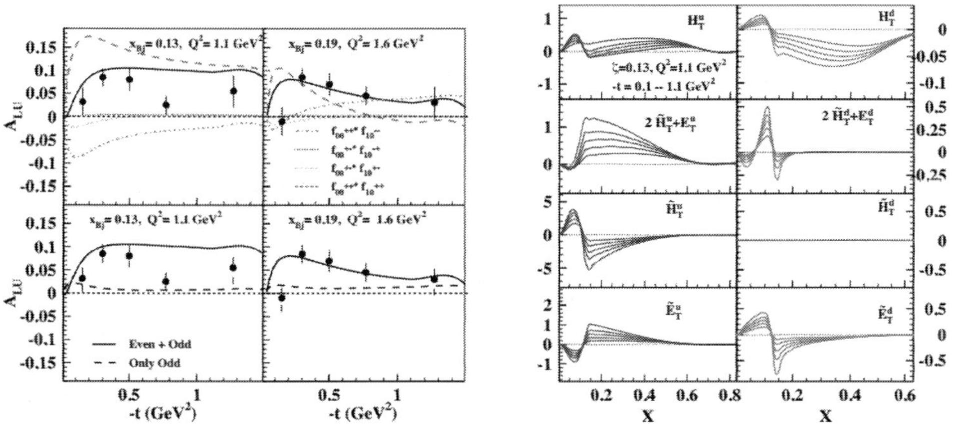

Figure 2. (Color online) Left (a): Beam spin asymmetry, A_{LU}, plotted vs. $-t$ for two different kinematics. Experimental data from Ref.[7]. Adapted from Ref. [8]. Right (b): The chiral-odd GPDs evaluated using the model described in the text plotted vs. X at $x_{Bj} = \zeta = 0.13$, $Q^2 = 2$ GeV^2. The range in $-t$ is: $0.1 \leq -t \leq 1.1$ GeV^2. Curves with the largest absolute values correspond to the lowest t. Adapted from Ref. [8]

We see that there are 8 quark GPDs per flavor (at leading twist). Correspondingly, the quark-nucleon amplitudes have the form N→ q :: q′ → N′ with 8 independent helicity amps. There are two questions to address: How to model the 8 GPDs? How to measure them? DVCS accesses Chiral Even GPDs through various cross sections and asymmetries. The GPDs, or their corresponding Compton Form Factors, enter linearly via Bethe-Heitler ⊗ DVCS interference. DVπ^0S accesses 2 Chiral Even + 4 Chiral Odd GPDs that enter bilinearly via $d\sigma/d\Omega$ & polarization asymmetries. The result of experimental observations that $d\sigma_T > d\sigma_L$ is that the chiral odd GPDs dominate.

In Ref.[12], after showing how DVπ^0P can be described in terms of chiral-odd GPDs, we estimated all of their contributions to the various observables with particular attention to the ones which were sensitive to the values of the tensor charge. Subsequently, an extension of the Flexible Reggeized Diquark model in the chiral-even sector in Refs.[13, 14] to the chiral odd sector was accomplished through the use of Parity and Charge Conjugation symmetry relations obeyed by the various helicity structures in the model.The chiral even GPDs integrate to the nucleon form factors, which constrains the GPD t-dependence,

$$\int_0^1 H^q(X,\zeta,t) = F_1^q(t), \int_0^1 E^q(X,\zeta,t) = F_2^q(t), \int_0^1 \widetilde{H}^q(X,\zeta,t) = G_A^q(t), \int_0^1 \widetilde{E}^q(X,\zeta,t) = G_P^q(t). \quad (3)$$

where $F_1^q(t)$ and $F_2^q(t)$ are the Dirac and Pauli form factors for the quark q components in the nucleon. $G_A^q(t)$ and $G_P^q(t)$ are the axial and pseudoscalar form factors. Furthermore, $H(x,0,0) = h_1(x)$ and $\widetilde{H}(x,0,0) = g_1(x)$.

3 Helicity, model parameters

The connection of the correlator, Eq.(2), with the helicity amplitudes proceeds by introducing [8, 12],

$$f_{\Lambda_\gamma 0}^{\Lambda\Lambda'}(\zeta,t) \quad = \quad \sum_{\lambda,\lambda'} g_{\Lambda_\gamma 0}^{\lambda\lambda'}(X,\zeta,t,Q^2) \otimes A_{\Lambda'\lambda',\Lambda\lambda}(X,\zeta,t), \quad (4)$$

where the helicities of the virtual photon and the initial proton are, Λ_γ, Λ, and the helicities of the produced pion and final proton are 0, and Λ', respectively. This describes a factorization into a "hard part", $g^{\lambda\lambda'}_{\Lambda_\gamma 0}$ for the partonic subprocess $\gamma^* + q \to \pi^0 + q$, and a "soft part" given by the quark-proton helicity amplitudes, $A_{\Lambda',\lambda';\Lambda,\lambda}$ that contain the GPDs. The amplitudes $A_{\Lambda'\lambda',\Lambda\lambda}$ implicitly contain an integration over the unobserved quark's transverse momentum, k_T, and are functions of $x_{Bj} = Q^2/2M\nu \approx \zeta, t$ and Q^2. The convolution integral in Eq.(4) is given by $\otimes \to \int_{-\zeta+1}^{1} dX$.

The expressions for the chiral-odd helicity amplitudes in terms of GPDs [11] are of the form

$$A_{++,--} = \sqrt{1-\xi^2}\left[H_T + \frac{t_0 - t}{4M^2}\widetilde{H}_T - \frac{\xi^2}{1-\xi^2}E_T + \frac{\xi}{1-\xi^2}\widetilde{E}_T\right], \ldots \tag{5}$$

where we use the symmetric notation for the kinematic variables. Analogous forms have been written for the chiral even and odd sectors [11].

For a transverse photon, inserting the expressions for g^{+-}_{10} and the A's into Eqs.(4) we obtain a set of helicity amplitudes that enter the observables. They include

$$f^{+-}_{10} = \frac{g^{V,odd}_\pi(Q) + g^{A,odd}_\pi(Q)}{2}\sqrt{1-\xi^2}\left[\mathcal{H}_T + \frac{t_0 - t}{4M^2}\widetilde{\mathcal{H}}_T - \frac{\xi^2}{1-\xi^2}\mathcal{E}_T + \frac{\xi}{1-\xi^2}\widetilde{\mathcal{E}}_T\right] \tag{6}$$

and analogous terms ([8, 12]), where $\mathcal{H}_T$, etc., are the Compton form factors. In the t-channel picture, which has its roots in a Regge analysis of this process [15], one separates the $J^{PC} = 1^{--}$ and $J^{PC} = 1^{+-}$ contributions to the amplitudes for transverse and longitudinal virtual photons, respectively, thus generating two different types of Q^2 dependence at the π^0 vertex [16].

For a longitudinal photon one has the convolution of g^{+-}_{00} with the A helicity amplitudes . There are contributions for the longitudinal photon from the chiral even $\mathcal{H}$, $\mathcal{E}$ as well (see Ref. [8]). Following from the chiral-even case, in π^0 electroproduction one obtains longitudinal photon amplitudes [17] that are significant. (see Ref. [8, 13]).

Our model for evaluating the chiral-odd GPDs extends our reggeized diquark model, which was already configured for chiral-even GPDs, to the chiral-odd sector. The parameters were initially fixed by a fit applied recursively first to PDFs, and to the nucleon form factors. The model was shown to reproduce data on different observables in DVCS (charge [18], longitudinal [19] and transverse [18] single spin asymmetries). A comparison with data from more recent analysis has also been shown in Ref.[20]. Recently [14], we presented a new fit that uses the form factor flavor separated data from Ref.[21].

The basic structures in our model are the quark-proton scattering amplitudes at leading order with proton-quark-diquark vertices. The quark parton helicity amplitudes describe a two body process, $q'(k')P \to q(k)P'$, where $q(k)$ corresponds to the "struck quark". The intermediate diquark system, X, can have $J^P = 0^+$ (scalar), or $J^P = 1^+$ (axial vector). We start from the region $X \geq \zeta$. At fixed M_X, the amplitudes read for the Scalar diquark (see [12, 13] for the full set of relations):

$$A^{(0)}_{\Lambda'\lambda',\Lambda\lambda} = \int d^2k_\perp \phi^*_{\Lambda'\lambda'}(k', P')\phi_{\Lambda\lambda}(k, P). \tag{7}$$

Next we consider reggeization that is, we extend the diquark model formalism to low X by allowing the spectator system's mass to vary up to very large values. This is accomplished by convoluting the GPD structures obtained in Eqs.(7) with a "spectral function", $\rho(M_X^2)$, where M_X^2 is the spectator's mass,

$$F^q_T(X,\zeta,t) = N_q \int_0^\infty dM_X^2 \rho(M_X^2) F_T^{(m_q,M^q_\Lambda)}(X,\zeta,t;M_X). \tag{8}$$

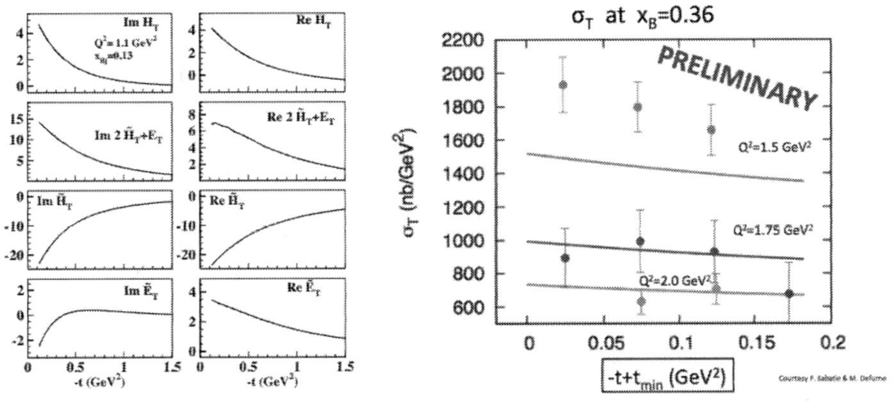

Figure 3. Left (a): Chiral-odd CFFs entering the process $\gamma^* p \to \pi^0 p'$. The various CFFs are plotted vs. $-t$ for the kinematic bin $x_{Bj} = 0.13$, $Q^2 = 1.1$ GeV^2. Adapted from Ref. [8]. Right (b): Transverse differential cross section vs. $(-t + t_{min})$ for $Q^2 = 1.5, 1.75, 2.0$ GeV^2 at $x_{Bj} = 0.36$. Preliminary data from Hall A, courtesy F. Sabatie & M. Defurne [23].

The spectral function was constructed in Refs.[13, 14] so that it approximately behaves as $(M_X^2)^\alpha$ for $M_X^2 \to \infty$ and $\delta(M_X^2 - \overline{M}_X^2)$ for M_X^2 at a few GeV^2, where $0 < \alpha < 1$, and $\overline{M}_X$ is in the GeV range. We parametrized this in a practical form as,

$$F_T^q(X,\zeta,t) \approx N_q X^{-\alpha_q + \alpha_q'(X)t}\, F_T^{(m_q,M_\Lambda^q)}(X,\zeta,t;\overline{M}_X) = R_{p_q}^{\alpha_q,\alpha_q'}(X,\zeta,t)\, G_{M_X,m}^{M_\Lambda}(X,\zeta,t) \tag{9}$$

where $\alpha_q'(X) = \alpha_q'(1-X)^{p_q}$. The functions $G_{M_X,m_q}^{M_\Lambda^q}$ and $R_{p_q}^{\alpha_q,\alpha_q'}$ are the quark-diquark and Regge contributions, respectively.

4 Observables & Data

The fitting procedure of GPDs is quite complicated owing to its many different steps: 1) the construction of chiral-odd helicity amplitudes; 2) the connection of these amplitudes to the chiral-even ones using Parity relations within spectator models; 3) the fixing of chiral-even parameters at an initial scale, Q_o^2, using the nucleon form factors and PQCD evolution to match DIS data; 4) the determination of chiral-odd GPDs; 5) the construction of the corresponding Compton form factors, and pseudoscalar meson electroproduction observables. In the following figures we show the GPDs at various kinematic ranges and some of the CFFs. A more detailed description of the other transversity functions including the first moment of $h_1^\perp \equiv 2\widetilde{H}_T^q + E_T^q$ is given in [8].

In Fig.2 (b) we show the t-dependent chiral odd GPDs that enter the helicity amplitudes evaluated in Eqs. 5 in a kinematical bin ($x_{Bj} = 0.13, Q^2 = 1.1$ GeV^2) consistent with the Jefferson Lab kinematical coverage.

In Fig.3(a) we show the proton CFFs, Eq.(6), which enter the $\gamma^* p \to \pi^0 p'$ reaction. The various GPDs enter the helicity amplitudes and those, in turn, determine all the cross section terms for π^0 electroproduction. The transverse and longitudinal cross sections have been separated experimentally at small t [23]. Some preliminary data compare favorably with our predictions in Fig 3(b).
(1) All GPD contributions should be considered separately. In particular, H_T, $\widetilde{H}_T$, and E_T are dominating; $\widetilde{E}_T$ is non zero in our model but small. Although the combination $2\widetilde{H}_T + E_T$ might be considered

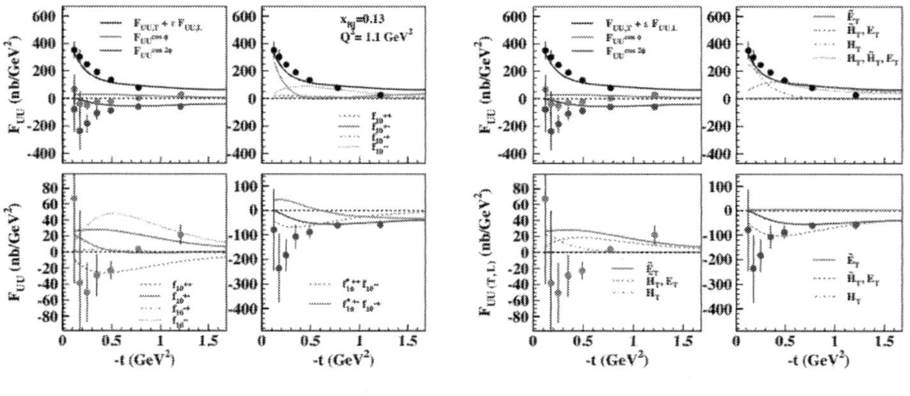

Figure 4. (Color online) LEFT: Unpolarized cross section components, $F_{UU,T} + \epsilon F_{UU,L}$, $F_{UU}^{\cos 2\phi}$, and $F_{UU}^{\cos\phi}$ in the kinematical bin, $x_{Bj} = 0.13$, $Q^2 = 1.2$ GeV^2. The upper left panel shows all components along with the data from Ref.[22]. The other panels show the contributions from the various helicity amplitudes. The right upper panel similarly shows $F_{UU,T} + \epsilon F_{UU,L}$. See Ref. [8], from which these are adapted, for details.

more fundamental in that its spin structure corresponds to the Boer-Mulders function [11]. $\widetilde{H}_T$, and E_T appear separately, and are multiplied by different factors in the amplitudes. $2\widetilde{H}_T + E_T$ should just be viewed as a forward limit.
(2) The behavior of f_{10}^{++} and f_{10}^{--} is determined by $\widetilde{H}_T$, and E_T. As a consequence of point (1), f_{10}^{--} is sensibly different from f_{10}^{++}. In particular, because of the different multiplicative factors, $f_{10}^{--} < f_{10}^{++}$.
(3) f_{10}^{+-} is determined by H_T at small $|t|$, and by E_T at large $|t|$.
(4) f_{10}^{-+} is determined by $\widetilde{H}_T$ only, but it is small due to the $|t|$ factor suppression.
(5) The longitudinal photon contributions, f_{00}^{+-}, and f_{00}^{++} are suppressed in the chiral-odd case.

In Fig. 4 we show unpolarized cross section components, $F_{UU,T} + \epsilon F_{UU,L}$, $F_{UU}^{\cos 2\phi}$, and $F_{UU}^{\cos\phi}$ as functions of t, for the kinematics $x_{Bj} = 0.13$, $Q^2 = 1.2$ GeV^2. The caption indicates the different contributions.

The unpolarized $\sin\phi$ modulation, $F_{LU}^{\sin\phi}$ describes the beam asymmetry, A_{LU},

$$A_{LU} = \sqrt{\epsilon(1-\epsilon)}\,\frac{F_{LU}^{\sin\phi}}{F_{UU,T} + \epsilon F_{UU,L}} \tag{10}$$

A_{LU} is shown in Fig.2(a) for two of the Jefferson Lab Hall B kinematical bins along with the different amplitudes contributions, in this case the products: $(f_{10}^{++*} f_{00}^{++})$, $(f_{10}^{--*} f_{00}^{++})$, $(f_{10}^{-+*} f_{00}^{+-})$ and $(f_{10}^{+-*} f_{00}^{+-})$. Notice that the longitudinally polarized amplitudes receive contributions from both the chiral-even and chiral-odd GPDs From the graph (lower panels) one can see a definite dominance of the chiral-even GPDs. We deduce that A_{LU} is not favored for the extraction of chiral-odd GPDs. The A_{UT} in Fig. 1(b) is quite sensitive to the transversity, however.

5 Conclusions & Outlook

The transversity parton distributions in the nucleon can be accessed through deeply virtual exclusive π^0 meson production. We addressed the feasibility of experimental extraction. This represents a consistent quantitative step with respect to our previous work [12]. In particular, only H_T and the combination $2\widetilde{H}_T + E_T$, were considered, but now they are separated. A similar simplified approach was taken also in Ref.[24] - we differ in the importance attached to the skewedness dependence of E_T, $\tilde{E}_T$.

We see the results of our extended approach for the many measured and measurable observables. What is especially gratifying is that certain asymmetries constrain the GPDs well enough to separately determine H_T, and consequently transversity through the limit $H_T(x, 0, 0)$, and the combination $2\widetilde{H}_T + (1 \pm \xi)E_T$.

Acknowledgements

We are grateful to the organizers of POETIC 6 for a productive meeting in an excellent setting. Work of S.L. supported by U.S. D.O.E. grant DE-FG02-01ER4120.

References

[1] A. Bacchetta, A. Courtoy and M. Radici, JHEP **1303**, 119 (2013).
[2] M. Anselmino, *et al.* Phys. Rev. D **87**, 094019 (2013).
[3] A. Airapetian *et al.* [HERMES Collaboration], Phys. Rev. Lett. **94**, 012002 (2005).
[4] COMPASS collaboration, E. Ageev *et al.*, Nucl. Phys. **B765**, 31 (2007).
[5] Belle collaboration, K. Abe *et al.*, Phys. Rev. Lett. **96**, 232002 (2006).
[6] G. R. Goldstein, J. .O. Gonzalez Hernandez and S. Liuti, arXiv:1401.0438 [hep-ph].
[7] R. De Masi, *et al.*, Phys. Rev. **C77**, 042201 (2008)
[8] G. R. Goldstein, J. .O. Gonzalez Hernandez and S. Liuti, Phys. Rev. D **91**, 114013 (2015).
[9] X. D. Ji, Phys. Rev. D **55**, 7114 (1997)
[10] P. Hoodbhoy and X. Ji, Phys. Rev. D 58, 054006 (1998).
[11] M. Diehl, Eur. Phys. Jour. C **19**, 485 (2001); *ibid*, Phys. Rept. **388**, 41 (2003).
[12] S. Ahmad, G. R. Goldstein and S. Liuti, Phys. Rev. D**79**, 054014 (2009).
[13] G. R. Goldstein, J. O. Hernandez and S. Liuti, Phys. Rev. D **84**, 034007 (2011).
[14] J. O. Gonzalez-Hernandez, S. Liuti, G. R. Goldstein and K. Kathuria, Phys. Rev. C **88**, 065206 (2013).
[15] G. R. Goldstein and J. F. Owens, Phys. Rev. D **7** (1973) 865.
[16] G. R. Goldstein, J. O. Hernandez and S. Liuti, J. Phys. G: Nucl. Part. Phys. **39** (2012) 115001.
[17] M . Vanderhaeghen, P. A. M. Guichon and M. Guidal, Phys. Rev. D **60**, 094017 (1999).
[18] A. Airapetian *et al.* [HERMES Collaboration], JHEP **0806**, 066 (2008); *ibid*, JHEP **0911**, 083 (2009).
[19] F. X. Girod *et al.* [CLAS Collaboration], Phys. Rev. Lett. **100**, 162002 (2008).
[20] E. Seder, *et al.* [CLAS], Phys. Rev. Lett. **114**, 032001 (2015); S. Pisano, *et al.* [CLAS], Phys. Rev. **D91**, 052014 (2015).
[21] G. D. Cates, C. W. de Jager, S. Riordan and B. Wojtsekhowski, Phys. Rev. Lett. **106**, 252003 (2011).
[22] I. Bedlinskiy, *et al.*, Phys.Rev.Lett. 109, 112001(2012).
[23] F. Sabatie, "DVCS & π^0 electroproduction at Hall A, Jefferson Lab", presentation at CIPANP2015, Vail, Colorado (May 2015).
[24] S. V. Goloskokov, P. Kroll, Eur. Phys. J. **A47**, 112 (2011).

EPJ Web of Conferences **112**, 01010 (2016)
DOI: 10.1051/epjconf/201611201010

Husimi distribution for nucleon tomography

Yoshitaka Hatta[1,a] and Yoshikazu Hagiwara[2]

[1] *Yukawa Institute for Theoretical Physics, Kyoto University, Kyoto 606-8502, Japan*
[2] *Department of Physics, Kyoto University, Kyoto 606-8502, Japan*

Abstract. We define the QCD Husimi distribution as the phase space distribution of partons inside the nucleon. Compared to the more well-known Wigner distribution, the Husimi distribution is better behaved and positive. It thus allows for a probabilistic interpretation and can be used to define the 'entropy' of the nucleon as a measure of complexity of the partonic structure. A possible connection to the Color Glass Condensate approach at small-x is also discussed.

1 Introduction

Over the past two decades, the study of the nucleon structure has literally expanded its dimensions. In addition to the ordinary parton distribution function (PDF) $f(x)$ which describes the one-dimensional (longitudinal) structure of the nucleon, the transverse-momentum-dependent distribution (TMD) $T(x, \vec{k}_\perp)$ and (the Fourier transform of) the generalized parton distribution (GPD) $G(x, \vec{b}_\perp)$ have become common theoretical tools to reveal the three-dimensional imaging of the nucleon. Moreover, the five-dimensional Wigner distribution $W(x, \vec{k}_\perp, \vec{b}_\perp)$ [1] has been conceived as the phase space distribution of quarks and gluons inside the nucleon [2]. The Wigner distribution has the favorable property that, upon integration over one variable, it reduces to the known distribution in the other variable: $\int d^2\vec{b}_\perp W(x, \vec{k}_\perp, \vec{b}_\perp) = T(x, \vec{k}_\perp)$, $\int d^2\vec{k}_\perp W(x, \vec{k}_\perp, \vec{b}_\perp) = G(x, \vec{b}_\perp)$. On the other hand, it cannot be literally interpreted as a probability distribution in phase space because it is not positive definite and often behaves badly. This dilemma is actually well-known in nonrelativistic quantum mechanics. A less known fact is that the Wigner distribution is not the unique phase space distribution in quantum mechanics, and there exists a positive definite distribution called the Husimi distribution [3]. In this contribution to the proceedings we discuss the potential use of the Husimi distribution in the context of the nucleon structure. For more details, see [4].

2 Wigner and Husimi distributions in quantum mechanics

Traditionally in quantum mechanics, the Wigner distribution [1] has been most commonly used as a quantum analog of the phase space distribution $f(q, p)$ in classical mechanics. For a state $|\psi\rangle$, it is defined by

$$f_W(q,p) = \int dx e^{-ipx/\hbar} \langle \psi | q - x/2 \rangle \langle q + x/2 | \psi \rangle \,. \tag{1}$$

[a] e-mail: hatta@yukawa.kyoto-u.ac.jp

It immediately follows that

$$\int \frac{dq}{2\pi\hbar} f_W(q,p) = |\langle\psi|p\rangle|^2 , \qquad \int \frac{dp}{2\pi\hbar} f_W(q,p) = |\langle\psi|q\rangle|^2 . \tag{2}$$

From these properties, it is tempting to interpret $f_W(q,p)$ as the probability distribution in phase space as in classical mechanics. However, this is not possible because f_W is not positive definite. We illustrate this point in the one-dimensional harmonic oscillator case for which the Wigner distribution can be analytically computed

$$f_W(q,p) = 2(-1)^n e^{-\frac{2H}{\hbar\omega}} L_n\left(\frac{4H}{\hbar\omega}\right) , \tag{3}$$

where n is the excited level, L_n is the Laguerre polynomial and $H = \frac{p^2}{2m} + \frac{m\omega^2 q^2}{2}$ is the classical Hamiltonian. Except for the ground state $n = 0$, (3) is not positive definite, and as n increases it oscillates more and more frequently. This is shown in Fig. 1(left) for $n = 4$. While the oscillations encode the quantum interference effects, the naive interpretation as a phase space density drastically fails. The source of this difficulty is the uncertainty principle

$$\Delta q \Delta p \gtrsim \frac{\hbar}{2} , \tag{4}$$

which tells that q and p cannot be determined simultaneously. Thus, in quantum systems, the very notion of 'phase space distribution' is ill-defined from the outset.

Nevertheless, it is physically reasonable to speak of a probability distribution in some averaged sense, as long as the averaging is done over a region in phase space larger than (4). A concrete realization of this idea is the Husimi distribution [3] which is obtained by the Gaussian smearing of the Wigner distribution

$$f_H(q,p) = \frac{1}{\pi\hbar} \int dq' dp' \, e^{-m\omega(q'-q)^2/\hbar - (p'-p)^2/m\omega\hbar} f_W(q',p') , \tag{5}$$

where the Gaussian widths in q' and p' are inversely related such that the product $\Delta q' \Delta p' = \hbar/2$ satisfies the minimal uncertainty. Remarkably, the Husimi distribution thus defined is positive (semi)definite. One can prove that

$$f_H(q,p) = |\langle\psi|\lambda\rangle|^2 \geq 0 , \tag{6}$$

where $|\lambda\rangle$ is the so-called coherent state which is the eigenstate of the annihilation operator $\hat{a}|\lambda\rangle = \lambda|\lambda\rangle$ and $\lambda = \frac{m\omega q + ip}{\sqrt{2\hbar m\omega}}$ is the eigenvalue. Together with the normalization condition $\int \frac{dqdp}{2\pi\hbar} f_H(q,p) = 1$, the Husimi distribution can thus be legitimately interpreted as a probability distribution. Since its invention in 1940, the Husimi distribution has been used for numerous applications in various branches of modern physics [5]. In Fig. 1(right), we plot the Husimi distribution of the harmonic oscillator

$$f_H(q,p) = \frac{1}{n!} e^{-\frac{H}{\hbar\omega}} \left(\frac{H}{\hbar\omega}\right)^n , \tag{7}$$

for $n = 4$. The peak is localized around the classical trajectory $H \approx 4\hbar\omega$, which makes perfect sense. It is very difficult to get this physical picture from the corresponding Wigner distribution. In the left figure of Fig. 1, the first thing one notices is the peak in the center, and one might wonder if there is important physics going on there. But it is just an artifact of the particular definition of the Wigner distribution.

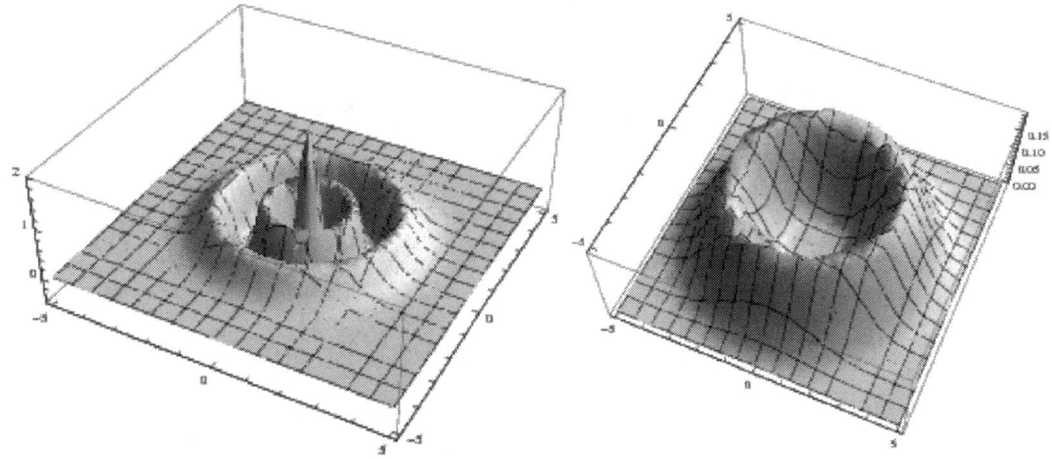

Figure 1. The Wigner distribution (left) and the Husimi distribution (right) in the (q, p) plane of the 4th excited state of the one-dimensional harmonic oscillator ($m = \omega = \hbar = 1$).

3 Wigner and Husimi distributions in QCD

Let us now turn to QCD. The quark Wigner distribution is defined by [2, 6]

$$W(x, \vec{b}_\perp, \vec{k}_\perp) = \int \frac{dz^- d^2z_\perp}{16\pi^3} \frac{d^2\Delta_\perp}{(2\pi)^2} e^{i(xP^+z^- - \vec{k}_\perp \cdot \vec{z}_\perp)} \langle P + \Delta/2 | \bar{\psi}(b - z/2)\gamma^+ \mathcal{L} \psi(b + z/2) | P - \Delta/2 \rangle \,, \quad (8)$$

where $|P\rangle$ is the nucleon state and $\mathcal{L}$ is the U-shaped Wilson line which connects the two points $b \pm z/2$ via the light-cone infinity $z^- = \pm\infty$. The gluon Wigner distribution can be similarly defined. An important difference with respect to the nonrelativistic case (1) is that the initial and final states are different due to the momentum recoil $\vec{\Delta}_\perp$ which inevitably occurs when probing a relativistic system. Integrating over $\vec{b}_\perp$, one find the TMD distribution

$$\int d^2\vec{b}_\perp W(x, \vec{b}_\perp, \vec{k}_\perp) = \int \frac{dz^- d^2z_\perp}{16\pi^3} e^{i(xP^+z^- - \vec{k}_\perp \cdot \vec{z}_\perp)} \langle P | \bar{\psi}(-z/2)\gamma^+ \mathcal{L} \psi(z/2) | P \rangle = T(x, \vec{k}_\perp) \,. \quad (9)$$

The Wigner distribution has been computed in many models [2, 6–12]. In models based on constituent quarks without gluons, one typically obtains a smooth positive distribution. However, once gluons are included, it behaves very badly (see below). This motivates us to define the QCD Husimi distribution

$$\begin{aligned} H(x, \vec{b}_\perp, \vec{k}_\perp) &\equiv \frac{1}{\pi^2} \int d^2b'_\perp d^2k'_\perp e^{-\frac{1}{\ell^2}(\vec{b}_\perp - \vec{b}'_\perp)^2 - \ell^2(\vec{k}_\perp - \vec{k}'_\perp)^2} W(x, \vec{b}'_\perp, \vec{k}'_\perp) \quad (10) \\ &= \int \frac{dz^- d^2z_\perp}{16\pi^3} \frac{d^2\Delta_\perp}{(2\pi)^2} e^{i(xp^+z^- - \vec{k}_\perp \cdot \vec{z}_\perp)} e^{-\ell^2 \frac{\Delta_\perp^2}{4} - \frac{z_\perp^2}{4\ell^2}} \langle P + \Delta/2 | \bar{\psi}(b - z/2)\gamma^+ \mathcal{L} \psi(b + z/2) | P - \Delta/2 \rangle \,. \end{aligned}$$

As before, the widths of the two Gaussian factors are inversely related to each other. The value of ℓ is in principle arbitrary, but there is a natural choice. To see this, note that the $\vec{b}_\perp$-integral of the Husimi distribution does *not* reduce to the TMD distribution

$$\int d^2\vec{b}_\perp H(x, \vec{b}_\perp, \vec{k}_\perp) = \int \frac{dz^- d^2z_\perp}{16\pi^3} e^{i(xp^+z^- - \vec{k}_\perp \cdot \vec{z}_\perp)} e^{-\frac{z_\perp^2}{4\ell^2}} \langle P | \bar{\psi}(-z/2)\gamma^+ \mathcal{L} \psi(z/2) | P \rangle \,. \quad (11)$$

The extra Gaussian factor $e^{-z_\perp^2/4\ell^2}$ may seem worrisome, but actually it is not something completely unfamiliar. It can be viewed as an effective nonperturbative cutoff on the $z_\perp$-integral which mimics confinement. (See also a discussion at the end.) Indeed, if we approximate $\ell \gtrsim z_\perp \approx 0$ in the matrix element, we find

$$\int d^2\vec{b}_\perp H(x,\vec{b}_\perp,\vec{k}_\perp) \approx \frac{1}{\pi\langle k_\perp^2\rangle} e^{-\vec{k}_\perp^2/\langle k_\perp^2\rangle} f(x)\,, \tag{12}$$

where $f(x)$ is the ordinary PDF and we identified $1/\ell^2 = \langle k_\perp^2\rangle$, the average transverse momentum squared. (12) agrees with the factorized ansatz for the TMD which has been successfully employed in phenomenology [13]. This suggests that $1/\ell$ should be of the order of the typical transverse momentum of partons.

The double moment of the Husimi distribution is the ordinary PDF.

$$\int d^2b_\perp d^2k_\perp H(x,\vec{b}_\perp,\vec{k}_\perp) = \int d^2b_\perp d^2k_\perp W(x,\vec{b}_\perp,\vec{k}_\perp) = f(x)\,. \tag{13}$$

Similarly, the canonical orbital angular momentum can be computed both from the Husimi and Wigner distributions [6, 14]

$$L_{can} = \int dx d^2b_\perp d^2k_\perp(\vec{b}_\perp \times \vec{k}_\perp)H(x,\vec{b}_\perp,\vec{k}_\perp) = \int dx d^2b_\perp d^2k_\perp(\vec{b}_\perp \times \vec{k}_\perp)W(x,\vec{b}_\perp,\vec{k}_\perp)\,. \tag{14}$$

4 One-loop example

We now come to the issue of positivity. Actually, in the QCD case one cannot rigorously prove that the Husimi distribution as defined in (10) is positive definite because it is not the forward matrix element due to the momentum recoil $\Delta_\perp$. However, the Gaussian factor $e^{-\ell^2\Delta_\perp^2/4}$ suppresses large values of $\Delta_\perp$, and there is a good chance that (10) can be treated as a positive function for practical purposes. Here we demonstrate this in a one-loop calculation. The one-loop Wigner distribution for a single quark (or an electron in QED, up to the Casimir factor) is [8]

$$W(x,\vec{b}_\perp,\vec{k}_\perp) = \frac{\alpha_s C_F}{2\pi^2}\int \frac{d^2\Delta_\perp}{(2\pi)^2} e^{-i\vec{\Delta}_\perp\cdot\vec{b}_\perp} \frac{\vec{q}_+\cdot\vec{q}_- P_{qq}(x) + m^2(1-x)^3}{(q_+^2 + m^2(1-x)^2)(q_-^2 + m^2(1-x)^2)}\,, \tag{15}$$

where $P_{qq}(x) = \frac{1+x^2}{1-x}$ is the splitting function and $\vec{q}_\pm \equiv \vec{k}_\perp \pm \frac{\vec{\Delta}_\perp}{2}(1-x)$. The $\Delta_\perp$-integral can be done only numerically, but one already sees that (15) is a badly-behaved function. Firstly, the $\Delta_\perp$-integral is divergent when $b_\perp = 0$. One thus needs a cutoff and the result depends rather strongly on the cutoff. Second, the coefficient of P_{qq} is negative when $k_\perp < \Delta_\perp$, and this makes W itself negative for small $k_\perp$. Furthermore, the factor $e^{-i\vec{\Delta}_\perp\cdot\vec{b}_\perp}$ oscillates rapidly at large $b_\perp$ and this leads to oscillations in W.

All these problems can be resolved by switching to the Husimi distribution (10). The Gaussian factor provides an effective cutoff $\Delta_\perp < 1/\ell$, so there is no problem of convergence. Moreover, the widths of the two Gaussian factors are deliberately related such that the smearing is done in a region larger than the region where the coefficient of P_{qq} is negative. In Fig. 2, we plot the numerical result in the $\vec{k}_\perp$-space at fixed $\vec{b}_\perp = (0.5\text{GeV}^{-1}, 0)$ (see also [8]). As expected, there is a sharp negative peak in the small-$k_\perp$ region in the Wigner distribution (left). This is hard to interpret intuitively because one is supposedly calculating the phase space density of the quark number $\sim \bar{\psi}\gamma^+\psi$, but there is no antiquark in this one-loop example. Thanks to the Gaussian smearing, even a function like this can be converted to a smooth positive Husimi distribution as shown in the right plot. We have surveyed different sets of parameters and found no evidence of negative regions as far as we could see. Thus, our construction is working, and the QCD Husimi distribution can be interpreted as the probability distribution of partons inside the nucleon.

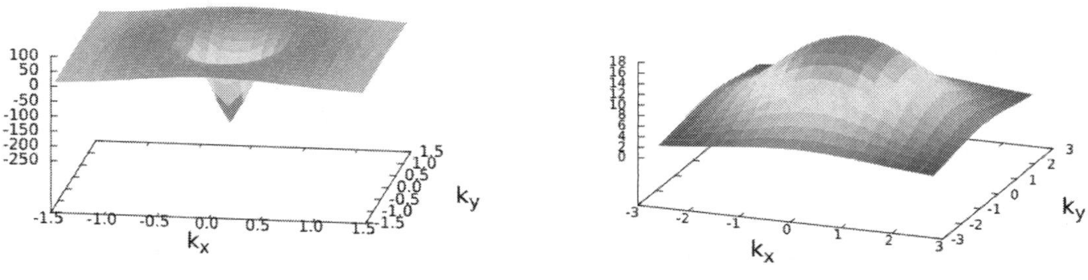

Figure 2. The one-loop Wigner distribution (left) and the Husimi distribution (right) for a single quark in the $\vec{k}_\perp$-space. We set $x = 0.5$, $m^2 = 0.1\text{GeV}^2$ and $\ell = 1\text{GeV}^{-1}$. The units of $k_{x,y}$ are in GeV^{-1}. The Wigner distribution is computed with a cutoff $\Delta_\perp^{max} = 5\text{GeV}$.

5 Discussions

We conclude with two speculative remarks which may be worth pursuing in future work.

5.1 Entropy

In quantum mechanics, since the Husimi distribution f_H is positive definite, one can take the logarithm and define an entropy called the Husimi-Wehrl entropy [15]

$$S \equiv -\int \frac{dqdp}{2\pi\hbar} f_H \ln f_H\,. \tag{16}$$

This is the classical counterpart of the quantum (von Neumann) entropy $S_q = -\text{tr}\hat{\rho}\ln\hat{\rho}$ ($\hat{\rho}$ is the density matrix), but unlike the latter, it is nonzero even for a pure state $\hat{\rho} = |\psi\rangle\langle\psi|$. It thus serves as the quantitative measure of complexity (or 'chaoticity') of pure quantum states. It is interesting to similarly define the 'entropy of the nucleon' from the QCD Husimi distribution

$$S(x) \equiv -\int d^2b_\perp d^2k_\perp H \ln H\,. \tag{17}$$

This characterizes quantitatively how 'complex' the nucleon wavefunction is due to the internal distribution of quarks and gluons at a given value of x. See also [16–19] for related discussions.

5.2 Relation to Color Glass Condensate

We have argued that the parameter $1/\ell$ should be of the order of the average transverse momentum $\langle k_\perp\rangle$ in the nucleon. When $x \sim O(1)$, $\langle k_\perp\rangle$ is a nonperturbative scale, of the order of Λ_{QCD}. However, at small-x and/or in a large nucleus, there arises a new perturbative scale, called the *saturation momentum* $Q_s(x)$, which sets the typical transverse momentum scale [20]. An effective theory of the nucleon structure in this regime is the Color Glass Condensate (CGC). It is then natural to choose $1/\ell = Q_s(x)$. One then notices that the Gaussian factor in (11) becomes formally identical to the so-called dipole S-matrix

$$S(z_\perp) = e^{-z_\perp^2 Q_s^2(x)/4}\,, \tag{18}$$

which is frequently encountered in the calculational framework of the CGC. This may not be a coincidence. The idea of the CGC is that at small-x there are so many gluons that they can be effectively treated as a classical, coherent field. On the other hand, the Husimi distribution is the coherent state expectation value of the density matrix. In a sense, the Husimi distribution aims to maximize the classical aspects of a given quantum state. Thus the two approaches can be related. As the parton density grows at small-x, the classical description of the nucleon/nucleus becomes more appropriate. The matrix elements computed within the quasiclassical/CGC approach could be interpreted as the Husimi distribution.

References

[1] E. Wigner, Phys. Rev. **40**, 749 (1932).
[2] A. V. Belitsky, X. Ji and F. Yuan, Phys. Rev. D **69**, 074014 (2004).
[3] K. Husimi, Proc. Phys. Math. Soc. Jpn. **22**, 264 (1940).
[4] Y. Hagiwara and Y. Hatta, Nucl. Phys. A **940**, 158 (2015) [arXiv:1412.4591 [hep-ph]].
[5] H. W. Lee, Phys. Rep. **259**, 147 (1995).
[6] C. Lorce and B. Pasquini, Phys. Rev. D **84**, 014015 (2011).
[7] C. Lorce, B. Pasquini, X. Xiong and F. Yuan, Phys. Rev. D **85**, 114006 (2012).
[8] A. Mukherjee, S. Nair and V. K. Ojha, Phys. Rev. D **90**, 014024 (2014).
[9] K. Kanazawa, C. Lorce, A. Metz, B. Pasquini and M. Schlegel, Phys. Rev. D **90**, 014028 (2014).
[10] A. Mukherjee, S. Nair and V. K. Ojha, Phys. Rev. D **91**, no. 5, 054018 (2015) [arXiv:1501.03728 [hep-ph]].
[11] T. Liu and B. Q. Ma, Phys. Rev. D **91**, 034019 (2015) [arXiv:1501.07690 [hep-ph]].
[12] A. Courtoy, G. R. Goldstein, J. O. G. Hernandez, S. Liuti and A. Rajan, arXiv:1412.0647 [hep-ph].
[13] M. Anselmino, in these proceedings; M. Anselmino, M. Boglione, J. O. Gonzalez Hernandez, S. Melis and A. Prokudin, JHEP **1404**, 005 (2014) [arXiv:1312.6261 [hep-ph]].
[14] Y. Hatta, Phys. Lett. B **708**, 186 (2014) [arXiv:1111.3547 [hep-ph]].
[15] A. Wehrl, Rev. Mod. Phys. **50**, 221 (1978).
[16] K. Kutak, Phys. Lett. B **705**, 217 (2011) [arXiv:1103.3654 [hep-ph]].
[17] R. Peschanski, Phys. Rev. D **87**, no. 3, 034042 (2013) [arXiv:1211.6911 [hep-ph]].
[18] H. Tsukiji, H. Iida, T. Kunihiro, A. Ohnishi and T. T. Takahashi, PTEP **2015**, no. 8, 083A01 (2015) [arXiv:1505.04698 [hep-ph]].
[19] A. Kovner and M. Lublinsky, Phys. Rev. D **92**, no. 3, 034016 (2015) [arXiv:1506.05394 [hep-ph]].
[20] L. D. McLerran and R. Venugopalan, Phys. Rev. D **49**, 2233 (1994).

EPJ Web of Conferences **112**, 01011 (2016)
DOI: 10.1051/epjconf/201611201011

Wide Angle Compton Scattering within the SCET factorization framework

Nikolay Kivel[a]

[1] *Helmholtz Institut Mainz, Johannes Gutenberg-Universität, D-55099 Mainz, Germany*

Abstract. Existing data for the electromagnetic proton form factors and for the cross section of the wide angle Compton scattering (WACS) show that the hard two-gluon exchange mechanism (collinear factorization) is still not applicable in the kinematical region where Mandelstam variables $s \sim -t \sim -u$ are about few GeV^2. On the other hand these observables can be described in phenomenological models where spectator quarks are soft which assumes a large contribution due to the soft-overlap mechanism. It turns out that the simple QCD factorization picture is not complete and must also include the soft-overlap contribution which can be described as a certain matrix element in the soft collinear effective theory (SCET). Then the leading power contribution to WACS amplitude is described as a sum of the hard- and soft-spectator contributions. The existing experimental data allows one to check certain conclusions based on the assumption about dominant role of the soft-spectator mechanism.

1 Introduction

The first data for the differential cross section of WACS has already been obtained long time ago [1]. New and more precise measurements were carried out at JLab [2]. Double polarization observables for a polarized photon beam and by measuring the polarization of the recoiling proton were also measured at Jefferson Lab (JLab) [3]. New measurements of various observables at higher energies are planned at the new JLAB 12 GeV facility, see e.g. [4].

The asymptotic limit of the WACS cross section, as predicted by QCD factorization, has been studied in many theoretical works [5–8]. It was found that the leading-twist contribution described by the hard two-gluon exchange between three collinear quarks predicts much smaller cross sections than is observed in experiments. One of the most promising explanations of this problem is that the kinematical region of the existing data is still far away from the asymptotic limit where the hard two-gluon exchange mechanism is predicted to dominate. Hence one needs to develop an alternative theoretical approach which is more suitable for the kinematic range of existing experiments.

Several phenomenological considerations, including the large value of the asymmetry $K_{\rm LL}$ [3] indicate that the dominant contribution in the relevant kinematic range can be provided by the so-called soft-overlap mechanism. In this case the underlying quark-photon scattering is described by the handbag diagram with one active quark while the other spectator quarks are assumed to be soft.

[a]e-mail: kivel@kph.uni-mainz.de

Various models have been considered in order to implement such scattering picture within a theoretical framework: diquarks Refs.[13], GPD-models [9–12] and constituent quarks [14] .

An attempt to develop a systematic approach within the soft collinear effective theory (SCET) framework was discussed in Refs.[15–17]. The description can be considered as a natural extension of the collinear factorization to the case with soft spectators.

In Refs [15, 16] the factorization of the three leading power amplitudes has been studied and a phenomenological analysis was made. The three amplitudes describing Compton scattering which involve a nucleon helicity-flip are power suppressed and they were neglected. In Ref.[17] these amplitudes were also included into description, together with all kinematical power corrections. For that purpose the factorization of helicity-flip amplitudes has been discussed assuming that it can be described as a sum of hard- and soft-spectator contributions. Corresponding soft contributions are described as matrix elements of the appropriate subleading SCET operators. Assuming that such soft contributions are dominant we estimate their possible numerical impact on the cross section and asymmetries.

2 SCET factorization for the WACS amplitudes

The amplitude of Compton scattering $\gamma(q) + N(p) \to \gamma(q') + N(p')$ can be parametrized as [18]

$$M^{\gamma p\to\gamma p} = -e^2\, \varepsilon^{*\mu}(q')\varepsilon^{\nu}(q)\bar{N}(p')\mathcal{A}^{\mu\nu}N(p), \tag{1}$$

$$\mathcal{A}^{\mu\nu} = \left\{-\mathcal{T}^{\mu\nu}_{12}(T_1 + \not{K}T_2) - \mathcal{T}^{\mu\nu}_{34}(T_3 + \not{K}T_4) + \mathcal{T}^{\mu\nu}_{5} i\gamma_5\, T_5 + \mathcal{T}^{\mu\nu}_{6}\, i\gamma_5 \not{K} T_6\right\}, \tag{2}$$

where e denotes the electromagnetic charge of the proton, $N(p)$ is the nucleon spinor. In Eq.(2) we introduced the orthogonal tensor structures

$$\mathcal{T}^{\mu\nu}_{12} = -\frac{P'^{\mu}P'^{\nu}}{P'^2},\quad \mathcal{T}^{\mu\nu}_{34} = \frac{N^{\mu}N^{\nu}}{N^2},\quad \mathcal{T}^{\mu\nu}_{5} = \frac{P'^{\mu}N^{\nu} - P'^{\nu}N^{\mu}}{P'^2K^2},\quad \mathcal{T}^{\mu\nu}_{6} = \frac{P'^{\mu}N^{\nu} + P'^{\nu}N^{\mu}}{P'^2K^2}, \tag{3}$$

with

$$P = \frac{1}{2}(p + p'),\ K = \frac{1}{2}(q + q'),\ P' = P - K\frac{(P\cdot K)}{K^2}, N_{\mu} = \varepsilon_{\mu\alpha\beta\gamma}P^{\alpha}\frac{1}{2}(p - p')^{\beta}K^{\gamma}. \tag{4}$$

The scalar amplitudes $T_i \equiv T_i(s,t)$ are functions of the Mandelstam variables. In the WACS kinematics $s \sim -t \sim -u \sim Q^2 \gg \Lambda_{QCD}$ these amplitudes include various contributions associated with the hard and long distance QCD dynamics. Using the SCET framework one can perform a systematic analysis of the various contributions including one which are associated with the soft spectator scattering (soft-overlap).

It turns out that in the case of nucleon target the soft spectator scattering mechanism is not suppressed comparing the hard spectator one. Therefore the complete factorization description must include both of them. The factorisation description of the soft spectator contribution is more complicate then the hard one because it involves hard and hard-collinear scales. The hard scale appears due to hard subprocesses describing partons with the virtualities of order Q^2, the hard-collinear scale is associated with the scattering of the hard-collinear particles with the virtualities of order ΛQ where we introduced a generic soft scale $\Lambda \sim \Lambda_{QCD}$. Assuming that hard, hard-collinear and soft scales are well separated $Q^2 \gg \Lambda Q \gg \Lambda^2_{QCD}$ the factorisation of the hard modes and hard-collinear modes can be carried out. Factorising the hard modes one reduces QCD to the effective theory known as SCET-I which still includes the hard-collinear modes. At the next step one integrate out these modes reducing description to the SCET-II. This effective theory describes the collinear and soft sectors. Corresponding particles have soft virtualities of order Λ^2 and therefore their dynamics is nonperturbative. The

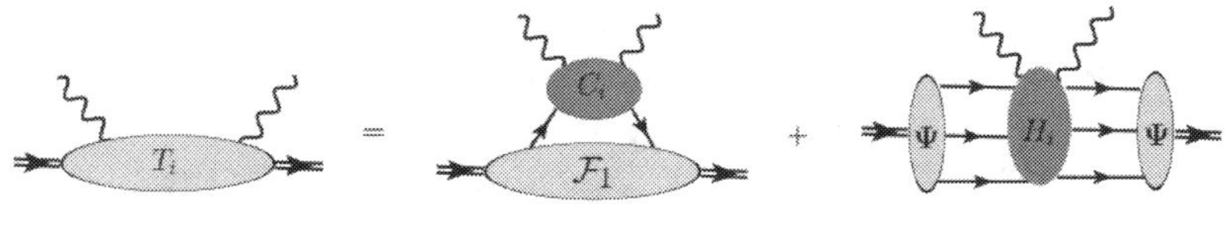

Figure 1. The reduced diagrams describing the factorization formula in Eq.(5).

matrix elements of operators constructed from the SCET-II fields do not depend on the hard scale and must be considered as a nonperturbative.

The kinematical region, where present data have been collected, corresponds to the relatively small value of the hard-collinear scale $\mu_{hc} < 1$ GeV. Therefore in this case one has to stop after factorization of the hard modes. The SCET-I matrix elements in such situation must be also considered as nonpertubative input despite they formally depend on the hard scale Q.

In Refs.[15, 16] it was shown that the complete leading power factorization formula can be written as

$$T_i(s,t) = C_i(s,t)\mathcal{F}_1(t) + \boldsymbol{\Psi} * H_i(s,t) * \boldsymbol{\Psi}, \quad i = 2,3,4. \tag{5}$$

Here the first term on the *rhs* describes the soft-spectator contribution while the second term corresponds to the well known hard-spectator mechanism. For illustration, in Fig.1 we show the different contributions in Eq.(5) as appropriate reduced diagrams. In the large Q limit *both* contributions in Eq.(5) behave as Q^{-6} up to logarithmic corrections and give

$$T_{2,4,6}(s,t) \sim Q^{-6}, \quad Q \to \infty. \tag{6}$$

For large values of Q the soft-spectator scattering is strongly suppressed due to the so-called Sudakov logarithms and therefore the hard-spectator contribution becomes dominant. But for moderate values of Q the effect of the Sudakov suppression is still weak therefore the soft-spectator contribution is quite large or even dominant.

The expression in Eq.(5) does not include the amplitudes $T_{1,3,5}$. These amplitudes describe the scattering when the helicity of nucleon is not conserved. Corresponding contribution in the cross section is suppressed by power $\mathcal{O}(1/Q^2)$ comparing the helicity conserving amplitudes. However the factorization of the amplitudes $T_{1,3,5}$ can also be describes in the same way as in Eq.(5) but the soft spectator contribution will be described by the set of the subleading SCET-I operators. More detailed consideration the SCET factorization for the amplitudes $T_{1,3,5}$ is given in Ref.[17].

The asterisks in Eq.(5) denote the convolution with respect to the collinear quark fractions, $\boldsymbol{\Psi}$ denotes the nucleon distribution amplitudes which describe the non-perturbative dynamics in the hard-spectator contributions. The hard coefficient functions $H_i(s,t)$ in Eq.(5) define the full dependence of the hard-spectator contribution on the Mandelstam variables. The leading-order approximation for these functions are defined by the two-gluon exchange diagrams and therefore they are of order $\alpha_s^2(Q^2)$.

The hard coefficient functions $C_{2,4,6}$ in the soft spectator term in Eq.(5) can be computed systematically order by order in pQCD. The leading-order expressions read

$$C_2(s,t) = -C_4(s,t) = \frac{s-u}{su}, \quad C_6(s,t) = \frac{t}{su}. \tag{7}$$

The NLO expressions have also been computed in Ref.[16].

The non-perturbative dynamics in the soft-spectator contribution in Eq.(5) is described by the SCET form factor (FF) $\mathcal{F}_1(t)$. In the SCET framework it is defined as

$$\langle p'| \sum e_q^2 \left\{ \bar{\xi}_n(0) W_n \gamma_\perp^\sigma W_{\bar{n}}^\dagger \xi_{\bar{n}}(0) - \bar{\xi}_{\bar{n}}(0) W_{\bar{n}} \gamma_\perp^\sigma W_n^\dagger \xi_n(0) \right\} |p\rangle_{\text{SCET}} = \bar{N}_n \gamma_\perp^\sigma N_{\bar{n}}\, \mathcal{F}_1(t), \tag{8}$$

where n and $\bar{n}$ denote the auxiliary light-like vectors associated with the nucleon momenta

$$p = \sqrt{-t}\tfrac{\bar{n}}{2} + \tfrac{m_N^2}{\sqrt{-t}}\tfrac{n}{2} \simeq \sqrt{-t}\tfrac{\bar{n}}{2}, \quad p' = \tfrac{m_N^2}{\sqrt{-t}}\tfrac{\bar{n}}{2} + \sqrt{-t}\tfrac{n}{2} \simeq \sqrt{-t}\tfrac{n}{2}, \tag{9}$$

The fields ξ_n and $\xi_{\bar{n}}$ in Eq.(8) denotes the hard-collinear quark fields in SCET-I (with the appropriate flavor) and W_n denotes hard-collinear Wilson line

$$W_n(z) = \mathrm{P}\exp\left\{ ig \int_{-\infty}^{0} ds\, \bar{n} \cdot A_{hc}^{(n)}(z + s\bar{n}) \right\}, \tag{10}$$

and similarly for the fields with the label $\bar{n}$.

This SCET FF depends only from the large momentum transfer t. However this dependence is associated only with the hard-collinear modes which can not be factorized for small values of the hard-collinear scale. One can see from Eq.(5) that the energy (s) dependence is completely defined by the hard coefficient function C_i and therefore can be computed in the perturbation theory. SCET analysis allows one to show that at large values of $-t$ one expect that $\mathcal{F}_1(t) \sim (-t)^{-2}$. Therefore at large Q the soft spectator term decreases with the same power as the hard spectator contribution.

The FF $\mathcal{F}_1(t)$ and the hard coefficient functions C_i also depend on the factorization scale μ_F which is not shown for simplicity. This scale defines the separation between the hard and hard-collinear regions. This scale dependence can be computed from the renormalization of the SCET operator in Eq.(8). In the NLO one finds [16]

$$\mu \frac{d}{d\mu} C_i(s, t; \mu^2) = \frac{\alpha_s}{4\pi} C_F \left\{ 4 \ln[-t/\mu^2] - 6 + O(\alpha_s) \right\} C_i(s, t; \mu^2), \tag{11}$$

Solving this equation one can resum the large logarithms to all order in perturbation theory.

3 Phenomenology

In order to apply the factorization formula (5) in a phenomenological analysis one must define the unknown nonperturbative form factor $\mathcal{F}_1$. This task can only be solved by using a nonperturbative approach within the SCET framework. At present time such approach has not yet developed. Meanwhile a useful phenomenological consideration can be carried out. This is possible due to the universality of the definition of the SCET FF $\mathcal{F}_1$ in the factorization approach and due to its specific properties. This one unknown quantity describes the three independent amplitudes $T_{2,4,6}$. Therefore the idea is to re-express this quantity in terms of any one amplitude and then use this expression for the remaining two amplitudes. This allows one to establish the relation between the three amplitudes up to well defined hard-spectator corrections.

In order to be specific let us use Eq.(5) and write for $\mathcal{F}_1$ the following expression

$$\mathcal{F}_1(t) = \mathcal{R}(s, t) - \mathbf{\Psi} * H_2(s, t) * \mathbf{\Psi}/C_2(s, t). \tag{12}$$

Here we define the ratio

$$\mathcal{R}(s, t) = \frac{T_2(s, t)}{C_2(s, t, \mu^2 = -t)}. \tag{13}$$

Note that the *rhs* of Eq.(12) does not depend on the total energy s. Substituting this equation in the expressions for the amplitudes $T_{4,6}$ (5) we obtain

$$T_i(s',t) = C_i(s',t)\mathcal{R}(s,t) + \Psi * \left\{ H_i(s',t) - C_i(s',t)\frac{H_2(s,t)}{C_2(s,t)} \right\} * \Psi. \tag{14}$$

On the left side of this equation we have a well defined physical amplitude therefore the right side must also be well defined. This means that the potential end-point singularities in the hard-spectator corrections must cancel in the difference on the *rhs* of Eq.(14). We assume that all the hard coefficient functions in Eq.(14) are defined at $\mu^2 = -t$. We also used the different values of the total energy s' in Eq.(14) in order to stress that the substitution (12) does not depend on the energy s.

If the values of the hard-spectator contributions in Eq.(14) are small, then such terms can be neglected and we obtain

$$T_i(s',t) \simeq C_i(s',t)\mathcal{R}(s,t) \Leftrightarrow \frac{T_i(s',t)}{C_i(s',t)} \simeq \mathcal{R}(s,t). \tag{15}$$

Notice that this formula is valid to all orders in α_s for the coefficient C_i but at order α_s^2 one has to take into account the hard-spectator corrections.

Obviously the choice of the amplitude T_i for the definition of the ratio $\mathcal{R}$ in (13) does not play any essential role. In the definition (13) we also used a freedom to fix the factorization scale and chose $\mu^2 = -t$ as simplest realization.

If the bulk contribution to the amplitudes $T_i(s,t)$ is provided by the soft-overlap term then one expects (this is also implied by Eq.(15)) that the ratio $\mathcal{R}$ depends only very weakly on the energy s

$$\frac{d}{ds}\mathcal{R}(s,t) \simeq 0, \tag{16}$$

This picture can be verified when comparing with the data. Consider the unpolarised WACS cross section. Using expressions (15) for the amplitudes $T_{2,4,6}$ and neglecting the helicity flip amplitudes $T_{1,3,5} \simeq 0$ and power suppressed terms $\sim m/Q$ we obtain

$$\frac{d\sigma}{dt} = \frac{\pi\alpha^2}{s^2}\,|\mathcal{R}(s,t)|^2(-su)\left(\frac{1}{2}|C_2(s,t)|^2 + \frac{1}{2}|C_4(s,t)|^2 + |C_6(s,t)|^2\right). \tag{17}$$

If the ratio $\mathcal{R}(s,t)$ depends mostly on the momentum transfer t then the energy dependence of the cross section in Eq.(17) is defined only by the hard coefficient functions. Using Eq.(17) one can extract the absolute value $|\mathcal{R}(s,t)|$ and to check the equation Eq.(16) using the experimental data for the cross section. In Fig.2 we show the results obtained for the ratio $\mathcal{R}$ where we also take into account the power suppressed kinematical corrections. From Fig.2 we see that this approximate scaling behavior is observed in the region where $-u \geq 2.5$ GeV^2. Hence we can adopt this value as a phenomenological lower limit of applicability of the described approach. For smaller values of u the extracted values of $\mathcal{R}$ (shown by the open squares) demonstrate already a clear sensitivity to s. Thus one can observe that for $-u = 1.3$ GeV^2 ($-t = 3.7$ GeV^2) the obtained value of $\mathcal{R}$ is about a factor 2 larger than the scaling curve. This observation clearly demonstrates that the given approach cannot describe the cross section data at small values of u.

References

[1] M. A. Shupe, R. H. Milburn, D. J. Quinn, J. P. Rutherfoord, A. R. Stottlemyer, S. S. Hertzbach, R. Kofler and F. D. Lomanno *et al.*, Phys. Rev. D **19** (1979) 1921.

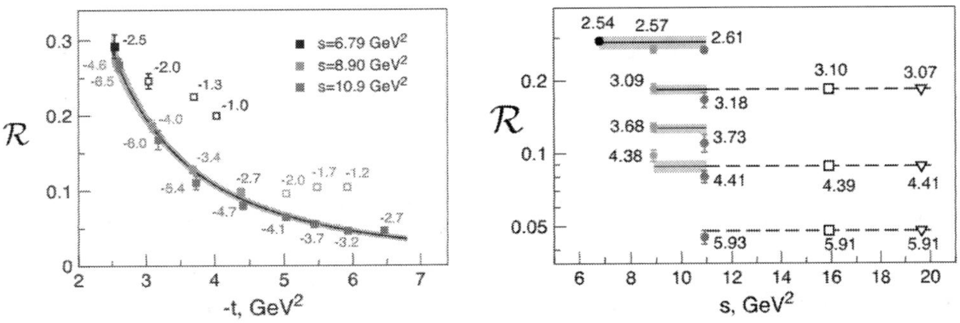

Figure 2. Left: The ratio $\mathcal{R}$ as function of t. The corresponding values of u are shown by the numbers next to the symbols. The open squares mark out the points with $-u < 2.5\text{GeV}^2$. The solid line represents an empirical fit, the gray bands show the 99% confidence interval. **Right:** The ratio $\mathcal{R}$ as a function of energy s at fixed values of t in the region where $-u > 2.5\text{GeV}^2$. The corresponding values of $-t$ are shown by the numbers next to the symbols. The open squares and triangles mark out the predicted points which will be checked in cross section measurements at JLab with the 12GeV beam facility.

[2] A. Danagoulian *et al.* [Hall A Collaboration], Phys. Rev. Lett. **98** (2007) 152001 [nucl-ex/0701068 [NUCL-EX]].
[3] D. J. Hamilton *et al.* [Jefferson Lab Hall A Collaboration], Phys. Rev. Lett. **94** (2005) 242001 [nucl-ex/0410001].
[4] D. Hamilton *et al.* [Jefferson Lab Hall C Collaboration], E1214003, "Wide-angle Compton Scattering at 8 and 10 GeV Photon Energies".
[5] A. S. Kronfeld and B. Nizic, Phys. Rev. D **44** (1991) 3445 [Erratum-ibid. D **46** (1992) 2272].
[6] M. Vanderhaeghen, P. A. M. Guichon and J. Van de Wiele, Nucl. Phys. A **622** (1997) 144C.
[7] T. C. Brooks and L. J. Dixon, Phys. Rev. D **62** (2000) 114021 [hep-ph/0004143].
[8] R. Thomson, A. Pang and C. -R. Ji, Phys. Rev. D **73** (2006) 054023 [hep-ph/0602164].
[9] A. V. Radyushkin, Phys. Rev. D **58** (1998) 114008 [hep-ph/9803316].
[10] M. Diehl, T. Feldmann, R. Jakob and P. Kroll, Eur. Phys. J. C **8** (1999) 409 [hep-ph/9811253].
[11] M. Diehl, T. Feldmann, R. Jakob and P. Kroll, Phys. Lett. B **460** (1999) 204 [hep-ph/9903268].
[12] M. Diehl, T. Feldmann, H. W. Huang and P. Kroll, Phys. Rev. D **67** (2003) 037502 [hep-ph/0212138].
[13] P. Kroll, M. Schurmann and W. Schweiger, Int. J. Mod. Phys. A **6** (1991) 4107.
[14] G. A. Miller, Phys. Rev. C **69** (2004) 052201 [nucl-th/0402092].
[15] N. Kivel and M. Vanderhaeghen, JHEP **1304** (2013) 029 [arXiv:1212.0683 [hep-ph]].
[16] N. Kivel and M. Vanderhaeghen, Nucl. Phys. B **883** (2014) 224 [arXiv:1312.5456 [hep-ph]].
[17] N. Kivel and M. Vanderhaeghen, Eur. Phys. J. C **75** (2015) 10, 483 [arXiv:1504.00991 [hep-ph]].
[18] D. Babusci, G. Giordano, A. I. L'vov, G. Matone and A. M. Nathan, Phys. Rev. C **58** (1998) 1013 [hep-ph/9803347].

EPJ Web of Conferences **112**, 01012 (2016)
DOI: 10.1051/epjconf/201611201012

Description and interpretation of DVCS measurements*

Krešimir Kumerički[1,a] and Dieter Müller[2,3,b]

[1] *Physics Department, University of Zagreb, 10000 Zagreb, Croatia*
[2] *Theoretical Physics Division, Rudjer Bošković Institute, 10002 Zagreb, Croatia*
[3] *Department of Physics, University of Cape Town, 7701 Rondebosch, South Africa*

Abstract. Having in mind the well-known limitations of certain models of generalized parton distributions (GPDs), we show that the allegedly universal GPDs, describing both deeply virtual Compton scattering (DVCS) and deeply virtual meson production (DVMP) data, fail to describe measurements of deep inelastic scattering. We present a new global DVCS fit that describes reasonably the world data set, which includes now also new measurements from Hall A and CLAS collaborations. We also explicitly illustrate that Compton form factors (CFFs) cannot be unambiguously extracted from photon electroproduction measurements off unpolarized proton alone and we argue that it is more reliable to interpret such measurements in terms of certain CFF combinations, where still some care is needed in order to estimate the propagated error.

1 Introduction

The deeply virtual Compton scattering (DVCS) process off a nucleon target is considered to be the theoretically cleanest process allowing access to generalized parton distributions (GPDs) from experimental measurements. Motivated by this, deeply virtual Compton scattering has been studied in the last decade quite extensively by HERA and JLab collaborations. Thereby, cross section measurements on an unpolarized proton have been performed in collider experiments H1 and ZEUS as well as now in fixed target experiments by the CLAS and Hall A collaborations, while HERMES provided an almost (over-)complete set of asymmetry measurements. On the other hand, considering possible photon and proton helicities, there are twelve DVCS amplitudes where, loosely spoken, the four photon helicity conserved amplitudes or Compton form factors (CFFs) $\mathcal{H}, \mathcal{E}, \widetilde{\mathcal{H}}, \widetilde{\mathcal{E}}$ appear in the $1/Q^2$ expansion as leading contributions. In the leading order and leading twist approximation these CFFs are given in terms of quark GPDs $H, E, \widetilde{H}, \widetilde{E}$,

$$\mathcal{H}(x_B, t, Q^2) \stackrel{\text{LO}}{=} \int_{-1}^{1} dx \left[\frac{1}{\xi - x - i\epsilon} - \frac{1}{\xi + x - i\epsilon}\right] \sum_{q=u,d,s,\cdots} e_q^2 H^q(x, \xi, t, Q^2), \tag{1}$$

where $\xi \approx x_B/(2 - x_B)$, x_B is Bjorken's scaling variable, t is the square of the momentum transfer, and the factorization scale has been equated to the (negative) photon virtuality $q^2 = -Q^2$. The proton

*Joint proceedings contribution, based on two talks given by the authors at the 6th International conference on Physics Opportunities at Electron-Ion colliders (POETIC6)

[a] e-mail: kkumer@phy.hr
[b] e-mail: dieter.mueller@irb.hr

helicity flip CFFs $\mathcal{E}$ and $\widetilde{\mathcal{E}}$ might be kinematically suppressed in the DVCS kinematics, *e.g.*, by $t/4M^2$ (M being the target mass); however, $\widetilde{\mathcal{E}}$ might also be enhanced at small $-t$ by the presence of the pion pole. The rather analogous set of four longitudinal photon helicity CFFs is formally suppressed by $1/Q$. The remaining four transverse helicity photon flip CFFs appear in the partonic description at next-to-leading order accuracy via the gluon transversity GPDs which might also be to a larger extent contaminated by $1/Q^2$ power suppressed contributions. Apart from the desire to access GPDs in the deeply virtual region where the photon virtuality is considered to be large, we require $Q^2 \gtrsim 1.5\mathrm{GeV}^2$, and the momentum transfer $-t$ with $-t/Q^2 \lesssim 1/4$ is small, virtual Compton scattering has its own interest. A unifying framework was proposed in Ref. [1] in which also the low energy limit, *i.e.*, the access to generalized polarizabilities was discussed.

The DVCS process is measurable in the leptoproduction of a photon on a nucleon (or nucleus) target and it interferes with the Bethe-Heitler process, which is parameterized in terms of the electromagnetic form factors. For a transversally polarized target the differential cross section is five-fold and is considered as function of x_{B}, t, Q^2, the azimuthal angle ϕ between the lepton and reaction plane, and the azimuthal angle φ of the transverse components of the polarization vector. Otherwise, the cross section is four-fold and for a massless electron reads

$$\frac{d^4\sigma}{dx_{\mathrm{B}}dQ^2dtd\phi} = \frac{\alpha^3 x_{\mathrm{B}} y^2}{8\pi Q^4\sqrt{1+\epsilon^2}}\left|\frac{\mathcal{T}}{e^3}\right|^2, \tag{2}$$

where $\alpha = e^2/4\pi \approx 1/137$ is the electromagnetic fine structure constant, $y = Q^2/x_{\mathrm{B}}\left(s - M^2\right)$ with $s = (p_1 + q_1)^2$, and $\epsilon^2 = 4x_{\mathrm{B}}^2 M^2/Q^2$ is an abbreviation. The electroproduction amplitude, entering as square of its absolute value into the cross section,

$$|\mathcal{T}|^2 = |\mathcal{T}^{\mathrm{BH}}|^2 + \mathcal{I} + |\mathcal{T}^{\mathrm{VCS}}|^2, \quad \mathcal{I} = 2\Re\mathrm{e}\mathcal{T}^{\mathrm{BH}}\mathcal{T}^{\dagger\,\mathrm{VCS}} \tag{3}$$

is the sum of the charge-odd Bethe-Heitler $\mathcal{T}^{\mathrm{BH}}$ amplitude and the charge-even virtual Compton scattering amplitude $\mathcal{T}^{\mathrm{VCS}}$ amplitude. The three different contributions might be decomposed in Fourier harmonics, *e.g.*, for an unpolarized nucleon

$$|\mathcal{T}^{\mathrm{BH}}|^2 = \frac{e^6}{x_{\mathrm{B}}^2 y^2(1+\epsilon^2)^2 t\,\mathcal{P}_1(\phi)\mathcal{P}_2(\phi)}\sum_{n=0}^{2} c_{n,\mathrm{unp}}^{\mathrm{BH}}\cos(n\phi), \tag{4a}$$

where the product of (rescaled) electron propagators $1/\mathcal{P}_1(\phi)\mathcal{P}_2(\phi)$ implies an additional ϕ dependence. The coefficients $c_{n,\mathrm{unp}}^{\mathrm{BH}}$ depend on the electromagnetic proton form factors for which we use the Kelly parametrization [2]. For the interference term we have

$$\mathcal{I} = \frac{\pm e^6}{x_{\mathrm{B}} y^3 t\mathcal{P}_1(\phi)\mathcal{P}_2(\phi)}\left[\sum_{n=0}^{3} c_{n,\mathrm{unp}}^{\mathrm{I}}\cos(n\phi) + \sum_{n=1}^{2} s_{n,\mathrm{unp}}^{\mathrm{I}}\sin(n\phi)\right], \tag{4b}$$

where the + (−) sign refers to electron (positron) scattering and

$$|\mathcal{T}^{\mathrm{VCS}}|^2 = \frac{e^6}{y^2 Q^2}\left[\sum_{n=0}^{2} c_{n,\mathrm{unp}}^{\mathrm{VCS}}\cos(n\phi) + s_{1,\mathrm{unp}}^{\mathrm{VCS}}\sin(\phi)\right], \tag{4c}$$

for the squared DVCS amplitude. The harmonics of the interference term are given in terms of linear CFF combinations

$$C_{\mathrm{unp}}^{\mathcal{I}} = F_1\mathcal{H} + \xi(F_1 + F_2)\widetilde{\mathcal{H}} - \frac{t}{4M^2}F_2\mathcal{E}, \tag{5}$$

$$C_{\mathrm{LP}}^{\mathcal{I}} \approx F_1\widetilde{\mathcal{H}} + \xi(F_1 + F_2)\mathcal{H} - \left(\xi F_1 + \frac{t}{4M^2}F_2\right)\xi\widetilde{\mathcal{E}}. \tag{6}$$

while those of the squared DVCS term are given in terms of bilinear (linear in both CFFs and their complex conjugates) combinations. Projecting first the experimental data onto (weighted) harmonics, see [3], is a standard filter procedure we utilize first in both global GPD model fitting and local CFF extraction.

The main emphasis of the present contribution is to spell out some phenomenological inconsistencies in the description of DVCS data which are not very much clarified in the literature. In Sec. 2 we show that universality of the GPD framework is violated in a certain model class and we provide a description of the world data DVCS set that is obtained by global fitting. In Sec. 3 we spell out that the interpretation of some results that arise from so-called quasi model-independent extraction of CFFs from CLAS cross section measurements might be misleading. Finally, we conclude.

2 Model predictions versus global fitting

Based on Regge theory arguments, one might assume that in collider experiments only one CFF contributes, giving access to the unpolarized quark GPD H^q combination, *e.g.*, in LO approximation by means of the convolution formula (1). The situation is different for fixed target kinematics, in particular for the JLab experiments with a 6 GeV electron beam and unpolarized or longitudinally polarized proton target. Additionally to model predictions, global fitting attempts were undertaken to extract the values of CFFs. It is clear from the outset that also in the case of a complete or overcomplete measurement the extraction of such information is a challenging task that can only be resolved within some physical or model-motivated assumptions. The common assumption most authors adopt is the dominance of leading twist (LT) in a perturbative leading order (LO) description, *i.e.*, the number of complex-valued Compton form factors

$$\mathcal{F} \in \{\mathcal{H}, \mathcal{E}, \widetilde{\mathcal{H}}, \widetilde{\mathcal{E}}, \cdots\}$$

reduces from twelve to four. This restriction might be not justified outside the DVCS kinematics, *e.g.*, $-t \sim Q^2/2$ or so.

The new CLAS beam spin asymmetry $A_{\rm LU}$ and longitudinally polarized target spin asymmetries $A_{\rm UL}$ and $A_{\rm LL}$ are measured in five $\{x_{\rm B}, Q^2\}$–bins versus t [4], where four bins cover DVCS kinematics

$$\{0.185, 1.63\,{\rm GeV}^2\},\ \{0.245, 1.79\,{\rm GeV}^2\},\ \{0.245, 2.12\,{\rm GeV}^2\},\ \{0.335, 2.78\,{\rm GeV}^2\}. \tag{7}$$

The Hall A collaboration provided high precision cross section measurements with a polarized electron beam and unpolarized proton target [5], which were superseded more recently for five $-t/{\rm GeV}^2 \in \{0.17, 0.23, 0.28, 0.33, 0.37\}$ values within five kinematical settings [6],

$$\begin{aligned}
{\rm Kin1}:&\quad 0.345 \le x_{\rm B} \le 0.399\,,\quad 1.453 \le Q^2/{\rm GeV}^2 \le 1.633\,,\\
{\rm Kin2}:&\quad 0.343 \le x_{\rm B} \le 0.381\,,\quad 1.820 \le Q^2/{\rm GeV}^2 \le 1.999\,,\\
{\rm Kin3}:&\quad 0.345 \le x_{\rm B} \le 0.373\,,\quad 2.218 \le Q^2/{\rm GeV}^2 \le 2.375\,,\\
{\rm KinX2}:&\quad 0.378 \le x_{\rm B} \le 0.401\,,\quad 2.012 \le Q^2/{\rm GeV}^2 \le 2.091\,,\\
{\rm KinX3}:&\quad 0.336 \le x_{\rm B} \le 0.342\,,\quad 2.161 \le Q^2/{\rm GeV}^2 \le 2.193\,,
\end{aligned} \tag{8}$$

where last two settings have subset of the same data re-binned for measuring the x_B-dependence. Somewhat less precise measurements of cross-section, but covering much larger kinematic space

$$0.1 \le x_{\rm B} \le 0.58\,,\quad 1 \le Q^2/{\rm GeV}^2 \le 4.6\,,\quad 0.17 \le -t/{\rm GeV}^2 \le 0.37\,, \tag{9}$$

were published by CLAS [7].

2.1 Model predictions

It is known since more than a decade that certain GPD models such as Radyushkin's double distribution ansatz (RDDA) or the leading SO(3)-partial wave expansion of conformal GPD moments fail to globally describe the DVCS data at leading twist and leading order of perturbation theory. While the failure of the latter model is well-accepted, problems of the former one are often forgotten or hidden in the literature, see, e. g., Ref. [8], where a good description of DVCS data by RDDA-based model has been reported.

It is rather popular to decorate the RDDA with t-dependence in such a manner,

$$H^q(x,\eta,t) = \int_0^1 dy \int_{-1+y}^{1-y} dz\, \delta(x-y-z\eta) \frac{q(y,t)}{1-y} \Pi\left(\frac{z}{1-y}\right) + \theta(|x| \le |\eta|) D\left(\frac{x}{|\eta|}, t\right), \tag{10}$$

that the t-dependence is entirely contained in the non-forward parton density $q(x,t)$. The model bias in such a parametrization is that the profile function is t-independent and, thus, for valence quarks it can be adjusted from form factor measurements where one relies on the functional ansatz. To complete polynomiality in the charge even sector, one might add a so-called D-term. Such an ansatz is the basis of older models, where partially a factorized t-dependence was utilized, the GK model [9], and it is also implemented in the VGG code [10].

The phenomenological problem with such models is that the skewness effect at small x_B is too large and so their predictions substantially overshoot the DVCS cross section measurements in collider kinematics, as pointed out for some time already in Ref. [11]. Thus, it is at first surprising that the GK model correctly describes the H1 and ZEUS DVCS data. However, the reason for that is the reduction to $n_F = 3$ of the flavour content of the PDF parametrization, adopted from CTEQ6 [12], see convolution formula (1). Such exclusion of the charm quark contribution from the model leads then to the large underestimate of the deep inelastic structure function $F_2(x_B, Q^2)$. This is displayed by the short dashed line in the left panel of Fig. 1 to be compared with the LO CTEQ6L1 (dash-dotted), NLO CTEQ6M (dashed), and our LO parametrization (solid). Hence, we conclude that the universality of the GPD framework in the GK model (and other GPD models based on RDDA) is not implemented.

On the other hand the GK model describes the asymmetry measurements of HERMES that were obtained within the missing mass technique, except for the beam spin asymmetry $A_{\mathrm{LU}}^{\sin(1\phi)}$, which is a key observable, compare full squares with short dashed line in the upper right panel of Fig. 1. On the other hand the recoil detector data of the HERMES collaboration (empty squares), which are consistent with the missing mass technique data, are described rather well by the GK model. Unfortunately, the two different data sets are not combined and the missing mass technique data suffer from a larger experimental uncertainty. As we will see in the next section the GK model quantitatively overestimates the beam spin asymmetry and the beam spin dependent cross section of JLab measurements.

2.2 Predictions of new JLab results and a new global fit

To fit globally the world data set of DVCS measurements from HERA and JLab collaborations, we employ a rather simple hybrid model. It is based on a double partial wave expansion for the sea-quark and gluon GPD, including the LO QCD evolution, and a dispersion integral representation for the valence quark content that does not evolve. Details of the model can be found in [14], and last published incarnation of the model (KMM12) [17] was obtained by a standard least-squares fitting procedure to the statistically independent set of 95 DVCS data points[1]. Total $\chi^2/n_{\mathrm{d.o.f.}}$ was 123.5/80,

[1] Actually, only 92 points were independent; we took all 6 CLAS $A_{UL}^{\sin\phi}$ measurements [18], where only 3 are independent. This is corrected in the new KM15 fit.

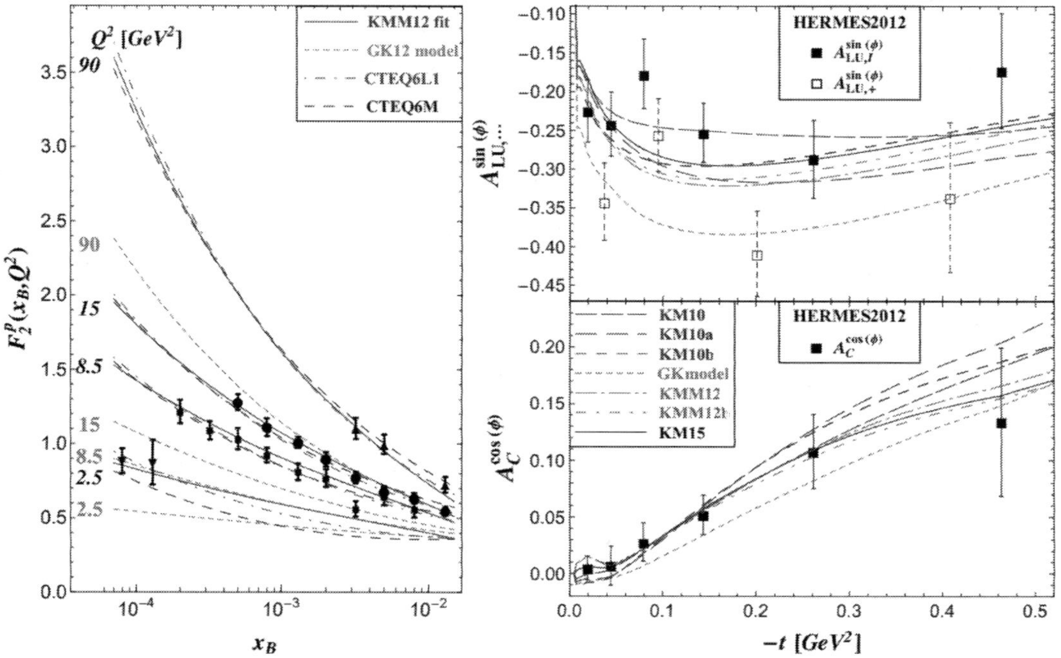

Figure 1. Left: The DIS structure function F_2^p measurements [13] are confronted with the GK model (short–dashed), based on the next-to-leading order PDF parametrization CTEQ6M (dashed), the leading order PDF parametrization CTEQ6L1 (dot–dashed) as well as our DIS fit (solid) from Ref. [14]. Right: Beam spin (upper panel) and beam charge (lower panel) asymmetry from the HERMES collaboration: measurements of $A_{\mathrm{LU},+}^{\sin(\phi)}$ with the recoil detector (empty squares) [15] and obtained from measurements of $A_{\mathrm{LU,I}}^{\sin(\phi)}$ and $A_{\mathrm{C}}^{\cos(\phi)}$ within the missing mass technique (filled squares) [16] versus the GK model prediction (short–dashed) and various fits, `KM10` (long-dashed), `KM10a` (dashed), `KM10b` (shorter dashed), `KMM12` (dash–dotted), `KMM12b` (shorter dash–dotted), and `KM15` (solid), to the HERMES data (filled squares) as solid curves.

where most problematic data was (i) HERMES data with longitudinally polarized target (and to some extent beam), and (ii) the very precise Hall A cross-section data (especially t-dependence of the first cosine harmonic).

On Table 1 we list all data used in fits, as well as χ^2/n_{pts} values for particular datasets, for the `KMM12` model, and for the new `KM15` model to be discussed below. To give additional information about compatibility of the fit with the data, we list also the values of the "pull" of each dataset, defined for dataset consisting of N measurements as

$$\mathrm{pull} \equiv \frac{1}{\sqrt{N}} \sum_{i=1}^{N} \frac{f(x_i) - y_i}{\Delta y_i} , \tag{11}$$

where $f(x_i)$ is model value for a given kinematics x_i, while y_i and Δy_i are the corresponding experimental value and uncertainty, respectively. For a perfect model, pull (11) would be a random variable distributed according to a Gaussian distribution with zero mean and unit standard deviation.

Adding 2015 JLab data to the world set, and refitting the same model, results in a model parameter values listed on Table 2. The key difference is brought by using 2015 Hall A cross section data instead

Table 1. Observables used for KMM12 and KM15 fits, together with corresponding values of $\chi^2/n_{\rm pts}$ and "pulls" (11).

Collaboration	Observable	Ref.	$n_{\rm pts}$	KMM12		KM15	
				$\chi^2/n_{\rm pts}$	pull	$\chi^2/n_{\rm pts}$	pull
ZEUS	$\sigma_{\rm DVCS}$	[19, 20]	11	0.49	-1.76	0.51	-1.74
ZEUS,H1	$d\sigma_{\rm DVCS}/dt$	[19, 21, 22]	24	0.97	0.85	1.04	1.37
HERMES	$A_{\rm C}^{\cos 0\phi}$	[23]	6	1.31	0.49	1.24	0.29
HERMES	$A_{\rm C}^{\cos\phi}$	[23]	6	0.24	-0.56	0.07	-0.20
HERMES	$A_{\rm LU,I}^{\sin\phi}$	[23]	6	2.08	-2.52	1.34	-1.28
CLAS	$A_{\rm LU}^{\sin\phi}$	[24]	4	1.28	2.09		
CLAS	$A_{\rm LU}^{\sin\phi}$	[4, 25]	13			1.24	0.63
CLAS	$\Delta\sigma^{\sin\phi,w}$	[7]	48			0.41	-1.66
CLAS	$d\sigma^{\cos 0\phi,w}$	[7]	48			0.16	-0.21
CLAS	$d\sigma^{\cos\phi,w}$	[7]	48			1.16	6.36
Hall A	$\Delta\sigma^{\sin\phi,w}$	[5]	12	1.06	-2.55		
Hall A	$d\sigma^{\cos 0\phi,w}$	[5]	4	1.21	2.14		
Hall A	$d\sigma^{\cos\phi,w}$	[5]	4	3.49	-0.26		
Hall A	$\Delta\sigma^{\sin\phi,w}$	[6]	15			0.81	-2.84
Hall A	$d\sigma^{\cos 0\phi,w}$	[6]	10			0.40	0.92
Hall A	$d\sigma^{\cos\phi,w}$	[6]	10			2.52	-2.42
HERMES,CLAS	$A_{\rm UL}^{\sin\phi}$	[18, 26]	10	1.90	-1.89	1.10	-1.94
HERMES	$A_{\rm LL}^{\cos 0\phi}$	[26]	4	3.44	2.17	3.19	1.99
HERMES	$A_{\rm UT,I}^{\sin(\phi-\phi_S)\cos\phi}$	[27]	4	0.90	0.61	0.90	0.71
CLAS	$A_{\rm UL}^{\sin\phi}$	[4]	10			0.76	0.38
CLAS	$A_{\rm LL}^{\cos 0\phi}$	[4]	10			0.50	-0.22
CLAS	$A_{\rm LL}^{\cos\phi}$	[4]	10			1.54	2.40

Table 2. Valence (top) and sea quark (bottom) related hybrid model parameters, as obtained by the global DVCS fit KM15 to 290 data points with χ^2/d.o.f. = 240./275. For detailed description of the model and parameters see Refs. [14, 17]. Parameters of the old KMM12 fit [17] are also given for comparison.

	$M^{\rm val}$	$r^{\rm val}$	$b^{\rm val}$	C	M_C	$\tilde{M}^{\rm val}$	$\tilde{r}^{\rm val}$	$\tilde{b}^{\rm val}$	r_π	M_π
KM15	0.789	0.918	0.4	2.768	1.204	3.993	0.881	0.4	2.646	4.
KMM12	0.95	1.12	0.4	1.	2.08	3.52	1.3	0.4	3.84	4.

	$(M^{\rm sea})^2$	$s_2^{\rm sea}$	$s_4^{\rm sea}$	$s_2^{\rm G}$	$s_4^{\rm G}$
KM15	0.482	1.071	-0.366	-2.991	0.905
KMM12	0.46	0.31	-0.14	-2.77	0.94

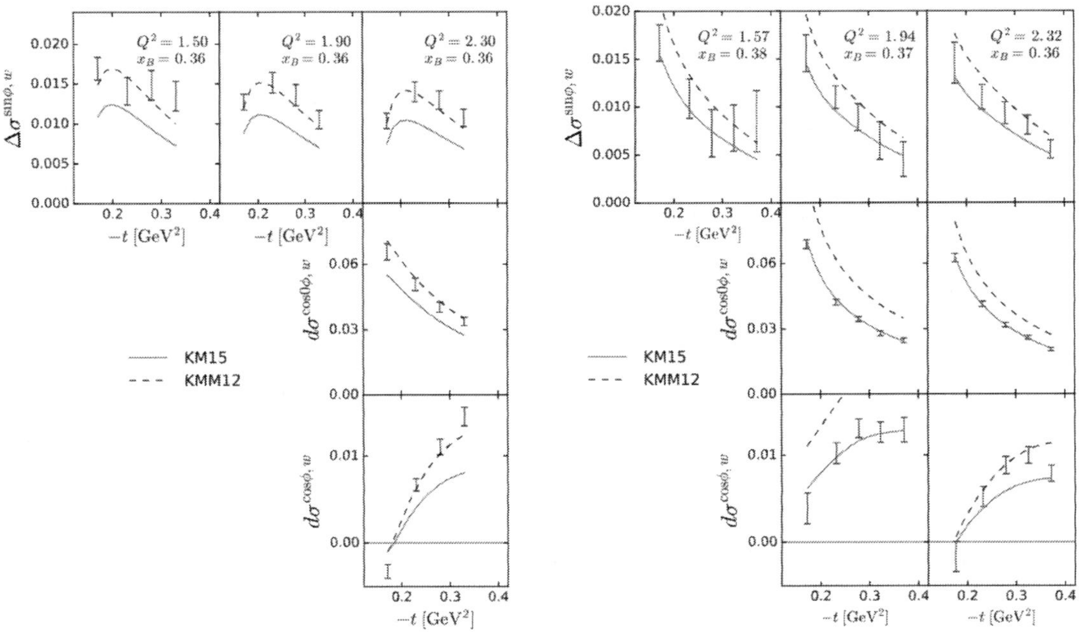

Figure 2. (Weighted) Fourier harmonics of Hall A 2006 [5] (left) and 2015 [6] (right) cross-section data, together with KMM12 (dashed) and KM15 (solid) models.

of superseded 2006 one. Old and new Hall A data, and KMM12 and KM15 models are confronted on Fig. 2. One notices that the cross-section went down a bit and the strong t-dependence of the first cosine harmonic ($d\sigma^{\cos\phi, w}$, lowest panels) seen in 2006 data is significantly milder now. All this made possible the decrease of skewness parameters r^{val} and $\tilde{r}^{\text{val}}$, *i.e.*, the decrease of $\mathcal{H}$ and $\tilde{\mathcal{H}}$, as well as the decrease of the pion pole contribution, thus improving the fit of most of the other observables, see Table. 1. In particular, description of HERMES $A_{\text{LU,I}}^{\sin\phi}$ is significantly improved, and overshooting (large positive pull) of CLAS $A_{\text{LU}}^{\sin\phi}$ is not present any more.

Looking at longitudinally polarized target data, χ^2 of $A_{\text{UL}}^{\sin\phi}$ is also improved, while very large χ^2 contribution of HERMES $A_{\text{LL}}^{\cos 0\phi}$ is mostly due to a single outlier point so we don't consider this a problem. First cosine harmonic of the cross section remains the most problematic observable to describe. As discussed above, in the Hall A case the tension is relaxed but still exists, and although χ^2 value for new CLAS measurement is fine, the pull of 6.36 is unacceptably large, showing consistent overshooting of the data.

3 Local versus global extraction of CFFs

It is a standard problem in the phenomenology of scattering processes that if only an incomplete measurement is performed one can extract only certain combinations of squared helicity amplitudes at given kinematics. In the photon leptoproduction the interference between Bethe-Heitler bremsstrahlung process and deeply virtual Compton scattering offers various possibilities to access

CFFs. To utilize the azimuthal angular dependence in the separation procedure, it is of advantage to perform first a Fourier analysis of the azimuthal angular dependence. In the following it is assumed that the eight sub-CFFs,

$$\mathcal{F} \in \{\Im\mathrm{m}\mathcal{H}, \Im\mathrm{m}\mathcal{E}, \Im\mathrm{m}\widetilde{\mathcal{H}}, \Im\mathrm{m}\widetilde{\mathcal{E}}, \Re\mathrm{e}\mathcal{H}, \Re\mathrm{e}\mathcal{E}, \Re\mathrm{e}\widetilde{\mathcal{H}}, \Re\mathrm{e}\widetilde{\mathcal{E}}\}, \tag{12}$$

appearing in LO and LT approximation (1) or, equivalently, in the hand-bag approach, dominate the twist-two associated first and zeroth asymmetry harmonics, where the latter one might be more contaminated by higher twist contributions than the former ones. Various methods for local extraction have been tested — we will discuss here only two: map of random numbers and a so-called quasi model-independent fitter code that is based on reference GPD models.

3.1 Mapping of random numbers

The mathematical problem to extract CFFs locally, *i.e.*, at fixed values of x_{B}, Q^2, and t, from experimental measurements, can be simply understood as a map from the space of observables into the space of CFFs [3]. For asymmetry measurements with both kinds of leptons available this map of random numbers has been described in Ref. [17]. For cross section measurements with a polarized electron beam and/or polarized nucleon target it can be abstractly formulated as a set of linear and quadratic equations,

$$\boldsymbol{X} = \boldsymbol{b}(F_1, F_2|\mathcal{G}) + \boldsymbol{c}^{-1}(F_1, F_2; \mathcal{G})\cdot\boldsymbol{\mathcal{F}} + \boldsymbol{\mathcal{F}}^{\dagger}\cdot\boldsymbol{d}\cdot\boldsymbol{\mathcal{F}}, \quad \boldsymbol{X} = \begin{pmatrix} X_{\mathrm{unp}} \\ \vdots \\ \vdots \\ X_{h_n S_m} \end{pmatrix}, \quad \boldsymbol{\mathcal{F}} = \begin{pmatrix} \Im\mathrm{m}H \\ \vdots \\ \Re\mathrm{e}H \\ \vdots \end{pmatrix}, \tag{13}$$

where the n-dimensional vector $\boldsymbol{X}$ contains the experimental measurements of (weighted) harmonics at given kinematics, $\boldsymbol{\mathcal{F}}$ is a n-dimensional vector of sub-CFFs (12) under the assumption that the remaining $24 - n$ dimensional sub-CFF vector $\boldsymbol{\mathcal{G}}$ can be considered as fixed. The constant vector

$$\boldsymbol{b}(F_1, F_2; \mathcal{G}) = \boldsymbol{b}^{\mathrm{BH}}(F_1, F_2) + \boldsymbol{b}^{\mathrm{INT}}(F_1, F_2)\cdot\boldsymbol{\mathcal{G}} + \boldsymbol{\mathcal{G}}^{\dagger}\cdot\boldsymbol{b}^{\mathrm{VCS}}\cdot\boldsymbol{\mathcal{G}}$$

contains contributions from the Bethe-Heitler term, interference term, and/or squared VCS amplitude, while the linear coefficient matrix $\boldsymbol{c}^{-1}(F_1, F_2; \mathcal{G})$ arises from the interference term and/or squared VCS amplitude. The (multi–valued) solution of the (quadratic) equations can be found by diagonalization.

Another local extraction method relies on the finding that apart from some power suppressed contributions, which are considered small, only certain linear and bilinear CFF combinations enter the interference term and the squared VCS amplitude, respectively. Strictly spoken, for each kinematical point we have different functional forms of these combinations which also depend on the center-of-mass energy or energy loss variable y. However, in the leading twist approximation the center-of-mass energy dependence approximately separates and Eq. (13) might be written as

$$\boldsymbol{X} \approx \boldsymbol{b}(F_1, F_2) + \boldsymbol{k}^{-1}\cdot\boldsymbol{C}(\mathcal{F}|F_1, F_2) + \boldsymbol{K}\cdot\boldsymbol{C}(\mathcal{F}^{\dagger}, \mathcal{F}), \tag{14}$$

where now $\mathcal{F}$ contains the full set of (sub-)CFFs and $\boldsymbol{k}^{-1}$ and $\boldsymbol{K}$ are matrices of dimensions n. For cross section measurements with an unpolarized nucleon target and a polarized electron beam [6, 7] the information which can be extracted are the first (weighted) harmonic of the beam helicity dependent cross section $\Delta(\phi) = \frac{1}{2}(d\sigma^{\uparrow} - d\sigma^{\downarrow})/dx_{\mathrm{B}}Q^2 dt d\phi$,

$$\Delta^{\sin(\phi)} \approx \frac{\mathcal{N}}{x_{\mathrm{B}} c_0^{\mathcal{P}} t}\frac{x_{\mathrm{B}} S_{++}(1)}{y}\Im\mathrm{m}\mathcal{C}^{\mathcal{I}}_{\mathrm{unp}}(\mathcal{F}) \tag{15a}$$

and zeroth and first (weighted) harmonics of the unpolarized cross section $\Sigma(\phi) = \frac{1}{2}(d\sigma^{\uparrow} + d\sigma^{\downarrow})/dx_{\rm B}Q^2dtd\phi$,

$$\Sigma^{\cos(0\phi)} \approx \frac{\mathcal{N}}{x_{\rm B}c_0^{\mathcal{P}}t}\left[\frac{c_{0,\rm unp}^{\rm BH}}{(1+\epsilon^2)^2} + \frac{x_{\rm B}C_{++}(0)}{y}\Re{\rm e}C_{\rm unp}^{\mathcal{I}}(\mathcal{F}) + \frac{2x_{\rm B}^2c_0^{\mathcal{P}}t}{Q^2}\frac{2-2y+y^2+\frac{\epsilon^2}{2}y^2}{1+\epsilon^2}C_{\rm unp}^{\rm VCS}(\mathcal{F},\mathcal{F}^*)\right], \tag{15b}$$

$$\Sigma^{\cos(\phi)} \approx \frac{-\mathcal{N}}{x_{\rm B}c_0^{\mathcal{P}}t}\left[\frac{c_{1,\rm unp}^{\rm BH}}{(1+\epsilon^2)^2} + \frac{x_{\rm B}C_{++}(1)}{y}\Re{\rm e}C_{\rm unp}^{\mathcal{I}}(\mathcal{F}) + \frac{2w_1x_{\rm B}^2c_0^{\mathcal{P}}t}{Q^2}\frac{2-2y+y^2+\frac{\epsilon^2}{2}y^2}{1+\epsilon^2}C_{\rm unp}^{\rm VCS}(\mathcal{F},\mathcal{F}^*)\right], \tag{15c}$$

allowing us to map directly the experimental data to the imaginary and real part of the CFF combination $C_{\rm unp}^{\mathcal{I}}(\mathcal{F})$ and the bilinear form $C_{\rm unp}^{\rm VCS}(\mathcal{F},\mathcal{F}^*)$. For precise definition of various symbols see Ref. [1].

In the practical application of the strategy (13) one might straightforwardly calculate the sub-CFFs $\mathcal{F}(\boldsymbol{X}; F_1, F_2|\boldsymbol{\mathcal{G}})$, *i.e.*, their mean and variance, as function of the experimental measurements, the form factor parametrization, and some given values of the remaining CFFs $\boldsymbol{\mathcal{G}}$, whose fixed values can be varied in some model-dependent or theory-inspired region. Thus, the uncertainty of the extracted sub-CFFs $\mathcal{F}(\boldsymbol{X}; F_1, F_2|\boldsymbol{\mathcal{G}})$ contains essentially two main sources: the experimental uncertainty of the DVCS and the form factor measurements, including also the uncertainties of QED corrections, and the estimated uncertainty of the unknown CFF contributions. The strategy (14) has the advantages that within this assumption the uncertainties of the extracted linear and bilinear CFF combinations are governed only by the experimental uncertainties. However, some care is needed to have control over the uncertainties that are induced by the assumed approximation. A consistency check between the two methods is provided by evaluating the set of (approximate) equalities

$$C(\mathcal{F}|F_1,F_2) \approx C(\mathcal{F}(\boldsymbol{X};F_1,F_2|\boldsymbol{\mathcal{G}})|F_1,F_2), \quad C(\mathcal{F}^{\dagger},\mathcal{F}) \approx C(\mathcal{F}^{\dagger}(\boldsymbol{X};F_1,F_2|\boldsymbol{\mathcal{G}}^{\dagger}),\mathcal{F}(\boldsymbol{X};F_1,F_2|\boldsymbol{\mathcal{G}})), \tag{16}$$

which have to be (almost) independent on the values of $\boldsymbol{\mathcal{G}}$.

3.2 Quasi model-independent method based on reference GPD models

Another strategy, based on the method of least squares fitting, was suggested in Ref. [28] and utilized to locally extract GPD information in LO and LT approximation (1) from experimental measurements [29–31]. Here, the parameters are eight multipliers, where their values are normalized to reference GPD predictions of the VGG code, reading here in our notation as

$$\Im{\rm m}\mathcal{F}(x_{\rm B},t,Q^2) = a_{\mathcal{F}}^{\Im{\rm m}}\Im{\rm m}\mathcal{F}^{\rm VGG}(x_{\rm B},t,Q^2), \quad \Re{\rm e}\mathcal{F}(x_{\rm B},t,Q^2) = a_{\mathcal{F}}^{\Re{\rm e}}\Re{\rm e}\mathcal{F}^{\rm VGG}(x_{\rm B},t,Q^2), \tag{17}$$

where $\mathcal{F}^{\rm VGG}$ are predicted by the VGG code via the convolution formula (1) and the multipliers $a_{\mathcal{H}}^{\Im{\rm m}},\cdots,a_{\widetilde{\mathcal{E}}}^{\Re{\rm e}}$ are the eight new model-independent fitting parameters. Further reduction might be made by setting some CFFs to zero, *e.g.*, the imaginary part of the helicity non-conserved CFF $\widetilde{\mathcal{E}}$ whose smallness is suggested by the GPD findings in the DVMP analyses of π^+ electroproduction [32]. Often the method was/is employed to an incomplete set of observables, *i.e.*, the fitting problem is ill-posed and to reach convergency one constrains the seven parameters $|a_{\mathcal{F}}^{\Im{\rm m}}| \lesssim 5$ or so, including the more unreliable model estimates $|a_{\mathcal{F}}^{\Re{\rm e}}| \lesssim 5$ for the real part of CFFs. This method has been recently extended by generating an ensemble of fits with randomly distributed start values, where it is hoped that in this statistical manner also ill-posed fitting problems can be managed [33].

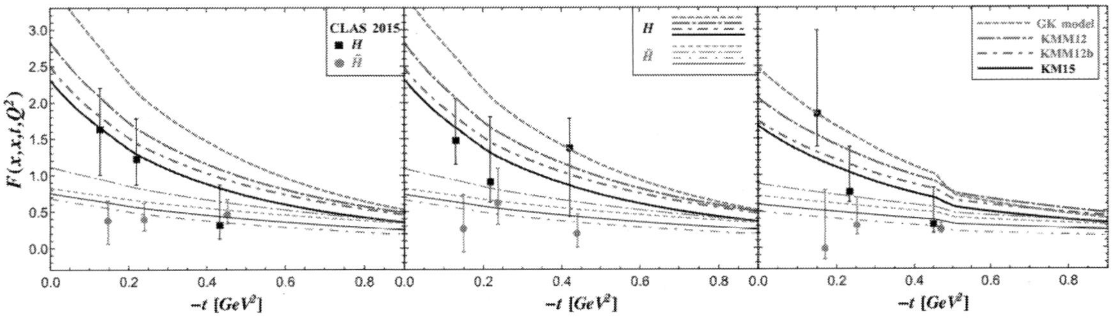

Figure 3. GPD H (filled squares, slightly shifted to the l.h.s., and thick curves) and $\widetilde{H}$ (filled circles, slightly shifted to the r.h.s., and thin curves) from DVCS asymmetry measurements in three $\{x_B, Q^2\}$-bins versus $-t$ [4]. They are locally extracted by the quasi model-independent method (squares and circles) [4], predicted by the GK model (dashed), KMM12 (dash-dotted) fit, KMM12b (shorter dash-dotted) fit, and from the new KM15 fit (solid).

3.3 Comparison of results

Various strategies to locally extract CFFs from the almost over-complete DVCS asymmetry measurements of the HERMES collaboration were studied in Ref. [17]. Namely, one-to-one mapping, stepwise regression, and least squares fitting methods. The quasi model-independent method, see Sec. 3.2, utilizes all available harmonics, *i.e.*, together with the stronger hypothesis that all observables are describable in the hand-bag approach noise is used to stabilize the fitting procedure. The results for $\{\Im\mathrm{m}\mathcal{H}, \Im\mathrm{m}\widetilde{\mathcal{H}}, \Re\mathrm{e}\mathcal{H}\}$ of Ref. [33] are in good agreement with our findings [17]. Note that in Ref. [17] all local CFF values were reported, including the negative one for $\widetilde{\mathcal{H}}$ at $x_B = 0.08$, $Q^2 = 1.9\,\mathrm{GeV}^2$, and $-t = 0.03\,\mathrm{GeV}^2$, however, we concluded that only the sub-CFFs $\Im\mathrm{m}\mathcal{H}$, $\Re\mathrm{e}\mathcal{H}$, $\Im\mathrm{m}\widetilde{\mathcal{H}}$, and maybe $\Im\mathrm{m}\mathcal{E}$ are accessible from HERMES measurements.

The new CLAS beam spin asymmetry A_{LU} and longitudinally polarized target spin asymmetry A_{UL} and A_{LL} measurements [4] were also analyzed with the quasi model-independent tool, shortly described in Sec. 3.2, were results for the imaginary parts of CFFs $\mathcal{H}$ and $\widetilde{\mathcal{H}}$ were presented in Fig. 25 of Ref. [4]. In contrast to the HERMES analysis, there are only three asymmetries that are sensitive to two linear combinations (5) and (6) of the imaginary part of CFFs and one to the real part of Eq. (6), where in addition the normalization of asymmetries, *i.e.*, the unpolarized cross section that is dominated by the Bethe-Heilter amplitude, is nonlinearly governed by all CFFs. We emphasise that CLAS data in the DVCS region do not indicate (apart from noise) the appearance of higher harmonics and that there might be some tension among longitudinal target spin asymmetry measurements from HERMES and CLAS collaborations. The results from the quasi model-independent approach have naturally large error bars, which provide a statistical representation of both the propagated experimental uncertainties and the model-dependent constraints on the non-extractable degree of freedom. As we would have expected, the locally extracted results in the DVCS kinematics are predicted from our older KMM12b and new KM15 fit (solid curves), shown for three DVCS bins in Fig. 3 while the GK (dashed) model prediction for GPD H is too large. Note that all models have a rather similar GPD $\widetilde{H}$ prediction. The t-dependence of GPD $\widetilde{H}$ comes out flatter than for GPD H, which has a natural explanation in GPD models that incorporate form factor constraint. Namely, from the common GPD modeling and the fact that the axial nucleon form factor possesses a flatter t-dependence than the Pauli form factor, the above interpretation can be considered a model-dependent GPD prediction and only by means of the GPD framework one might loosely link the CFF $\widetilde{\mathcal{H}}$ to the axial-charge distribution, quantified by the axial form factor.

As explained in Sec. 3.1, the information which can be extracted from the new Hall A [6] and CLAS [7] cross section measurements are the imaginary and real parts of the CFF combination $\mathcal{C}^{\mathcal{I}}_{\rm unp}(\mathcal{F})$ and the bilinear form $\mathcal{C}^{\rm VCS}_{\rm unp}(\mathcal{F},\mathcal{F}^*)$, see Eq. (15), while extraction of sub-CFFs is an ill-posed problem. We utilized the method (15), where uncertainties were added in quadrature and propagated via the transformation of the covariance matrix, see [17]. Note that in the propagation of experimental uncertainties we ignored the fact that systematic ones are asymmetric; however, for the purpose of an illustration at twist-two level this is not an important issue. Our results, shown as filled squares in the upper row of Fig. 4 coincide for the imaginary and real part of the linear combination $\mathcal{C}^{\mathcal{I}}_{\rm unp}$ and for the bilinear form $\mathcal{C}^{\rm VCS}$ with those that were obtained by the Hall A collaboration by fits to the ϕ–dependent cross sections, see Fig. 22 of Ref. [6], where the net uncertainties are dominated by the systematic errors. The same method applied to the previous Hall A data [5], shown by empty squares, reveals some differences between the data sets (here we assume that these CFF combinations are smooth functions that vary only slightly within the small kinematical difference). However, as we will explain now, the adopted approximation implies an additional uncertainty and so the net errors of these results might be underestimated.

On the other hand we might map the experimental data to the CFF $\mathcal{H}(\mathcal{E},\widetilde{\mathcal{H}})$ and the bilinear form $\mathcal{C}^{\rm VCS}(\mathcal{F}^*,\mathcal{F})$ by utilizing the "exact" formulae for the restricted set of four CFFs. As expected, the extracted mean values for the linear combination

$$\Im\mathrm{m}\,\mathcal{C}^{\mathcal{I}}_{\rm unp}(\Delta|\mathcal{E}=\widetilde{\mathcal{H}}=0)\ \ \text{and}\ \ \Re\mathrm{e}\,\mathcal{C}^{\mathcal{I}}_{\rm unp}(\Sigma|\mathcal{E}=\widetilde{\mathcal{H}}=0)$$

only slightly deviate from the findings within the direct extraction procedure in which power suppressed corrections are neglected. In average we find that the results deviate on the 0.2–0.3% level where the maximal deviation does not exceed 0.5% and 8% for the imaginary and real parts, respectively. Taking values for the unconstrained CFFs from the GK model prediction (filled circles) or our KM15 fit (stars) will only slightly change the mean values for this linear CFF combinations. In particular for the extraction of the bilinear form power suppressed corrections play a more important role and, thus, the two different methods provide results where in average the differences are ~ 7%, however, can be up to ~ 15% in some bins. The results where the unconstrained CFFs $\mathcal{E}$, $\widetilde{\mathcal{H}}$, and $\widetilde{\mathcal{E}}$ are set to the GK model and the KM15 fit findings are shown as filled circles and stars, respectively. For both choices the values for $\widetilde{\mathcal{H}}$ are very similar and since the sensitivity to $\mathcal{E}$ is much more suppressed, we find rather similar results which are also very similar to those obtained in the $\mathcal{E}=\widetilde{\mathcal{H}}=0$ case. However, varying the unconstrained CFFs in the region

$$|\Im\mathrm{m}\mathcal{E},\Im\mathrm{m}\widetilde{\mathcal{H}}|\le 2\ \ \text{and}\ \ |\Re\mathrm{e}\mathcal{E},\Re\mathrm{e}\widetilde{\mathcal{H}}|\le 4 \tag{18}$$

generates for the real part of the linear combination and for the bilinear form a larger variation of the extracted mean values inside the gray rectangles, which reflects the importance of power suppressed contributions. Again the differences are much more pronounced for the extracted values of the bilinear CFF coefficient $\mathcal{C}^{\rm VCS}_{\rm unp}$. Taking into account this additional uncertainty, we can conclude that our previous KM10b ([35], dashed curves) and KMM12b (dash-dotted curves) fits in which only the relative ϕ-modulation of the superseded dataset [5] was utilized predicts the locally extracted values of the new data [6] while the new global KM15 fit (solid curves) is entirely consistent with the locally extracted values. Compared to the older data set (empty squares), the GK model prediction (short dashed curves and dotted curves that adds a D-term contribution from Ref. [34]) gets more disfavored with respect to the new data set (filled squares).

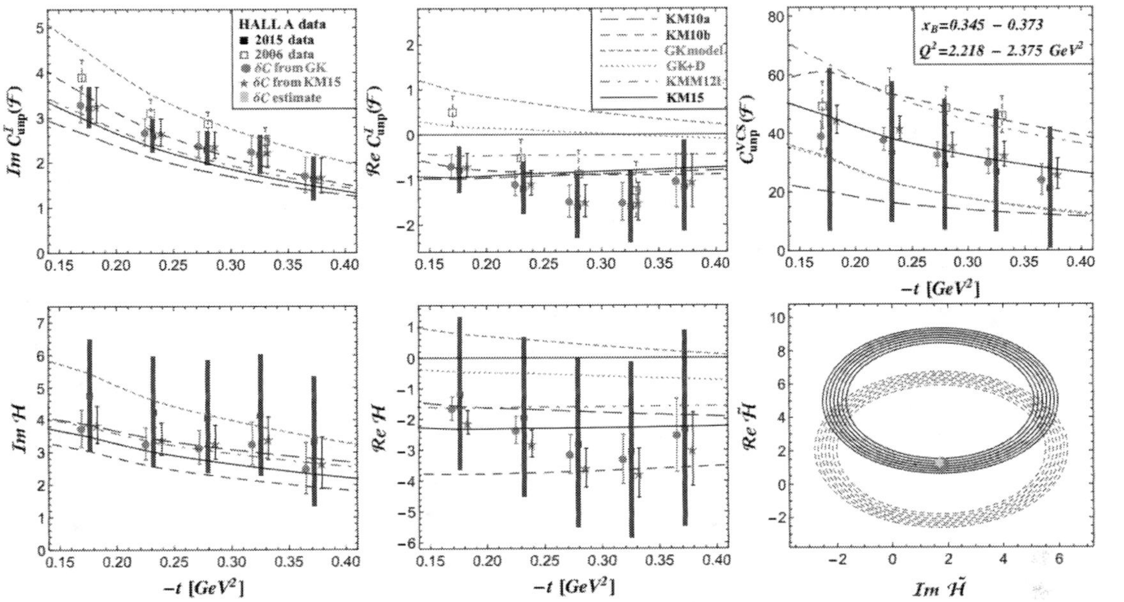

Figure 4. Upper row: imaginary part [left panel] and real part [middle panel] of the linear CFF combination $C^{I}_{\text{unp}}(\mathcal{F})$, defined in Eq. (5), as well as of the bilinear CFF combination $C^{\text{VCS}}_{\text{unp}}(\mathcal{F}^{*}_{\text{unp}}, \mathcal{F})$ [right panel] are mapped from the Hall A cross section measurements [6] for Kin3 kinematics (8) within the Hall A approximation (filled squares) and utilizing the CFF $\mathcal{E}$ and $\widetilde{\mathcal{E}}$ values from the GK model (filled circles) and the KM15 fit (stars). The bands show the model uncertainties of power suppressed contributions in the range $|\Im\mathrm{m}\mathcal{E}, \Im\mathrm{m}\widetilde{\mathcal{H}}| \leq 2$ and $|\Re\mathrm{e}\mathcal{E}, \Re\mathrm{e}\widetilde{\mathcal{H}}| \leq 4$. The empty squares sketch the first Hall A measurements [5] at slightly different kinematics. Lower row: possible region for $\Im\mathrm{m}\mathcal{H}$ [left panel] and $\Re\mathrm{e}\mathcal{H}$ [left panel] within the same CFF constraints as well as the 1-σ ellipses for the $\widetilde{\mathcal{H}}$ constraint at $t = -0.232\,\text{GeV}^2$ [right panel] for fixed CFF $\mathcal{E}$ and $\widetilde{\mathcal{E}}$ values taken from the GK model [dashed curves] and the KM15 fit [solid curves]. Model predictions are shown as curves, see Fig. 1. The dotted curves show a RDDA based GPD H, taken from the GK model, and the D-term form factor parametrization of Ref. [34].

Clearly, if one asks for the values of the CFF $\mathcal{H}$ the uncertainties depend much on the constraints for the remaining CFFs. This is illuminated in the first two panels of the lower row in Fig. 4 where due to the variation (18) the mean values of $\mathcal{H}(\mathcal{E} = 0, \widetilde{\mathcal{H}} = 0)$ (filled squares) can move in the region of the rectangles. The findings for $\Im\mathrm{m}\mathcal{H}$ would be consistent with the values from the RDDA based GK model, however, taking values for $\widetilde{\mathcal{H}}$ and $\mathcal{E}$ from the GK model (circles) or the KM15 fit (stars) we realize that $\mathcal{H}$ is consistent with the KM15 fit and to some extent with our model predictions. In the right panel of the lower row we show the correlation between the real and imaginary part of the CFF $\widetilde{\mathcal{H}}$ within the 1-σ band and with remaining CFFs fixed by the GK model (dashed curves) or the KM15 fit (solid curves). The model values for $\widetilde{\mathcal{H}}$ are again displayed as filled circle and star and one realizes that the GK model violates this constraint while the KM15 fit respects it.

The CLAS collaboration used the quasi model-independent tool and tried to extract the CFF $\mathcal{H}$ from the cross section measurements [7], where only the two CFFs $\mathcal{H}$ and $\widetilde{\mathcal{H}}$ were taken as four fitting parameters. Apart from noise effects this is essentially a three parameter fit problem, see Eq. (15), where, however, four constrained parameters were used. We utilized a similar method, shown for the $\{x_{\text{B}} = 0.185, Q^2 = 1.63\ \text{GeV}^2\}$-bin in Fig. 5 by the empty circles and squares. In the lower panel

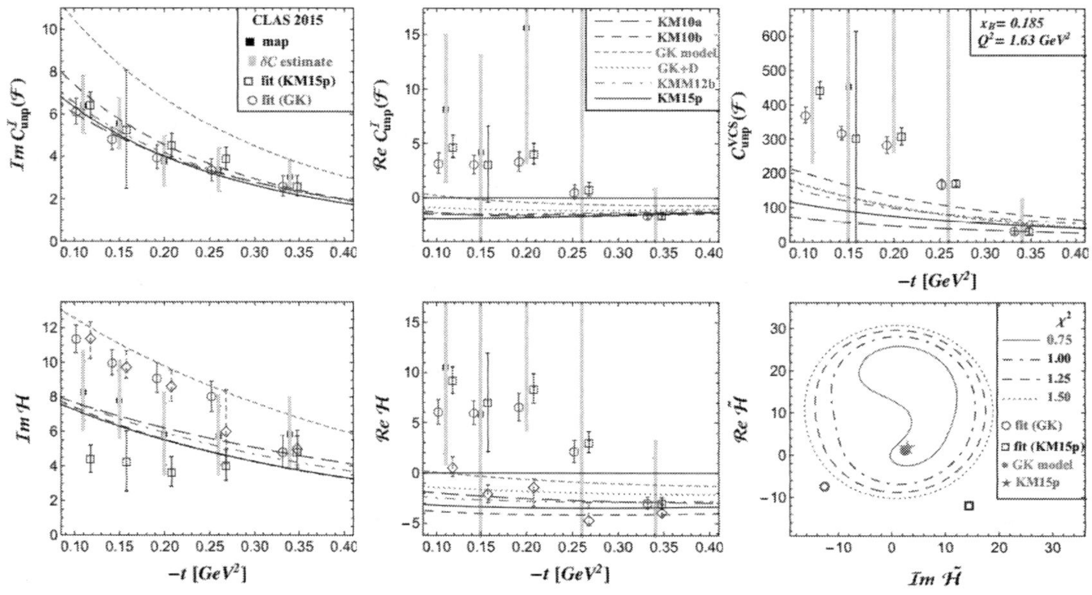

Figure 5. Imaginary [upper left panel] and real [middle panel] parts of linear CFF combination $C^{\mathcal{I}}_{\mathrm{unp}}$ as well as the bilinear combination $C^{\mathrm{VCS}}_{\mathrm{unp}}$ [upper right panel] for the $\{x_{\mathrm{B}} = 0.185, Q^2 = 1.63\ \mathrm{GeV}^2\}$-bin are locally extracted from CLAS data [7] by a map (filled squares), where rectangles show the variation of mean values by utilizing the constraint (18), and from a constrained four parameter, $\Im\mathrm{m}\mathcal{H}$, $\Im\mathrm{m}\widetilde{\mathcal{H}}$, $\Re\mathrm{e}\mathcal{H}$, $\Re\mathrm{e}\widetilde{\mathcal{H}}$, fit with start values from the GK model (empty circles) and KM15 fit (empty squares). Model predictions are shown as curves and are described in Fig. 4.

row we show that the resulting values, in particular for $\Im\mathrm{m}\mathcal{H}$ (left panel), might depend on the initial values and we might reproduce the CLAS result (see empty rhombi) if we take for the initial values the GK model predictions, however, we might find another solution if we utilize the KM15 result (empty squares) for the start values. Compared to our mapping procedure, providing us $\mathcal{H}(\mathcal{E}, \widetilde{\mathcal{H}})$ (filled squares and rectangles), the uncertainties from the fitting procedures are underestimated. The reason is that the unconstrained degrees of freedom tend to be on the boundary, see $t = -0.11\ \mathrm{GeV}^2$ example in the lower right panel, and thus the error propagation is not reliable. Such large values for $\widetilde{\mathcal{H}}$ can also be considered as inconsistent with the findings from the new CLAS asymmetry measurements, described above. Plugging our CFF fit results into the CFF combinations, see upper panel row, provides results that are consistent with our mapping procedure, however, the uncertainties from the fits might still be underestimated. Our model predictions and fit results are consistent with the CFF values from the map. In contrast to the CLAS statement, we conclude that the locally extracted $\mathcal{H}$ values disfavour a RDDA based GPD model (here the GK model).

4 Conclusions

We compared model predictions with the new DVCS measurements from the CLAS and Hall A collaborations where we utilize a (weighted) Fourier transform as filter to strengthen the discriminating power of predictions. We presented a new global DVCS fit KM15 showing that superseding 2006 Hall A cross section data with new one relieves some tensions, but description of first cosine harmonic of cross section remains imperfect. We showed that our global fit predictions KM10b and KMM12b, which ignored the normalization of the first Hall A cross section measurements, predict the new measurements reasonably well while a RDDA based model, *e.g.*, the GK model, becomes more disfavored by the new data sets. These conclusions are partially in disagreement with the first analyses of the data by collaborations themselves. In particular, the new cross section measurements support the previous findings that even in fixed target kinematics such models provide a GPD H that is too large. We further provided within our simple GPD ansatz from previous fits a good description of the new global DVCS data set and showed that the locally extracted CFF values are consistent with the global description. This analysis includes the GPD $\widetilde{H}$ which is rather compatible with RDDA based model predictions with respect to normalization and t-dependence which comes out flatter than for GPD H. Due to the new cross section measurements the indication that the GPD H has a rather small skewness effect becomes more reliable.

Acknowledgements

We like to thank M. Guidal for a comparison of local fitting results that were obtained from our and his fitting analysis and we are indebted to H. Avakian, M. Garcon, and S. Niccolai for discussions. This work has been supported in part by the Croatian Science Foundation under the project number 8799, Croatian Ministry of Science, Education and Sport (MSES) under the NEWFELPRO grant agreement no. 54, and by the NRF of South Africa under CPRR grant no. 90509.

References

[1] A.V. Belitsky, D. Müller, Y. Ji, Nucl.Phys. **B878**, 214 (2014), 1212.6674
[2] J.J. Kelly, Phys. Rev. **C70**, 068202 (2004)
[3] A.V. Belitsky, D. Müller, A. Kirchner, Nucl. Phys. **B629**, 323 (2002), hep-ph/0112108
[4] S. Pisano et al. (CLAS), Phys. Rev. **D91**, 052014 (2015), 1501.07052
[5] C.M. Camacho et al. (Jefferson Lab Hall A), Phys. Rev. Lett. **97**, 262002 (2006), nucl-ex/0607029
[6] M. Defurne et al. (Hall A) (2015), 1504.05453
[7] H. Jo et al. (the CLAS) (2015), 1504.02009
[8] P. Kroll, H. Moutarde, F. Sabatie, Eur.Phys.J. **C73**, 2278 (2013), 1210.6975
[9] S.V. Goloskokov, P. Kroll, Eur. Phys. J. **C53**, 367 (2008), 0708.3569
[10] M. Vanderhaeghen, P.A.M. Guichon, M. Guidal, Phys. Rev. **D60**, 094017 (1999), hep-ph/9905372
[11] A. Freund, M. McDermott, M. Strikman, Phys. Rev. **D67**, 036001 (2003), hep-ph/0208160
[12] J. Pumplin et al., JHEP **07**, 012 (2002), hep-ph/0201195
[13] S. Aid et al. (H1 Collaboration), Nucl.Phys. **B470**, 3 (1996), hep-ex/9603004
[14] K. Kumerički, D. Müller, Nucl. Phys. **B841**, 1 (2010), 0904.0458
[15] A. Airapetian et al. (HERMES), JHEP **1210**, 042 (2012), 1206.5683
[16] A. Airapetian et al. (HERMES), JHEP **11**, 083 (2009), 0909.3587
[17] K. Kumerički, D. Müller, M. Murray, Phys. Part. Nucl. **45**, 723 (2014), 1301.1230
[18] S. Chen et al. (CLAS), Phys. Rev. Lett. **97**, 072002 (2006), hep-ex/0605012
[19] S. Chekanov et al. (ZEUS), JHEP **05**, 108 (2009), 0812.2517
[20] S. Chekanov et al. (ZEUS), Phys. Lett. **B573**, 46 (2003), hep-ex/0305028
[21] F. Aaron et al. (H1), Phys.Lett. **B681**, 391 (2009), 0907.5289

[22] A. Aktas et al. (H1), Eur. Phys. J. **C44**, 1 (2005), hep-ex/0505061
[23] A. Airapetian et al. (HERMES Collaboration), JHEP **1207**, 032 (2012), 1203.6287
[24] F.X. Girod et al. (CLAS), Phys. Rev. Lett. **100**, 162002 (2008), 0711.4805
[25] G. Gavalian et al. (CLAS), Phys. Rev. **C80**, 035206 (2009), 0812.2950
[26] A. Airapetian et al. (HERMES), JHEP **06**, 019 (2010), 1004.0177
[27] A. Airapetian et al. (HERMES), JHEP **06**, 066 (2008), 0802.2499
[28] M. Guidal, Eur. Phys. J. **A37**, 319 (2008), 0807.2355
[29] M. Guidal, H. Moutarde, Eur. Phys. J. **A42**, 71 (2009), 0905.1220
[30] M. Guidal, Phys. Lett. **B689**, 156 (2010), 1003.0307
[31] M. Guidal, Phys. Lett. **B693**, 17 (2010), 1005.4922
[32] C. Bechler, D. Müller (2009), 0906.2571
[33] M. Boër, M. Guidal, J. Phys. **G42**, 034023 (2015), 1412.4651
[34] B. Pasquini, M. Polyakov, M. Vanderhaeghen, Phys.Lett. **B739**, 133 (2014), 1407.5960
[35] K. Kumerički, T. Lautenschlager, D. Müller, K. Passek-Kumericki, A. Schaefer, M. Meskauskas (2011), 1105.0899

EPJ Web of Conferences **112**, 01013 (2016)
DOI: 10.1051/epjconf/201611201013

The gauge-invariant canonical energy-momentum tensor

Cédric Lorcé[1,a]

[1] *Centre de Physique Théorique, Ecole polytechnique, CNRS, Université Paris-Saclay, F-91128 Palaiseau, France*

Abstract. The canonical energy-momentum tensor is often considered as a purely academic object because of its gauge dependence. However, it has recently been realized that canonical quantities can in fact be defined in a gauge-invariant way provided that strict locality is abandoned, the non-local aspect being dictacted in high-energy physics by the factorization theorems. Using the general techniques for the parametrization of non-local parton correlators, we provide for the first time a complete parametrization of the energy-momentum tensor (generalizing the purely local parametrizations of Ji and Bakker-Leader-Trueman used for the kinetic energy-momentum tensor) and identify explicitly the parts accessible from measurable two-parton distribution functions (TMDs and GPDs). As by-products, we confirm the absence of model-independent relations between TMDs and parton orbital angular momentum, recover in a much simpler way the Burkardt sum rule and derive three similar new sum rules expressing the conservation of transverse momentum.

1 Introduction

Following Noether's procedure, one obtains a canonical energy-momentum (EMT) $T^{\mu\nu}$ which is usually neither symmetric[1] nor gauge invariant. These properties are often considered as pathological and can be cured by adding to the EMT a superpotential term of the form $\partial_\alpha f^{[\alpha\mu]\nu}$ [2–4], where the square brackets stand for antisymmetrization. This amounts to a redefinition of the local density of energy and momentum [5, 6] while leaving the total linear and angular momenta unchanged.

Superpotential terms have also been used to decompose the angular momentum into spin and orbital contributions [7, 8]. According to standard textbooks like *e.g.* [9, 10], one cannot perform this decomposition for the gauge field in a gauge-invariant way. Nonetheless, it appears that the photon spin and orbital angular momentum (OAM) are routinely measured, see *e.g.* [11] and references therein. Similarly, a gauge-invariant quantity called ΔG, interpreted in the light-front gauge as the gluon spin [7], has been measured in polarized deep inelastic and proton-proton scatterings, see [12] for a recent analysis. Since physical observables are gauge invariant, Chen *et al.* [13] claimed that the textbooks were wrong and proposed a formal gauge-invariant decomposition of the photon and

[a] e-mail: cedric.lorce@polytechnique.edu

[1] We stress that the symmetry requirement essentially comes from General Relativity where torsion is assumed to vanish. In more general theories of gravitation like *e.g.* Einstein-Cartan theory and metric-affine gauge theory, torsion is allowed to be nonzero leading to asymmetric EMTs and a natural coupling between gravitation and spin. These effects are however expected to be extremely small and to show up only under extreme conditions, see *e.g.* [1] and references therein.

gluon angular momentum. Their work received strong criticisms and triggered numerous theoretical papers on the subject, summarized in recent reviews [14, 15]. It is now understood that textbooks implicitly referred to local quantities only, whereas the formal construction of Chen *et al.* and the above-mentioned quantities extracted from experimental data are intrinsically non-local [16–19].

Typical examples of measurable non-local quantities in Quantum ChromoDynamics (QCD) are parton distribution functions (PDFs) whose gauge invariance is ensured by a Wilson line along a path determined by the factorization theorems [20]. Generalized Transverse-Momentum dependent Distributions (GTMDs) [21] are natural generalizations of the PDFs and provide a natural way to access the canonical OAM [22–24] and other angular momentum correlations [25]. Unfortunately, it is not known so far how to access them directly in experiments. They can however be accessed indirectly using *e.g.* realistic models [22, 26–29], or lattice QCD in the infinite-momentum limit [30–33].

We present here the first complete parametrization of the EMT obtained in [34] which can be applied to virtually any form of the EMT discussed in the literature, provided that the non-locality appears only along the light-front (LF) direction n.

2 A basis of gauge-invariant energy-momentum tensors

Most of the EMTs that appeared in the literature can be decomposed in a basis of five gauge-invariant tensors

$$\begin{aligned} T_1^{\mu\nu} &= \overline{\psi}\gamma^\mu \tfrac{i}{2}\overleftrightarrow{D}^\nu\psi, & T_2^{\mu\nu} &= -2\mathrm{Tr}[G^{\mu\alpha}G^\nu{}_\alpha] + g^{\mu\nu}\tfrac{1}{2}\mathrm{Tr}\big[G^{\alpha\beta}G_{\alpha\beta}\big], \\ T_3^{\mu\nu} &= -\overline{\psi}\gamma^\mu g A^\nu_{\text{phys}}\psi, & T_4^{\mu\nu} &= \tfrac{1}{4}\,\epsilon^{\mu\nu\alpha\beta}\partial_\alpha\big[\overline{\psi}\gamma_\beta\gamma_5\psi\big], \\ T_5^{\mu\nu} &= -2\partial_\alpha\mathrm{Tr}\big[G^{\mu\alpha}A^\nu_{\text{phys}}\big], \end{aligned} \tag{1}$$

where $\frac{i}{2}\overleftrightarrow{D}^\mu = \frac{i}{2}\overleftrightarrow{\partial}^\mu + gA^\mu$ is the hermitian covariant derivative with $\overleftrightarrow{\partial}^\mu = \overrightarrow{\partial}^\mu - \overleftarrow{\partial}^\mu$, and $\epsilon_{0123} = +1$. For example, the kinetic form of the quark and gluon EMTs are given by $T_1^{\mu\nu}$ and $T_2^{\mu\nu}$, respectively, and the corresponding canonical forms are given by $T_1^{\mu\nu} + T_3^{\mu\nu}$ and $T_2^{\mu\nu} - T_3^{\mu\nu} + T_5^{\mu\nu}$, respectively. The superpotentials $T_4^{\mu\nu}$ and $T_5^{\mu\nu}$ together with the QCD equations of motion

$$\begin{aligned} \overline{\psi}(r)\gamma^{[\mu}i\overleftrightarrow{D}^{\nu]}\psi &= -\epsilon^{\mu\nu\alpha\beta}\partial_\alpha\big[\overline{\psi}\gamma_\beta\gamma_5\psi\big], \\ 2\big[\mathcal{D}_\alpha G^{\alpha\beta}\big]^c_{\ c'} &= -g\,\overline{\psi}_{c'}\gamma^\beta\psi^c, \end{aligned} \tag{2}$$

where c, c' are color indices in the fundamental representation and $\mathcal{D}_\mu = \partial_\mu - ig[A_\mu,\quad]$ is the adjoint covariant derivative, can be used to relate the various EMTs [14, 34]. Because of the first identity in Eq. (2), we have $T_4^{\mu\nu} = -\frac{1}{2}\,T_1^{[\mu\nu]}$ which can then be discarded in the following discussions.

In order to obtain gauge-invariant canonical EMTs, one needs to decompose the gauge potential A_μ into pure-gauge and "physical" (or covariant) contributions

$$A_\mu^{\text{pure}} \equiv \tfrac{i}{g}\,\mathcal{W}\partial_\mu\mathcal{W}^{-1}, \qquad A_\mu^{\text{phys}} \equiv A_\mu - A_\mu^{\text{pure}}, \tag{3}$$

where $\mathcal{W}$ is a non-integrable phase factor transforming as $\mathcal{W} \mapsto U\mathcal{W}$ under gauge transformations. In the gauge where $\mathcal{W}(r) = 1$, the gauge-invariant canonical decomposition formally reduces to the Jaffe-Manohar decomposition [7], and can therefore be considered as a gauge-invariant extension of the latter [14, 17, 35, 36].

3 General parametrization

Since our aim is to relate the matrix elements of the EMT to measurable parton distributions, we identify the non-local phase factor $\mathcal{W}$ with a straight Wilson line running along the LF direction n to $r_n = r + \infty\eta n$, and then in the transverse direction to $\infty_\perp$. It is then clear that, beside the average target momentum $P = (p' + p)/2$ and the momentum transfer $\Delta = p' - p$, the matrix elements of the generic LF EMT also depend in principle on $N = \frac{M^2 n}{P\cdot n}$ for a target of mass M, and on the direction of the Wilson line $\eta = \pm 1$. The scalar functions parametrizing the generic LF EMT are complex-valued functions of $\xi = -(\Delta\cdot N)/2(P\cdot N)$ and $t = \Delta^2$, which are the only two independent scalars that can be formed with P, Δ and N.

In Ref. [34], we used the techniques presented in the Appendix A of Ref. [21] and found that the generic LF EMT for a spin-1/2 target can be parametrized as $\langle p', S'|T^{\mu\nu}_a(0)|p, S\rangle = \overline{u}(p', S')\Gamma^{\mu\nu}_a(P, \Delta, N; \eta)u(p, S)$ with $a = 1, \cdots, 5$ and the matrix $\Gamma^{\mu\nu}_a$ given by

$$\begin{aligned}\Gamma^{\mu\nu}_a &= Mg^{\mu\nu}A^a_1 + \frac{P^\mu P^\nu}{M}A^a_2 + \frac{\Delta^\mu\Delta^\nu}{M}A^a_3 + \frac{P^\mu i\sigma^{\nu\Delta}}{2M}A^a_4 + \frac{P^\nu i\sigma^{\mu\Delta}}{2M}A^a_5\\
&+ \frac{N^\mu N^\nu}{M}B^a_1 + \frac{P^\mu N^\nu}{M}B^a_2 + \frac{P^\nu N^\mu}{M}B^a_3 + \frac{N^\mu i\sigma^{\nu\Delta}}{2M}B^a_4 + \frac{N^\nu i\sigma^{\mu\Delta}}{2M}B^a_5 + \frac{\Delta^\mu i\sigma^{\nu N}}{2M}B^a_6 + \frac{\Delta^\nu i\sigma^{\mu N}}{2M}B^a_7\\
&+ \left[Mg^{\mu\nu}B^a_8 + \frac{P^\mu P^\nu}{M}B^a_9 + \frac{\Delta^\mu\Delta^\nu}{M}B^a_{10} + \frac{N^\mu N^\nu}{M}B^a_{11} + \frac{P^\mu N^\nu}{M}B^a_{12} + \frac{P^\nu N^\mu}{M}B^a_{13}\right]\frac{i\sigma^{N\Delta}}{2M^2}\\
&+ \frac{P^\mu\Delta^\nu}{M}B^a_{14} + \frac{P^\nu\Delta^\mu}{M}B^a_{15} + \frac{\Delta^\mu N^\nu}{M}B^a_{16} + \frac{\Delta^\nu N^\mu}{M}B^a_{17} + \frac{M}{2}i\sigma^{\mu\nu}B^a_{18} + \frac{\Delta^\nu i\sigma^{\mu\Delta}}{2M}B^a_{19}\\
&+ \frac{P^\mu i\sigma^{\nu N}}{2M}B^a_{20} + \frac{P^\nu i\sigma^{\mu N}}{2M}B^a_{21} + \frac{N^\mu i\sigma^{\nu N}}{2M}B^a_{22} + \frac{N^\nu i\sigma^{\mu N}}{2M}B^a_{23}\\
&+ \left[\frac{P^\mu\Delta^\nu}{M}B^a_{24} + \frac{P^\nu\Delta^\mu}{M}B^a_{25} + \frac{\Delta^\mu N^\nu}{M}B^a_{26} + \frac{\Delta^\nu N^\mu}{M}B^a_{27}\right]\frac{i\sigma^{N\Delta}}{2M^2}.\end{aligned} \tag{4}$$

For convenience, we introduced the notation $i\sigma^{\mu b} \equiv i\sigma^{\mu\alpha}b_\alpha$ and the factors of i such that the real part of the scalar functions is η-even while the imaginary part is η-odd

$$X^a_j(\xi, t; \eta) = X^{e,a}_j(\xi, t) + i\eta X^{o,a}_j(\xi, t) \tag{5}$$

as a consequence of time-reversal symmetry. The hermiticity property then implies that the real part of B^a_j with $j \geq 14$ is ξ-odd and the imaginary part is ξ-even. For the other functions, it is the opposite. The fact that only 32 independent structures exist can be obtained from a naive simple counting : the EMT $T^{\mu\nu}_a$ has $4\times 4 = 16$ components; the target state polarizations $\pm S$ and $\pm S'$ bring another factor of $2\times 2 = 4$; time-reversal and hermiticity having been used to fix the factors of i and the ξ-dependence, we are left with parity which reduces the number of independent polarization configurations by a factor 2. As announced, this leads to a total of 32 independent complex-valued amplitudes.

Manifestly, the EMTs $T^{\mu\nu}_1$ and $T^{\mu\nu}_2$ do not depend on N or η. All the corresponding scalar functions must then vanish except for $A^{e,a}_j(0, t)$ with $a = 1, 2$ and $j = 1, \cdots, 5$, which are linearly related to the standard energy-momentum form factors [8, 14, 37] as follows

$$\begin{aligned}
A_q(t) &= A^{e,1}_2(0,t), & A_G(t) &= A^{e,2}_2(0,t),\\
B_q(t) &= A^{e,1}_4(0,t) + A^{e,1}_5(0,t) - A^{e,1}_2(0,t), & B_G(t) &= A^{e,2}_4(0,t) + A^{e,2}_5(0,t) - A^{e,2}_2(0,t),\\
C_q(t) &= A^{e,1}_3(0,t), & C_G(t) &= A^{e,2}_3(0,t),\\
\bar{C}_q(t) &= A^{e,1}_1(0,t) + \tfrac{t}{M^2}A^{e,1}_3(0,t), & \bar{C}_G(t) &= A^{e,2}_1(0,t) + \tfrac{t}{M^2}A^{e,2}_3(0,t),\\
D_q(t) &= A^{e,1}_4(0,t) - A^{e,1}_5(0,t), & 0 &= A^{e,2}_4(0,t) - A^{e,2}_5(0,t).
\end{aligned} \tag{6}$$

4 Linear and angular momentum constraints

The parametrization (4) is only constrained by space-time symmetries. We briefly present here the additional constraints arising from the conservation of total linear and angular momentum.

The average four-momentum in the LF form of dynamics can be obtained by contracting the EMT with $\frac{1}{2M^2} N_\mu$ in the forward limit $\Delta \to 0$

$$\langle p^\nu_a \rangle \equiv \frac{1}{2M^2} \langle P, S | T^{N\nu}_a(0) | P, S \rangle = P^\nu A^{e,a}_2 + N^\nu (A^{e,a}_1 + B^{e,a}_2) + \delta^{a3} \frac{\eta}{2} \epsilon^{\nu S}_T (B^{o,3}_{18} - B^{o,3}_{20}). \tag{7}$$

Interestingly, the last term in Eq. (7) originating from the potential EMT $T^{\mu\nu}_3$ is η-odd and can be interpreted as the spin-dependent contribution to the momentum arising from initial and/or final-state interactions, see *e.g.* [38] and references therein. It comes with the structure $\epsilon^{\nu S}_T \equiv \epsilon^{\nu\mu\alpha\beta} S_\mu n_\alpha \bar{n}_\beta$ where $\bar{n}$ is another lightlike four-vector satisfying $n \cdot \bar{n} = 1$ and such that $P^\mu = (P \cdot n)\bar{n}^\mu + (P \cdot \bar{n}) n^\mu$, which means that a transverse target polarization is required. The total four-momentum being P^ν, we recover from the sum over all partons the well-known momentum constraints

$$\sum_{a=1,2} A^{e,a}_1(0,0) = \sum_{a=q,G} \bar{C}_a(0) = 0, \qquad \sum_{a=1,2} A^{e,a}_2(0,0) = \sum_{a=q,G} A_a(0) = 1. \tag{8}$$

Thanks to the complete parametrization (4), we can easily compute the matrix elements of the corresponding OAM tensors $L^{\mu\nu\rho}_a = r^\nu T^{\mu\rho}_a - r^\rho T^{\mu\nu}_a$. Because of the explicit factors of r, the matrix elements of the generic LF OAM tensor need to be handled with care [14, 37]. For a longitudinally polarized target, we found a simple expression for the longitudinal component of OAM

$$\langle L^a_L \rangle \equiv \frac{\epsilon_{T\alpha\beta}}{2M^2} \left[i \frac{\partial}{\partial \Delta_\alpha} \langle p', S' | T^{N\beta}_a(0) | p, S \rangle \right]_{\Delta=0} = A^{e,a}_4(0,0). \tag{9}$$

Similarly, the quark and gluon spin contributions $S^{\mu\nu\rho}_1 = \frac{1}{2} \epsilon^{\mu\nu\rho\sigma} \overline{\psi} \gamma_\beta \gamma_5 \psi$ and $S^{\mu\nu\rho}_2 = -2\mathrm{Tr}\left[G^{\mu[\nu} A^{\rho]}_{\text{phys}}\right]$ can be expressed in terms of $L^{\mu\nu\rho}_4$ and $L^{\mu\nu\rho}_5$, respectively. We then found

$$\begin{aligned} \langle S^q_L \rangle &\equiv \frac{1}{2M^2} \langle P, S | \tfrac{1}{2} \epsilon_{T\alpha\beta} S^{N\alpha\beta}_1(0) | P, S \rangle = -\frac{1}{2} \left[A^{e,1}_4(0,0) - A^{e,1}_5(0,0) \right], \\ \langle S^G_L \rangle &\equiv \frac{1}{2M^2} \langle P, S | \tfrac{1}{2} \epsilon_{T\alpha\beta} S^{N\alpha\beta}_2(0) | P, S \rangle = -A^{e,5}_4(0,0), \end{aligned} \tag{10}$$

where Eq. (2) has been used to express $L^{\mu\nu\rho}_4$ in terms of $T^{\mu\nu}_1$. Adding together the spin and OAM contributions, we naturally recover the Ji relation for total angular momentum [8]

$$\begin{aligned} \langle J^q_L \rangle &= \langle S^q_L \rangle + \langle L^q_L \rangle = \tfrac{1}{2} \left[A^{e,1}_4(0,0) + A^{e,1}_5(0,0) \right] = \tfrac{1}{2} \left[A_q(0) + B_q(0) \right], \\ \langle J^G_L \rangle &= \langle S^G_L \rangle + \langle L^G_L \rangle = \tfrac{1}{2} \left[A^{e,2}_4(0,0) + A^{e,2}_5(0,0) \right] = \tfrac{1}{2} \left[A_G(0) + B_G(0) \right]. \end{aligned} \tag{11}$$

Finally, combining the fact that the total angular momentum is $1/2$ and the momentum constraints (8), we recover also the anomalous gravitomagnetic moment sum rule [39, 40]

$$\sum_{a=1,2} [A^{e,a}_4(0,0) + A^{e,a}_5(0,0) - A^{e,a}_2(0,0)] = \sum_{a=q,G} B_a(0) = 0. \tag{12}$$

5 Link with measurable parton distributions

The matrix elements of the EMT we are interested in can easily be expressed in terms of the GTMD correlator [18]

$$\langle p', S'|T^{\mu\nu}(0)|p, S\rangle = \int \mathrm{d}^4k\, k^\nu\, W^\mu_{S'S}. \tag{13}$$

By considering appropriate projections [21, 41, 42], we can at the end relate some of the scalar functions appearing in our generic parametrization (4) to measurable parton distributions, like *e.g.* Generalized Parton Distributions (GPDs) accessed in exclusive scatterings [43] and Transverse-Momentum dependent Distributions (TMDs) accessed in semi-inclusive scatterings [20]. The detailed relations between the EMT scalar functions and two-parton GPDs and TMDs of any twist can be found in [34].

Among the interesting results, let us just mention that we derived the following sum rules

$$\sum_{a=q,G} \int \mathrm{d}x\, \mathrm{d}^2k_T\, \tfrac{k_T^2}{2M^2}\, F^a(x, k_T^2) = 0 \tag{14}$$

with $F^a = f_{1T}^{\perp a}, f^{\perp a}, f_L^{\perp a}, f_{3T}^{\perp a}$. They express the fact that the total transverse momentum (w.r.t. the target momentum) has to vanish. The leading-twist relation (*i.e.* the one involving the Sivers function $f_{1T}^{\perp a}$) is known as the Burkardt sum rule [44, 45]. The other three are new and much harder to test experimentally, but it would be very interesting to test them using phenomenological models, Lattice QCD and perturbative QCD.

6 Conclusions

It is possible to define a gauge-invariant canonical energy-momentum tensor once one relaxes the assumption of strict locality. We argued that the canonical energy-momentum tensor can be considered as a physical object and *a priori* measured experimentally *via* particular moments of parton distributions extracted from various physical processes.

We presented here a complete parametrization for the matrix elements of the generic light-front energy-momentum tensor and discussed some of the constraints arising from linear and angular momentum conservation. We showed that this energy-momentum tensor can be related to particular moments of the generalized and transverse-momentum dependent parton distributions. Among the interesting results, we rederived in a simpler way the Burkardt sum rule and obtained three new sum rules involving higher-twist distributions, all expressing basically the conservation of transverse momentum. We expect in a near future exciting new developments in these matters coming from new experimental data obtained in existing and future facilities, and explicit investigations using phenomenological models, Lattice QCD and perturbative QCD.

References

[1] Y. N. Obukhov, A. J. Silenko and O. V. Teryaev, Phys. Rev. D **90**, no. 12, 124068 (2014).
[2] F. J. Belinfante, Physica **6**, 887 (1939).
[3] F. J. Belinfante, Physica **7**, 449 (1940).
[4] L. Rosenfeld, Mém. Acad. Roy. Belg. **18**, 1 (1940).
[5] F. W. Hehl, Rept. Math. Phys. **9**, 55 (1976).
[6] K. F. Liu and C. Lorcé, arXiv:1508.00911 [hep-ph].
[7] R. L. Jaffe and A. Manohar, Nucl. Phys. B **337**, 509 (1990).

[8] X. D. Ji, Phys. Rev. Lett. **78**, 610 (1997).
[9] C. Cohen-Tannoudji, J. Dupont-Roc, and C. Grynberg, *Photons and Atoms*, John Wiley and Sons Inc., 1989, New York.
[10] J. W. Simmons and M. J. Guttman, *States, Waves, and Photons: A modern introduction to light*, Addison-Wesley, 1970, Reading.
[11] K. Y. Bliokh, J. Dressel and F. Nori, New J. Phys. **16**, no. 9, 093037 (2014).
[12] D. de Florian, R. Sassot, M. Stratmann and W. Vogelsang, Phys. Rev. Lett. **113**, no. 1, 012001 (2014).
[13] X. S. Chen, X. F. Lu, W. M. Sun, F. Wang and T. Goldman, Phys. Rev. Lett. **100**, 232002 (2008).
[14] E. Leader and C. Lorcé, Phys. Rept. **541**, 163 (2014).
[15] M. Wakamatsu, Int. J. Mod. Phys. A **29**, 1430012 (2014).
[16] Y. Hatta, Phys. Rev. D **84**, 041701 (2011).
[17] C. Lorcé, Phys. Rev. D **87**, no. 3, 034031 (2013).
[18] C. Lorcé, Phys. Lett. B **719**, 185 (2013).
[19] C. Lorcé, Nucl. Phys. A **925**, 1 (2014).
[20] J. Collins, *Foundations of perturbative QCD*, Cambridge Univeristy Press, 2011.
[21] S. Meissner, A. Metz and M. Schlegel, JHEP **0908**, 056 (2009).
[22] C. Lorcé and B. Pasquini, Phys. Rev. D **84**, 014015 (2011).
[23] C. Lorcé, B. Pasquini, X. Xiong and F. Yuan, Phys. Rev. D **85**, 114006 (2012).
[24] Y. Hatta, Phys. Lett. B **708**, 186 (2012).
[25] C. Lorcé and B. Pasquini, arXiv:1512.06744 [hep-ph].
[26] K. Kanazawa, C. Lorcé, A. Metz, B. Pasquini and M. Schlegel, Phys. Rev. D **90**, no. 1, 014028 (2014).
[27] A. Mukherjee, S. Nair and V. K. Ojha, Phys. Rev. D **90**, no. 1, 014024 (2014).
[28] A. Mukherjee, S. Nair and V. K. Ojha, Phys. Rev. D **91**, no. 5, 054018 (2015).
[29] T. Liu and B. Q. Ma, Phys. Rev. D **91**, 034019 (2015).
[30] B. U. Musch, P. Hägler, J. W. Negele and A. Schäfer, Phys. Rev. D **83**, 094507 (2011).
[31] B. U. Musch, P. Hägler, M. Engelhardt, J. W. Negele and A. Schäfer, Phys. Rev. D **85**, 094510 (2012).
[32] X. Ji, Phys. Rev. Lett. **110**, 262002 (2013).
[33] X. Ji, J. H. Zhang and Y. Zhao, Phys. Lett. B **743**, 180 (2015).
[34] C. Lorcé, JHEP **1508**, 045 (2015).
[35] X. Ji, X. Xiong and F. Yuan, Phys. Rev. D **88**, no. 1, 014041 (2013).
[36] C. Lorcé, Phys. Rev. D **88**, 044037 (2013).
[37] B. L. G. Bakker, E. Leader and T. L. Trueman, Phys. Rev. D **70**, 114001 (2004).
[38] D. Boer, C. Lorcé, C. Pisano and J. Zhou, Adv. High Energy Phys. **2015**, 371396 (2015).
[39] O. V. Teryaev, hep-ph/9904376.
[40] S. J. Brodsky, D. S. Hwang, B. Q. Ma and I. Schmidt, Nucl. Phys. B **593**, 311 (2001).
[41] C. Lorcé, B. Pasquini and M. Vanderhaeghen, JHEP **1105**, 041 (2011).
[42] C. Lorcé and B. Pasquini, JHEP **1309**, 138 (2013).
[43] M. Diehl, Phys. Rept. **388**, 41 (2003).
[44] M. Burkardt, Phys. Rev. D **69**, 057501 (2004).
[45] M. Burkardt, Phys. Rev. D **69**, 091501 (2004).

EPJ Web of Conferences **112**, 01014 (2016)
DOI: 10.1051/epjconf/201611201014

The 3D structure of QCD and the roots of the Standard Model

P.J. Mulders[1,a]

[1] *Nikhef Theory Group and Vrije Universiteit Amsterdam, De Boelelaan 1081, 1081 HV Amsterdam, the Netherlands*

Abstract. For many phenomenological applications involving hadrons in high energy processes the hadronic structure can be taken care of by parton distribution functions (PDFs), in which only the collinear momenta of quarks and gluons are important. In principle the transverse structure, however, provides interesting new phenomenology. Taking into account transverse momenta of partons one works with transverse momentum dependent PDFs (TMDs), These allow all spin-spin correlations and also spin-orbit correlations that have a time reversal odd character and lead to new observables. In many theoretical developments the link to the collinear treatment is used. In this talk I will speculate on a novel view of the 3-dimensional (3D) structure of QCD, which fits in a broader study looking at the roots of the Standard Model of particle physics.

1 Introduction

In the abstract of this talk I emphasized the role of transverse components of the partonic momenta in parton distribution functions (PDFs), hence generalizing them to transverse momentum dependent (TMD) PDFs or in short TMDs. This extends upon the highly successful collinear description of high energy scattering processes where collinear PDFs $f(x)$ only depending on the momentum fraction x in $p = x\,P + \ldots$ describe the partonic (quark and gluon) content of a proton with momentum P. One can think about jet physics, scale dependence, the role of Collins-Soper-Sterman (CSS) formalism to study transverse momenta in the perturbative regime [1], or soft collinear effective theories (SCET [2]). While collinear PDFs only can accomodate unpolarized and spin-spin correlations between hadron and partons, TMDs can accomodate spin-momentum correlations, important in accounting for single spin asymmetries and understanding azimuthal asymmetries in high energy scattering processes.

Most TMDs still have quite naturally the interpretation of momentum densities, now of the type $f(x, p_T^2)$, including the transverse component in $p = x\,P + p_T + \ldots$, which incorporate for instance longitudinal-transverse spin correlations. One of the important issues in dealing with TMDs is that their operator structure necessarily involves (covariant) transverse derivatives, hence gluon fields. This leads for some of the TMDs, in particular the time reversal odd ones (important for single spin asymmetries) to process dependence and the occurrence of multiple functions, for instance for the socalled quark pretzelocity function and the linear gluon polarization. For these issues, I want to refer to contributions in the proceedings of DIS2015 [3] or QCD evolution 2015 [4] and references therein.

---

[a] e-mail: p.j.g.mulders@vu.nl

In this contribution I will focus on the question if there possibly could be more fundamental reasons for the success of collinear physics in QCD linked to the transition from 1D to 3D with the warning that the material in the next sections is very preliminary and at many points speculative [5]. I want to argue that the standard model and its particle content may naturally emerge in a world with an $IO(1,1)\otimes SU(3)$ spacetime-color symmetry broken down to a threefold $IO(1,3)\otimes[SU(2)\times U(1)]$ spacetime-electroweak symmetry for the asymptotic fields leaving only the electroweak charges for these asymptotic particle states. The bosonic excitations are the electroweak gauge bosons, the massless photon and massive W and Z bosons, as well as the Higgs boson. Fermionic excitations come in three families of colorless (leptonic) excitations living in $E(1,3)$ or they form three families of colored (quark) excitations living in $E(1,1)$. Many features of the Standard Model emerge naturally. Taking a supersymmetric starting point solves the naturalness problem. There is an underlying left-right symmetry leading to custodial symmetry in the electroweak sector. In the spectrum one naturally has Dirac-type charged leptons and Majorana-type neutrinos. The electroweak behavior of the naturally confined quarks, accomodates fractional electric charges and the doublet and singlet structure of left- and right-handed quarks, respectively. Most prominent feature is the link between the number of colors, families and space directions.

2 A 1D supersymmetric starting point

As already indicated in the introduction, we take the 1D world serious as a starting point with operators H, P (combined into P^μ or combined in the combinations $P^\pm = H \pm P$) and the boost operator K generating the 2-dimensional Poincaré symmetry group $IO(1,1)$,

$$[H,P]=0,\ [K,H]=iP \text{ and } [K,P]=iH, \quad \text{or} \quad [P^+,P^-]=0 \text{ and } [K,P^\pm]=\pm iP^\pm, \tag{1}$$

with Casimir operator $P^2 = P^\mu P_\mu = H^2 - P^2 = P^+P^-$. The $IO(1,1)$ symmetry can be combined with an $SO(N)$ symmetry to obtain the $IO(1,N)$ space-time symmetry. with as generators H, P^i, K^i and $J^{[ij]}$ (combined into P^μ and $J^{\mu\nu}$, of course after also including discrete space- and time-reversal symmetries.

Positive and negative energy (particle and anti-particle) eigenstates of the momentum operator have a space-time phase dependence $\phi_k(x) = e^{ik\cdot x}$ with $(k^0)^2 - \boldsymbol{k}^2 = k^-k^+ - \boldsymbol{k}_T^2 = M^2$. These plane waves are solutions of the Klein-Gordon equation $(\partial^\mu\partial_\mu + M^2)\phi = 0$ or, with the appropriate Clifford algebra, the Dirac equation $(i\partial\!\!\!/ - M)\xi = 0$. For massless excitations in 1D one can distinguish independent right-movers, depending on x^+, and left-movers, depending on x^-. Expanded in these modes right- and left-handed fields can be defined satisfying $[P^-,\phi_R] = i\partial_+\phi_R = 0$ and $[P^+,\phi_L] = i\partial_-\phi_L = 0$. For massive fields left and right modes become coupled, while the other derivatives $[P^+,\phi_R] = i\partial_-\phi_R$ and $[P^-,\phi_L] = i\partial_+\phi_L$ acquire roles as (front form) canonical momenta [6]. For $M = 0$ the fermion fields in 1D satisfy $\gamma^-\xi_R = \gamma^+\xi_L = 0$ and $\xi_{R/L}$ are independent good fields [7]. Massive fermion fields satisfy the constraints

$$[P^-,\xi_R] = i\partial_+\xi_R = -iM\xi_L \quad \text{and} \quad [P^+,\xi_L] = i\partial_-\xi_L = iM\xi_R. \tag{2}$$

As an additional ingredient, the Poincaré algebra can be extended to a supersymmetric algebra [8] with fermionic operators $Q_{R/L}$,

$$\{Q_R,Q_R^\dagger\} = 2P^+ = 2\,P\cdot\bar{n}, \quad \{Q_L,Q_L^\dagger\} = 2P^- = 2\,P\cdot n \quad \text{and} \quad \{Q_R,Q_L\} = 0, \tag{3}$$

$$[P^\pm,Q_{R/L}] = 0 \quad \text{and} \quad [K,Q_{R/L}] = \pm\tfrac{1}{2}i\,Q_{R/L}. \tag{4}$$

The supersymmetry charges connect the fields,

$$[Q_{R/L},\phi_{R/L}] = \xi_{R/L} \quad \text{and} \quad \{Q_{R/L},\xi^\dagger_{R/L}\} = [P^\pm,\phi_{R/L}] = i\partial_\mp\phi_{R/L}. \tag{5}$$

The contact with the real world is made by considering the excitations of three real scalar and real fermionic (Majorana) fields ϕ and ξ, still in a single space dimension, however. If their masses are zero one can distinguish right-movers (R) and left-movers (L) as independent degrees of freedom, which for bosons is a simple doubling but for fermions coincides with right- and left-handed fermions. Including left-right coupling via a single mass M and a single Yukawa coupling between the boson fields, linked in the supersymmetric situation to the coupling of fermions and bosons, we have basically the Wess-Zumino model [9] in two dimensions,

$$\begin{aligned}\mathscr{L} &= \tfrac{1}{2}\partial_-\phi_R\,\partial_+\phi_R + \tfrac{1}{2}\partial_+\phi_L\,\partial_-\phi_L + \tfrac{i}{2}\xi_R\partial_+\xi_R + \tfrac{i}{2}\xi_L\partial_-\xi_L - V(\phi,\xi) \\ &= \tfrac{1}{2}\partial^\mu\phi_S\,\partial_\mu\phi_S + \tfrac{1}{2}\partial^\mu\phi_P\,\partial_\mu\phi_P + i\,\overline{\psi}\partial\!\!\!/\psi - V(\phi,\xi), \end{aligned} \tag{6}$$

where (real) right and left fields for bosons can be combined into (real) scalar (CP-even) and pseudoscalar (CP-odd) fields $\phi_{S/P} = (\phi_R \pm \phi_L)/\sqrt{2}$. Real fermion fields can be combined in a (self-conjugate) spinor $\psi = (\xi_R, -i\xi_L)/\sqrt{2}$. Supersymmetry strongly restricts the interaction terms. The most compact expression is in terms of the scalar and pseudoscalar fields containing a mass term coupling left and right fields and a single Yukawa coupling that also governs the fermion-boson coupling,

$$V(\phi,\xi) = \tfrac{1}{2}(M + g\phi_S)^2(\phi_S^2 + \phi_P^2) + \tfrac{1}{2}g^2\phi_P^2(\phi_S^2 + \phi_P^2) + \overline{\psi}(M + g\phi_S + g\phi_P\gamma^1)\psi + \lambda F \tag{7}$$

using $\gamma_5 = \gamma^0\gamma^1$. The constraint is given by

$$\lambda F = \frac{\lambda}{4g}\left((M + 2g\phi_R\sqrt{2})(M + 2g\phi_L\sqrt{2}) - M^2\right) = \frac{\lambda}{g}\left((g\phi_S + M/2)^2 - g^2\phi_P^2 - M^2/4\right). \tag{8}$$

Defining $M/2g \equiv v$, we introduce fields $\phi_S + v \equiv v\Phi_S$ and $v\Phi_P \equiv \phi_P$ which can be re-defined as $\Phi_S = \cosh\eta$ and $\Phi_P = \sinh\eta$ or if one likes one can use an *imaginary* representation for Φ_P by writing $\eta = i\theta$. The bosonic part of the potential including constraint becomes

$$V(\Phi) = \frac{v^2M^2}{2}\Phi_S^2\Phi_P^2 = \frac{v^2M^2}{2}\Phi_S^2(\Phi_S^2 - 1) \tag{9}$$

or $V(\Phi) = \frac{1}{8}M^2v^2\sinh^2(2\eta)$. Defining $|\Phi|^2 \equiv (\Phi_S^2 + \Phi_P^2)/2 = \Phi_R^2 + \Phi_L^2$ we have $|\Phi|^2 = \cosh(2\eta)$ and we have $\Phi_S^2 = |\Phi|^2 + 1/2$ and $\Phi_P^2 = |\Phi|^2 - 1/2$. Looking at the minimum of the potential ($\eta = 0$ or $\theta = 0$) we see that the boson field acquires a vacuum expectation value which is right-left symmetric, $\Phi_R = \Phi_L = |\Phi| = 1/\sqrt{2}$ (or $\Phi_S = 1$ and $\Phi_P = 0$). The real excitations around the vacuum are Majorana modes $\Psi = \Psi^c = (\xi, -i\xi)/\sqrt{2}$ and real scalar bosonic modes $\Phi_S/\sqrt{2} = \Phi = \Phi^c = (1 + H)/\sqrt{2}$. Note that $\phi_S = vH$. The 1D pseudoscalar field ϕ_P can be identified as a vector field writing $i\partial_\mu\Phi_{R/L} = (i\partial_\mu \pm gA_\mu)\phi_S/\sqrt{2}$. In the ground state $A_\mu = 0$ and around the vacuum one has $A^\mu \approx \phi_P(n^\mu - \bar{n}^\mu)$ or $A^+ = -A^- \approx \phi_P$. This suggests working with a complex field Φ rather than left and right fields that are CP symmetric, $\Phi_R = \Phi_L^*$. For a single field a global $U(1)$ symmetry is not relevant and local symmetries don't lead to dynamics either, but taking multiple scalar fields the symmetry pattern becomes much richer.

3 Symmetric extension to three fields

The symmetric extension to N real boson and fermion fields (we take $N = 3$) has interesting consequences for the dynamics, which is studied by looking at the possible fluctuations around the vacuum, in the symmetric basis $\langle\Phi\rangle^T = (1,1,1)/\sqrt{3}$. Including complex phases we consider $SU(3)$ fluctuations, although the lagrangian is only invariant under $SO(3)$ transformations of the fields, which is also the symmetry of the groundstate. We propose to use the $SO(3)$ symmetry in combination with

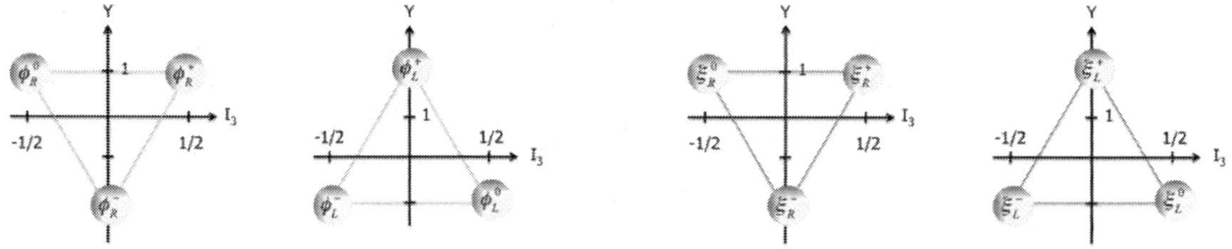

Figure 1. SU(3) quantum number assignments for bosonic (ϕ) and fermionic (ξ) excitations. The right-moving positively charged (ϕ_R^+, ξ_R^+) and neutral (ϕ_R^0, ξ_R^0) fields are in an isospin doublet with $Y = +1$, while the right-moving negatively charged (ϕ_R^-, ξ_R^-) fields are in a weak isosinglet with $Y = -2$. The left-moving negatively charged (ϕ_L^-, ξ_L^-) and neutral (ϕ_L^0, ξ_L^0) fields are in an isospin doublet with $Y = -1$ while the left-moving positively charged (ϕ_L^+, ξ_L^+) fields are in a weak isosinglet with $Y = +2$. The asymptotic fermion fields become the right- and left-handed leptonic fields, $\xi_{R/L}^\pm \to e_{R/L}^\pm$ and $\xi_{R/L}^0 \to \nu_{R/L}^0$.

inversion and time reversal symmetry, to extend the $d = 2$ Poincaré symmetry to a $d = 4$ Poincaré symmetry. Implemented in Weyl mode, the asymptotic fields become real representations of $IO(1,3)$ living in $E(1,3)$.

At this stage part of the freedom in fluctuations around the vacuum has been incorporated. The already accounted for real $SO(3)$ rotations are identified with the subalgebra generated by the $SU(3)$ generators λ_2, $-\lambda_5$ and λ_7, constituting the algebra of the factor group of the subgroup $G' = SU(2) \times U(1)$ with generators $\lambda_1/2$, $\lambda_2/2$, $\lambda_3/2$ and $\lambda_8/2$. This subgroup contains the $SU(3)$ Cartan subalgebra consisting of $I_3 = \lambda_3/2$ and $Y = \lambda_8\sqrt{3}$ that will serve as electroweak charge labels. Labelling the (massless) bosonic states using this Cartan subalgebra, gives fields $\phi_{R/L}^{[I_3,Y]}(\boldsymbol{x},t)$ living in $E(1,3)$. For two fields this would have been just a $U(1)$ charge assigment after combining two real fields into complex fields. The basic recoupled bosonic and fermionic starting point for the three fields and their electroweak quantum numbers is illustrated in Fig. 1.

The fluctuations around the vacuum in space-time and internal space is also reflected in the covariant derivative. We consider the possibilities

$$E(1,1): \qquad iD_\mu\Phi^i = i\partial_\mu\Phi^i + g\sum_{a\in G} A_\mu^a(T_a)^i_{\ j}\Phi^j, \tag{10}$$

$$E(1,3): \qquad iD_\mu\Phi^i = i\partial_\mu\Phi^i + g\sum_{a\in G'} A_\mu^a(T_a)^i_{\ j}\Phi^j. \tag{11}$$

The first expression for the covariant derivative applies to the fields in $E(1,1)$ and accounts for local $SU(3)$ gauge invariance. It involves eight (color) gauge fields also living in $E(1,1)$. The second expression for the covariant derivative is relevant for (asymptotic) fields in $E(1,3)$, coupling for the *real* continuous $SO(3)$ transformations field and space rotations in such a way that there are no gauge fields for that part, and account for the complex fluctuations through four (electroweak) gauge fields living in $E(1,3)$. Such a transition from 1D to 3D, in whatever way we implement it, does require that the Poincaré symmetry and internal symmetry are direct products [10].

To achieve this, we note that the embedding of $SO(3)$ directions into $SU(3)$ is not unique. The discrete symmetry group A_4 governs the possible oriented embeddings. For singlet representations of the embedding group (A_4) one can consider $SU(3) \supset SO(3) \times A_4 \times [SU(2) \otimes U(1)] \to SO(3) \otimes [SU(2) \otimes U(1)]$, decoupling space-time and internal symmetries. The unitary transformation matrix for these singlet states [11–14] is the matrix W. Its specific role now is that it rotates the *symmetric* embedding of the vacuum into the *electroweak* embedding,

$$W^\dagger \begin{pmatrix} \sqrt{1/3} \\ \sqrt{1/3} \\ \sqrt{1/3} \end{pmatrix} = \begin{pmatrix} 0 \\ 1 \\ 0 \end{pmatrix} \quad \text{with} \quad W = \frac{1}{\sqrt{3}} \begin{pmatrix} 1 & 1 & 1 \\ \omega^2 & 1 & \omega \\ \omega & 1 & \omega^2 \end{pmatrix}, \tag{12}$$

where $\omega = \exp(i\,2\pi/3)$. Since the starting point only had $SO(3)$ as a symmetry group, the vacuum indeed is not invariant under $SU(2) \otimes U(1)$ transformations, but it is neutral for $Q = I_3 + Y/2$. The symmetry pattern and its breaking thus is summarized as

$$IO(1,1) \otimes SU(3) \supset \underbrace{IO(1,1) \times SO(3)}_{IO(1,3)} \otimes \underbrace{SU(2)_I \otimes U(1)_Y}_{\to U(1)_Q}.$$

All bosons and fermions, however, still do originate as (finite dimensional) representations of the basic $SU(3)$ symmetry group, which will become important later. There are three families of particles corresponding to singlets of A_4. In 3D the interaction changes from a confining potential to a $1/r$ (or Yukawa) potential between the (electroweak) charges, which thus can be *free*, in contrast to the (color) $SU(3)$ charges that live in 1D.

4 Bosons and Fermions in the Standard Model

After the introduction of the covariant derivatives in Eqs 10 or 11, part of the potential is included in the term

$$D^\mu\Phi^* D_\mu\Phi \quad = \quad \partial^\mu\Phi^*\partial_\mu\Phi + \frac{N\,C_A}{2}\,g^2 A^\mu A_\mu\,\Phi^*\Phi. \tag{13}$$

With $C_A(G = SU(3)) = 4/3$, the second term in Eq. 10 is precisely $-V(\Phi)$ and we are left with the $1+1$ dimensional QCD lagrangian (without a Higgs mass-term),

$$\mathscr{L} \quad = \quad \tfrac{1}{2}\partial^\mu H \partial_\mu H - \frac{1}{4}\,F^{\mu\nu}F_{\mu\nu} + \overline{\Psi}(i\not{D} - M - gv\,H)\Psi, \tag{14}$$

but with a scalar field, which does not seem harmful [15]. We return to the electroweak and family structue of the colored fermions (quarks) in the next section.

For the covariant derivative in Eq. 11 we have $C_A(G') = 2/3$ and the second term is only $-V(\Phi)/2$. This leaves the scalar field massive with $M_H = M/\sqrt{2}$. This is an experimentally interesting scenario for the standard model if the fermion mass is identified with the top quark mass, $M = M_t$. For the bosons, we have (in principle arbitrarily) assigned right to the triplet and left to the anti-triplet. The fields can be rotated into a single scalar field with a nonzero vacuum expectation value as is done in the usual standard model treatment, even if they formed triplets at the start,

$$\Phi_L \quad = \quad \frac{1}{\sqrt{2}}\exp\Big(-\tfrac{i}{2}\sum_{a=1,2,3}\theta^a\lambda_a\Big)\begin{pmatrix} 0 \\ 1+H \\ 0 \end{pmatrix},$$

$$\Phi_R \quad = \quad \frac{1}{\sqrt{2}}\exp\Big(+\tfrac{i}{2}\sum_{a=1,2,3}\theta^a\lambda_a\Big)\begin{pmatrix} 1+H \\ 0 \\ 0 \end{pmatrix}.$$

The electroweak charges and corresponding generators of gauge transformations are identified with the $SU(2)_I \otimes U(1)_Y$ transformations but with a single coupling constant within $SU(3)$. The charged fields are neither $I_3 = \lambda_3/2$ or $Y = \lambda_8\sqrt{3}$ eigenstates but they are eigenstates of $Q = I_3 + Y/2$. The breaking of the $SU(2)_I \times U(1)_Y$ symmetry to $U(1)_Q$ after the choice of ground state being neutral, produces three massive and one massless gauge boson. As discussed in a slightly different context [16], the $SU(3)$ embedding gives a weak mixing angle, $\sin\theta_W = 1/2$ after rewriting in

$$iD_\mu\Phi \quad = \quad i\partial_\mu\Phi + \frac{g}{2}\Big(\sum_{i=1}^{3} W^i_\mu\lambda_i + B_\mu\lambda_8\Big)\Phi \tag{15}$$

the neutral combination $gW^0_\mu I_3 + (g/2\sqrt{3})B_\mu Y$ in terms of Z^μ and A^μ. One obtains (using the dimensionful coupling in $d = 2$) $e = g/2$ and masses $M_W^2 = M^2/4$, $M_Z^2 = M_W^2/\cos^2\theta_W = M^2/3$ and $M_A^2 = 0$. In zeroth order, the weak mixing is fine and the Higgs mass and gauge boson masses are related and they are of the right order with $M = M_t$. Taking $v = M/2g = 1$, one even is tempted to compare $e/M = 1/4$ with $\sqrt{4\pi\alpha} \approx 0.3$. Besides providing a global zeroth order picture for electroweak bosons, we note that the left-right symmetric starting point also ensures custodial symmetry [17, 18].

For the fermionic excitations, the starting $SU(3)$ triplets ξ_R and anti-triplets ξ_L in 1D match those of the bosons, implying the underlying supersymmetry of the elementary fermionic and bosonic $d = 2$ excitations. Also in this case one fixes one direction for the $SU(3)$ representations (the $SO(3)$ embedding) and uses the (remaining) symmetry to fix the electroweak structure as an $SU(3)$ triplet or anti-triplet. The fermions then have electroweak charges corresponding to isospin doublets and singlets as shown in Fig. 1. We like to identify the three possible singlets of A_4 with three families. For this we note that the off-diagonal charge operator in terms of real fields ($\xi^\pm \sim \xi_1 \pm i\xi_3$) in the *electroweak* basis is

$$Q_{\rm ew} = \begin{pmatrix} 0 & 0 & -i \\ 0 & 0 & 0 \\ i & 0 & 0 \end{pmatrix} = V_Q \begin{pmatrix} 1 & 0 & 0 \\ 0 & 0 & 0 \\ 0 & 0 & -1 \end{pmatrix} V_Q^\dagger \quad \text{with} \quad V_Q = \begin{pmatrix} \sqrt{1/2} & 0 & \sqrt{1/2} \\ 0 & 1 & 0 \\ i\sqrt{1/2} & 0 & -i\sqrt{1/2} \end{pmatrix}, \tag{16}$$

while in the *symmetric* basis for the fields one has

$$Q_s \quad = \quad \frac{1}{\sqrt{3}}\begin{pmatrix} 0 & i & -i \\ -i & 0 & i \\ i & -i & 0 \end{pmatrix} = W \begin{pmatrix} 1 & 0 & 0 \\ 0 & 0 & 0 \\ 0 & 0 & -1 \end{pmatrix} W^\dagger. \tag{17}$$

They are related by $Q_s = WV_Q^\dagger\, Q_{\rm ew}\, V_Q W^\dagger = U_{\rm HPS}\, Q_{\rm ew}\, U_{\rm HPS}^\dagger$ via the tribimaximal mixing matrix [19]

$$U_{\rm HPS} \quad = \quad WV_Q^\dagger = \begin{pmatrix} \sqrt{2/3} & \sqrt{1/3} & 0 \\ -\sqrt{1/6} & \sqrt{1/3} & -\sqrt{1/2} \\ -\sqrt{1/6} & \sqrt{1/3} & \sqrt{1/2} \end{pmatrix}. \tag{18}$$

This looks like a promising zeroth order description for leptons providing arguments for the role of the discrete symmetry group A_4, which is a subgroups of both $SO(3)$ *and* $SU(3)$, in the structuring of families and the mixing matrices. The details of this, however, need to be worked out. Finally note that (without looking at the role of the masses) the extension of 1D fermion fields leads to 3D 'good' light-front fields $\Psi = (\xi_R, -\epsilon\xi_L)$ with two-component spinors $\xi_{R/L}$ and $\epsilon = i\sigma^2$. The rotations are represented by $\boldsymbol{J} = \boldsymbol{\sigma}/2$, boosts by $\boldsymbol{K} = \pm i\boldsymbol{\sigma}/2$ for right and left fields, respectively. The coupling of fermions to the pseudoscalar fields, combined into a 3D vector field, becomes the $\overline{\Psi}\slashed{A}\Psi$ coupling.

5 Quarks and Gluons

The fermionic modes ξ can also just live in $E(1,1)$ and be arranged in three families of $SU(3)_C$ color triplets, which are identified as colored quarks but living in E(1,1) where color is confined via

Table 1. Fermionic excitations with their assigments in $SO(1,1) \times SU(3)_C$ and $SO(1,3) \times SU(2)_I \times U(1)_Y$ symmetry schemes. The column labeled space L indicates one of the basic (left/right) modes of the 1D theory (the colored fermionic modes). The columns labeled $(T_1\ T_2)$ contain charge eigenstates of two basic modes that can be combined with the L-mode, giving allowed I_3 states within $SU(3)$.

	space			electroweak			charge	color
	L	$(T_1$	$T_2)$	I	I_3	Y	Q	$\underline{c}$
ν_L	ξ_L^0	ξ_L^0	ξ_L^0	1/2	+1/2	−1	0	$\underline{1}$
e_L^-	ξ_L^-	ξ_L^-	ξ_L^-	1/2	−1/2	−1	−1	$\underline{1}$
e_L^+	ξ_L^+	ξ_L^+	ξ_L^+	0	0	+2	+1	$\underline{1}$
ν_R	ξ_R^0	ξ_R^0	ξ_R^0	1/2	−1/2	+1	0	$\underline{1}$
e_R^+	ξ_R^+	ξ_R^+	ξ_R^+	1/2	+1/2	+1	+1	$\underline{1}$
e_R^-	ξ_R^-	ξ_R^-	ξ_R^-	0	0	−2	−1	$\underline{1}$
u_L	ξ_L^0	$(\xi_R^+$	$\xi_R^+)$	1/2	+1/2	+1/3	+2/3	$\underline{3}$
d_L	ξ_L^-	$(\xi_R^0$	$\xi_R^0)$	1/2	−1/2	+1/3	−1/3	$\underline{3}$
$\bar{u}_L$	ξ_L^0	$(\xi_L^-$	$\xi_R^-)$	0	0	−4/3	−2/3	$\underline{3}^*$
$\bar{d}_L$	ξ_L^+	$(\xi_L^0$	$\xi_R^0)$	0	0	+2/3	+1/3	$\underline{3}^*$
$\bar{u}_R$	ξ_R^0	$(\xi_L^-$	$\xi_L^-)$	1/2	−1/2	−1/3	−2/3	$\underline{3}^*$
$\bar{d}_R$	ξ_R^+	$(\xi_L^0$	$\xi_L^0)$	1/2	+1/2	−1/3	+1/3	$\underline{3}^*$
u_R	ξ_R^0	$(\xi_L^+$	$\xi_R^+)$	0	0	+4/3	+2/3	$\underline{3}$
d_R	ξ_R^-	$(\xi_L^0$	$\xi_R^0)$	0	0	−2/3	−1/3	$\underline{3}$

the instantaneous confining linear potential of the gauged $SU(3)$ symmetry. In order to study the electroweak structure of quarks (their *valence* nature) one has to study their interactions with the electroweak gauge bosons. This is achieved by mapping the structure of the excitations into three spatial directions in a *frozen color* scheme in which we just consider fermions of one particular color (say r). Take the case of all ξ_R states with color r and all ξ_L being $\bar{r}$. Taking a step back and looking at what was done in order to find leptons where the frozen colors were in essence space dimensions. The one-dimensional state would be labeled by a single momentum component, which is extended to states labeled by a 3-dimensional momentum vector in $E(1,3)$. For two space dimensions, the fermions could be labeled by their helicity in $E(1,2)$, charge eigenstates being $(\xi^-\xi^-)$, $(\xi^+\xi^+)$ and $(\xi^0\xi^0)$. For leptons in three space dimensions ξ_L^0 was combined with $(\xi_L^0\xi_L^0)$ to find an asymptotic charge eigenstate with $(Q, I_3) = (0, +1/2)$, which we already discussed as the left-handed Majorana neutrino ν_L. For colored eigenstates we specify how states are 'viewed' in 3 dimensions by combining the (frozen) anti-red ξ_L^0 state with the (frozen) rr combinations (ξ_R^0, ξ_R^0), $(\xi_R^+\xi_R^+)$ or $(\xi_R^-\xi_R^-)$. Then only the combination $(\xi_R^+\xi_R^+)$ leads to acceptable $SU(3)$ quantum numbers (roots), being an asymptotic acceptable $SU_I(2)$ weak eigenstate with $I_3 = 1/2$, which has $U_Q(1)$ charge $Q = +2/3$, identified as the weak iso-doublet quark state u_L with color r belonging to a color triplet. Combining the (frozen) color $\bar{r}$ state ξ_L^0 with the (frozen) $r\bar{r}$ combination $(\xi_L^0\xi_R^0)$, $(\xi_L^+\xi_R^+)$ or $(\xi_L^-\xi_R^-)$ gives only for $(\xi_L^-\xi_R^-)$ an acceptable (frozen) color $\bar{r}$ state with $(Q, I_3) = (-2/3, 0)$, the weak iso-singlet antiquark state $\bar{u}_L$. The full set of electroweak assignments of quarks as viewed in three space dimensions is shown in Table 1 and illustrated in Fig. 2. The resulting allowed $SU(3)$ quantum numbers are for each family a left-handed quark doublet and right-handed antiquark doublet and two singlets of opposite handedness. The way in which the electroweak structure emerges resembles the rishon model [20], but rather than

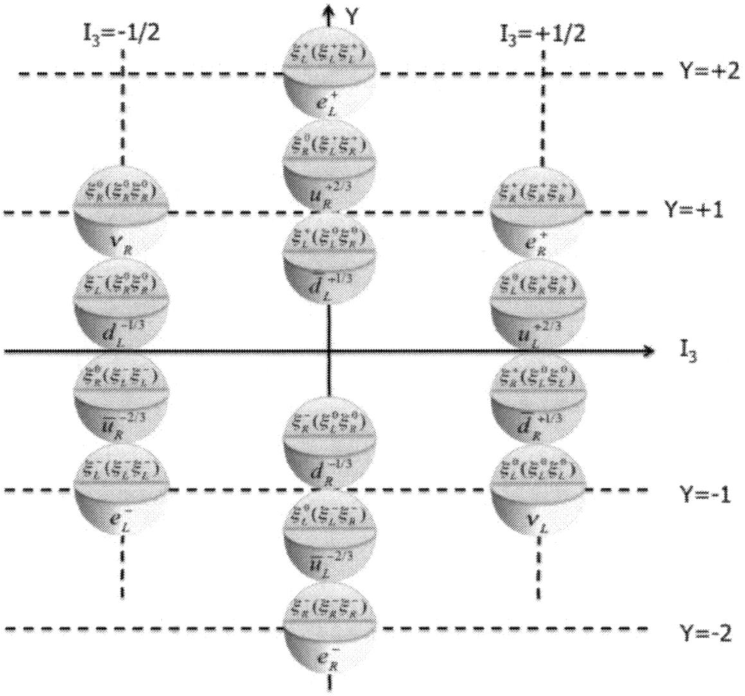

Figure 2. Fermionic excitations and their electroweak quantum numbers for leptons and for quarks (of a given color). Combinations such as $\xi^0_R(\xi^0_R\xi^0_L)$ with $(I_3, Y) = (-1/6, 1/3)$ or $\xi^0_R(\xi^+_R\xi^+_L)$ with $(I_3, Y) = (1/6, 1)$ don't fit in this scheme.

having two fractionally charged fields (V and T) in 3D, our basic modes are charged or neutral fields living in 1D. The family mixing would also for quarks originate from symmetries in fixing a direction, but in zeroth order there is only a single heavy quark, the top quark (with $M_t = M$), so the mixing would be simpler. The complete mechanism for masses and mixing for quarks and leptons, however, requires further study.

6 Conclusions

Instead of extending the standard model of particle physics, I have described an attempt to start at a more basic level with just a single space dimension (1D) and as starting point a fully supersymmetric set of three real fields describing bosonic and fermionic excitations. With this supersymmetric, superrenormalizable starting point, there is no naturalness or hierarchy problem. The $SO(3)$ symmetry of the classical ground state, including parity and time reversal, is then in Weyl mode realized as excitations living in 3D. The bosonic degrees of freedom are rearranged into the Higgs particle and the electroweak gauge bosons, while fermions are arranged in three families with two charged (Dirac) and one neutral (Majorana) lepton arranged in left-handed weak isospin doublets and singlets and corresponding right-handed antileptons. All these excitations appear as asymptotic states in 3D. The excitations of the fields also can live in 1D. The $SU(3)$ gauge theory has an instantaneous confining interaction and no physical gauge degrees of freedom. But this is not how these degrees of freedom show up asymptotically. We argue that the quarks reveal themselves in 3D as good (front form) components of fractionally charged Dirac fields arranged in a lefthanded weak isospin doublet and two righthanded singlets (and corresponding right- and left-handed antiparticles).

In this way a minimal scenario is created to obtain the standard model of particle physics with also in 3D *elementary* fields and confinement of color being implicit. Most prominent is that it links the number of colors, families and space directions. The Higgs or top quark mass are the natural basic scales for wave-lengths of the one-dimensional excitations producing the right orders of magnitude for masses of top quark, Higgs particle and gauge bosons. There are many details that need to be investigated to see if the proposed scheme can be made consistent, the embedding mechanism for the family structure, the origin of mixing matrices and the emergence of the scale of QCD, where the 1D and 3D descriptions meet. The conjectures as put forward here will likely not invalidate the existing highly successful field theoretical framework for the standard model. Hopefully it could lead to the calculation of parameters as deviations from a zeroth order description. For the QCD part, it also may provide insights why and to what extent collinear effective theories or the many effective theories for QCD at low energies, work. It might provide handles on universality breaking effects such as the 'proton radius puzzle', the reason being that atomic Hydrogen involves all degrees of freedom of just *one* family while muonic Hydrogen is different in this respect. It could also be interesting to look at more (or maybe less) than three fields, which may also be relevant in the context of the evolution of our universe, in which the world above hadronic scales, i.e. the visible part at nuclear, atomic, molecular scales up to astronomical scales, lives in three space dimensions.

Acknowledgements

I acknowledge useful discussions with several colleagues at Nikhef, in particular Tomas Kasemets. This research is part of the FP7 EU "Ideas" programme QWORK (Contract 320389).

References

[1] J.C. Collins, *Foundations of perturbative QCD* (Cambridge University Press, 2011)

[2] T. Becher, A. Broggio, A. Ferroglia (2014), arXiv:1410.1892

[3] P.J. Mulders, *Spin Physics and Transverse Structure*, in *23rd International Workshop on Deep-Inelastic Scattering and Related Subjects (DIS 2015) Dallas, Texas, United States, April 27-May 1, 2015* (2015), arXiv:1508.04244

[4] P.J. Mulders, *Operator Structure of TMDs*, in *Proceedings, QCD Evolution Workshop (QCD 2015)* (2015), arXiv:1510.05871

[5] P.J. Mulders, *The Roots of the Standard Model of Particle Physics*, arXiv:1601.00300

[6] P.A. Dirac, Rev.Mod.Phys. **21**, 392 (1949)

[7] J.B. Kogut, D.E. Soper, Phys. Rev. **D1**, 2901 (1970)

[8] S.P. Martin, Adv.Ser.Direct.High Energy Phys. **21**, 1 (2010),

[9] J. Wess, B. Zumino, Phys.Lett. **B49**, 52 (1974)

[10] S.R. Coleman, J. Mandula, Phys.Rev. **159**, 1251 (1967)

[11] N. Cabibbo, Phys. Lett. **B72**, 333 (1978)

[12] L. Wolfenstein, Phys. Rev. **D18**, 958 (1978)

[13] E. Ma, G. Rajasekaran, Phys. Rev. **D64**, 113012 (2001),

[14] G. Altarelli, F. Feruglio, Nucl. Phys. **B741**, 215 (2006),

[15] D.B. Kaplan (2013), ArXiv:1306.5818

[16] S. Weinberg, Phys.Rev. **D5**, 1962 (1972)

[17] M. Veltman, Nucl.Phys. **B123**, 89 (1977)

[18] P. Sikivie, L. Susskind, M.B. Voloshin, V.I. Zakharov, Nucl.Phys. **B173**, 189 (1980)

[19] P.F. Harrison, D.H. Perkins, W.G. Scott, Phys. Lett. **B530**, 167 (2002),

[20] H. Harari, N. Seiberg, Nucl.Phys. **B204**, 141 (1982)

EPJ Web of Conferences **112**, 01015 (2016)
DOI: 10.1051/epjconf/201611201015

Dispersion representation of the D-term form factor in deeply virtual Compton scattering

B. Pasquini[1,a]

[1] *Dipartimento di Fisica, Università degli Studi di Pavia, Pavia, Italy, and Istituto Nazionale di Fisica Nucleare, Sezione di Pavia, Pavia, Italy*

Abstract. We review the dispersion relation formalism for the deeply virtual Compton scattering amplitudes, and derive a dispersive representation of the D-term form factor, using unsubtracted t-channel dispersion relations. Results for the D-term form factor as function of t as well as at $t = 0$ are shown in comparison with available model predictions and phenomenological parametrizations.

1 Introduction

Dispersion relations (DRs) provide an useful framework to connect different observables and to extract nucleon structure quantities. In particular, DRs for Compton scattering processes, with both real and virtual photons, had wide applications for the prediction and extraction from experimental data of low-energy properties of hadron systems [1–5]. More recently, the dispersion formalism has been extended to describe the virtual Compton scattering process in the deep inelastic region [6–12]. In particular, it was shown that the amplitudes for deeply virtual Compton scattering (DVCS) satisfy subtracted DRs at fixed t with the subtraction function defined by the D-term form factor [6–8]. The D term was originally introduced to complete the parametrization of the generalized parton distributions (GPDs) in hard exclusive reactions in terms of double distributions, and restore the polynomiality property of the singlet moments of unpolarized GPDs [13]. This term turned out to be a crucial contribution in the phenomenological description of DVCS observables, where different forms have been assumed with parameters tuned to DVCS data [9, 14].

Here we review the dispersive representation of the D-term form factor in terms of DRs in the t-channel, as recently proposed in Ref. [15]. The advantage of this dispersive representation is to provide a microscopic interpretation of the physical content of the D-term form factor in terms of t-channel exchanges with the appropriate quantum numbers.

In Section 2, we review the derivation of the s-channel subtracted dispersion relations for the DVCS amplitudes. In Section 3, we derive t-channel DRs for the D-term form factor, and give the ingredients for the explicit calculation, by saturating the unitarity relation for the t-channel amplitudes with two-pion intermediate states. We then discuss the dispersive predictions for the D-term form factor in Section 4, and we conclude summarizing our results.

[a] e-mail: barbara.pasquini@unipv.it

2 Subtracted dispersion relations in the s-channel

We consider the DVCS process

$$\gamma^*(q)\,N(p) \to \gamma(q')\,N(p'), \tag{1}$$

where the variables in brackets denote the four-momenta of the participating particles. The familiar Mandelstam variables are $s = (p+q)^2$, $t = (q-q')^2$, $u = (q-p')^2$, and are constrained by $s+u+t = 2M_N^2 - Q^2$, with M_N the nucleon mass and $Q^2 = -q^2$. We will consider the Bjorken regime, where the photon virtuality Q^2 and s are large, and $-t \ll s, Q^2$.
The DVCS amplitude reads

$$T_{\lambda'_\gamma\lambda'_N,\lambda_\gamma\lambda_N} = \varepsilon_\mu(q,\lambda_\gamma)\varepsilon^*_\nu(q',\lambda'_\gamma)H^{\mu\nu}_{\lambda'_N,\lambda_N}, \tag{2}$$

where $\varepsilon_{\lambda_\gamma}$ ($\varepsilon^*_{\lambda'_\gamma}$) is the polarization vector of the incoming virtual (outgoing real) photon and the DVCS tensor is defined as the nucleon matrix element of the T-product of two electromagnetic currents:

$$H^{\mu\nu}_{\lambda'_N,\lambda_N} = -i\int \mathrm{d}^4x\, e^{-i(q\cdot x)}\langle N(p',\lambda'_N)|T[J^\mu(x)J^\nu(0)|N(p,\lambda_N)\rangle, \tag{3}$$

where λ_N (λ'_N) is the helicity of the incoming (outgoing) nucleon.

The DVCS amplitude for unpolarized nucleon and at leading order in Q can be parametrized as

$$T_{\lambda'_\gamma\lambda'_N,\lambda_\gamma\lambda_N} = \varepsilon_\mu(q,\lambda_\gamma)\varepsilon^*_\nu(q',\lambda'_\gamma)\frac{(-g_\perp^{\mu\nu})}{2}$$
$$\times\left[\bar u(p',\lambda'_N)\gamma\cdot n\, u(p,\lambda_N)\sum_q e_q^2C^q - \bar u(p',\lambda'_N)u(p,\lambda_N)\frac{1}{M_N}\sum_q e_q^2F^q\right], \tag{4}$$

where we introduced the lightlike vector $n^\mu = 1/(\sqrt{2}P^+)(1,0,0,-1)$, with $P = (p+p')/2$, and the symmetric tensor $g_\perp^{\mu\nu} = g^{\mu\nu} - n^\mu\tilde p^\nu - n^\nu\tilde p^\mu$, with $\tilde p^\mu = P^+/\sqrt{2}(1,0,0,1)$. Furthermore, the light-front component for a generic four-vector a^μ is defined as $(a^0+a^3)/\sqrt{2}$. In Eq. (4), the invariant amplitudes C^q and F^q are given by

$$C^q(\xi,t) = \int_{-1}^{1}\mathrm{d}x\frac{H^{(+)}(x,\xi,t)+E^{(+)}(x,\xi,t)}{x-\xi+i\epsilon}, \qquad F^q(\xi,t) = \int_{-1}^{1}\mathrm{d}x\frac{E^{(+)}(x,\xi,t)}{x-\xi+i\epsilon}, \tag{5}$$

with the skewedness variable defined as $\xi = Q^2/(2s+Q^2)$. $H^{(+)}(x,\xi,t) = H^q(x,\xi,t) - H^q(-x,\xi,t)$ denotes the singlet ($C=+1$) combination of nucleon helicity-conserving GPDs, and analogously for the nucleon helicity-flip GPD $E^{(+)}$. The invariant amplitudes and the GPDs in Eq. (5) depend also on the renormalization scale μ^2, which is not explicitly displayed, and it is identified with the hard scale of the process Q^2.
In the following we will consider the invariant amplitude F^q in the $\nu-t$ plane at fixed Q^2, with $\nu = (s-u)/4M_N = Q^2/4M_N\xi$. In this plane, F^q satisfies the following fixed-t subtracted relation [6, 11]

$$F^q(\nu,t) = F^q(0,t) + \frac{\nu^2}{\pi}\int_{\nu_0}^{\infty}\frac{\mathrm{d}\nu'^2}{\nu'^2}\frac{\mathrm{Im}F^q(\nu',t)}{\nu'^2-\nu^2}, \tag{6}$$

where the lower limit of integration is $\nu_0 = Q^2/4M_N$ and the nucleon pole term residing in this point may be considered separately. Following Refs. [6, 8], the subtraction function $F^q(0,t)$ can be related to the D-term form factor $D^q(t)$ [13] as follows

$$F^q(0,t) = 2\int_{-1}^{+1}\mathrm{d}z\frac{D^q(z,t)}{1-z} = 4D^q(t). \tag{7}$$

3 t-channel dispersion relations for the D-term form factor

The dispersive representation for the D-term form factor $D^q(t)$ of Eq. (7) is obtained by applying unsubtracted DRs, this time in the variable t:

$$F^q(0,t) = \frac{1}{\pi}\int_{4m_\pi^2}^{+\infty} dt' \frac{\mathrm{Im}_t F^q(0,t')}{t'-t} + \frac{1}{\pi}\int_{-\infty}^{-a} dt' \frac{\mathrm{Im}_t F^q(0,t')}{t'-t}. \tag{8}$$

The imaginary part in the integral from $4m_\pi^2 \to +\infty$ in Eq. (8) is saturated by the possible intermediate states for the t-channel process, which lead to cuts along the positive-t axis. For low values of t, the t-channel discontinuity is dominated by $\pi\pi$ intermediate states.
The second integral in Eq. (8) extends from $-\infty$ to $-a = -2(m_\pi^2 + 2M_N m_\pi) - Q^2$. As we are interested in evaluating Eq. (8) for large Q^2 values and small (negative) values of t ($|t| \ll a$), the integral from $-\infty \to -a$ is suppressed, and will be neglected in this work. Consequently, we shall saturate the integral in Eq. (8) by the contribution of $\pi\pi$ intermediate states, which turns out to be a good approximation for small t.

We start by decomposing the t-channel helicity amplitude for $\gamma^*\gamma \to \bar{N}N$ into a partial wave series,

$$T^t_{\lambda_{\bar{N}}\lambda_N,\,\lambda_\gamma\lambda'_\gamma}(\nu,t) = \sum_J \frac{2J+1}{2}\, T^{J(\gamma^*\gamma\to N\bar{N})}_{\lambda_N\lambda_{\bar{N}},\,\lambda'_\gamma\lambda_\gamma}(t)\, d^J_{\Lambda_N\Lambda_\gamma}(\theta_t)\,, \tag{9}$$

where $\Lambda_\gamma = \lambda'_\gamma - \lambda_\gamma$, $\Lambda_N = \lambda_N - \lambda_{\bar{N}}$, and $d^J_{\Lambda_N\Lambda_\gamma}$ are Wigner d-functions. As described in Ref. [15], the unitarity relation for the the partial-wave amplitudes in the t-channel, with only the $\pi\pi$ intermediate states, reads

$$2\,\mathrm{Im}T^{J\,(\gamma^*\gamma\to N\bar{N})}_{\lambda_{\bar{N}}\lambda_N,\,\lambda_\gamma\lambda'_\gamma}(t) = \frac{1}{(8\pi)}\frac{p_\pi}{\sqrt{t}}\left[T^{J\,(\gamma^*\gamma\to\pi\pi)}_{\Lambda_\gamma}(t)\right]\left[T^{J\,(\pi\pi\to N\bar{N})}_{\Lambda_N}(t)\right]^*, \tag{10}$$

where $T^{J\,(\gamma^*\gamma\to\pi\pi)}$ and $T^{J\,(\pi\pi\to N\bar{N})}$ are the partial-wave amplitudes for the $\gamma^*\gamma \to \pi\pi$ and the $\pi\pi \to N\bar{N}$ subprocesses, respectively.

The partial-wave amplitudes $T^{J\,(\pi\pi\to N\bar{N})}_{\Lambda_N=0}$ are related to the amplitudes $f^J_+(t)$ of Frazer and Fulco [16] by the relation

$$T^{J\,(\pi\pi\to N\bar{N})}_{\Lambda_N=0}(t) = \frac{16\pi}{p_t}(p_t\, p_\pi)^J f^J_+(t)\,. \tag{11}$$

The reaction $\gamma^*\gamma \to \pi\pi$ at large Q^2 and small t can be described in a factorized form [17, 18], as the convolution of a short-distance contribution, $\gamma^*\gamma \to q\bar{q}$, perturbatively calculable, and nonperturbative matrix elements describing the exclusive fragmentation of a $q\bar{q}$ pair into two pions. These nonperturbative functions correspond to two-pion generalized distribution amplitudes (GDAs). In particular, one finds

$$T^{J\,(\gamma^*\gamma\to\pi\pi)}_{\Lambda_\gamma=0}(t) = \sum_q e_q^2\, T^{J\,(\gamma^*\gamma\to q\bar{q})}_{\Lambda_\gamma=0}(t) \tag{12}$$

with

$$T^{J\,(\gamma^*\gamma\to q\bar{q})}_{\Lambda_\gamma=0}(t) = \frac{6}{2J+1}\sum_{\substack{n=\max(1,J-1)\\ \text{odd}}}\int_0^1 dz\,(2z-1)\tilde{B}_{nJ}(t) C_n^{(3/2)}(2z-1), \tag{13}$$

where $C_n^{(3/2)}$ are Gegenbauer polynomials, and $\tilde{B}^q_{nl}$ are expansion coefficients of the GDA amplitude [18–20]. Inserting the partial wave expansion of Eqs. (11) and (12) into Eq. (10), we can finally express the 2π t-channel contribution to $\mathrm{Im}_t F^q(\nu = 0, t)$ as

$$\mathrm{Im}_t F^{q\,(\pi\pi)} = -\frac{M_N p_\pi}{\sqrt{t}\, p_t^2} \sum_{\substack{J \\ \text{even}}} \frac{2J+1}{2} (-1)^{J/2} \frac{(J-1)!!}{J!!} (p_t p_\pi)^J \, T^{J\,(\gamma^*\gamma\to q\bar{q})}_{\Lambda_\gamma=0} f_+^{J*}(t). \tag{14}$$

For the numerical estimate, we restrict ourselves to the S- and D-wave contributions in Eq. (14). The partial-wave amplitudes of the $\pi\pi \to N\bar{N}$ subprocess are taken from [21]. The two-pion GDAs are calculated through DRs using the Omnès representation [19, 20, 22–24]. The results for the S- and D-wave coefficients, for $n = 1$ in the Gegenbauer expansion of (13), read

$$\tilde{B}_{10}(t) = -B_{12}(0)\frac{3C - \beta^2}{2} f_0(t), \quad \tilde{B}_{12}(t) = \beta^2 B_{12}(0) f_2(t), \tag{15}$$

where the Omnès functions $f_{0,2}$ can be related to $\pi\pi$ phase-shifts $\delta^0_{0,2}(t)$ using Watson theorem and dispersion relations [19]. In Eq. (15), the constant C is taken from Ref. [20], using the estimate from the instanton model [25] at low energies, $C = 1 + bm_\pi^2 + O(m_\pi^4)$ with $b \approx -1.7$ GeV^{-2}, while the coefficient $B_{12}(0)$ is obtained using the crossing relations between the quark 2πDA's and the corresponding parton distributions in the pion, i.e.

$$B_{12}(0) = \frac{10}{9} \int \mathrm{d}x\, x \frac{1}{N_f} \sum_f [q_\pi^f(x) + \bar{q}_\pi^f(x)]. \tag{16}$$

As final result, taking into account only the contribution with $J = 0$ and $J = 2$, Eq. (14) simplifies to

$$\mathrm{Im}_t F^{q\,(\pi\pi)} = \frac{3M_N p_\pi}{2\sqrt{t}\, p_t^2} B_{12}(0) \left[(3C - \beta^2) f_0(t) f_+^{0*}(t) + (p_\pi p_t)^2 \beta^2 f_2(t) f_+^{2*}(t) \right]. \tag{17}$$

In Eq. (17), the dependence on the renormalization scale enters only through the coefficient B^q_{12} evaluated at $t = 0$, and therefore is disjoined from the t dependence of the amplitude. Furthermore, the coefficients B^q_{12} evolve in the same way as the quark momentum fraction in the pion, in accordance with Eq. (16).

We note that the restriction to $n = 1$ in the Gegenbauer expansion of the GDAs corresponds to a similar restriction for the D-term form factor. As a matter of fact, the D-term $D(z, t)$ in Eq. (7) can be expanded in terms of Gegenbauer polynomials $C_n^{3/2}$, leading to the following representation for the D-term form factor $D^q(t) = \sum_{\substack{n=1 \\ n\,\text{odd}}}^{\infty} d_n^q(t)$. Accordingly, the numerical results shown in the next section are for the contribution $d_1^q(t)$.

4 Results

In Fig. 1 we present the dispersive predictions for $d_1^Q = \sum_q d_1^q(t)$ as function of t, with the sum over flavors restricted to up and down quarks. The solid and dashed curves are obtained using as input in Eq. (16) the parametrization of the pion distributions at $Q^2 = 4$ GeV^2 from Refs. [26] and [27], respectively. The different inputs for the pion distributions change the results by an overall normalization factor, without affecting the t dependence. As outlined above, the Q^2 dependence enters only through the quark momentum fraction of the pion, which changes only by a few percent in the range of $Q^2 = [1, 10]$ GeV^2. At $t = 0$, we find $d_1^Q = -1.59$ and $d_1^Q(0) = -1.92$ for the solid and dashed curve in Fig. 1, respectively. These values compare with the results obtained, at a low normalization scale, in the χQSM [28], $d_1^Q(0) = -2.35$, in the Skyrme model [29], $d_1^Q(0) = -4.48$, and in a recent calculation

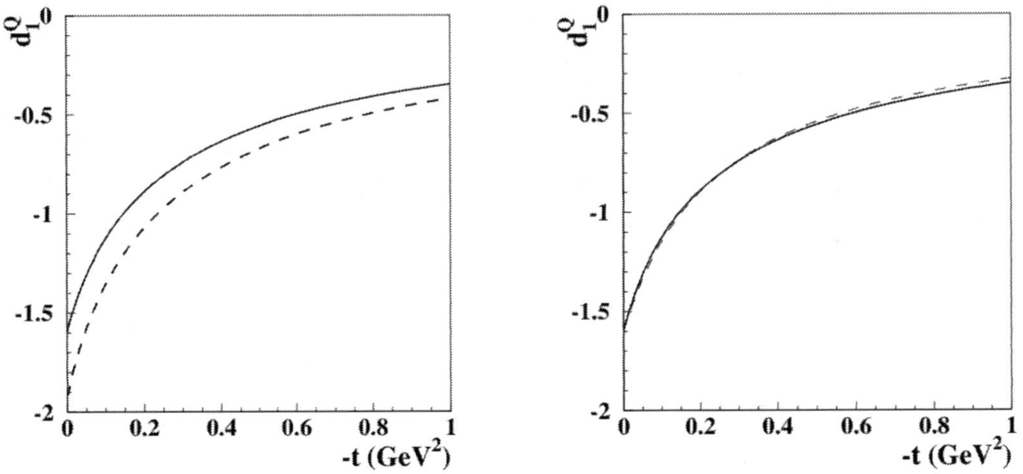

Figure 1. d_1^Q as function of t, at $Q^2 = 4\,\mathrm{GeV}^2$, obtained with the quark distributions in the pion q_π^f from Ref. [26] (solid curve) and Ref. [27] (dashed curve). Right panel: t dependence of the DR results for $d_1^Q(t)$ with q_π^f from Ref. [26] (solid curve) in comparison with the function in Eq. (18) (dashed curve).

with effective light-front wave functions from a Regge-improved diquark model [30], $d_1^q(0) = -2.01$.
$d_1^Q(t)$ is particularly interesting, as it enters in the parametrization of the quark part of the energy momentum tensor of QCD, and provides information on how strong forces are distributed and stabilized in the nucleon [31]. In all theoretical studies so far, as well as in the present dispersive calculation, $d_1^Q(t)$ at zero-momentum transfer $t = 0$ is found to have a negative sign. The negative value of this constant has a deep relation to the spontaneous breaking of the chiral symmetry in QCD [31, 32], and has also an appealing connection with the criterion of stability of the nucleon [28].
In most of phenomenological studies of DVCS, the t dependence of D-term form factor is parametrized by a dipole function [14]. However, the dispersive results favor a different functional form, as shown in the right panel of Fig. 1 where we compare the result for d_1^Q as function of t with the following parametrization

$$F_D = \frac{d_1^Q(0)}{[1 - t/(\alpha M_D^2)]^\alpha}, \text{ with } M_D = 0.487\,\mathrm{GeV} \text{ and } \alpha = 0.841. \tag{18}$$

5 Conclusions

We presented a dispersive representation for the quark contribution to the D-term form factor in hard exclusive reactions in terms of unsubtracted t-channel dispersion relations. The unitarity relation for the t-channel amplitudes is saturated with two-pion intermediate states, taking into account the contribution from S- and D-wave intermediate states. The input for the imaginary part of the dispersion relation are the two-pion GDAs, determined through the first-x moment of the flavor-singlet pion PDFs, the $\pi\pi$ phase shifts, and the partial waves for the $\pi\pi \to N\bar{N}$ amplitudes obtained from dispersion theory by analytical continuation of πN scattering amplitudes. We found that the t and

Q^2 dependence of the D-term form factor are disjoined. The t-dependence is not trivial and does not follow a dipole behavior as normally assumed in phenomenological parametrizations. The value at $t = 0$ is also compatible with estimates in chiral-quark soliton model and a Regge-improved diquark model. The Q^2 dependence enters only in the normalization point at $t = 0$, which is proportional to the first x-moment of the flavor-singlet pion PDFs.

References

[1] D. Drechsel, B. Pasquini and M. Vanderhaeghen, Phys. Rept. **378**, 99 (2003).
[2] B. Pasquini, M. Gorchtein, D. Drechsel, A. Metz and M. Vanderhaeghen, Eur. Phys. J. A **11**, 185 (2001); Phys. Rev. C **62**, 052201 (2000).
[3] D. Drechsel, M. Gorchtein, B. Pasquini and M. Vanderhaeghen, Phys. Rev. C **61**, 015204 (1999).
[4] B. Pasquini, D. Drechsel and M. Vanderhaeghen, Phys. Rev. C **76**, 015203 (2007).
[5] B. Pasquini, D. Drechsel and S. Scherer, Phys. Rev. C **77**, 065211 (2008).
[6] I. V. Anikin and O. V. Teryaev, Phys. Rev. D **76**, 056007 (2007).
[7] O. V. Teryaev, hep-ph/0510031.
[8] A. V. Radyushkin, Phys. Rev. D **83**, 076006 (2011).
[9] K. Kumericki, D. Müller and K. Passek-Kumericki, Nucl. Phys. B **794**, 244 (2008).
[10] D. Müller and K. M. Semenov-Tian-Shansky, Phys. Rev. D **92**, no. 7, 074025 (2015).
[11] M. Diehl and D. Y. Ivanov, Eur. Phys. J. C **52**, 919 (2007).
[12] M. V. Polyakov and M. Vanderhaeghen, arXiv:0803.1271 [hep-ph].
[13] M. V. Polyakov and C. Weiss, Phys. Rev. D **60**, 114017 (1999).
[14] M. Guidal, H. Moutarde and M. Vanderhaeghen, Rept. Prog. Phys. **76**, 066202 (2013).
[15] B. Pasquini, M. V. Polyakov and M. Vanderhaeghen, Phys. Lett. B **739**, 133 (2014).
[16] W. R. Frazer and J. R. Fulco, Phys. Rev. **117**, 1603 (1960).
[17] M. Diehl, T. Gousset, B. Pire and O. Teryaev, Phys. Rev. Lett. **81**, 1782 (1998).
[18] M. Diehl, T. Gousset and B. Pire, Phys. Rev. D **62**, 073014 (2000).
[19] M. V. Polyakov, Nucl. Phys. B **555**, 231 (1999).
[20] N. Kivel, L. Mankiewicz and M. V. Polyakov, Phys. Lett. B **467**, 263 (1999).
[21] G. Höhler, Pion-Nucleon Scattering, Landolt-Börnstein, Vol. I/9b2, edited by H. Schopper Springer, Berlin, 1983.
[22] B. Lehmann-Dronke, P. V. Pobylitsa, M. V. Polyakov, A. Schäfer and K. Goeke, Phys. Lett. B **475**, 147 (2000).
[23] B. Lehmann-Dronke, A. Schäfer, M. V. Polyakov, K. Goeke, Phys. Rev. D **63**, 114001 (2001).
[24] N. Warkentin, M. Diehl, D. Y. Ivanov and A. Schafer, Eur. Phys. J. A **32**, 273 (2007).
[25] M. V. Polyakov and C. Weiss, Phys. Rev. D **59**, 091502 (1999).
[26] J. F. Owens, Phys. Rev. D **30**, 943 (1984).
[27] M. Gluck, E. Reya and A. Vogt, Z. Phys. C **53**, 651 (1992).
[28] K. Goeke, J. Grabis, J. Ossmann, M. V. Polyakov, P. Schweitzer, A. Silva and D. Urbano, Phys. Rev. D **75**, 094021 (2007).
[29] C. Cebulla, K. Goeke, J. Ossmann and P. Schweitzer, Nucl. Phys. A **794**, 87 (2007).
[30] D. Müller and D. S. Hwang, arXiv:1407.1655 [hep-ph].
[31] M. V. Polyakov, Phys. Lett. B **555**, 57 (2003).
[32] M. V. Polyakov and A. G. Shuvaev, hep-ph/0207153.

EPJ Web of Conferences **112**, 01016 (2016)
DOI: 10.1051/epjconf/201611201016

On the universality of the $J = 0$ fixed pole contribution in DVCS

D. Müller[1,a] and K. Semenov-Tian-Shansky[2,b]

[1] Theoretical Physics Division, Rudjer Bošković Institute, HR-10002 Zagreb, Croatia
[2] Petersburg Nuclear Physics Institute, Gatchina, 188300, St.Petersburg, Russia

Abstract. S. Brodsky, F.J. Llanes-Estrada and A. Szczepaniak formulated the $J = 0$ fixed pole universality hypothesis for (deeply) virtual Compton scattering. We show that in the Bjorken limit this hypothesis is equivalent to the validity of the inverse moment sum rule for the D-term form factor. However, any supplementary D-term added to a generalized parton distribution (GPD) results in an additional $J = 0$ fixed pole contribution that violates universality. Unfortunately, one can not provide any reliable theoretical argument excluding the existence of such supplementary D-term. Moreover, the violation of $J = 0$ fixed pole universality was revealed in field theoretical GPD models. Therefore, $J = 0$ fixed pole universality hypothesis remains just an external assumption and probably will never be proven theoretically.

1 Introduction

Studies of Compton scattering off a nucleon $\gamma^{(*)}(q_1) + N(p_1) \to \gamma^{(*)}(q_2) + N(p_2)$ with both photons being real ($q_1^2 = q_2^2 = 0$) or with one ($q_1^2 = -Q_1^2$, $q_2^2 = 0$) or two ($q_1^2 = -Q_1^2$, $q_2^2 = -Q_2^2$) virtual space-like photons allow to probe nucleon structure in low and high energy regimes. The customary theoretical framework which helps to address Compton scattering in high energy regime is the Regge theory. Assuming the existence of the leading Regge trajectory $\alpha(t)$ ($t \equiv (p_2 - p_1)^2$), the high energy leading asymptotic behavior $\sim s^{\alpha(t)}$ of the amplitude originates from a moving pole in the plane of complex cross-channel angular momentum J. Besides these common moving poles (and/or cuts) there also might exist so-called *fixed pole* singularities (see *e.g.* Chapter I of Ref. [1]), that do not slide with the change of t and can not be revealed by means of the analytic continuation in J. A $J = J_0$ fixed pole singularity may arise from a cross channel exchange with a non-Reggeized (elementary) particle of spin J_0 in the cross channel (or from a contact interaction term). It appears then as the Kronecker-δ singularity in the complex J-plane.

The discussion on the manifestation of the $J = 0$ fixed pole singularity in Compton scattering, which results in constant contribution in high-energy asymptotic limit, lasts since the late sixties (see the pioneering paper [2] and Ref. [3] as well as references therein). S. Brodsky, F. Close and J. Gunion identified this contribution at the diagrammatic level. They argue it originates from "seagull" diagrams corresponding to local two photon interaction. More recently, in [4] S. Brodsky, F. J. Llanes-Estrada and A. Szczepaniak emphasized that such $J = 0$ fixed pole contribution in virtual Compton scattering

---

[a] e-mail: dieter.mueller@irb.hr
[b] e-mail: cyrstsh@gmail.com

is universal (*i.e.* independent of photon virtualities) and can be expressed by the inverse moment of the corresponding t-dependent parton distribution function (PDF).

On the other hand, the $J = 0$ fixed pole contribution in deeply virtual Compton scattering (DVCS) is closely related with the D-term form factor, which arises as the subtraction constant for deeply virtual Compton scattering amplitude within the partonic picture. The D-term [5] was originally introduced to complement the polynomiality condition for GPDs within the double distribution (DD) representation. Later it was realized that it could be implemented as an inherent part of GPD both within the modified DD representation and within the representations based on conformal partial wave (PW) expansion of GPDs.

Below, following Refs. [6, 7], we review the relation between the $J = 0$ fixed pole contribution and the D-term form factor. We tie together the J-analytic properties of cross channel SO(3) PWs and those of GPD conformal moments in the conformal spin j. We show that the $J = 0$ fixed pole universality hypothesis generally remains unproven and only can be accepted as external principle for building up phenomenological GPD models.

2 Dispersive approach for Compton amplitude

We focus on the Compton form-factor (CFF) $\mathcal{H}(\nu \equiv \frac{s-u}{4}, t|Q_1^2, Q_2^2)$ occurring in the parametrization of the transverse non-flip photon helicity amplitude of Compton scattering. This CFF has even signature ($P = +1$; $C = +1$) and is the analogue of the usual Dirac electromagnetic form factor. In the forward kinematics the imaginary part of $\mathcal{H}$ corresponds to the deep inelastic structure function F_1.

A fixed-t dispersion relation for the CFF $\mathcal{H}$ can be obtained adopting usual assumptions on the analytic properties of CFFs. Neglecting the Born term (which is irrelevant in the Bjorken limit) we consider the deformed integration contour surrounding cuts on the real axis starting at the pion production threshold $\nu_{\text{cut}} = \frac{Q_1^2+Q_2^2+t+(M+2m_\pi)^2-M^2}{4M}$ (see Fig. 1). Assuming that $\mathcal{H}$ vanishes at infinity ($\lim_{|\nu|\to\infty} \mathcal{H}(\nu, t|Q_1^2, Q_2^2) = 0$), we obtain the unsubtracted DR in the standard form,

$$\mathcal{H}(\nu, t|Q_1^2, Q_2^2) = \frac{1}{\pi} \int_{\nu_{\text{cut}}}^{\infty} d\nu' \frac{2\nu' \, \text{Im}\, \mathcal{H}(\nu', t|Q_1^2, Q_2^2)}{\nu'^2 - \nu^2 - i\epsilon}. \quad (1)$$

Once $\mathcal{H}$ does not vanish at infinity one still can formally consider the unsubtracted DR by specifying a regularization procedure both for the divergent integral along the real axis and (either divergent) contributions from large semicircles of the contour on right panel of Fig. 1. A possible choice is the analytic regularization (see *e.g.* Ref. [8]). The dispersive integral is replaced by the loop integral in the complex plane that includes the point $\nu = \infty$, denoted as (∞). The unsubtracted DR for CFF $\mathcal{H}$ then reads

$$\mathcal{H}(\nu, t|Q_1^2, Q_2^2) = \mathcal{H}_\infty(t|Q_1^2, Q_2^2) + \frac{1}{\pi} \int_{\nu_{\text{cut}}}^{(\infty)} d\nu' \frac{2\nu' \, \text{Im}\, \mathcal{H}(\nu', t|Q_1^2, Q_2^2)}{\nu'^2 - \nu^2 - i\epsilon}, \quad (2)$$

where the constant $\mathcal{H}_\infty$, arises from the analytic regularization at $\nu = \infty$. Within the Regge–pole expansion of the amplitude it is interpreted as the $J = 0$ fixed pole contribution.

An alternative form of DR employed within the deeply virtual (d.v.) regime involves one subtraction at the unphysical point $\nu = 0$:

$$\mathcal{H}(\nu, t|Q_1^2, Q_2^2) \overset{\text{d.v.}}{=} \mathcal{H}_0(t|Q_1^2, Q_2^2) + \frac{1}{\pi} \int_{\nu_{\text{cut}}}^{\infty} \frac{d\nu'}{\nu'} \frac{2\nu^2 \, \text{Im}\, \mathcal{H}(\nu', t|Q_1^2, Q_2^2)}{\nu'^2 - \nu^2 - i\epsilon}. \quad (3)$$

Generally, the subtraction constant $\mathcal{H}_0(t|Q_1^2, Q_2^2)$ represents an unknown quantity. However, in the deeply virtual regime one can formulate the external principle allowing to fix the subtraction constant

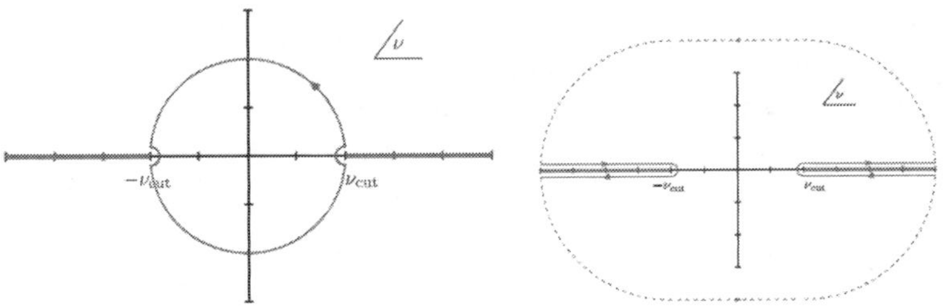

Figure 1. Left panel: Initial integration contour in the complex ν plane to express the CFF through the Cauchy integral. Right panel: Deformation of the integration contour in eq. (1).

from the absorptive part of the amplitude. Relying on the operator product expansion one can check that for equal photon virtualities $Q_1^2 = Q_2^2 = Q^2$ (DIS kinematics) due to the current conservation the subtraction constant vanishes to the leading twist accuracy. Within the kinematics of DVCS the subtraction constant in (3) corresponds to the D-term form factor $\mathcal{H}_0(t|Q_1^2 = Q^2, Q_2^2 = 0) \overset{\text{d.v.}}{=} 4\mathcal{D}(t)$.

The dispersion relations (2) and (3) are supposed to represent the same function. Therefore, the $J = 0$ fixed pole contribution $\mathcal{H}_\infty$ could be related to the subtraction constant $\mathcal{H}_0$. Plugging in the algebraic decomposition $\frac{2\nu'}{\nu'^2-\nu^2-i\epsilon} = \frac{1}{\nu'}\frac{2\nu^2}{\nu'^2-\nu^2-i\epsilon} + \frac{2}{\nu'}$ of the Cauchy kernel into (2) and comparing it with (3), we read off the sum rule

$$\mathcal{H}_\infty(t|Q_1^2, Q_2^2) = \mathcal{H}_0(t|Q_1^2, Q_2^2) - \frac{2}{\pi}\int_{\nu_{cut}}^{(\infty)} \frac{d\nu}{\nu}\, \mathrm{Im}\mathcal{H}(\nu, t|Q_1^2, Q_2^2)\,,$$

expressing the $J = 0$ fixed pole contribution through the subtraction constant and the analytically regularized inverse moment of the absorptive part of the amplitude.

In the deeply virtual kinematics regime the convenient kinematical variable are the Bjorken-like variable $\xi = \frac{Q^2}{(p_1+p_2)\cdot\frac{1}{2}(q_1+q_2)}$, where $Q^2 = -\frac{(q_1+q_2)^2}{4}$; and the photon asymmetry parameter[1] $\vartheta = \frac{q_1^2-q_2^2}{q_1^2+q_2^2} + O(t/Q^2)$. Note that for $t = 0$ $\vartheta = 0$ one recovers the usual DIS kinematics, while $\vartheta = 1$ corresponds to the DVCS set up.

Within these scaling variables the analytically regularized DR (2) and the subtracted one (3) read as follows:

$$\mathcal{H}(\xi, t|\vartheta) = \frac{1}{\pi}\int_{(0)}^{1} \frac{d\xi'}{\xi'} \frac{2\xi^2\, \mathrm{Im}\mathcal{H}(\xi', t|\vartheta)}{\xi^2 - \xi'^2 - i\epsilon} + \mathcal{H}_\infty(t|\vartheta); \tag{4}$$

$$\mathcal{H}(\xi, t|\vartheta) = \frac{1}{\pi}\int_{0}^{1} d\xi' \frac{2\xi'\, \mathrm{Im}\mathcal{H}(\xi', t|\vartheta)}{\xi^2 - \xi'^2 - i\epsilon} + \mathcal{H}_0(t|\vartheta)\,. \tag{5}$$

Here, the lower integration limit, $\xi = 0$, corresponds to $\nu = \infty$ and the upper integration limit $\xi_{cut} = \frac{Q^2}{2M\nu_{cut}}$, has been set to $\xi_{cut} = 1$ in the (generalized) Bjorken limit. The equivalence of the two DRs (4), (5) is ensured by the sum rule (4), which now reads

$$\mathcal{H}_\infty(t|\vartheta) = \mathcal{H}_0(t|\vartheta) - \frac{2}{\pi}\int_{(0)}^{1} \frac{d\xi}{\xi}\, \mathrm{Im}\mathcal{H}(\xi, t|\vartheta)\,. \tag{6}$$

Within these notations, the $J = 0$ fixed pole universality hypothesis of [4] reads:

$$\mathcal{H}_\infty(t|\vartheta) = -\frac{2}{\pi}\int_{(0)}^{1} \frac{d\xi}{\xi}\, \mathrm{Im}\mathcal{H}(\xi, t|\vartheta = 0)\,. \tag{7}$$

[1]Various definitions of ϑ encountered in literature differ by $O(1/Q^2)$ terms vanishing in the generalized Bjorken limit.

3 Fixed pole contribution in GPD framework

In this section we address the issue of the subtraction constant for the dispersive representation of CFF within the partonic picture. We would now like to point out the origin of additional fixed pole contribution $\Delta\mathcal{H}_\infty$, which violates the universality conjecture within the GPD framework.

Within the collinear factorization approach to the leading order (LO) accuracy, the CFF $\mathcal{H}(\xi, t|\vartheta)$ arises from the elementary amplitude

$$\mathcal{H}(\xi, t|\vartheta) \stackrel{\text{LO}}{=} \int_0^1 dx \, \frac{2x}{\xi^2 - x^2 - i\epsilon} H^{(+)}(x, \eta = \vartheta\xi, t) \,, \tag{8}$$

where $H^{(+)}(x, \eta, t) \equiv H(x, \eta, t) - H(-x, \eta, t)$ stands for the antisymmetric charge even quark GPD combination. Note that for all allowed values of $|\vartheta| \leq 1$ the imaginary part of the CFF is given by the GPD in the outer region $\xi \geq \eta$.

Inserting the imaginary part of (8) into the sum rule (6) allows to express the $J = 0$ fixed pole contribution $\mathcal{H}_\infty(t|\vartheta)$ to the LO accuracy by the GPD in the outer region:

$$\mathcal{H}_\infty(t|\vartheta) \stackrel{\text{LO}}{=} \mathcal{H}_0(t|\vartheta) - 2\int_{(0)}^1 \frac{dx}{x} H^{(+)}(x, \vartheta x, t) \,. \tag{9}$$

By plugging the imaginary part of (8) into the subtracted DR (5) and equating it with the LO convolution formula (8) for the CFF, one can work out the GPD sum rule [13, 14]:

$$\mathcal{H}_0(t|\vartheta) \equiv 4\mathcal{D}(t|\vartheta) \stackrel{\text{LO}}{=} \int_0^1 dx \frac{2x}{x^2 - \xi^2} \left[H^{(+)}(x, \vartheta x, t) - H^{(+)}(x, \vartheta\xi, t)\right] \tag{10}$$

for the D-term form factor, which corresponds to the subtraction constant $\mathcal{H}_0$. It seems that by formally taking the high energy limit $\xi \to 0$ one can provides a regular proof for the $J = 0$ fixed pole universality conjecture (7). We argue that one has to be prudent while performing this limiting procedure and separate the integration region in (9) into central $x \in [0; \theta\xi]$ and outer $x \in [\theta\xi; 1]$ regions. Omitting the contribution from the central region means ignoring possible separate D-term addenda to GPD. As an illustration, we considered the specific form of a DD representation for GPD:

$$H^{(+)}(x, \eta) = \int_0^1 dy \int_{-1+y}^{1-y} dz \left[(1 - ax)\delta(x - y - z\eta) - \{x \to -x\}\right] h(y, z) \,, \tag{11}$$

where $h(y, z)$ is DD, symmetric in z and antisymmetric in y. The case $a = 0$ corresponds to the usual DD representation (without a D-term). For $a \neq 0$ GPD (11) includes an inherent D-term part and the polynomiality condition is respected in its complete form. The GPD sum rule (10) is also respected and the $J = 0$ fixed pole universality conjecture (7) is valid.

However, adding a supplementary separate D-term contribution

$$H^{(+)}(x, \eta) \to H^{(+)}(x, \eta) + \theta(|x| \leq |\eta|) d^{\text{f.p.}}(x/|\eta|) \tag{12}$$

leads to breakdown of the $J = 0$ fixed pole universality conjecture by the $J = 0$ fixed pole contribution $4\mathcal{D}^{\text{f.p.}}(t|\vartheta) = \int_0^1 dx \, \frac{2x\vartheta^2}{1-\vartheta^2 x^2} \, d^{\text{f.p.}}(x, t)$:

$$\mathcal{H}_\infty(t|\vartheta) \stackrel{\text{LO}}{=} -2\int_0^1 \frac{dx}{x} q^{(+)}(x, t) + \underbrace{4\mathcal{D}^{\text{f.p.}}(t|\vartheta)}_{\Delta\mathcal{H}_\infty(t|\vartheta)} \,. \tag{13}$$

A particular example of a field theoretical GPD model with a separate D-term contribution is provided by the calculation [9] of pion GPDs in the non-local chiral quark model [10]. In this model the universality conjecture (7) is not valid due to a supplementary D-term $d^{\text{f.p.}}(x,t)$ contribution, which has to be added to make GPD satisfy the soft pion theorem [11] fixing pion GPDs in the limit $\eta \to 1$. Another example is given by the photon GPD H_1, calculated to one-loop accuracy in [12], where a direct calculation of the CFF in the $\xi \to 0$ limit disproves the conjecture (7).

We conclude that the validity of the $J = 0$ fixed pole universality conjecture corresponds to the internal duality principle for GPDs, relating GPDs in the inner and outer support regions (see Ref. [13] for a detailed discussion).

It is extremely convenient to address this property within the approach [15] based on operator product expansion over the conformal basis. In this case there turns to be two kinds of analytical properties relevant for GPDs and associated CFFs:

- analyticity of CFFs in the cross channel angular momentum J;
- analyticity of GPD Gegenbauer/Mellin moments in the variable j, labeling the conformal spin $j+2$ of twist-2 quark conformal basis operators [2] $\mathcal{O}_j^a = \frac{\Gamma(3/2)\Gamma(1+j)}{2^j\Gamma(3/2+j)} (i\overleftrightarrow{\partial}_+)^j \, \bar{\psi}\lambda^a\gamma_+ \, C_j^{3/2}\left(\frac{\overleftrightarrow{D}_+}{\overleftrightarrow{\partial}_+}\right)\psi$.

In order to show explicitly the J-analytic properties we employ the Froissart-Gribov projection [16] of the cross channel SO(3) PWs of the CFF: $a_J(t|\vartheta) \equiv \frac{1}{2}\int_{-1}^{1} d(\cos\theta_t)\, P_J(\cos\theta_t)\mathcal{H}^{(+)}(\cos\theta_t, t|\vartheta)$, where (neglecting the threshold corrections) the cosine of the t-channel scattering angle reads: $\cos\theta_t = -1/(\vartheta\xi) + \mathcal{O}(1/Q^2)$.

For $J > 0$ PWs the Froissart-Gribov projection provides to the LO accuracy

$$a_{J>0}(t|\vartheta) \overset{\text{LO}}{=} 2\int_0^1 dx \frac{Q_J(1/x)}{x^2} H^{(+)}(x, \vartheta x, t)\,, \tag{14}$$

where $Q_J(1/x)$ denote the Legendre functions of the second kind. For $J = 0$ one obtains

$$a_{J=0}(t|\vartheta) \overset{\text{LO}}{=} 2\int_0^1 dx \left[\frac{Q_0(1/x)}{x^2} - \frac{1}{x}\right] H^{(+)}(x, \vartheta x, t) + 4\mathcal{D}(t|\vartheta)\,. \tag{15}$$

Indeed, as clearly seen from Eqs. (14) and (15), one can not recover $a_{J=0}(t|\vartheta)$ by means of the analytic continuation of $a_{J>0}(t|\vartheta)$ to $J = 0$. Therefore, analyticity in the cross channel angular momentum J is affected by the presence of a $J = 0$ fixed pole contribution

$$a_{J=0}^{\text{f.p.}}(t|\vartheta) \equiv \mathcal{H}_\infty(t|\vartheta) \overset{\text{LO}}{=} 4\mathcal{D}(t|\vartheta) - 2\int_{(0)}^1 \frac{dx}{x} H^{(+)}(x, \vartheta x, t)\,. \tag{16}$$

Within the conformal operator product expansion approach the analytic properties in the conformal spin j of the Gegenbauer moments of GPDs play the crucial role. In particular, the presence of a $J = 0$ fixed pole contribution (16) to the CFF $\mathcal{H}$ is a direct consequence of absence of non-singular conformal operators with the Lorentz spin $J \equiv j + 1 = 0$. Such a $j = -1$ contribution has to be subtracted from the $J = 0$ PW (see the $1/x$ moment in the integrand of Eq. (15)). The analogous cancelation appears also in the framework of dual parametrization of GPDs [6]. Moreover, assuming maximal analyticity in the conformal spin j (which corresponds to absence of the $j = -1$ fixed pole singularities manifest as the Kronecker $\delta_{j,-1}$ terms or singular operator contributions in the momentum

[2]Here the covariant derivative $\overleftrightarrow{D}$ and the total derivative $\overleftrightarrow{\partial}$ are contracted with the light-cone vector n, $C_j^{\frac{3}{2}}$ stand for the usual Gegenbauer polynomials with the index 3/2.

fraction representation) leads to the validity of the internal duality principle for GPDs (see [13]). This corresponds the implementation of $J = 0$ fixed pole universality hypothesis with a definite $J = 0$ fixed pole contribution given by the inverse moment sum rule

$$\mathcal{H}_\infty(t|\vartheta) \stackrel{\text{LO}}{=} -2 \int_{(0)}^{1} \frac{dx}{x} H^{(+)}(x, 0, t). \tag{17}$$

It is worth emphasizing that the inverse moment of PDF (17) can not be extracted form the sum rule (10) for D-term form factor as it simply cancels out. This sum rule is only potentially sensitive to possible non-universal contribution into $\mathcal{H}_\infty(t|\vartheta)$.

Conclusions

The $J = 0$ fixed pole universality hypothesis remains an attractive theoretical assumption for building up consistent GPD models intended for description of Compton scattering in deeply virtual regime. However, at the moment this hypothesis lacks theoretical proof. Phenomenological verification of this conjecture from the GPD sum rule (9) is in principle possible but is biased by the necessity of extrapolation of experimental results into high energy asymptotic regime and therefore will hardly be considered as decisive.

References

[1] V. de Alfaro, S. Fubini, G. Furlan, C. Rossetti, *Currents in Hadron Physics* (North-Holland, Amsterdam, 1973).

[2] M. J. Creutz, S. D. Drell and E. A. Paschos, Phys. Rev. **178**, 2300 (1969).

[3] S. J. Brodsky, F. E. Close and J. F. Gunion, Phys. Rev. D **5**, 1384 (1972); Phys. Rev. D **6**, 177 (1972).

[4] S. J. Brodsky, F. J. Llanes-Estrada and A. P. Szczepaniak, Phys.Rev. D **79**, 033012 (2009).

[5] M. V. Polyakov and C. Weiss, Phys. Rev. D **60**, 114017 (1999).

[6] D. Müller, M. V. Polyakov and K. M. Semenov-Tian-Shansky, JHEP **1503**, 052 (2015).

[7] D. Müller and K. M. Semenov-Tian-Shansky, Phys. Rev. D **92**, 074025 (2015).

[8] I. M. Gelfand and G. E. Shilov, *Generalized Functions Vol. I* (Academic Press, New York, 1964).

[9] K. M. Semenov-Tian-Shansky, Eur. Phys. J. A **36**, 303 (2008).

[10] M. Praszalowicz and A. Rostworowski, Acta Phys.Polon. B **34**, 2699 (2003).

[11] P. Pobylitsa, M. V. Polyakov and M. Strikman, Phys.Rev.Lett. **87**, 022001 (2001).

[12] S. Friot, B. Pire and L. Szymanowski, Phys. Lett. B **645**, 153 (2007).

[13] K. Kumerički, D. Müller and K. Passek-Kumerički, Eur. Phys. J. C **58**, 193 (2008).

[14] I. V. Anikin and O. V. Teryaev, Phys. Rev. D **76**, 056007 (2007).

[15] K. Kumerički, D. Müller and K. Passek-Kumerički, Nucl. Phys. B **794**, 244 (2008).

[16] V. N. Gribov, Zh. Eksp. Teor. Fiz. **41**, 667 (1961) [transl. Sov. Phys. JETP **14**, 478 (1962)]; M. Froissart, Phys. Rev. **123**, 1053 (1961).

EPJ Web of Conferences **112**, 01017 (2016)
DOI: 10.1051/epjconf/201611201017

Spin-Directed Momentum Transfers in SIDIS Baryon Production

Dennis Sivers
Portland Physics Institute
University of Michigan

Abstract

The measurement of transverse single-spin asymmetries for baryon production in the target fragmentation region of semi-inclusive deep-inelastic scattering (SIDIS), can produce important insight into those nonperturbative aspects of QCD directly associated with confinement and with the dynamical breaking of chiral symmetry. We discuss here, in terms of spin-directed momentum transfers, the powerful quantum field- theoretical constraints on the spin-orbit dynamics underlying these transverse spin observables. The A_τ-odd spin-directed momentum shifts, originating either in the target nucleon (δk_{TN}) or in the QCD jets (δp_{TN}) produced in the deep inelastic scattering process, represent significant quantum entanglement effects connecting information from current fragmentation with observables in target fragmentation.

Instrumentation of the "target fragmentation" region at an electron-ion collider (EIC) can provide a broad range of new tools for the study of nonperturbative dynamics in QCD. [1,2] The specific observables discussed here are the transverse single-spin asymmetry measurements

$$\begin{aligned} &A_N d\sigma(ep\uparrow \Rightarrow e'BX) \quad B = p, n, \Lambda, \Sigma, \Delta ... \\ &P d\sigma(ep \Rightarrow e'B\uparrow X) \quad B\uparrow = \Lambda\uparrow, \Sigma\uparrow, \Xi\uparrow ... \end{aligned} \tag{1.1}$$

The systematic, organized program of such measurements should also be done at the Jefferson Lab CEBAF 12 GeV machine [3] due to the importance of single-spin observables in understanding spin-orbit dynamics in a full range of kinematics. As described below, these baryon asymmetry measurements both supplement and enhance the understanding of measurements of target spin asymmetries for meson production

$$A_N d\sigma(ep\uparrow \Rightarrow e'MX) \quad M = \pi^\pm, \pi^0, K^\pm, K^0, \eta^0 ... \tag{1.2}$$

found in the "current fragmentation" region. It is convenient to parameterize the baryon observables of (1.1) in terms of the transverse-momentum dependent (TMD) "fractured functions" introduced in references [1] and [2]. For the target-spin asymmetries we can define the Fractured Orbital Functions (FOF) $\Delta^N M^q_{B/(q,q):p\uparrow}(x,k_{TN};z,\vec{p}^B_T\cdot\vec{k}_T\;;Q^2)$, and the Fractured Collins, Heppelman, Ladinsky Functions (FCHLF) $\Delta^N M^{q\uparrow}_{B/\{q,q\}\uparrow:p\uparrow}(x,\vec{k}_T\cdot\vec{p}^B_T;z,p^B_{TN};Q^2)$. Correspondingly, for the final-state polarization observables we can define the Polarizing Fractured Functions (PFF) $\Delta^N M^q_{B\uparrow/(q,q):p}(x,\vec{k}_T\cdot\vec{p}^B_T;z,p^B_{TN};Q^2)$ and the Fractured Boer Mulders Functions (FBMF) $\Delta^N M^{q\uparrow}_{B\uparrow/\{q,q\}\uparrow:p}(x,k_{TN};z,\vec{k}_T\cdot\vec{p}_T;Q^2)$. The content of the expressions for these functions is dense and the labeling complicated so the names are attached to connect these transverse spin-odd fractured functions involving quantum diquark structures to the classification of transverse-spin odd distributions and fragmentation functions for quarks created by Mulders and Tangerman. [4] The four fractured functions presented here describe the spin-dependent differences of the conjoint probabilities for detecting a quark jet with kinematics defined by $x = x_{bj} = Q^2/2p.q$ and transverse momentum $\vec{k}_T$ and a final-state baryon with kinematics defined by the Feynman variable $z = z_B = p\cdot p_B/p\cdot q$ and transverse momentum $\vec{p}_T$ in the same deep-inelastic scattering event. A more complete discussion of these fractured functions can be found in references [1,2]. For FOF and FCHLF the Δ^N symbol indicates a target transverse spin asymmetry,

$$\Delta^N M^q_{B/(,):p\uparrow} = M^q_{B/(,):p\uparrow} - M^q_{B/(,):p\downarrow}. \tag{1.3}$$

For the fractured functions PFF and FBMF the symbol Δ^N defines a final-state polarization asymmetry from an unpolarized target,

$$\Delta^N M^q_{B\uparrow/(,):p} = M^q_{B\uparrow/(,):p} - M^q_{B\downarrow/(,):p}. \tag{1.4}$$

The set of conjoint probability densities $[M^q_{B\uparrow\downarrow/(,):p}, M^q_{B/(,):p\uparrow\downarrow},\ldots]$ can capture the full range of quantum information contained in the measurements as constrained in the Bell's inequalities [5,6] and used in the study of quantum entanglement. This connection is important when viewed in the context of the powerful quantum-field-theoretical superselection principles applicable in the measurement of single-spin asymmetries involving the hard scattering of light quarks. [7,8]. For valence SIDIS these arguments also introduce quantum entanglement connecting the target fragmentation region to information in current fragmentation.

All single-spin asymmetries are odd under an operator, O, such that

$$\mathrm{O}\{\vec{k}_i;\vec{\sigma}_j\}\mathrm{O}^{-1} = \{\vec{k}_i;-\vec{\sigma}_j\}, \tag{1.5}$$

where $\vec{k}_i$ are momentum 3-vectors and $\vec{\sigma}_j$ are spin axial vectors. This operator can be seen to be the 3-D Hodge dual of the parity operator, P,

$$P\{\vec{k}_i;\vec{\sigma}_j\}P^{-1} = \{-\vec{k}_i;\vec{\sigma}_j\}. \tag{1.6}$$

The operator product $\mathrm{A}_\tau = P\mathrm{O}$ then has the action

$$\mathrm{A}_\tau\{\vec{k}_i;\vec{\sigma}_j\}\mathrm{A}_\tau^{-1} = \{-\vec{k}_i;-\vec{\sigma}_j\}. \tag{1.7}$$

The operator A_τ now designated "naïve time reversal" [9,10] is odd for transverse, parity conserving single-spin asymmetries. In this note we will denote the A_τ-odd spin-directed momentum,

$$k_{TN} = \overset{\text{1}}{k}_T \cdot (\hat{\sigma} \times \hat{Q}), \tag{1.8}$$

as the observable that defines a single-spin observable in SIDIS and other hard-scattering processes.

The explicit connection of the spin-directed momentum shift that can be produced by one unit of h and the asymmetries defined in Eq. (1.1) is indicated in the sketches of Fig. 1

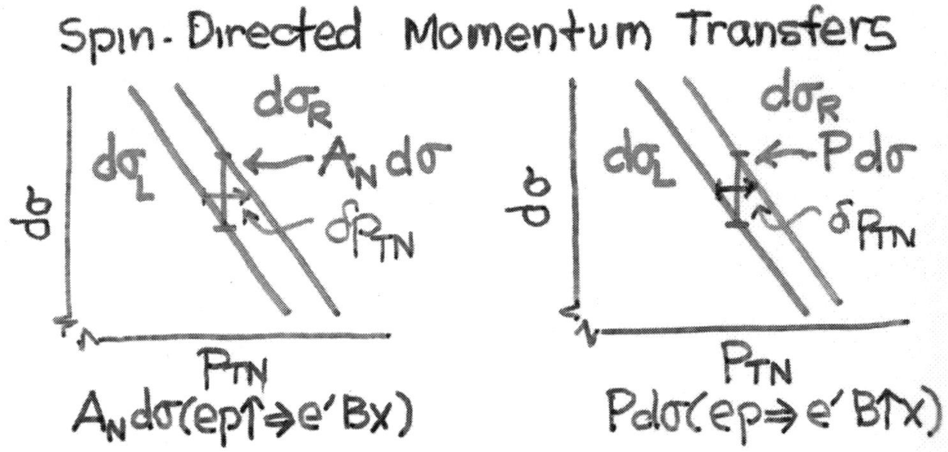

Fig. 1 Spin-directed momentum transfers provide an alternate, rigorous, definition of single-spin asymmetries.

In the discussion below we will designate a spin-directed momentum shift generated within the target nucleon by $\langle \delta k_{TN}(x, \mu^2) \rangle$ and shift generated within a QCD jet fragmentation by $\langle \delta p_{TN}(x, \mu^2) \rangle$ and treat them as directly observable asymmetries. The observation that transverse single-spin asymmetries are prohibited as $m_q \to 0$ in hard-scattering processes of perturbative QCD [11] can be designated Kane-Pumplin-Repko [12] factorization. For processes such as SIDIS this allows the use of field-theoretical idempotent projection operators,

$$\Pi_A^{\pm} = \left(\frac{1 \pm A_\tau}{2} \right). \tag{1.9}$$

These operators create superselection rules that isolate the spin-directed momentum shift. In processes involving the hard scattering of massless quarks A_τ-odd dynamical effects can only be produced in the wave function of the target nucleon or in the color re-arrangement that incurs in the fragmentation dynamics leading to final-state hadrons. The superselection rules allow for the calculation of these shifts within the nucleon by lattice gauge simulations [13]. The application of these operators to a valence-quark SIDIS process is indicated in Fig. 2,

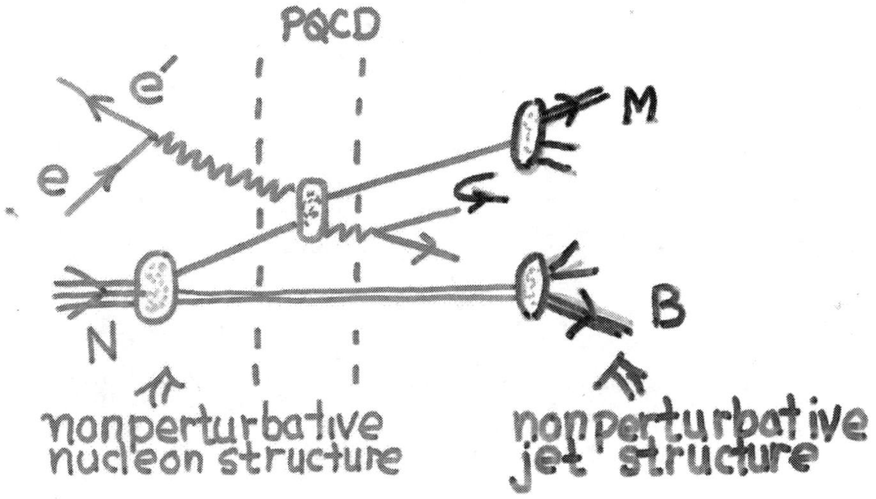

Fig. 2 The hard scattering in valence SIDIS separates nonperturbative jet structure from nonperturbative dynamics in the target nucleon.

The hard scattering process $eq \Rightarrow eqX$ in the central part of this diagram that is calculable in QCD perturbation theory for a valence u or d quark is necessarily A_τ-even so that the A_τ-odd dynamics leading to the spin-directed momentum shift must appear in the wave function of the target nucleon or in the color-rearrangement of the fragmentation processes leading to final-state hadrons. The phenomenological study of spin-directed momentum transfers in the current fragmentation region of this diagram [14,15] has already proved to be very instructive in isolating interesting dynamics. The diagram in Fig. 2 does not explicitly show the final-state interaction between the struck quark and the diquark structure that creates the process dependence of A_τ-odd distribution functions. The quantum entanglement produced by these final-state interactions can be isolated by the projection operator Π_A^- of (1.9) The fact that the SIDIS process demonstrates TMD factorization [16] combined with the fact that SIDIS kinematics allows for a distinction between A_τ-odd dynamics occurring in the target nucleon from A_τ-odd effects in fragmentation. [17] The same kinematic restrictions occur for the production of baryons in the target fragmentation region of SIDIS as can be demonstrated by the three intersecting planes shown in Fig. 3

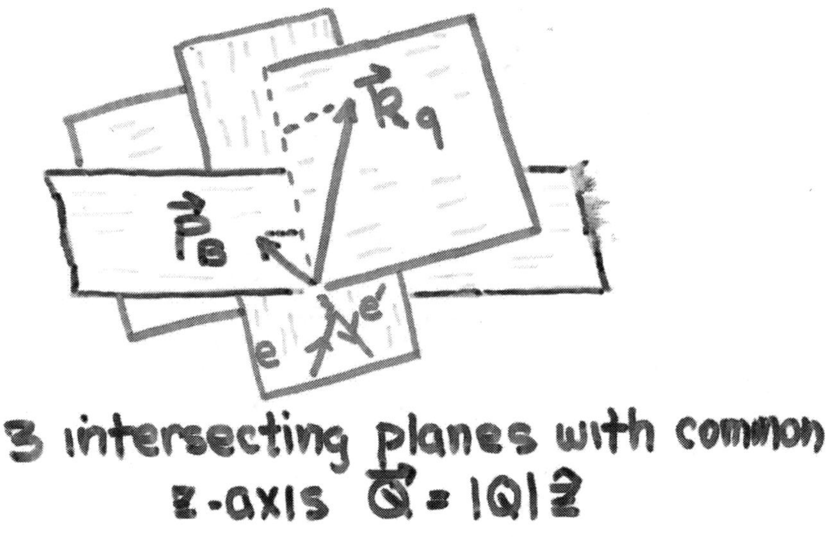

Fig. 3 The three independent planes in SIDIS can be used to differentiate between A_T-odd dynamics in the nucleon and in jet fragmentation.

A significant feature of spin-observables involving baryons is that polarization asymmetries and target spin asymmetries can be studied in tandem. For example the comparison of the FOF $\Delta^N M^u_{\Lambda/[u,d];p\uparrow}$ with the PFF $\Delta^N M^u_{\Lambda\uparrow/[u,d];p}$ in the same range of kinematics allows for the calibration of the $\langle \delta k_{TN}(x,\mu^2)\rangle$ associated with the Fractured Orbital Function with the $\langle \delta p_{TN}(z,\mu^2)\rangle$ generated in the Polarizing Fractured Function. Since these momentum shifts do not evolve under TMD evolution [11,18], the data collected from spin asymmetries can be used to study the evolution of spin-averaged TMD's.

Most of the familiar examples of quantum entanglement involve spin observables measured in widely separated kinematical regions. [5,19] The hard scattering of a valence quark in a polarized proton as pictured in Fig. 2 can be seen to resolve the structure of a complicated internal quantum state. The Fractured Orbital Function, FOF, resolves this state in terms of an orbiting configuration involving quark-diquark and virtual meson degrees of freedom confined within the polarized proton. The final-state interactions of the struck quark that lead to a spin-directed momentum transfer involve the attractive chromo-electric force between the struck quark and the diquark spectator. This force leads to a fraction of the transverse momentum shift of the quark to be transmitted to the diquark,

$$\left\langle \delta k_{TN}(x,\mu^2)\right\rangle^{(O)}_{\{u,d\}} + \left\langle \delta k_{TN}(x,\mu^2)\right\rangle^{(O)}_{[u,d]} = -2\eta_u \left\langle \delta k_{TN}(x,\mu^2)\right\rangle^{(O)}_{u}$$
$$\left\langle \delta k_{TN}(x,\mu^2)\right\rangle^{(O)}_{\{u,u\}} = -\eta_d \left\langle \delta k_{TN}(x,\mu^2)\right\rangle^{(O)}_{d}. \quad (1.10)$$

In these equations the parameters $0 \le \eta_u, \eta_d \le 1$ provide a measure of the interaction of the quark-diquark system with the remaining internal constituents. These equations allow the prediction of $\Delta^N M^u_{p/[u,d]:p\uparrow}$, $\Delta^N M^u_{n/[u,d]:p\uparrow}$ and $\Delta^N M^u_{\Lambda/[u,d]:p\uparrow}$ from existing data on pion spin asymmetries. This type of long-range correlation tests the underlying understanding of orbital distributions in the same manner that the Collins conjugation prediction [20] relating the asymmetry in SIDIS and the Drell-Yan process,

$$\left\langle \delta k_{TN}(x,\mu^2)\right\rangle^{(O)}_{SIDIS} = -\left\langle \delta k_{TN}(x,\mu^2)\right\rangle^{(O)}_{DY}, \quad (1.11)$$

when expressed in terms of spin-directed momentum transfers.

The spin state of a diquark by itself represents a quantum-entangled system and the rank dependence of the fragmentation process in the FCHLF $\Delta^N M^{q\uparrow}_{B/\{q,q'\}\uparrow:p\uparrow}$ is closely related to that of the Collins Function, $\Delta^N D_{M/q\uparrow}$. That is to say, the A_τ-odd flux rupture leading to $\{q,q'\}\uparrow \Rightarrow B + \bar{q}\uparrow$ necessarily involves

$$\left\langle \delta p_{TN}(z,\mu^2)\right\rangle_{\bar{q}\uparrow} = -\left\langle \delta p_{TN}(z,\mu^2)\right\rangle_{B}. \quad (1.12)$$

This means that there will be spin-asymmetries associated with the production of mesons within the target fragmentation region as well as baryons. We will not discuss here the formulation of dihadron fractured functions that can be used to study spin-oriented correlations between hadrons in a full range of kinematics. At this conference Harut Avakian presented Jefferson Lab data on spin-dependent correlations between baryons and mesons. [21] These data support the basic structure of the formalism presented here. The author is grateful for informative talks and advice from Gary Goldstein and Simonetta Liutti and would like to thank Nikolai Kochelev for directing attention to ref. [22] where many of these topics have previously been explored.

REFERENCES

1. D. Sivers, Phys. Rev. **D79**, 085008 (2009).
2. D. Sivers, Phys. Rev. **D80**, 034029 (2010).
3. See, for example, M Battaglieri et al. Acta Phys. Polon. **B46**, 2, 257 (2015).
4. R.D. Tangerman and P.J. Mulders, Phys. Rev. **D51**, 3357 (1995); Nucl. Phys. **B461**, 197 (1996).
5. J.S. Bell, Physics **1**, 195 (1964); Rev. Mod. Phys. **38**, 447 (1966).
6. P.H. Eberhard, Nuovo Cim. **B38** 75 (1977).
7. D. Sivers, Phys. Rev. **D74**, 094008 (2006).
8. D. Sivers, arXiV:0704.1791 (hep-ph) unpublished
9. R. Jaffe, in *Proceedings on the Workshop on Deep Inelastic Scattering off Polarized Targets, DEZY-Zeuthen, Germany 1997,* edited by J.Blumlein et al., 167.

10. D. Sivers, in *Proceedings of the Workshop on Deep Inelastic Scattering off Polarized Targets, DEZY-Zeuthen, Germany 1997,* edited by J. Blumlein et al., 383.
11. D. Sivers, Int. J. Mod. Phys. Conf. Ser. **25**, 1460002 (2014).
12. G. Kane, J. Pumplin and W. Repko, Phys. Rev. Lett. **41**, 1689 (1978).
13. B. U. Musch, et al., Phys. Rev. **D85**, 094510 (2012).
14. M. Anselmino et al. , arXiV:1107.4446 (2011).
15. Z.-B. Kang, et al., arXiV:1505.05589 (2015).
16. J. Collins, arXiV:1307.2920 (2013).
17. A. Bacchetta et al., Phys. Rev. **D70**, 117504 (2011).
18. D. Sivers, in EBJ Web. Conf. **85**, 2007 (2015).
19. A. Einstein, B Podolsky and N. Rosen, Phys. Rev. **D47**, 777 (1935).
20. J. Collins, Phys. Lett. **B536**, 43 (2002).
21. H. Avakian, this conference.
22. A.E. Dorokhov, et al.; Phys. Part. Nucl. Lett. **61**, 440 (2009)

EPJ Web of Conferences **112**, 01018 (2016)
DOI: 10.1051/epjconf/201611201018

Probing the transversity spin structure of a nucleon in neutrino-production of a charmed meson

B. Pire[1,a], L. Szymanowski[2,b], and J. Wagner[2,c]

[1] *Centre de physique théorique, École Polytechnique, CNRS, Université Paris-Saclay, 91128 Palaiseau, France*
[2] *National Centre for Nuclear Research (NCBJ), Warsaw, Poland*

Abstract. Including $O(m_c)$ terms in the coefficient functions and/or $O(m_D)$ twist 3 contributions in the heavy meson distribution amplitudes leads to a non-zero transverse amplitude for exclusive neutrino production of a D pseudoscalar charmed meson on an unpolarized target. We work in the framework of the collinear QCD approach where chiral-odd transversity generalized parton distributions (GPDs) factorize from perturbatively calculable coefficient functions.

1 Introduction

The now well established framework of collinear QCD factorization [1–3] for exclusive reactions mediated by a highly virtual photon in the generalized Bjorken regime describes hadronic amplitudes using generalized parton distributions (GPDs) which give access to a 3-dimensional analysis [4] of the internal structure of hadrons. Neutrino production is another way to access (generalized) parton distributions [5]. Although neutrino induced cross sections are orders of magnitudes smaller than those for electroproduction and neutrino beams are much more difficult to handle than charged lepton beams, they have been very important to scrutinize the flavor content of the nucleon and the advent of new generations of neutrino experiments opens new possibilities. In particular, the flavor changing character of the electroweak current allows charmed quark to be produced in processes involving light quark partonic distributions [6]. This in turn allows helicity flip hard amplitudes to occur at the $O(\frac{m_c}{Q})$ level where Q is the typical large scale allowing QCD collinear factorization. Such a coefficient function has to be attached to a chiral-odd generalized parton distribution, the elusive transversity GPDs [7–9]. The transverse character of these GPDs select the transverse polarization of the W−boson, which phenomenologically allows a separation of this interesting amplitude through the azimuthal distribution of the final state particles [6].

[a] e-mail: bernard.pire@polytechnique.edu
[b] e-mail: lech.szymanowski@ncbj.gov.pl
[c] e-mail: jakub.wagner@ncbj.gov.pl

2 Kinematics

For definiteness, we consider the exclusive production of a pseudoscalar D–meson through the reaction (see Fig. 1):

$$\nu_l(k) N(p_1) \quad \rightarrow \quad l^-(k') D^+(p_D) N(p_2), \tag{1}$$

where N is a proton or a neutron, in the kinematical domain where collinear factorization leads to a description of the scattering amplitude in terms of nucleon GPDs and the D–meson distribution amplitude, with the hard subprocesses:

$$W^+(\varepsilon, q) d \quad \rightarrow \quad D^+(p_D) d\,. \tag{2}$$

Our kinematical notations are as follows (m and M_D are the nucleon and D–meson masses, m_c will denote the charmed quark mass):

$$\begin{aligned} &q = k - k' \quad ; \quad Q^2 = -q^2 \quad ; \quad \Delta = p_2 - p_1 \quad ; \quad \Delta^2 = t; \\ &p_1^\mu = (1+\xi)p^\mu + \frac{1}{2}\frac{m^2 - \Delta_T^2/4}{1+\xi}n^\mu - \frac{\Delta_T^\mu}{2} \quad ; \quad p_2^\mu = (1-\xi)p^\mu + \frac{1}{2}\frac{m^2 - \Delta_T^2/4}{1-\xi}n^\mu + \frac{\Delta_T^\mu}{2}; \\ &q^\mu = -2\xi' p^\mu + \frac{Q^2}{4\xi'}n^\mu \quad ; \quad p_D^\mu = 2(\xi - \xi')p^\mu + \frac{M_D^2 - \Delta_T^2}{4(\xi - \xi')}n^\mu - \Delta_T^\mu, \end{aligned} \tag{3}$$

with $p^2 = n^2 = 0$ and $p.n = 1$. As in the double deeply virtual Compton scattering case [10], it is meaningful to introduce two distinct momentum fractions:

$$\xi = -\frac{(p_2 - p_1).n}{2} \quad , \quad \xi' = -\frac{q.n}{2}. \tag{4}$$

Neglecting the nucleon mass and Δ_T, the approximate values of ξ and ξ' are

$$\xi \approx \frac{Q^2 + M_D^2}{4p_1.q - Q^2 - M_D^2} \quad , \quad \xi' \approx \frac{Q^2}{4p_1.q - Q^2 - M_D^2}. \tag{5}$$

To unify the description of the scaling amplitude, we define a modified Bjorken variable $x_B^D \equiv \frac{Q^2+M_D^2}{2p_1.q}$ which allows to express ξ and ξ' in a compact form:

$$\xi \approx \frac{x_B^D}{2 - x_B^D} \quad , \quad \xi' \approx \frac{x_B}{2 - x_B^D}. \tag{6}$$

If the meson mass is the relevant large scale (for instance in the limiting case where Q^2 vanishes as in the timelike Compton scattering kinematics [11]) :

$$Q^2 \rightarrow 0 \;;\; \xi' \rightarrow 0; \xi \approx \frac{\tau}{2-\tau} \;;\; \tau = \frac{M_D^2}{s_{WN} - m^2}. \tag{7}$$

3 The transverse amplitude

In the Feynman gauge, the non-vanishing m_c–dependent part of the Dirac trace in the hard scattering part depicted in Fig. 1a reads:

$$Tr[\sigma^{pi}\gamma^\nu \hat{p}_D \gamma^5 \gamma^{\nu'} \frac{m_c}{D_1} \hat{\varepsilon}(1-\gamma^5)\frac{-g_{\nu\nu'}}{D_2}] = \frac{2(Q^2 + M_D^2)}{\xi}\varepsilon_\mu[\epsilon^{\mu pin} + i g_\perp^{\mu i}]\frac{m_c}{D_1}\frac{1}{D_2}, \tag{8}$$

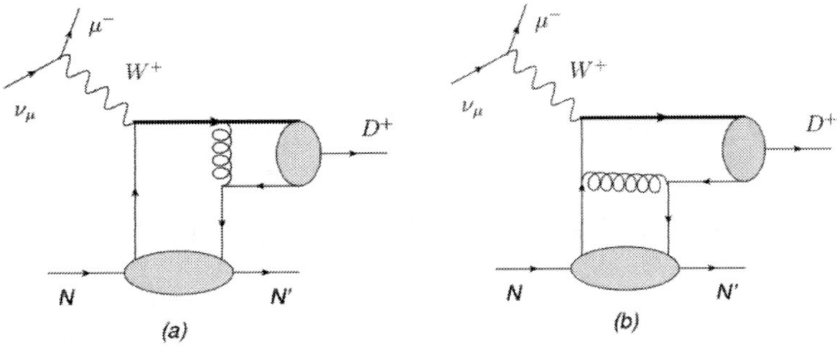

Figure 1. Feynman diagrams for the factorized amplitude for the $\nu_\mu N \to \mu^- D^+ N'$ or the $\nu_\mu N \to \mu^- D^0 N'$ process involving the quark GPDs; the thick line represents the heavy quark. In the Feynman gauge, diagram (a) involves convolution with both the transversity GPDs and the chiral even ones, whereas diagram (b) involves only chiral even GPDs.

where ε is the polarization vector of the $W^\pm$boson (we denote $\hat{p} = p_\mu \gamma^\mu$ for any vector p). The fermionic trace vanishes for the diagram shown on Fig. 1b thanks to the identity $\gamma^\rho \sigma^{\alpha\beta} \gamma_\rho = 0$. The denominators of the propagators read:

$$D_1 = k_c^2 - m_c^2 + i\epsilon = \frac{Q^2}{2\xi'}(x + \xi - 2\xi') - m_c^2 + i\epsilon = \frac{Q^2 + M_D^2}{2\xi}(x + \xi) - Q^2 - m_c^2 + i\epsilon , \quad (9)$$

$$D_2 = k_g^2 + i\epsilon = \bar{z}[\bar{z} m_D^2 + \frac{Q^2 + m_D^2}{2\xi}(x - \xi) + i\epsilon] ,$$

where k_c (k_g) is the heavy quark (gluon) momentum. The transverse amplitude is then written as ($\tau = 1 - i2$):

$$T_T = \frac{iC\xi(m_c - 2M_D)}{\sqrt{2}(Q^2 + M_D^2)} \bar{N}(p_2) \left[\mathcal{H}_T^\phi i\sigma^{n\tau} + \tilde{\mathcal{H}}_T^\phi \frac{\Delta^\tau}{m_N^2} + \mathcal{E}_T^\phi \frac{\hat{n}\Delta^\tau + 2\xi\gamma^\tau}{2m_N} - \tilde{\mathcal{E}}_T^\phi \frac{\gamma^\tau}{m_N} \right] N(p_1), \quad (10)$$

with $C = \frac{2\pi}{3} C_F \alpha_s V_{dc}$, in terms of transverse form factors that we define as :

$$\mathcal{F}_T^\phi = f_D \int \frac{\phi(z)dz}{\bar{z}} \int \frac{F_T^d(x, \xi, t)dx}{(x - \xi + \beta\xi + i\epsilon)(x - \xi + \alpha\bar{z} + i\epsilon)}, \quad (11)$$

where F_T^d is any d-quark transversity GPD, $\alpha = \frac{2\xi M_D^2}{Q^2 + M_D^2}$, $\beta = \frac{2(M_D^2 - m_c^2)}{Q^2 + M_D^2}$ and we shall denote $\bar{\mathcal{E}}_T^\phi = \xi \mathcal{E}_T^\phi - \tilde{\mathcal{E}}_T^\phi$. In the following, we shall put β to 0.

The prefactor in Eq.(10) shows the two sources of the transverse amplitude : m_c signals the contribution from the helicity changing part of the heavy quark propagator, while M_D signals the contribution from the twist 3 heavy meson distribution amplitude which we parametrize (omitting the Wilson lines) as :

$$\langle 0|\bar{c}(y)\,\gamma^5\, d(-y)|D^-(P_D)\rangle = -i\frac{f_D}{4}M_D\int_0^1 dz\, e^{i(2z-1)P_D.y}\,\phi_D^{(s)}(z)\,, \tag{12}$$

and for simplicity we identify ϕ^s with the leading twist 2 pseudoscalar charmed meson DA ϕ_D defined as:

$$\langle 0|\bar{c}(y)\,\gamma^\mu\gamma^5\, d(-y)|D^-(P_D)\rangle = -i\frac{f_D}{4}P_D^\mu\int_0^1 dz\, e^{i(2z-1)P_D.y}\phi_D(z)\,.$$

4 The azimuthal dependence of neutrinoproduction

The dependence of a leptoproduction cross section on azimuthal angles is a widely used way to analyze the scattering mechanism. This procedure is helpful as soon as one can define an angle φ between a leptonic and a hadronic plane, as for deeply virtual Compton scattering [12] and related processes. In the neutrino case, it reads :

$$\frac{d^4\sigma(\nu N\to l^-N'D)}{dx_B\,dQ^2\,dt\,d\varphi} = \tag{13}$$

$$\tilde{\Gamma}\Big\{\frac{1+\sqrt{1-\varepsilon^2}}{2}\sigma_{--} + \varepsilon\sigma_{00} + \sqrt{\varepsilon}(\sqrt{1+\varepsilon}+\sqrt{1-\varepsilon})(\cos\varphi\ \mathrm{Re}\sigma_{-0} + \sin\varphi\ \mathrm{Im}\sigma_{-0})\Big\},$$

with

$$\tilde{\Gamma} = \frac{G_F^2}{(2\pi)^4}\frac{1}{16x_B}\frac{1}{\sqrt{1+4x_B^2m_N^2/Q^2}}\frac{1}{(s-m_N^2)^2}\frac{Q^2}{1-\epsilon},$$

and the "cross-sections" $\sigma_{lm} = \epsilon_l^{*\mu}W_{\mu\nu}\epsilon_m^\nu$ are product of amplitudes for the process $W(\epsilon_l)N\to DN'$, averaged (summed) over the initial (final) hadron polarizations. In the anti-neutrino case, one gets a similar expression with $\sigma_{--}\to\sigma_{++}$, $\sigma_{-0}\to\sigma_{+0}$, $1+\sqrt{1-\varepsilon^2}\to 1-\sqrt{1-\varepsilon^2}$ and $\sqrt{1+\varepsilon}+\sqrt{1-\varepsilon}\to\sqrt{1+\varepsilon}-\sqrt{1-\varepsilon}$. We use the standard notations of deep exclusive leptoproduction, namely $y = p_1.q/p_1.k$ and $\epsilon\simeq 2(1-y)/[1+(1-y)^2]$. The azimuthal angle φ is defined in the initial nucleon rest frame as:

$$sin\ \varphi = \frac{\vec{q}\cdot[(\vec{q}\times\vec{p}_D)\times(\vec{q}\times\vec{k})]}{|\vec{q}||\vec{q}\times\vec{p}_D||\vec{q}\times\vec{k}|}\,, \tag{14}$$

while the final nucleon momentum lies in the xz plane ($\Delta^y = 0$).

The quantity σ_{-0} is directly related to the observables $< cos\ \varphi >$ and $< sin\ \varphi >$ through

$$< cos\,\varphi > = \frac{\int cos\,\varphi\, d\varphi\, d^4\sigma}{\int d\varphi\, d^4\sigma} = \frac{\sqrt{\varepsilon}(\sqrt{1+\varepsilon}+\sqrt{1-\varepsilon})\,\mathcal{R}e\sigma_{-0}}{2\epsilon\sigma_{00}+(1+\sqrt{1-\varepsilon^2})\sigma_{--}}\,,$$

$$< sin\,\varphi > = \frac{\int sin\,\varphi\, d\varphi\, d^4\sigma}{\int d\varphi\, d^4\sigma} = \frac{\sqrt{\varepsilon}(\sqrt{1+\varepsilon}+\sqrt{1-\varepsilon})\,\mathcal{I}m\sigma_{-0}}{2\epsilon\sigma_{00}+(1+\sqrt{1-\varepsilon^2})\sigma_{--}}\,. \tag{15}$$

Estimating the counting rates and the angular observables defined above is in progress [13]. As a first step, we calculate σ_{--} which is bilinear in transversity quark GPDs. At zeroth order in Δ_T, σ_{--} reads:

$$\sigma_{--} = \frac{4\xi^2C^2(m_c+2M_D)^2}{(Q^2+M_D^2)^2}\Big\{(1-\xi^2)|\mathcal{H}_T^\phi|^2 + \frac{\xi^2}{1-\xi^2}|\bar{\mathcal{E}}_T^\phi|^2 - 2\xi\mathcal{R}e[\mathcal{H}_T^\phi\bar{\mathcal{E}}_T^{\phi *}]\Big\}. \tag{16}$$

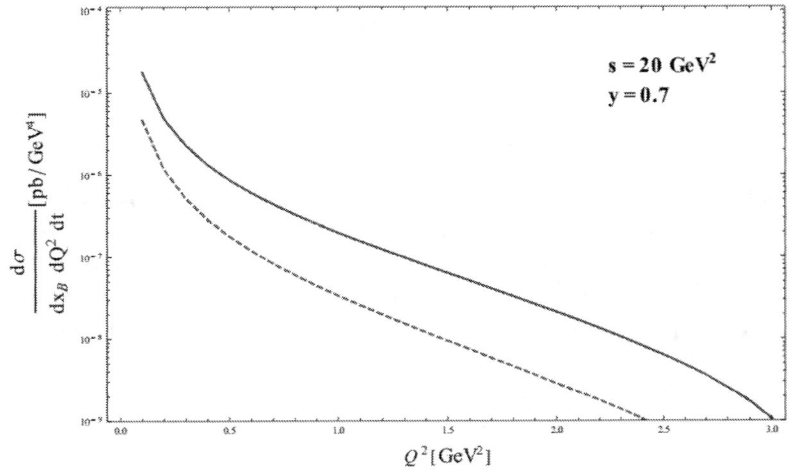

Figure 2. Contribution of σ_{--} to the differential cross section $\frac{d\sigma}{dx_B dQ^2 dt}$ of neutrino-production of a D^+ meson on a proton (dashed line) or a neutron (solid line) at $s = 20$ GeV^2 , $y = 0.7$ and $t = t_0$, as a function of Q^2.

Using the model of Ref [14] for the D^+ meson distribution amplitude and the parametrization of the dominant transversity GPD $H_T(x, \xi, t)$ from Ref [15] (and neglecting for the time being other chiral-odd GPDs contributions), we compute the contribution to the differential cross section given in Eq.(13) integrated over φ. The result is shown in Fig. 2 as a function of Q^2 for $s = 20$ GeV^2, $y = 0.7$ and $t = t_{min}$. Since the process selects the $d-$ quark contribution, the proton and neutron target cases allow to access H_T^d and H_T^u respectively. Although small, the cross-sections are of the same order of magnitude as those for the neutrino production of π or D_s mesons estimated in [5]. This shows that these processes should be measurable in intense neutrino beam facilities.

Let us remind the reader that we allow Q^2 to be quite small since the hard scale governing our process is $M_D^2 + Q^2$.

5 Conclusion

Collinear QCD factorization has allowed us to calculate neutrino production of $D-$mesons in terms of GPDs. Gluon and both chiral-odd and chiral-even quark GPDs contribute to the amplitude for different polarization states of the $W^\pm$ boson. The azimuthal dependence of the cross section allows to separate different contributions. Planned high energy neutrino facilities [16] which have their scientific program oriented toward the understanding of neutrino oscillations or elusive inert neutrinos may thus allow - without much additional equipment - some important progress in the realm of hadronic physics.

This work was partially supported by the COPIN-IN2P3 Agreement and by the french grant ANR PARTONS (Grant No. ANR-12-MONU-0008-01); L. Sz was partially supported by grant of National Science Center, Poland, No. 2015/17/B/ST2/01838.

References

[1] D. Müller *et al.*, Fortsch. Phys. **42**, 101 (1994).

[2] X. Ji, Phys. Rev. **D55**, 7114 (1997); A. V. Radyushkin, Phys. Rev. D **56**, 5524 (1997).
[3] J. C. Collins, L. Frankfurt, M. Strikman, Phys. Rev. D **56**, 2982 (1997).
[4] M. Burkardt, Phys. Rev. D **62**, 071503 (2000) [Phys. Rev. D **66**, 119903 (2002)]; J. P. Ralston and B. Pire, Phys. Rev. D **66**, 111501 (2002); M. Diehl and P. Hagler, Eur. Phys. J. C **44**, 87 (2005).
[5] B. Lehmann-Dronke and A. Schafer, Phys. Lett. B **521** (2001) 55; C. Coriano and M. Guzzi, Phys. Rev. D **71** (2005) 053002; P. Amore, C. Coriano and M. Guzzi, JHEP **0502** (2005) 038; A. Psaker, W. Melnitchouk and A. V. Radyushkin, Phys. Rev. D **75** (2007) 054001; B. Z. Kopeliovich, I. Schmidt and M. Siddikov, Phys. Rev. D **86**, 113018 (2012) and D **89**, 053001 (2014); G. R. Goldstein, O. G. Hernandez, S. Liuti and T. McAskill, AIP Conf. Proc. **1222** (2010) 248.
[6] B. Pire and L. Szymanowski, Phys. Rev. Lett. **115**, 092001 (2015); B. Pire and L. Szymanowski, arXiv:1510.01869.
[7] M. Diehl, Eur. Phys. J. C **19**, 485 (2001).
[8] M. Diehl *et. al.* Phys. Rev. D **59**, 034023 (1999); J. C. Collins *et. al.*, Phys. Rev. D **61**, 114015 (2000).
[9] D. Yu. Ivanov *et al.*, Phys. Lett. B **550**, 65 (2002); R. Enberg, B. Pire and L. Szymanowski, Eur. Phys. J. C **47**, 87 (2006).
[10] M. Guidal and M. Vanderhaeghen, Phys. Rev. Lett. **90**, 012001 (2003); A. V. Belitsky and D. Mueller, Phys. Rev. Lett. **90**, 022001 (2003).
[11] E. R. Berger, M. Diehl and B. Pire, Eur. Phys. J. C **23** (2002) 675.
[12] M. Diehl *et al.*, Phys. Lett. B **411** (1997) 193; A. V. Belitsky, D. Mueller and A. Kirchner, Nucl. Phys. B **629**, 323 (2002).
[13] B. Pire, L. Szymanowski and J. Wagner, work in progress.
[14] T. Kurimoto, H. n. Li and A. I. Sanda, Phys. Rev. D **65**, 014007 (2002); T. Kurimoto, H. n. Li and A. I. Sanda, Phys. Rev. D **67**, 054028 (2003).
[15] M. El Beiyad *et. al.*, Phys. Lett. B **688**, 154 (2010).
[16] D. S. Ayres *et al.* [NOvA Collaboration], hep-ex/0503053; J. Mousseau *et al.* [MINERvA Collaboration], arXiv:1601.06313 [hep-ex]; see also M. L. Mangano, S. I. Alekhin, M. Anselmino *et al.*, CERN Yellow Report CERN-2004-002, pp.185-257 [hep-ph/0105155].

EPJ Web of Conferences **112**, 01019 (2016)
DOI: 10.1051/epjconf/201611201019

Overview of HERMES results

Charlotte Van Hulse[1,a]

[1] *University of the Basque Country – UPV/EHU, Spain.*

Abstract. The HERMES experiment has collected a wealth of deep-inelastic scattering data using the 27.6 GeV polarized lepton beam at HERA and various pure gas targets, both unpolarized and polarized. This allowed for a series of diverse and unique measurements. Among them are measurements that provide information on the three-dimensional structure of the nucleon, both in momentum space and in position space. Results of measurements of exclusive ω production on an unpolarized and transversely polarized nucleon target, sensitive to the distribution in transverse-position and longitudinal-momentum space, are discussed as well as the three-dimensional extraction of azimuthal asymmetries measured in semi-inclusive deep-inelastic scattering, sensitive to twist-2 and twist-3 distributions in three-dimensional momentum space.

1 Introduction

The HERMES experiment was a fixed-target experiment located around the HERA accelerator at DESY in Hamburg (Germany). It took data from 1995 until 2007 by scattering longitudinally polarized electrons or positrons off a stationary gas target, internal to the beam pipe. The various types of targets used include hydrogen, deuterium, helium and heavier targets, with hydrogen longitudinally or transversely polarized or unpolarized, deuterium and helium longitudinally polarized or unpolarized, and the heavier targets unpolarized. The scattered lepton and particles produced in the collision were detected by a forward spectrometer. Here tracking was performed using multi-wire proportional chambers, while lepton-hadron particle identification was obtained using a preshower hodoscope, a calorimeter, and a transition-radiation detector, and hadron discrimination into pions, kaons and protons was achieved using a ring-imaging Cherenkov detector. The last two years of data collection got complemented by the installation of a recoil detector, located around the target, in order to detect low-energetic particles, outside of the acceptance of the forward spectrometer.

In the here presented proceedings, the focus is placed on the three-dimensional structures of the nucleon. In section 2 results sensitive to the distribution of quarks as a function of their longitudinal momentum fraction with respect to the momentum of the nucleon, where longitudinal refers to the direction of the probe used to investigate the nucleon, and their transverse position are presented. More in particular, measurements of exclusive ω production on transversely polarized and unpolarized hydrogen as well as on unpolarized deuterium are discussed.

In section 3, measurements of asymmetries that provide access to the three-dimensional structure of the nucleon in terms of the distribution of quarks as a function of their longitudinal momentum frac-

[a] e-mail: cvhulse@mail.desy.de

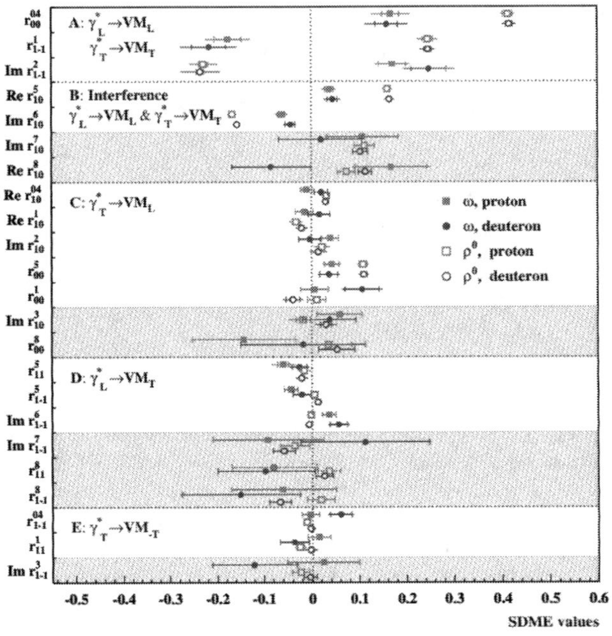

Figure 1. Extracted ω SDMEs (closed symbols) and ρ SDMEs (open symbols) for data collection on proton (red squares) and deuteron (blue circles) in the entire kinematic range. The unpolarized SDMEs are represented in the white bands, while the polarized SDMEs are represented in the grey bands. The different classes of SDMEs are labeled in the figure.

tion and transverse momentum are presented. The focus lies here on the three-dimensional extraction of the relevant asymmetry amplitudes, previously extracted one-dimensionally.

2 Exclusive ω production

The study of exclusive ω production is sensitive to generalized parton distributions (GPDs) and thus offers a means to study the three-dimensional structure of the nucleon. The Fourier transform in impact-parameter space of GPDs provides information on the distribution of quarks inside the nucleon as a function of their longitudinal momentum fraction and their transverse position. In addition, through the Ji relation, GPDs offer access to the total angular momentum of the quarks [1]. Here, the two relevant GPDs are the GPDs H and E, which both describe the distribution with parton-helicity conservation and with respectively nucleon-helicity conservation and nucleon-helicity flip. Exclusive ω production is sensitive to both these GPDs. However, for exclusive ω production on an unpolarized nucleon, the contribution of the GPD E is suppressed, and one is mainly sensitive to the GPD H. In contrast, exclusive ω production on a transversely polarized nucleon is sensitive to the interference of the GPDs H and E, thus opening access to the GPD E. In the following results on exclusive ω

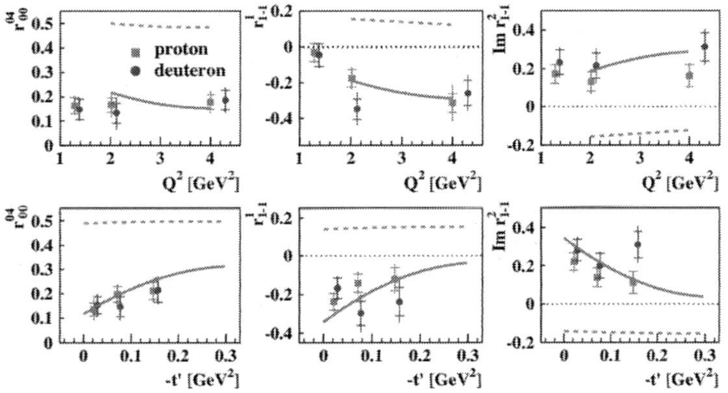

Figure 2. The class-A ω SDMEs as a function of Q^2 (upper row) and $-t'$ (lower row). The predictions of the GK model are also represented, with (red, full line) and without (green, dashed line) the pion-pole contribution.

production off unpolarized protons and deuterons as well as off transversely polarized protons are presented.

The analysis of exclusive ω production off unpolarized hydrogen and deuterium is performed through the study of the three-dimensional angular distribution of the ω-meson decay products. The ω meson decays with a branching ratio of 89.1% as $\omega \to \pi^+\pi^-\pi^0$, with $\pi^0 \to \gamma\gamma$. The angular distribution of the decay products is parametrized in terms of spin density matrix elements (SDMEs). The SDMEs themselves can be written in terms of helicity amplitudes that describe the transition of the helicity states of the initial nucleon and the virtual photon, mediating the lepton-nucleon interaction, to the helicity states of the final-state nucleon and the produced ω meson. The selection of exclusive ω production is performed using a missing-mass technique without the detection of the recoiling nucleon and the SDMEs are extracted through a maximum-likelihood fit to the angular distribution of the ω-meson decay products. For an unpolarized lepton beam fifteen "unpolarized" SDMEs are extracted and for a longitudinally polarized lepton beam eight "polarized" SDMEs are extracted.

The extracted SDMEs are presented in figure 1 for the entire kinematic range [2], in comparison with the SDMEs extracted from the exclusive production of ρ mesons [3], as measured by the HERMES collaboration as well. There are five classes of SDMEs, corresponding to the transition of a longitudinally (transversely) polarized photon to a longitudinally (transversely) polarized meson (class A), the interference of both transitions (class B), the transition of a transversely polarized photon to a longitudinally polarized meson (class C), the transition of a longitudinally polarized photon to a transversely polarized meson (class D), and finally the transition of a transversely polarized photon to a transversely polarized meson with helicity flip (class E). The conservation of s-channel helicity, i.e., the property that the photon and meson have the same helicity state, is observed for classes A and B, while for classes C and D a slight s-channel-helicity violation is observed at the 2-to-3 σ level for proton and deuteron. One can also see that the ω SDMEs r^1_{1-1} and $\Im r^2_{1-1}$ are respectively negative and positive and opposite in sign to the corresponding ρ SDMEs. This corresponds to a large unnatural-parity exchange for the ω meson and a large natural-parity exchange for the ρ meson. The term (un)natural-parity exchange refers to the quantum numbers of the particle exchanged between the virtual photon, which dissociates into a quark-antiquark pair, and the nucleon in Regge theory. In Regge phenomenology the interaction of the quark-antiquark pair with the nucleon proceeds through

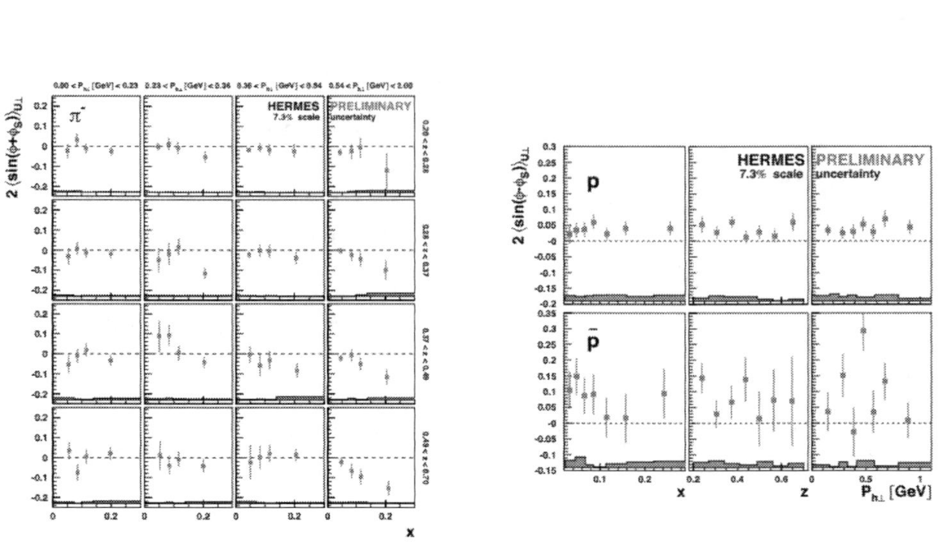

Figure 3. The Collins amplitude (left) for negative pions as a function of x for different bins in z (subdivision in rows) and $P_{h\perp}$ (subdivision in columns) and the Sivers amplitude (right) for protons and anti-protons as a function of x, z and $P_{h\perp}$.

the exchange of a pomeron or other reggeons, and depending on the quantum numbers of the particle, one talks about natural-parity exchange ($J^P = 0^+, 1^-,..$) or unnatural-parity exchange ($J^P = 0^-, 1^+$). The same large unnatural-parity exchange for ω mesons is also clearly visible from the analysis of linear combinations of specific SDMEs [3].

The extraction of the SDMEs was also performed in three kinematic bins in Q^2 and $-t'$, where $-Q^2$ represents the squared four-momentum of the virtual photon and $t' = t - t_{min}$, with t the Mandelstam variable and $-t_{min}$ the smallest kinematically allowed value of $-t$ at fixed virtual-photon energy and Q^2. As an example, the measurements for the class-A SDMEs are shown in figure 2. No pronounced kinematic dependence is observed, neither for class A nor for the other classes. Also shown in the figure are the results of the calculations from the "GK" model [4]. Two versions of the model are shown: one including the full contribution to the electromagnetic form factor from the pion (full, red line) and one taking only into account the relatively small perturbative contribution to this form factor (green, dashed line). As can be seen, the inclusion of the pion pole results in good agreement between the experimental data and the model.

From the data collected on transversely polarized protons, five sine and two cosine asymmetry amplitudes were extracted in the entire kinematic range as well as in two bins of Q^2 and $-t'$ [5]. Within the GK model, the experimental data favours the version of the model with pion-pole contribution with a positive $\pi\omega$ transition form factor above the versions without pion-pole contribution and with pion-pole contribution with a negative $\pi\omega$ transition form factor.

3 Asymmetries in semi-inclusive deep-inelastic scattering

Semi-inclusive deep-inelastic scattering (DIS) offers the possibility to study the three-dimensional distribution of the nucleon in terms of transverse-momentum-dependent (TMD) parton distribution functions (PDFs). These describe the distribution of partons as a function of their longitudinal and

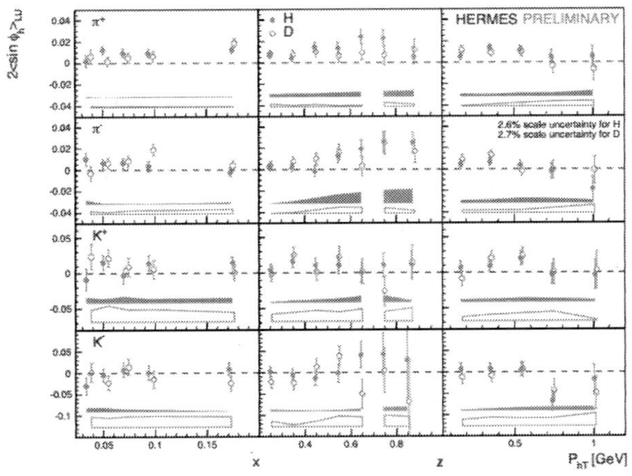

Figure 4. Lepton-beam asymmetries for pions (upper two rows) and kaons (lower two rows) as a function of x, z and $P_{h\perp}$ (labeled as P_{hT}). Note that for the projections in x and $P_{h\perp}$, the highest two z bins, indicated in grey, are not included.

transverse momenta. At leading-twist, eight TMD PDFs are needed to describe the full semi-inclusive DIS cross section, each of them relating to a specific spin-state configuration of the quark and the nucleon. For example, the Sivers TMD PDF describes the distribution of unpolarized quarks in a transversely polarized nucleon, the transversity distribution describes the distribution of transversely polarized quarks in a transversely polarized nucleon and the Boer-Mulders distribution describes the distribution of transversely polarized quarks in an unpolarized nucleon. The TMD PDFs enter the cross section in convolution with a fragmentation function, which describes the fragmentation of a quark into a final-state hadron. The two fragmentation functions of relevance are the spin-independent fragmentation function and the Collins fragmentation function, which describes the fragmentation of a transversely polarized quark into an unpolarized hadron. The TMD PDF, in combination with the fragmentation function, can be accessed through the extraction of asymmetries, given that each combination enters the cross section with a characteristic angular modulation. Besides the leading-twist contributions, also higher-twist contributions enter the cross section.

The one-dimensional extraction for pions and kaons of the Collins amplitude, sensitive to the transversity distribution and the Collins fragmentation function, and of the Sivers amplitude, sensitive to the Sivers PDF and the spin-independent fragmentation function, had been performed previously by the HERMES collaboration as a function of x, z and $P_{h\perp}$ using data collected on a transversely polarized hydrogen target [6–8]. Here, x represents the Bjorken-x variable, z the fractional energy of the produced hadron with respect to the virtual-photon energy and $P_{h\perp}$ the transverse momentum of the produced hadron. The analysis has been extended by performing a three-dimensional extraction of all amplitudes related to a transversely polarized target and including also results for protons and anti-protons, where for the latter due to limited statistics only a one-dimensional extraction was possible.

In figure 3, left, the Collins amplitude is shown for negative pions as a function of x for different bins in z (subdivision in rows) and $P_{h\perp}$ (subdivision in columns). The asymmetry amplitude is on average negative. In the one-dimensional extraction, a rise of the magnitude of the asymmetry amplitude with increasing x is observed. In the present three-dimensional extraction, this rise is seen to be concentrated at high values of $P_{h\perp}$. For positive pions, an analogous behaviour in the three-dimensional extraction is not clearly observed, although in the one-dimensional extraction a rise of the asymmetry amplitude, which on average is positive in sign, with increasing x is also seen.

In figure 3, right, the one-dimensional extraction of the Sivers amplitude is shown for protons and anti-protons as a function of x, z and $P_{h\perp}$. The data for the anti-protons is clearly limited by statistics, while the protons show a clear positive asymmetry amplitude, without a clear dependence on any of the kinematic variables. For protons, also the three-dimensional extraction was performed. From this extraction, no clear dependence on any of the kinematic variables is observed neither.

As stated earlier, besides the leading-twist contributions, also higher-twist contributions enter the cross section. Among those extracted, the twist-3 contribution to the cross section for longitudinally polarized beam and unpolarized nucleon is discussed here. This twist-3 contribution is sensitive to a combination of twist-3 PDFs, e and $g^{\perp}$, and twist-2 fragmentation functions, the Collins and spin-independent fragmentation functions, and a combination of twist-2 PDFs, the spin-independent PDF and the Boer-Mulders PDF, and twist-3 fragmentation functions, $\tilde{G}^{\perp}$ and $\tilde{E}$ [9]. The extraction of the twist-3 contribution has been performed one- and three-dimensionally in x, z and $P_{h\perp}$ for pions, kaons and protons for data collected on a hydrogen and deuterium target. In figure 4, the asymmetries extracted in one dimension as a function of x, z and $P_{h\perp}$ are shown for pions and kaons. Note that the projections in x and $P_{h\perp}$ only contain the z region of $z < 0.7$. The asymmetries for pions are positive. They increase with increasing z up to z of 0.7 and decrease with increasing $P_{h\perp}$. Also the positive kaons show a positive asymmetry amplitude, but no specific kinematic dependence is observed. The asymmetries for negative kaons are consistent with zero. The same observation is true for protons and anti-protons.

References

[1] X. Ji, Phys. Rev. Lett. **78**, 610 (1997).

[2] A. Airapetian et al., Eur. Phys. J. C **74**, 3110 (2014).

[3] A. Airapetian et al., Eur. Phys. J. C **62**, 659 (2009).

[4] S. Goloskokov and P. Kroll, Eur. Phys. J **A** 50, 146 (2014).

[5] A. Airapetian et al., Eur. Phys. J. C **75**, 600 (2015).

[6] A. Airapetian et al., Phys. Rev. Lett. **94**, 012002 (2005).

[7] A. Airapetian et al., Phys. Lett. B **693**, 11-16 (2010).

[8] A. Airapetian et al., Phys. Rev. Lett. **103**, 152002 (2009).

[9] A. Bacchetta et al., JHEP **0702**, 093 (2007).

EPJ Web of Conferences **112**, 01020 (2016)
DOI: 10.1051/epjconf/201611201020

GPDs in heavy meson production and Compton scattering

D.Yu. Ivanov[1], B. Pire[2], L. Szymanowski[3], and J. Wagner[3]

[1] *Sobolev Institute of Mathematics and Novosibirsk State University, 630090 Novosibirsk, Russia*
[2] *Centre de physique théorique, École Polytechnique, CNRS, Université Paris-Saclay, 91128 Palaiseau, France*
[3] *National Centre for Nuclear Research (NCBJ), Warsaw, Poland*

Abstract. Exclusive processes of heavy meson production and spacelike and timelike deeply virtual Compton scattering allow us to investigate the hadron structure in terms of Generalized Parton Distributions (GPDs). We review recent developments in the NLO description of such processes.

1 DVCS & TCS

The studies of deeply virtual Compton scattering (DVCS) $\gamma^*(Q^2)p \to \gamma p$ in electron-proton collisions at JLAB, HERA are the primary source of our knowledge on GPDs [1]. The related timelike Compton scattering (TCS) process [2] shares all its virtues for accessing correlated information on the light cone momentum fraction and the transverse location of partons in hadrons [3]. Both reactions, illustrated in Fig.1, can be seen as limiting cases of the double virtual Compton scattering process,

$$\gamma^*(q_{in})N(p) \to \gamma^*(q_{out})N'(p') \,. \tag{1}$$

The relevant light-cone ratios describing the processes of interest in the generalized Bjorken limit are the scaling variable ξ and skewness $\eta > 0$:

$$\xi = -\frac{q_{out}^2 + q_{in}^2}{q_{out}^2 - q_{in}^2}\eta \,, \quad \eta = \frac{q_{out}^2 - q_{in}^2}{(p+p')\cdot(q_{in}+q_{out})} \,. \tag{2}$$

The scattering amplitude is written in a factorized form as :

$$\mathcal{A}^{\mu\nu}(\xi,\eta,t) = -e^2 \frac{1}{(P+P')^+}\bar{u}(P')\Big[g_T^{\mu\nu}\left(\mathcal{H}(\xi,\eta,t)\gamma^+ + \mathcal{E}(\xi,\eta,t)\frac{i\sigma^{+\rho}\Delta_\rho}{2M}\right) + i\epsilon_T^{\mu\nu}\left(\widetilde{\mathcal{H}}(\xi,\eta,t)\gamma^+\gamma_5 + \widetilde{\mathcal{E}}(\xi,\eta,t)\frac{\Delta^+\gamma_5}{2M}\right)\Big]u(P) \,, \tag{3}$$

where the Compton form factors are defined as :

$$\mathcal{H}(\xi,\eta,t) = +\int_{-1}^{1} dx \left(\sum_q T^q(x,\xi,\eta)H^q(x,\eta,t) + T^g(x,\xi,\eta)H^g(x,\eta,t)\right) \,, \tag{4}$$

$$\widetilde{\mathcal{H}}(\xi,\eta,t) = -\int_{-1}^{1} dx \left(\sum_q \widetilde{T}^q(x,\xi,\eta)\widetilde{H}^q(x,\eta,t) + \widetilde{T}^g(x,\xi,\eta)\widetilde{H}^g(x,\eta,t)\right) \,,$$

and similarly for $\mathcal{E}(\xi, \eta, t)$ and $\widetilde{\mathcal{E}}(\xi, \eta, t)$.

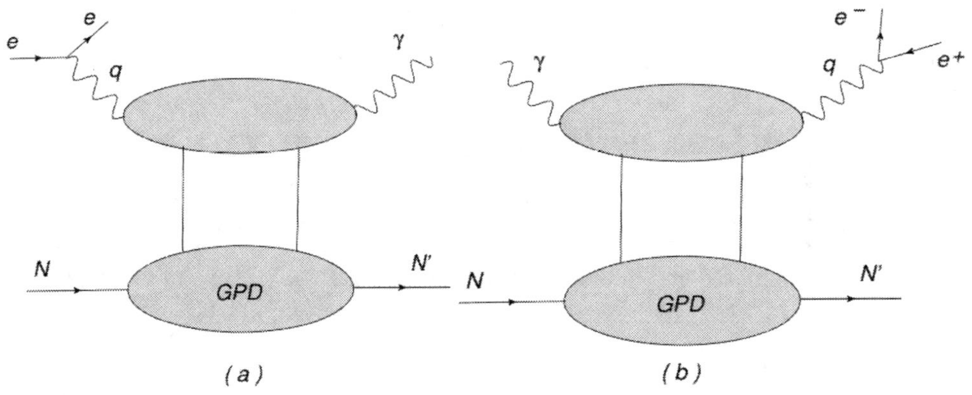

Figure 1. (left)Deeply Virtual Compton Scattering (DVCS) : $lN \to l'N'\gamma$ (right) Timelike Compton Scattering (TCS): $\gamma N \to l^+l^-N'$.

The study of $O(\alpha_s)$ corrections to the DVCS and TCS amplitudes turns out to be full of surprises. Without entering a detailed analysis [4] the renormalized coefficient functions for DVCS are given by

$$\begin{aligned}
T^q(x) &= \left[C_0^q(x) + C_1^q(x) + \ln\left(\frac{Q^2}{\mu_F^2}\right) \cdot C_{coll}^q(x)\right] - (x \to -x), \\
T^g(x) &= \left[C_1^g(x) + \ln\left(\frac{Q^2}{\mu_F^2}\right) \cdot C_{coll}^g(x)\right] + (x \to -x), \\
\widetilde{T}^q(x) &= \left[\widetilde{C}_0^q(x) + \widetilde{C}_1^q(x) + \ln\left(\frac{Q^2}{\mu_F^2}\right) \cdot \widetilde{C}_{coll}^q(x)\right] + (x \to -x), \\
\widetilde{T}^g(x) &= \left[\widetilde{C}_1^g(x) + \ln\left(\frac{Q^2}{\mu_F^2}\right) \cdot \widetilde{C}_{coll}^g(x)\right] - (x \to -x).
\end{aligned}$$

Thanks to the analytic structure (in Q^2) of the amplitude, the results for DVCS and TCS cases are simply related [5]:

$$^{TCS}T(x, \eta) = \pm \left(^{DVCS}T(x, \xi = \eta) + i\pi \cdot C_{coll}(x, \xi = \eta) \right)^* ,$$

where + (−) sign corresponds to vector (axial) case. This difference has very important phenomenological consequences [6]. $O(\alpha_s)$ corrections are not small and one may question the relevance of phenomenological studies based on $O(\alpha_s)$ coefficient functions. Moreover, the factorization scale dependence turns out to be rather large, as can be seen on Fig. 2 and Fig. 3. Resumming higher order corrections may help to stabilize this unwanted feature. In any case, this is a needed improvement of the theoretical description of DVCS and TCS (see for instance [7]), in particular for the high energy domain which is already accessible thanks to the ultra peripheral reactions in hadron colliders [8]. This will be a central issue for the phenomenology of the electron-ion collider [9].

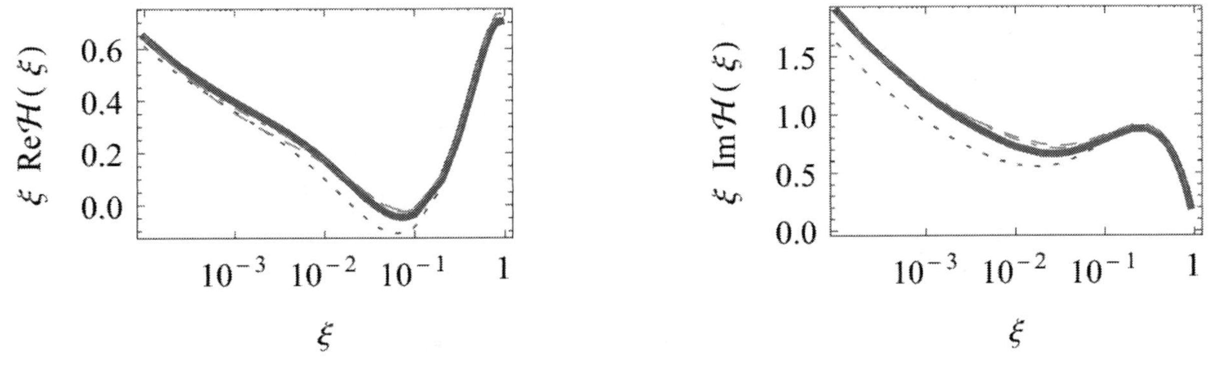

Figure 2. Full NLO result for DVCS Compton Form Factors, as a function of ξ. Left column - $\xi \cdot Re(\mathcal{H}(\xi))$, right column - $\xi \cdot Im(\mathcal{H}(\xi))$, $Q^2 = 4GeV^2$, $\mu_F^2 = Q^2, Q^2/2, Q^2/3, Q^2/4$ (dotted, solid, dash-dotted, dashed lines respectively).

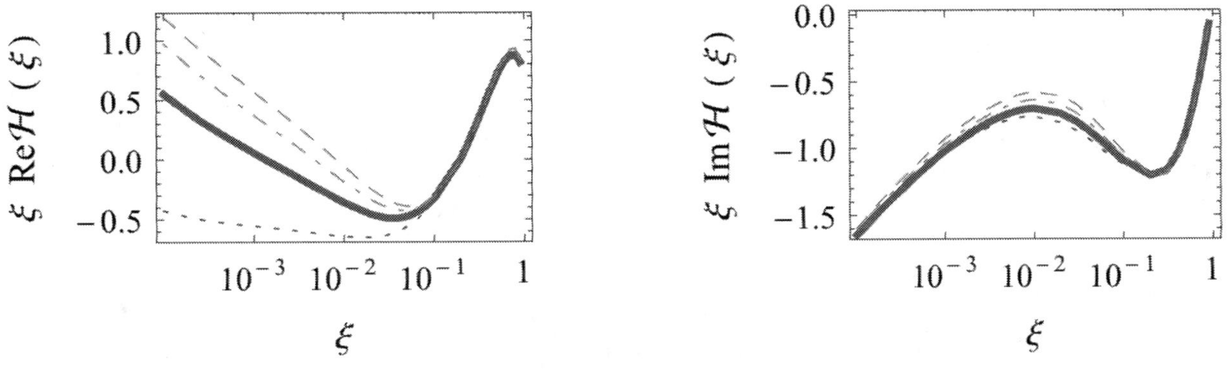

Figure 3. Full NLO result for TCS Compton Form Factors, as a function of ξ. Left column - $\xi \cdot Re(\mathcal{H}(\xi))$, right column - $\xi \cdot Im(\mathcal{H}(\xi))$, $Q^2 = 4\ GeV^2$, $\mu_F^2 = Q^2, Q^2/2, Q^2/3, Q^2/4$ (dotted, solid, dash-dotted, dashed lines respectively).

2 Heavy vector meson production

The photoproduction of a heavy vector meson:

$$\gamma p \to V p \tag{5}$$

is a subject of intense experimental [10] and theoretical [11] studies. One motivation of such studies is the possibility to explore gluon GPDs in the nucleon. We present here preliminary results [12] on the use of the collinear factorization approach at the next to leading order in α_s, which was developed in [13] , in the context of ultraperipheral collisions.

The amplitude $\mathcal{M}$ of the process (5) is given by the factorization formula:

$$\mathcal{M} \sim \left(\frac{\langle O_1\rangle_V}{m^3}\right)^{1/2} \int_{-1}^{1} dx \left[T_g(x,\xi)\, F^g(x,\xi,t) + T_q(x,\xi) \sum_{q=u,d,s} F^q(x,\xi,t) \right], \quad (6)$$

with $F^{g(q)}(x,\xi,t;\mu_F^2)$ the gluon (quark) GPDs; m is a pole mass of the heavy quark, and $\xi = M^2/(2W^2 - M^2)$ is the skewness parameter.

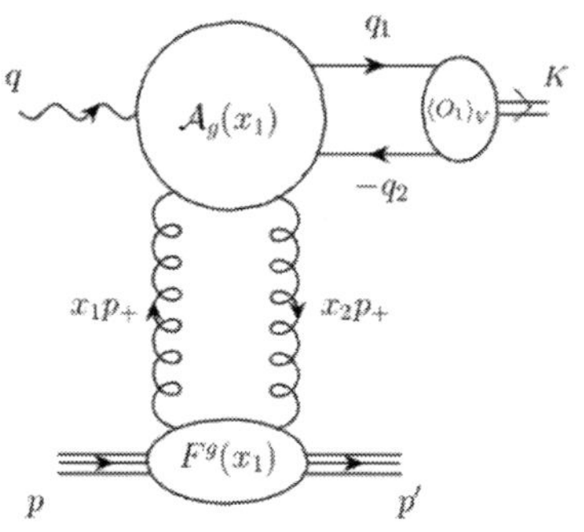

Figure 4. Kinematics of the heavy vector meson photoproduction.

All information about the quarkonium structure is encoded in the NRQCD [14] matrix element $\langle O_1\rangle_V$ which enters the leptonic decay rate

$$\Gamma[V \to l^+ l^-] = \frac{2e_q^2 \pi\alpha^2}{3} \frac{\langle O_1\rangle_V}{m^2} \left(1 - \frac{8\alpha_S}{3\pi}\right)^2 .$$

The coefficient functions read

$$T_g(x,\xi) = \frac{\xi}{(x-\xi+i\varepsilon)(x+\xi-i\varepsilon)} \mathcal{A}_g\left(\frac{x-\xi+i\varepsilon}{2\xi}\right),$$

$$T_q(x,\xi) = \mathcal{A}_q\left(\frac{x-\xi+i\varepsilon}{2\xi}\right).$$

At leading order $\mathcal{A}_g^{(0)}(y) = \alpha_S$, $\mathcal{A}_q^{(0)}(y) = 0$. The inclusion of NLO corrections has dramatic effects on the production cross section : NLO corrections are very big and the overall result depends very strongly on the choice of the factorization scale, especially for the high values of W.[1]

Why are NLO corrections so large in this case, where $\xi \ll 1$? The inspection of NLO hard-scattering amplitudes shows that the imaginary part of the amplitude dominates and that the leading contribution to the NLO correction originates from the broad integration region $\xi \ll x \ll 1$, where the gluonic part approximates ($\bar{\alpha}_s = 3\alpha_s/\pi$):

$$Im\mathcal{M}^g \sim H^g(\xi,\xi) + \bar{\alpha}_s \left[\log\frac{M_V^2}{\mu_F^2} - \log 4\right] \int_{\xi}^{1} \frac{dx}{x} H^g(x,\xi) . \quad (7)$$

[1] In our calculations, in both LO and NLO cases, we keep the value of renormalization scale fixed $\mu_R = M_{J/\psi}$.

Given the behavior of the gluon GPD at small x, $H^g(x,\xi) \sim xg(x) \sim$ const, we see that the NLO correction is parametrically large, $\sim \ln(1/\xi)$, and negative unless one chooses the value of the factorization scale sufficiently lower than the hard kinematic scale, $Q = M_V$.

The size of the corrections, and the sensitivity of the NLO result to the factorization scale choice, shows that some additional information about still higher order contributions is needed to provide reliable theoretical predictions. This may come from some strategy for the scale choice to minimalize the one-loop corrections . In our opinion the most promising approach is related with the resummation of the higher orders terms enhanced at small ξ by powers of large logarithms of energy, $\sim \bar{\alpha}_s^n \ln^n(1/\xi)$, see [15]:

$$\mathcal{I}m\mathcal{M}^g \sim H^g(\xi,\xi) + \int\limits_{\xi}^{1} \frac{dx}{x} H^g(x,\xi) \sum_{n=1} C_n(L) \frac{\bar{\alpha}_s^n}{(n-1)!} \log^{n-1}\frac{x}{\xi}, \tag{8}$$

where $C_n(L)$ are polynomials of $L = \ln(M_V^2/\mu_F^2)$ which maximum power is L^n. For DIS inclusive structure functions F_T and F_L corresponding $C_n(L)$ coefficients were calculated long time ago by Catani and Hautman [16]. Their method developed for inclusive DIS can be straightforwardly generalized to exclusive, nonforward processes.

Resummed coefficient functions, parameterized by $C_n(L)$ polynomials, can be calculated and conveniently represented using Mellin transformation. In the Mellin space the resummed coefficient function is a polynomial in the variable $z = \bar{\alpha}_s/N$. Conversely, the contributions proportional to the nth power of this variable generate terms $\sim \bar{\alpha}_s^n \ln^n(1/\xi)$ in the process amplitude. For meson photoproduction our result in the $\overline{MS}$ scheme reads

$$1 + z(L - \ln 4) + \frac{z^2}{6}\left(\pi^2 + 3\ln^2 4 + 3L(L - \ln 16)\right) + \ldots ,$$

where only two nontrivial terms of the expansion are shown. This leads to $C_1(L) = L - \ln 4$, in accordance with the found high energy asymptotic of NLO result in Eq. (7), $C_2(L) = \left(\pi^2 + 3\ln^2 4 + 3L(L - \ln 16)\right)/6$, and so on. On the Fig.5 we present the effect of such resummation of the imaginary part of amplitude (normalized in such way that the LO result equals $H^g(\xi,\xi,t,\mu_F)$), as a function of W for $\mu_F = M_V$.

Let us discuss the value of resummed coefficient and its dependence on the scale of factorization μ_F. Below we present results for two cases. a) μ_F is equal to kinematic hard scale $\mu_F = M_V$ ($L = 0$), and b) μ_F is chosen to vanish the value of the first high energy term C_1, it requires $\mu_F = M_V/2 = m$ ($L = \ln 4$):

$$\begin{aligned} a) \quad (\mu_F = M_V): &\quad 1 - 1.39\,z + 2.61\,z^2 + 0.481\,z^3 - 4.96\,z^4 + \ldots \\ b) \quad (\mu_F = M_V/2): &\quad 1 + 0.\,z + 1.64\,z^2 + 3.21\,z^3 + 1.08\,z^4 + \ldots . \end{aligned}$$

We see that almost all the high energy term coefficients, $C_n(L)$, have large absolute values. It shows that it is important to take into account not only large NLO effects but also contributions from still higher orders of QCD collinear expansion that are enhanced by the powers of large logarithms of energy. Another important observation is that it is not possible by appropriate choice of factorization scale μ_F to move all enhanced by powers of $\ln(1/\xi)$ contributions from the coefficient function into the GPD (through its μ_F- evolution). Such a strategy is promoted in [17]. We see that our results above, the case b), do not support this suggestion. The choice $\mu_F = M_V/2$ indeed eliminates the big part of NLO correction from the hard coefficient, but it can not allow to get rid of such big terms from the higher orders contributions. On the other hand, as illustrated on the Fig.6, the resummation up to the 6th order of the gluon GPD dependent coefficient function, greatly reduces the factorization scale dependence [2].

We believe that high energy resummation described above have to be incorporated in the analysis of the J/Ψ and Υ meson photoproduction processes, this work is now in progress.

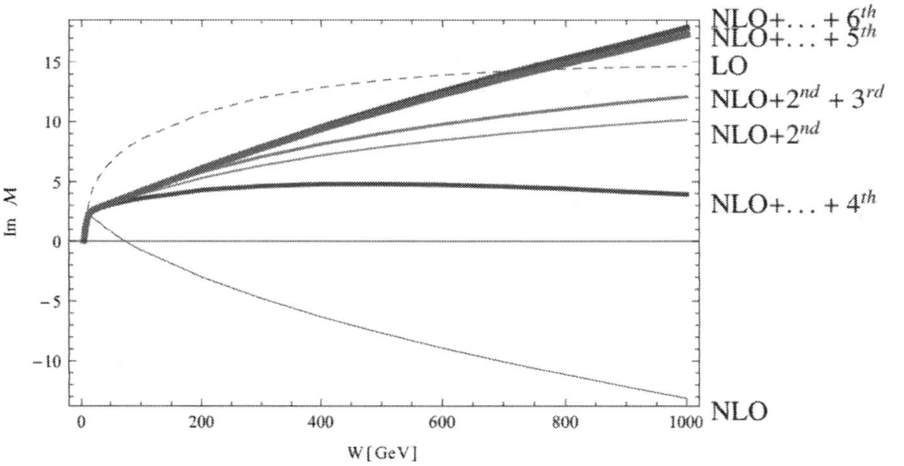

Figure 5. Resummation of the (gluonic GPDs dependent) imaginary part of amplitude (normalized in such way that the LO result equals $H^g(\xi, \xi, t, \mu_F)$) up to the sixth higher order term, as a function of W, for $\mu_F = M_V$.

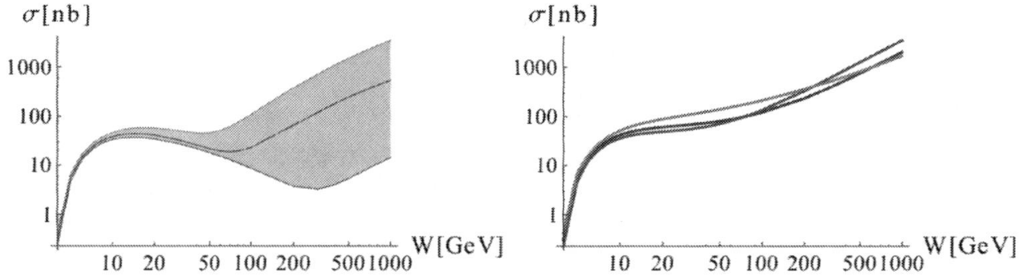

Figure 6. NLO(left panel) and resummed (right panel) photoproduction cross section (only gluonic GPDs included in both cases) as a function of $W = \sqrt{s_{\gamma p}}$ for $\mu_F^2 = M^2_{J/\psi} \times \{0.5, 1, 2\}$ (pink, blue and yellow lines respectively).

3 Summary

GDPs enter factorized amplitudes for hard exclusive reactions in a similar manner as PDFs enter factorized cross sections for inclusive DIS. Ultraperipheral collisions at hadron colliders open a new way to measure GPDs in TCS and photoproduction of heavy vector mesons at very small skewness parameters. NLO corrections turn out to be rather large for TCS and even more for vector meson

[2]In both cases (NLO and resummed) only the forward evolution of the PDFs, which enter the double distribution model for GPDs is performed.

photoproduction in the kinematics typical for experiments at the EIC collider. Various resummation techniques have been invoked to help to stabilize the perturbative expansion. In the heavy meson production case, high energy resummation techniques are suggested as a tool to provide reliable theoretical predictions in this kinematical domain.

Acknowledgments

This work is partly supported by the COPIN-IN2P3 Agreement, French grant ANR PARTONS (Grant No. ANR- 12-MONU-0008-01) and by the grant RFBR-15-02-05868, L.Sz. was partially supported by grant of National Science Center, Poland, No. 2015/17/B/ST2/01838.

References

[1] M. Diehl, Phys. Rept. **388** (2003) 41; A. V. Belitsky and A. V. Radyushkin, Phys. Rept. **418**, 1 (2005); S. Boffi and B. Pasquini, Riv. Nuovo Cim. **30**, 387 (2007).

[2] E. R. Berger, M. Diehl and B. Pire, Eur. Phys. J. C **23**, 675 (2002); M. Boer, M. Guidal and M. Vanderhaeghen, Eur. Phys. J. A **51**, no. 8, 103 (2015).

[3] M. Burkardt, Phys. Rev. D **62**, 071503 (2000) and Int. J. Mod. Phys. A **18**, 173 (2003); J. P. Ralston and B. Pire, Phys. Rev. D **66**, 111501 (2002); M. Diehl, Eur. Phys. J. C **25**, 223 (2002).

[4] B. Pire, L. Szymanowski and J. Wagner, Phys. Rev. D **83**, 034009 (2011).

[5] D. Mueller *et al.*, Phys. Rev. D **86**, 031502 (2012).

[6] H. Moutarde *et al.*, Phys. Rev. D **87**, no. 5, 054029 (2013); J. Wagner *et al.*, EPJ Web Conf. **66**, 06016 (2014).

[7] T. Altinoluk, B. Pire, L. Szymanowski and S. Wallon, JHEP **1210** (2012) 049.

[8] B. Pire, L. Szymanowski and J. Wagner, Phys. Rev. D **79** (2009) 014010.

[9] D. Boer *et al.*, arXiv:1108.1713 [nucl-th]; J. L. Abelleira Fernandez *et al.* [LHeC Study Group Collaboration], J. Phys. G **39**, 075001 (2012).

[10] C. Alexa *et al.* [H1 Collaboration], HERA," Eur. Phys. J. C **73** (2013) 2466, and references therein.

[11] S. P. Jones, A. D. Martin, M. G. Ryskin and T. Teubner, JHEP **1311** (2013) 085, and references therein.

[12] D.Yu. Ivanov, L. Szymanowski and J. Wagner, in preparation.

[13] D.Yu. Ivanov, A. Schafer, L. Szymanowski and G. Krasnikov, Eur. Phys. J. C **34** (2004) 3, 297 [Eur. Phys. J. C **75** (2015) 2, 75].

[14] G.T. Bodwin, E. Braaten and G.P. Lepage, Phys. Rev. D **51**, 1125 (1995) [Erratum-ibid. D **55**, 5853 (1997)].

[15] D.Yu. Ivanov, arXiv:0712.3193 [hep-ph].

[16] S. Catani and F. Hautmann, Nucl. Phys. B **427** (1994) 475.

[17] S. P. Jones, A. D. Martin, M. G. Ryskin and T. Teubner, arXiv:1507.06942 [hep-ph].

EPJ Web of Conferences **112**, 01021 (2016)
DOI: 10.1051/epjconf/201611201021

Non-dipolar gauge links for transverse-momentum-dependent pion wave functions

Yu-Ming Wang[1,a]

[1] *Fakultät für Physik, Universität Wien, Boltzmanngasse 5, 1090 Vienna, Austria*

Abstract. I discuss the factorization-compatible definitions of transverse-momentum-dependent (TMD) pion wave functions which are fundamental theory inputs entering QCD factorization formulae for many hard exclusive processes. I will first demonstrate that the soft subtraction factor introduced to remove both rapidity and pinch singularities can be greatly reduced by making the maximal use of the freedom to construct the Wilson-line paths when defining the TMD wave functions. I will then turn to show that the newly proposed TMD definition with non-dipolar Wilson lines is equivalent to the one with dipolar gauge links and with a complicated soft function, to all orders of the perturbative expansion in the strong coupling, as far as the infrared behavior is concerned.

1 Introduction

Hard exclusive processes have significantly sharpened our understanding of the strong interaction dynamics involved in the hadronic amplitudes, due to the establishment of various QCD factorization theorems which allow for separation of the hard scattering contributions and the (universal) long-distance QCD effects. Albeit the fact that factorization in a non-abelian gauge theory like QCD is by no means a trivial issue, collinear factorization for a number of interesting hard reactions, at leading power in $1/Q$ where Q is the typical hard scale for a given process, can be established with rigorous power counting schemes and standard assumptions. By contrast, transverse-momentum-dependent (TMD) factorization for hard exclusive processes based upon the theory of on-shell Sudakov form factor [1], with the original aim of avoiding the rapidity divergences in collinear factorization for the pion electromagnetic form factor at sub-leading power [2], turns out to be far more sophisticated conceptually due to the absence of a definite power counting scheme and to the emergence of light-cone singularities in the infrared subtraction, and this can be already partially recognized from the troubled history of TMD wave functions [3–14].

The primary difficulty of defining TMD wave functions lies in eliminating the rapidity divergence involved in the infrared subtraction, due to the TMD hard scattering kernels, without introducing additional pinch singularities. It is the subjective of this presentation to overview the current status of TMD factorization for hard exclusive processes, focusing on the comparison between collinear and TMD pion wave functions and the regularization of unwanted rapidity and pinch divergences. I will consider the simplest hard exclusive process $\gamma^* \to \pi\gamma$ (neglecting the involved sub-leading contributions [15] for the moment) through the discussion in order not to bother about the complication due

[a]e-mail: yu-ming.wang@univie.ac.at

to the Glauber/Coulomb gluon exchange [16] which are relevant to the construction of factorization theorems for more complicated processes, e.g., the pion electromagnetic form factor. It is certainly interesting to generalize the discussion presented here to understand the factorization properties of those processes taking into account the contribution from the Glauber gluons, which is also crucial to construct the Wilson line structures of TMD wave functions. In what follows, I will first explain the origin of light-cone singularities in the infrared subtraction determined by the convolution of TMD wave functions and the hard functions, and present new definitions of the TMD pion wave functions with dipolar and non-dipolar Wilson lines in section 2. Demonstration of the equivalence of different TMD definitions is then discussed with the aid of rapidity evolution equations in section 3.

2 Factorization-compatible TMD wave functions

To facilitate the comparison between the collinear and TMD pion wave functions, I will begin with the collinear factorization formula for the pion-photon transition form factor [3]

$$F_\pi(Q^2) = \frac{\sqrt{2} f_\pi}{3} \int_0^1 dx\, H(x, Q^2, \mu)\, \phi_\pi(x, \mu), \tag{1}$$

where the pion light-cone distribution amplitude (LCDA) $\phi_\pi(x, \mu)$ is given by

$$\langle 0|\bar{q}(0)\, W_{n_-}(0, t\, n_-)\, \not{n}_-\, \gamma_5\, q(t\, n_-)|\pi^+(p)\rangle = i\, f_\pi\, p_+ \int_0^1 dx\, e^{-ixt\, p_+}\, \phi_\pi(x, \mu), \tag{2}$$

with the collinear gauge link

$$W_{n_-}(0, t\, n_-) = P \exp\left[i\, g_s \int_0^t d\lambda\, T^a\, n_- \cdot A^a\, (\lambda\, n_-)\right]. \tag{3}$$

The hard function $H(x, Q^2, \mu)$ determined from the perturbative QCD matching is known to the next-to-next-to-leading order in the strong coupling [17]. Taking into account the Sudakov form factor to suppress the soft gluon contribution [2], the TMD factorization formula for the $\gamma^* \to \pi\gamma$ form factor can be written as

$$F_\pi(Q^2) = \frac{\sqrt{2} f_\pi}{3} \int_0^1 dx \int d^2\vec{b}\, H(x, \vec{b}, Q^2, \mu)\, \phi_\pi(x, \vec{b}, \mu)\, e^{-S(x, \vec{b}, Q)}, \tag{4}$$

where $S(x, \vec{b}, Q)$ is the perturbative Sudakov form factor and the non-perturbative Sudakov evolution is left out due to the strong model dependence. Naively, one may expect the following definition of the TMD pion wave function in momentum space

$$\begin{aligned} \phi_\pi^{\rm naive}(x, \vec{k}_T, \mu) \;\; &\stackrel{?}{=} \;\; \int \frac{dz_-}{2\pi} \int \frac{d^2 z_T}{(2\pi)^2}\, e^{i(x p_+ z_- - \vec{k}_T \cdot \vec{z}_T)} \\ &\quad \times \langle 0|\bar{q}(0) W_{n_-}^\dagger(+\infty, 0)\, \not{n}_-\, \gamma_5\, [\text{tr. link}]\, W_{n_-}(+\infty, z)\, q(z)|\pi^+(p)\rangle, \end{aligned} \tag{5}$$

with the coordinate $z = (0, z_-, \vec{z}_T)$, which can be constructed from a straightforward extension of (2). Unfortunately, the above-mentioned definition yields the light-cone divergence in the infrared subtraction, which can be readily understood from the one-loop correction to the quark-Wilson-line diagram

$$\phi_\pi^{(1)} \otimes H^{(0)} \propto \int [dl] \frac{1}{[(k+l)^2 + i0)][l_+ + i0][l^2 + i0]} \left[H^{(0)}(x + l_+/p_+, \vec{k}_T + \vec{l}_T) - H^{(0)}(x, \vec{k}_T)\right]. \tag{6}$$

In contrast to the collinear factorization, this loop integral develops an endpoint singularity from the $1/l_+$ singularity in the Eikonal propagator, due to the non-cancellation of two hard functions at $l_+ = 0$. Two different definitions were proposed [4] to regularize such rapidity divergence either by rotating the gauge links off the light cone or by introducing an additional subtraction factor

$$\begin{aligned}\phi_\pi^{C1}(x,\vec{k}_T,y_u,\mu) &\stackrel{?}{=} \int\frac{dz_-}{2\pi}\int\frac{d^2z_T}{(2\pi)^2}e^{i(xp_+z_- - \vec{k}_T\cdot\vec{z}_T)} \\ &\quad\times\langle 0|\bar{q}(0)W_u^\dagger(+\infty,0)\,\not{n}_-\gamma_5\,[\text{tr. link}]\,W_u(+\infty,z)\,q(z)|\pi^+(p)\rangle\,, \qquad (7)\\ \phi_\pi^{C2}(x,\vec{k}_T,y_u,\mu) &\stackrel{?}{=} \int\frac{dz_-}{2\pi}\int\frac{d^2z_T}{(2\pi)^2}e^{i(xp_+z_- - \vec{k}_T\cdot\vec{z}_T)} \\ &\quad\times\frac{\langle 0|\bar{q}(0)W_{n_-}^\dagger(+\infty,0)\,\not{n}_-\gamma_5\,[\text{tr. link}]\,W_{n_-}(+\infty,z)\,q(z)|\pi^+(p)\rangle}{\langle 0|W_{n_-}^\dagger(+\infty,0)W_u(+\infty,0)\,[\text{tr. link}]\,W_{n_-}(+\infty,z)\,W_u^\dagger(+\infty,z)|0\rangle}\,, \qquad (8)\end{aligned}$$

where the gauge vector $u = (u_+, u_-, \vec{0}_T)$ is away from the light cone.

Some years later, Bacchetta et al [18] found that the TMD definitions like (7) and (8) will induce the pinch singularity for a TMD parton density with space-like Wilson lines and hence they cannot be used to construct TMD factorization formulae for the hard inclusive reactions in practice. Even worse, the pinch singularity appears in TMD wave functions for any off-light-cone gauge vector, because the radiative gluon is not necessary to be on the mass shell [19]. To develop a better understanding of the resulting pinch singularity, one can compute the Wilson line self-energy graph explicitly

$$\phi_\pi \propto \int[dl]\frac{u^2}{[l+i0)][u\cdot l+i0][u\cdot l-i0]}\delta(x'-x+l_+/p_+)\,\delta^{(2)}(\vec{k}'_T-\vec{k}_T+\vec{l}_T)\,, \qquad (9)$$

which yields the linear divergence in the length of the Wilson lines in coordinate space. One then concludes that *off-light-cone dipolar Wilson lines regularize the rapidity divergence at the price of introducing an unwanted pinch singularity.*

The problem of defining factorization-compatible TMD definitions was resolved only recently by Collins [16] in the context of parton densities. Generalizing his proposal to the TMD wave functions one obtains

$$\begin{aligned}\phi_\pi^{C}(x,\vec{k}_T,y_2,\mu) &= \lim_{\substack{y_1\to+\infty\\ y_u\to-\infty}}\int\frac{dz_-}{2\pi}\int\frac{d^2z_T}{(2\pi)^2}e^{i(xp_+z_- - \vec{k}_T\cdot\vec{z}_T)} \\ &\quad\times\langle 0|\bar{q}(0)W_u^\dagger(+\infty,0)\,\not{n}_-\gamma_5\,[\text{tr. link}]\,W_u(+\infty,z)\,q(z)|\pi^+(p)\rangle \\ &\quad\times\sqrt{\frac{S(z_T;y_1,y_2)}{S(z_T;y_1,y_u)\,S(z_T;y_2,y_u)}}\,, \qquad (10)\end{aligned}$$

where the soft function is defined as

$$S(z_T;y_A,y_B) = \frac{1}{N_c}\langle 0|W_{n_B}^\dagger(\infty,\vec{z}_T)_{ca}\,W_{n_A}(\infty,\vec{z}_T)_{ad}\,W_{n_B}(\infty,0)_{bc}\,W_{n_A}^\dagger(\infty,0)_{db}|0\rangle, \qquad (11)$$

with n_A and n_B being the rapidities of the gauge-fixing vectors n_A and n_B. This definition is unique, up to the choice of the ultraviolet renormalization prescription, provided that the following criteria are satisfied [20].

- The unsubtracted wave function only involves light-cone Wilson lines.
- Each soft factor has at most one off-light-cone Wilson line.

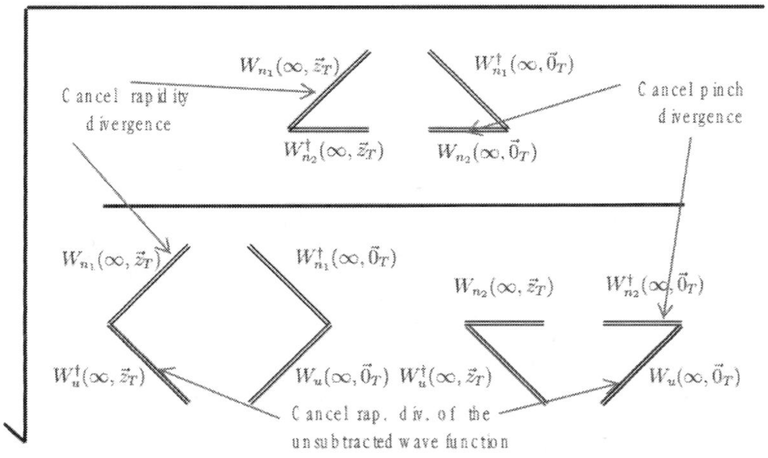

Figure 1. Diagrammatical representation of the soft subtraction in (10).

- The subtracted wave function is free of rapidity and pinch singularities and is defined as the product of the unsubtracted wave function and powers of the soft function in transverse coordinate space.

To demonstrate the cancellation mechanism in a transparent way, a detailed account of the rapidity and pinch structures of different pieces in the soft subtraction is display in figure 1.

At this point, one might be curious to ask whether there exists an alternative definition of the TMD pion wave function with a simpler soft subtraction function. The key to answering this question consists in identifying the very origin of pinch singularities, which is due to *the dipolar structure of the gauge links* defining the TMD wave functions or parton densities. It is then a straightforward task to propose the following definition [19]

$$\begin{aligned} \phi_\pi^{\text{new}}(x, \vec{k}_T, y_2, \mu) &= \int \frac{dz_-}{2\pi} \int \frac{d^2 z_T}{(2\pi)^2} e^{i(xp_+ z_- - \vec{k}_T \cdot \vec{z}_T)} \\ &\times \frac{\langle 0|\bar{q}(0) W_{n_2}^\dagger(+\infty, 0)\ \not{n}_-\ \gamma_5\ [\text{links}@\infty]\ W_v(+\infty, z)\, q(z)|\pi^+(p)\rangle}{[\text{color factor}]\ \langle 0|W_{n_2}^\dagger(+\infty, 0)\ [\text{links}@\infty]\ W_v(+\infty, 0)|0\rangle}, \end{aligned} \tag{12}$$

with two *distinct* off-light-cone gauge vectors n_2 and v. It is evident that the pinch singularity appeared in the naive definition (7) is now alleviated to a soft divergence in the unsubtracted wave function

$$\phi_\pi^{\text{new}} \supset \int [dl] \frac{n_2 \cdot v}{}$$

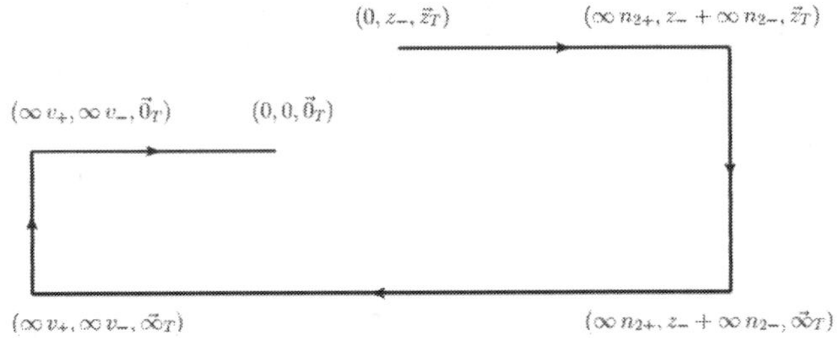

Figure 2. The Wilson-line path adopted to define the unsubtracted wave function (12).

subtraction will be needed in the TMD definition [19]

$$
\begin{aligned}
\phi_\pi^{\rm I}(x, \vec{k}_T, y_2, \mu) &= \int \frac{dz_-}{2\pi} \int \frac{d^2 z_T}{(2\pi)^2} e^{i(xp_+ z_- - \vec{k}_T \cdot \vec{z}_T)} \\
&\times \langle 0|\bar{q}(0) W_{n_2}^\dagger(+\infty, 0) \not{n}_- \gamma_5 \,[\text{links@}\infty]\, W_v(+\infty, z)\, q(z)|\pi^+(p)\rangle\,, \qquad (14)
\end{aligned}
$$

with

$$
n_2 = (e^{y_2}, e^{-y_2}, \vec{0}_T)\,, \qquad v = (-e^{y_2}, e^{-y_2}, \vec{0}_T)\,. \qquad (15)
$$

One can readily verify that both of the two definitions (10) and (14) result in the same collinear logarithm for the infrared subtraction (without the contribution from the LSZ terms)

$$
\phi_\pi^{C,\rm I} \otimes H^{(0)} = -\frac{\alpha_s C_F}{4\pi} \,[2\ \ln x + 3]\, \ln\left(\frac{k_T^2}{Q^2}\right) H^{(0)}(x, k_T) + (\text{IR finite terms})\,, \qquad (16)
$$

at one-loop approximation, which reproduces the collinear behaviour of the one-loop QCD amplitude for the partonic Green function corresponding to $\gamma^* \to \pi\gamma$ at leading Fock-state approximation. Alternatively, one can facilitate the TMD definition (12) with the antiparallel Wilson lines

$$
\begin{aligned}
\phi_\pi^{\rm II}(x, \vec{k}_T, y_2, \mu) &= \int \frac{dz_-}{2\pi} \int \frac{d^2 z_T}{(2\pi)^2} e^{i(xp_+ z_- - \vec{k}_T \cdot \vec{z}_T)} \\
&\times \frac{\langle 0|\bar{q}(0) W_{n_2}^\dagger(+\infty, 0) \not{n}_- \gamma_5 \,[\text{links@}\infty]\, W_{n_2}(-\infty, z)\, q(z)|\pi^+(p)\rangle}{[\text{color factor}]\ \langle 0|W_{n_2}^\dagger(+\infty, 0)\,[\text{links@}\infty]\, W_{n_2}(-\infty, 0)|0\rangle}\,, \qquad (17)
\end{aligned}
$$

which can be also shown to reproduce the very collinear structure in (16) and I will not expand the discussion for this definition.

3 Equivalence of TMD definitions

Now I will demonstrate the equivalence of different TMD definitions to all orders of the strong coupling in light of the arguments inspired from the rapidity evolution equations. First, I will explain that both of the definitions (10) and (14) will recover the naive definition (5) in the limit of vanishing regulators. One can readily verify that switching off the regulator in (10), i.e., $y_2 \to -\infty\ (\Rightarrow y_2 = y_u)$ implies that

$$
\sqrt{\frac{S(z_T; y_1, y_2)}{S(z_T; y_1, y_u)\, S(z_T; y_2, y_u)}} \to 1\,, \qquad (18)
$$

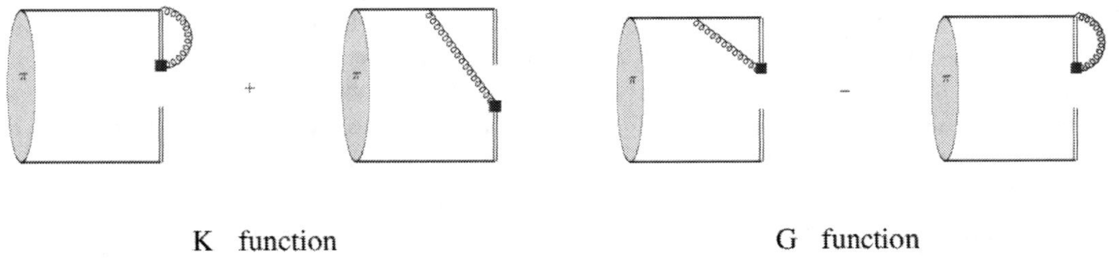

Figure 3. Graphic representation of the rapidity evolution kernels for the new TMD definition (14).

which evidently reduces the Collins' definition to (5). Similarly, setting $y_2 \to -\infty$ ($\Rightarrow n_2 = v = n_-$) in (14) trivially gives rise to

$$\phi_\pi^{\rm I}(x, \vec{k}_T, y_2, \mu) \to \phi_\pi^{\rm naive}(x, \vec{k}_T, \mu)\,. \tag{19}$$

We are then led to conclude that both Collins' and our proposals for the TMD definitions yield the same collinear divergence in the limit $y_2 = y_u \to -\infty$. The remaining task is to show that the two definitions can bring about the same collinear behaviour for an arbitrary rapidity y_2. To this end, we need to examine the rapidity evolution equations for both definitions of the TMD pion wave function, employing the standard resummation techniques in QCD [1, 22].

Noticing that the rapidity dependence in Collins' definition (10) comes from the soft subtraction entirely and applying the routine chain rule for the rapidity derivative one can write

$$\frac{d}{dy_2}\phi_\pi^C = \lim_{\substack{y_1\to+\infty \\ y_u\to-\infty}} \frac{1}{2}\Bigg[\underbrace{\frac{S'(z_T; y_1, y_2)}{S(z_T; y_1, y_2)}}_{K(z_T; y_1, y_2)} - \underbrace{\frac{S'(z_T; y_2, y_u)}{S(z_T; y_2, y_u)}}_{K(z_T; y_2, y_u)}\Bigg]\phi_\pi^C \approx \lim_{y_1\to+\infty} K(z_T; y_1, y_2)\,\phi_\pi^C\,, \tag{20}$$

where the primed soft functions S' are computed from the Wilson-line diagrams defining the soft functions, with the Eikonal vertex n_2^μ replaced by a special vertex given by Eq. (4.4) in [19], and only the soft function $K(z_T; y_2, y_u)$ enters the rapidity evolution of the TMD definition (10) in contrast to the original Collins-Soper-Sterman evolution [1, 22].

Constructing the rapidity evolution for the newly proposed definition (14) is slightly more involved due to the appearance of two gauge vectors dependent on the same rapidity parameter y_2 and to the rapidity dependence of the unsubtracted wave function. It is nevertheless without the conceptual difficulty to derive the following evolution equation

$$\frac{d}{dy_2}\phi_\pi^{\rm I} = \lim_{y_1\to+\infty} \left[K(z_T; y_1, y_2) + G(y_2)\right]\phi_\pi^{\rm I}\,, \tag{21}$$

where the soft and hard functions can be computed from the effective diagrams presented in figure 3. We can then readily conclude that the infrared behaviour of the TMD definition (14) is equivalent to that proposed by Collins (10), to all orders of the strong coupling, due to the infrared safety of the hard function $G(y_2)$.

4 Concluding discussion

To summarize, I presented a brief historic report on understanding the TMD factorization for hard exclusive processes, using the definition of TMD wave functions as an illustrative example. I provided a detailed account of justifying the new definition (12), with a simple soft function, applicable to a QCD description of $\gamma^* \to \pi\gamma$ at large momentum transfer. Furthermore, I demonstrated the equivalence of Collins' and our proposals for the TMD definitions in the infrared behaviours to all orders of perturbative expansion, with the aid of the rapidity evolution equations using the classical QCD resummation techniques. However, there remain many conceptual issues of constructing TMD factorization formulae for hard exclusive reactions as already emphasized above. Among these unsettled questions, the top priority should be to establish a definite power counting scheme for all the energy scales involved in a given problem, including the *intrinsic* transverse momentum. I therefore anticipate exciting developments of putting TMD factorization for hard exclusive processes on a solid ground, in addition to its phenomenological success.

Acknowledgements

I am grateful to Hsiang-nan Li for a very fruitful collaboration and to the organizers of POETIC6 for the generous finical support.

References

[1] J. Botts and G. F. Sterman, Nucl. Phys. B **325**, 62 (1989).
[2] H. n. Li and G. F. Sterman, Nucl. Phys. B **381**, 129 (1992).
[3] G. P. Lepage and S. J. Brodsky, Phys. Rev. D **22**, 2157 (1980).
[4] J. C. Collins, Acta Phys. Polon. B **34**, 3103 (2003) [hep-ph/0304122].
[5] S. Nandi and H. n. Li, Phys. Rev. D **76**, 034008 (2007) [arXiv:0704.3790 [hep-ph]].
[6] F. Feng, J. P. Ma and Q. Wang, Phys. Lett. B **674**, 176 (2009) [arXiv:0807.0296 [hep-ph]].
[7] H. n. Li, Y. L. Shen, Y. M. Wang and H. Zou, Phys. Rev. D **83**, 054029 (2011) [arXiv:1012.4098 [hep-ph]].
[8] H. n. Li, Y. L. Shen and Y. M. Wang, Phys. Rev. D **85**, 074004 (2012) [arXiv:1201.5066 [hep-ph]].
[9] W. F. Wang and Z. J. Xiao, Phys. Rev. D **86**, 114025 (2012) [arXiv:1207.0265 [hep-ph]].
[10] H. N. Li, Y. L. Shen and Y. M. Wang, JHEP **1302**, 008 (2013) [arXiv:1210.2978 [hep-ph]].
[11] H. N. Li, Y. L. Shen and Y. M. Wang, JHEP **1401**, 004 (2014) [arXiv:1310.3672 [hep-ph]].
[12] S. Cheng, Y. Y. Fan, X. Yu, C. D. Lü and Z. J. Xiao, Phys. Rev. D **89**, 094004 (2014) [arXiv:1402.5501 [hep-ph]].
[13] Y. L. Shen and Y. M. Wang, EPJ Web Conf. **80**, 00047 (2014) [arXiv:1409.1048 [hep-ph]].
[14] Y. M. Wang, Int. J. Mod. Phys. Conf. Ser. **37**, 1560049 (2015).
[15] S. S. Agaev, V. M. Braun, N. Offen and F. A. Porkert, Phys. Rev. D **83**, 054020 (2011) [arXiv:1012.4671 [hep-ph]].
[16] J. Collins, "*Foundations of perturbative QCD*," Cambridge monographs on particle physics, nuclear physics and cosmology, 32.
[17] B. Melic, D. Mueller and K. Passek-Kumericki, Phys. Rev. D **68**, 014013 (2003) [hep-ph/0212346].
[18] A. Bacchetta, D. Boer, M. Diehl and P. J. Mulders, JHEP **0808**, 023 (2008) [arXiv:0803.0227 [hep-ph]].
[19] H. n. Li and Y. M. Wang, JHEP **1506**, 013 (2015) [arXiv:1410.7274 [hep-ph]].
[20] J. Collins, Int. J. Mod. Phys. Conf. Ser. **4**, 85 (2011) [arXiv:1107.4123 [hep-ph]].
[21] I. O. Cherednikov, T. Mertens and F. F. Van der Veken, "*Wilson lines in quantum field theory*," De Gruyter Stud. Math. Phys. **24** (2014).
[22] J. C. Collins and D. E. Soper, Nucl. Phys. B **193**, 381 (1981) [Nucl. Phys. B **213**, 545 (1983)].

EPJ Web of Conferences **112**, 01022 (2016)
DOI: 10.1051/epjconf/201611201022

Electron–deuteron DIS with spectator tagging at EIC: Development of theoretical framework

W. Cosyn[1], V. Guzey[2], M. Sargsian[3], M. Strikman[4], and C. Weiss[5]

[1]*Ghent University, 9000 Ghent, Belgium*
[2]*Petersburg Nuclear Physics Institute, Gatchina, 188300, Russia*
[3]*Florida International University, Miami, FL 33199, USA*
[4]*Pennsylvania State University, University Park, PA 16802, USA*
[5]*Jefferson Lab, Newport News, VA 23606, USA*

Abstract. An Electron–Ion Collider (EIC) would enable next-generation measurements of deep-inelastic scattering (DIS) on the deuteron with detection of a forward-moving nucleon (p, n) and measurement of its recoil momentum ("spectator tagging"). Such experiments offer full control of the nuclear configuration during the high-energy process and can be used for precision studies of the neutron's partonic structure and its spin dependence, nuclear modifications of partonic structure, and nuclear shadowing at small x. We review the theoretical description of spectator tagging at EIC energies (light-front nuclear structure, on-shell extrapolation in the recoil nucleon momentum, final-state interactions, diffractive effects at small x) and report about on-going developments.

Measurements of deep-inelastic scattering and related processes on light ions ($A = 2, 3, ...$) address basic questions of high-energy nuclear physics and Quantum Chromodynamics (QCD), such as the partonic structure of the neutron and its spin decomposition [1], the nuclear modification of the nucleon's quark and gluon densities [2, 3], and the onset of nuclear shadowing and coherent phenomena at small x [4]. The common challenge in the analysis of such measurements is to account for the multitude of nuclear configurations present during the high-energy scattering process. The scattering can involve any of the constituent nucleons (p, n) in different states of their quantum–mechanical motion (momentum, spin), as well as non-nucleonic degrees of freedom excited by the nuclear binding. In the extraction of neutron structure one needs to isolate the cross section arising from scattering on the neutrons and eliminate the effects of nuclear binding (Fermi motion, non-nucleonic degrees of freedom). For neutron spin structure one must also infer the effective polarization of the neutron and account for the polarization of non-nucleonic degrees of freedom, particularly intrinsic Δ isobars in polarized ^{3}He [5]. In the study of nuclear modifications at $x \gtrsim 0.1$ (EMC effect, antishadowing) one would like to relate the modifications to the interactions taking place in a particular configuration (exchange mechanisms, short-range correlations). In the analysis of coherent phenomena at small x one wants to understand what configurations in the wave function build up the coherent response (onset of shadowing, saturation). While these challenges can partly be addressed by theoretical calculations, major progress could come from new experiments that provide information on the nuclear configurations during the high–energy process.

Deuteron and spectator tagging. High-energy scattering on the deuteron ($A = 2$) with detection of a proton or neutron in the nuclear fragmentation region (recoil momentum $p_R \sim$ few 10–100 MeV/c

in the deuteron rest frame) represents a unique method for performing DIS measurements in controlled nuclear configurations. The deuteron's pn wave function is simple and well known from low–energy measurements, including the light–front wave function entering in high–energy processes (see below). Because the deuteron has isospin $I = 0$, Δ isobars in the wave function are strongly suppressed (they can occur only in $\Delta\Delta$ configurations), so that the deuteron can be treated as a pn system for most of the configurations relevant to DIS [6]. Detection of the "spectator" nucleon and measurement of its momentum positively identifies the active nucleon and controls its momentum during the DIS process. Because of the simple geometry of high-energy scattering on the $A = 2$ system the possibilities for final–state interactions are limited, and in configurations where they occur they can be computed using theoretical models. Regarding polarization the deuteron has spin $S = 1$; its wave function is predominantly S–wave with a small admixture of D–wave, so that the effective nucleon polarization is well known.

DIS on the deuteron with recoil proton detection ("spectator tagging") was measured in the JLab CLAS BONuS experiment at 6 GeV beam energy at recoil momenta $p_R \gtrsim 70$ MeV/c [7] and will be explored further at 11 GeV. In such fixed–target experiments it is difficult to get the slow proton (or neutron) out of the target and measure its momentum with sufficient resolution. Much more suitable for this purpose are colliding–beam experiments, where the spectator nucleon moves on with approximately half the deuteron beam momentum and can be detected using forward detectors. An EIC with electron–nucleon center–of–mass energies $\sqrt{s_{eN}} \sim 15$–100 GeV and luminosity $L \sim 10^{33}$–10^{34} cm^{-2} s^{-1} is projected as a future facility for nuclear physics [8]. Both EIC designs presently developed have capabilities for forward nucleon detection [9]. The JLab EIC detector is designed to provide full coverage for spectator protons down to zero transverse momentum, and with a momentum resolution corresponding to $p_R \sim 20$ MeV/c in the rest frame, as well as forward neutron detection [10]. This setup would enable detailed measurements of DIS with spectator tagging over the entire (x, Q^2) range covered by the collider, which includes the region of sea quarks, gluons, and small–x phenomena. It would allow also tagged measurements on the polarized deuteron, which is potentially the most precise method for determining neutron spin structure. Altogether this setup would permit nuclear DIS measurements with full control of the nuclear configuration and enable a new level of understanding of nuclear effects in QCD.

To realize this potential it is necessary to develop the theoretical apparatus for describing DIS with spectator tagging at collider energies and with polarized beams. In this article we summarize the theoretical description and report about on-going developments [11]. A general review of nuclear physics with EIC is given in Ref. [12].

Cross section and structure functions. The basic observable is the invariant differential cross section for inclusive DIS on the deuteron with an identified nucleon with recoil momentum p_R in the final state (see Fig. 1a):

$$\frac{d\sigma(eD \to e'NX)}{dx\, dQ^2\, (d^3p_R/E_R)} = [\text{flux}] \left[F_{2D}(x, Q^2; \alpha_R, p_{RT}) - (1-\epsilon)F_{LD}(..) + \ldots\right], \tag{1}$$

where ϵ is the virtual photon polarization parameter. The "tagged" deuteron structure functions F_{2D} and F_{LD} depend on x and Q^2 as well as on the recoil nucleon momentum, which is parametrized here by the light-cone fraction $\alpha_R \equiv 2(E_R + p_R^z)/(E_D + p_D^z)$ and transverse momentum $p_{RT} \equiv |\boldsymbol{p}_{RT}|$ in a frame where the momentum transfer $\boldsymbol{q}$ and the deuteron momentum $\boldsymbol{p}_D$ are collinear and along the z–axis. Here $x \equiv Q^2/(p_D q) = 2x_B$ is the rescaled Bjorken variable with limits $0 < x < 2$. The ellipsis in Eq. (1) stands for azimuthal angle– (ϕ_R–) and spin–dependent terms which we left out for brevity. The full form of the cross section for scattering of a polarized electron on a polarized deuteron (vector, tensor) has been determined and will be given elsewhere [13]. We emphasize that the form of Eq. (1) is general and does not rely on any assumptions regarding composite structure of the deuteron.

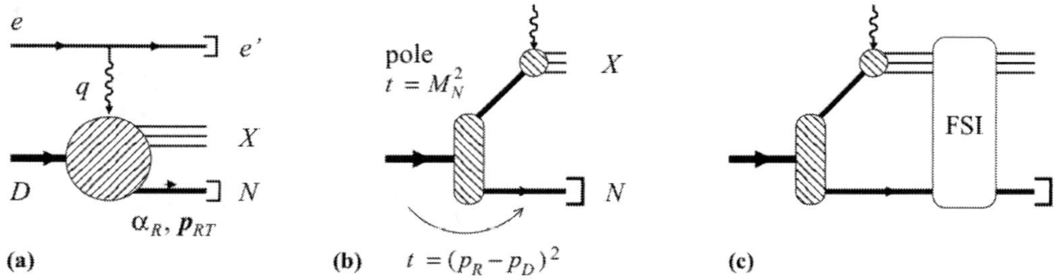

Figure 1. (a) Tagged DIS on the deuteron, $eD \to e' + N + X$. (b) Impulse approximation amplitude, containing the nucleon pole at $t = M_N^2$. (c) Final–state interaction amplitude.

Light-front nuclear structure and impulse approximation. The theoretical description aims to express the tagged structure functions of Eq. (1) in terms of nuclear structure and nucleon structure functions. High–energy processes such as DIS probe the nucleus at fixed light–front time, where the structure is described by the light–front wave function. The light–front description is unique in the sense that the off–shellness of the constituents remains finite as the scattering energy becomes large, which permits a composite description of the nucleus in DIS in terms of nucleon degrees of freedom [6]. It can be implemented using either non-covariant light–front quantization or a covariant scheme based on Feynman diagrams [14]. The starting point is the impulse approximation, in which the electromagnetic current interacts with a single nucleon and the DIS final state evolves independently of the other nucleon (see Fig. 1b). Neglecting terms $(\text{mass})^2/Q^2$ it gives

$$F_{2D}(x, Q^2; \alpha_R, p_{RT}) = \frac{|\Psi_D(\alpha_R, \boldsymbol{p}_{RT})|^2}{2 - \alpha_R} F_{2N}(\widetilde{x}, Q^2; \text{ off-shell}), \qquad \widetilde{x} \equiv \frac{x}{2 - \alpha_R}. \tag{2}$$

Here $2-\alpha_R$ is the LF momentum fraction of the active nucleon in the deuteron, which is kinematically fixed by that of the spectator, α_R. The nucleon structure function is evaluated at $\widetilde{x} = x/(2-\alpha_R)$. Ψ_D is the deuteron's NN light–front wave function, normalized such that $\int d\alpha_R d^2 p_{RT} |\Psi_D(\alpha_R, \boldsymbol{p}_{RT})|^2/[\alpha_R(2-\alpha_R)] = 1$, which can be obtained from the non-relativistic wave function for rest frame momenta $|\boldsymbol{p}_R| \ll M_N$ [6]. F_{2N} is the structure function of the active nucleon, which is generally modified by off-shell effects (see below). When integrated over the recoil momentum Eq. (2) satisfies the baryon number and light-cone momentum sum rules for the deuteron.

Equation (2) implies Bjorken scaling of the tagged deuteron structure function at $Q^2 \gg 1\,\text{GeV}^2$ for fixed α_R and p_{RT} (this holds even in the presence of modified nucleon structure and final-state interactions). This feature can be verified experimentally and represents a basic test of the theoretical framework. It represents a special case of the QCD factorization theorem for DIS with identified slow hadrons in the target fragmentation region [16].

Nucleon pole and on-shell extrapolation. An essential feature of the impulse approximation is that it captures the leading singularity of the tagged structure function in the invariant momentum transfer between the deuteron and the recoiling nucleon, $t \equiv (p_D - p_R)^2 = \text{function}(\alpha_R, p_{RT})$. The deuteron LF wave function has a pole at $t = M_N^2$ corresponding to the nucleon on-shell point. (In the covariant formulation this pole represents "nucleon exchange" between the deuteron and the electromagnetic current.) In the limit $t \to M_N^2$ one therefore has

$$F_{2D}(x, Q^2; \alpha_R, p_{RT}) \sim \frac{R}{(t - M_N^2)^2} F_{2N}(\widetilde{x}, Q^2) \qquad (t \to M_N^2), \tag{3}$$

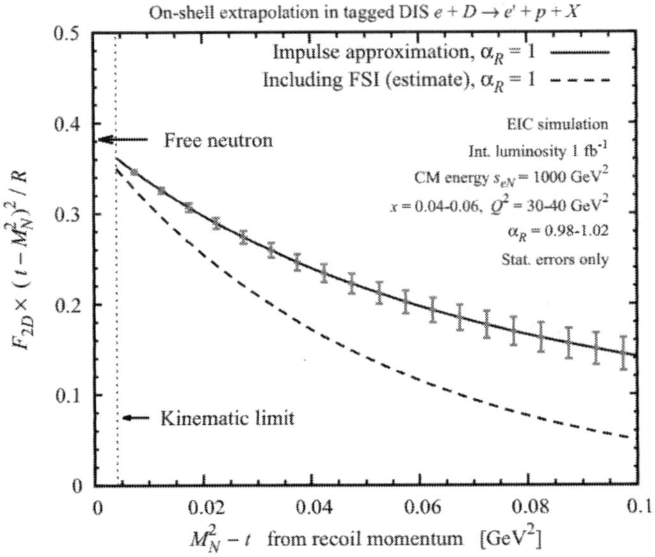

Figure 2. Neutron structure extraction by on-shell extrapolation $t \to M_N^2$ in tagged DIS $e + D \to e' + p + X$. The plot shows the tagged deuteron structure function F_{2D}, with the pole factor $R/(t - M_N^2)^2$ removed, as a function of the off-shellness $t - M_N^2$, at $\alpha_R = 1$. *Solid line:* Impulse approximation. *Dashed line:* Sum of impulse approximation and final state interactions (estimate) [15]. *Data:* Simulated measurement of t–dependence with EIC. Parameters are shown on the plot. Error bars indicate expected statistical errors; for systematic errors see text.

where $R \equiv R(\alpha_R)$ is a calculable residue and F_{2N} is the on-shell (free) structure function of the active nucleon. Theoretical arguments show that final–state interactions (see Fig. 1c) do not modify the leading singularity in t, as they involve an integral over the intermediate nucleon momentum [17]. One can thus obtain the on-shell nucleon structure function by measuring the deuteron structure function F_{2D} over a range of t at fixed α_R and performing the on-shell extrapolation to $t \to M_N^2$. The physical region is $t - M_N^2 < -\epsilon_D M_D = -0.0041\ \text{GeV}^2$ (ϵ_D is the deuteron binding energy), so that measurements can be done extremely close to the on-shell point. In the deuteron rest frame $t - M_N^2 = -2|\boldsymbol{p}_R|^2 - \epsilon_D M_D$, and the kinematic limit corresponds to zero recoil momentum $\boldsymbol{p}_R = 0$.

Simulations show that precise measurements of the free neutron structure function F_{2n} by proton tagging and on-shell extrapolation are feasible with EIC (see Fig. 2). One measures the tagged deuteron structure function F_{2D} over a range of t at fixed α_R, removes the pole factor in Eq. (3), and performs polynomial extrapolation to $t \to 0$. Both statistical and systematic data errors can be propagated through the extrapolation procedure. Systematic errors arise mainly from the uncertainty in the recoil momentum determination and are expected to result in a correlated F_{2n} uncertainty of $\lesssim 3\%$ with the JLab EIC beam and detector design [11].

The on-shell extrapolation can also be used to extract neutron spin structure functions from polarized tagged DIS on the deuteron. It is convenient to measure the double spin asymmetry of the tagged cross section as a function of the recoil momentum,

$$A_{\|}(x, Q^2; \alpha_R, p_{RT}) = \frac{\sigma(+\frac{1}{2}, +1) - \sigma(-\frac{1}{2}, +1)}{\sigma(+\frac{1}{2}, +1) + \sigma(-\frac{1}{2}, +1)}, \tag{4}$$

and perform the extrapolation to $t \to M_N^2$ for the asymmetry rather than the cross sections. [Here $\sigma(\lambda_e, \lambda_D)$ denotes the cross section for electron helicity $\lambda_e = \pm 1/2$ and deuteron helicity $\lambda_D = \pm 1$, i.e., longitudinal polarization along the respective beam directions.] The asymmetry exhibits only very weak t dependence. An interesting feature is that the D–wave in the deuteron wave function practically drops out in the on-shell extrapolation, since its wave function is proportional to the squared rest frame nucleon momentum $|\boldsymbol{p}_R|^2$ and the on-shell point corresponds to extremely small unphysical momenta

$|\boldsymbol{p}_R|^2 = -\epsilon_D M_D$. This means that at the on-shell point the neutron is effectively 100% polarized in the deuteron spin direction. Furthermore, kinematic factors and many systematic errors cancel in the cross section ratio. Simulations show that precise measurements of neutron spin structure with on-shell extrapolation are possible over most of the (x, Q^2) range covered by EIC with an integrated luminosity $\sim 100\ \text{fb}^{-1}$ [18].

Nuclear modifications and final–state interactions (FSI). Away from the nucleon pole $t = M_N^2$ the tagged deuteron structure function is modified both by possible off–shell dependence of the nucleon structure functions and by FSI. The off-shell modifications can be accommodated within the impulse approximation, by allowing the effective nucleon structure function to depend on $t - M_N^2$ in a certain range of virtualities ("virtual nucleon model"). Specific dynamical models for the off-shell dependence have been proposed.

FSI arise from the interactions of the spectator with the DIS final state produced by the active nucleon (see Fig. 1c). While they do not change the total (untagged) DIS cross section on the deuteron, such interactions can distort the recoil momentum spectrum and change the outgoing particle flux, and must therefore be accounted for in the analysis of tagged DIS. FSI can be estimated theoretically by modeling the composition of the nucleon DIS final state and its rescattering from the spectator. Theoretical arguments suggest that at $x \lesssim 0.1$ the dominant FSI arise from "slow" hadrons in the DIS final state, with rest frame momenta $\sim$ 1–2 GeV/c, as these hadrons are expected to form close to the interaction point and can rescatter from the spectator with hadronic cross sections [15]. "Fast" hadrons with momenta $\gg$ 1 GeV/c form outside the nucleus and interact marginally with the spectators, as evidenced by the absence of absorption of such hadrons in DIS on larger nuclei [19]. Estimates of the FSI effects in tagged DIS can be made in a schematic model, using empirical DIS hadron distributions and rescattering cross sections [15]. Preliminary results show noticeable modifications of the t–dependence of the tagged structure function away from the pole (see Fig. 2). The modifications vanish in the limit $t \rightarrow M_N^2$, in accordance with the findings of Ref. [17], and thus do not prevent the extraction of free neutron structure through on-shell extrapolation; their effect on the extraction errors remains to be investigated.[1]

An interesting question is how the effects of modified nucleon structure and FSI in tagged DIS could be separated experimentally in future measurements at EIC. Tagged DIS measurements at $\widetilde{x} \sim 0.2$, where nuclear modifications of the inclusive structure functions are negligible [2], could isolate the effects of FSI in the tagged DIS cross section. Direct evidence for the dominance of FSI in this region could come from observation of their distinctive angular dependence at larger rest frame recoil momenta $p_R \gtrsim 200$ MeV/c [15]. Tagged DIS measurements at $\widetilde{x} > 0.3$, where substantial nuclear modifications are seen in the inclusive data (EMC effect) [2], could then reveal the combined effects of modified nucleon structure and FSI. Further information on nuclear FSI will come from measurements of nuclear modification of hadron spectra in DIS on heavier nuclei. The experimental capabilities of EIC (kinematic range, recoil momentum coverage and resolution, variety of nuclear beams) would enable a program of comparative measurements designed to separate the two effects.

Diffraction in tagged DIS. In tagged DIS at $x \lesssim 0.01$ a new mechanism of FSI arises due to diffractive scattering on the active nucleon, in which the nucleon is left intact and appears in the DIS final state with a rest–frame momentum $\sim$ few 100 MeV/c (see Fig. 3). In DIS on the deuteron such diffractive scattering can happen on the proton as well as on the neutron, with quantum-mechanical interference of the two amplitudes in the cross section. In inclusive DIS $e + D \rightarrow e' + X$ this effect

[1] The FSI effect on the t–distribution depends on the recoil light-cone fraction α_R [15]. FSI is maximal for the situation shown in Fig. 2, $\alpha_R = 1$, which corresponds to an angle of $\sim 90°$ between the recoil momentum $\boldsymbol{p}_R$ and the $\boldsymbol{q}$–vector in the rest frame (sideways recoil). FSI becomes smaller for $\alpha_R < 1$ or > 1, which correspond to angles $< 90°$ or $> 90°$ (forward or backward recoil). However, for $\alpha_R \neq 1$ the kinematic limit in $t - M_N^2$ moves further away from zero, which is not favorable for on-shell extrapolation.

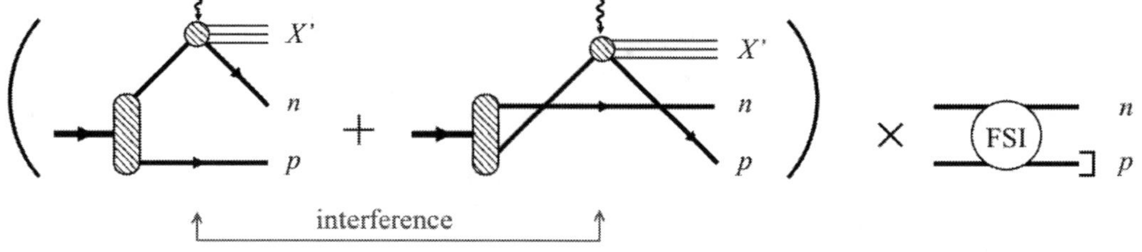

Figure 3. Diffraction in tagged DIS on the deuteron $e + D \to e' + p + X$. Diffractive scattering on the p and n enables interference of the respective amplitudes (shadowing), as well as strong FSI within the outgoing pn system.

gives rise to nuclear shadowing and has been the subject of extensive theoretical studies [4]. In tagged DIS $e+D \to e'+N+X$ the presence of the slow diffractive nucleon in the final state enables strong FSI between that nucleon and the spectator, resulting in considerable modification of the recoil momentum distribution compared to the impulse approximation. The distortion is particularly strong in the spin–1, isospin–0 partial wave of the pn system, where the outgoing pn scattering state must be orthogonal to the deuteron bound state in that channel (a similar effect happens in low–energy deuteron breakup reactions). Theoretical work aims to calculate this effect using methods analogous to those used for shadowing in inclusive DIS [20].

Experiments at EIC could perform detailed tests of the predicted distortion effects using single–nucleon tagging $e + D \to e' + p + X$, or double–nucleon tagging $e + D \to e' + p + n + X'$, which completely fixes the kinematics of the produced pn system. In this way they could elucidate the general mechanism of nuclear shadowing, which is essential for analyzing inclusive scattering on the deuteron (including polarization) and quantifying the approach to saturation at small x.

In summary, electron–deuteron DIS with spectator tagging at EIC would enable next–generation measurements of neutron spin structure, nuclear modifications of partonic structure, and small–x shadowing, with maximum control of the nuclear configurations participating in the high–energy process. Theoretical methods based on light–front nuclear structure and analytic properties provide a transparent and efficient description of such measurements, consistent with that of inclusive nuclear DIS. Work is in progress to adapt these methods to collider kinematics and polarized beams.

Notice: Authored by Jefferson Science Associates, LLC under U.S. DOE Contract No. DE-AC05-06OR23177. The U.S. Government retains a non-exclusive, paid-up, irrevocable, world-wide license to publish or reproduce this manuscript for U.S. Government purposes.

References

[1] C. A. Aidala, S. D. Bass, D. Hasch and G. K. Mallot, Rev. Mod. Phys. **85**, 655 (2013).
[2] S. Malace, D. Gaskell, D. Higinbotham and I. Cloet, Int. J. Mod. Phys. E **23**, 1430013 (2014).
[3] L. Frankfurt and M. Strikman, Phys. Rept. **160**, 235 (1988).
[4] L. Frankfurt, V. Guzey and M. Strikman, Phys. Rept. **512**, 255 (2012).
[5] L. Frankfurt, V. Guzey and M. Strikman, Phys. Lett. B **381**, 379 (1996); F. Bissey, V. Guzey, M. Strikman and A. Thomas, Phys. Rev. C **65**, 064317 (2002); C. Boros, V. Guzey, M. Strikman and A. Thomas, Phys. Rev. D **64**, 014025 (2001).
[6] L. Frankfurt and M. Strikman, Phys. Rept. **76**, 215 (1981).
[7] S. Tkachenko *et al.* [CLAS Collaboration], Phys. Rev. C **89**, 045206 (2014).

[8] U.S. Department of Energy Office of Science Nuclear Science Advisory Committee (NSAC) 2015 Long Range Plan for Nuclear Science, `http://science.energy.gov/np/nsac`

[9] For current information on the EIC machine designs, see: `https://eic.jlab.org/wiki/` (JLab) and `https://wiki.bnl.gov/eic/` (BNL).

[10] P. Nadel–Turonski, Presentation at Electron-Ion User Group Meeting 2016, UC Berkeley, 6–9 January 2016, `https://conferences.lbl.gov/event/56/`

[11] C. Weiss *et al.*, *"Physics potential of polarized light ions with EIC@JLab,"* Jefferson Lab 2014/15 Laboratory–directed R&D Project, `https://www.jlab.org/theory/tag/`

[12] A. Accardi, V. Guzey, A. Prokudin and C. Weiss, Eur. Phys. J. A **48**, 92 (2012).

[13] W. Cosyn, M. Sargsian, C. Weiss, in preparation.

[14] M. Sargsian, Int. J. Mod. Phys. E **10**, 405 (2001)

[15] M. Strikman and C. Weiss, in preparation.

[16] L. Trentadue and G. Veneziano, Phys. Lett. B **323**, 201 (1994); J. C. Collins, Phys. Rev. D **57**, 3051 (1998) [Erratum-ibid. D **61**, 019902 (2000)].

[17] M. Sargsian and M. Strikman, Phys. Lett. B **639**, 223 (2006).

[18] W. Cosyn *et al.*, J. Phys. Conf. Ser. **543**, 012007 (2014).

[19] J. Ashman *et al.* [European Muon Collaboration], Z. Phys. C **52**, 1 (1991); A. Airapetian *et al.* [HERMES Collaboration], Eur. Phys. J. A **47**, 113 (2011).

[20] V. Guzey et al., in preparation.

EPJ Web of Conferences **112**, 01023 (2016)
DOI: 10.1051/epjconf/201611201023

Results on longitudinal spin physics at COMPASS

Malte Wilfert[1,a,b] on behalf of the COMPASS Collaboration

[1] *Universität Mainz, Institut für Kernphysik, Johann-Joachim-Becher-Weg 45, D 55128 Mainz*

Abstract. The COMPASS experiment at the CERN SPS has taken data on deep inelastic scattering of polarised muons on a polarised NH_3 target in 2007 and 2011 and on a polarised LiD target in 2002-2004 and 2006. The new results on the longitudinal double spin asymmetry A_1^p and the spin-dependent structure function g_1^p obtained from the 2011 data set are presented. These results are used in a NLO QCD fit to the world data to obtain the polarised parton distributions. Also an update of the results on the Bjorken sum rule, connecting the integral of the non-singlet spin-dependent structure function with the ratio of the weak coupling constants, will be given.
Direct access to the gluon polarisation is possible via the photon gluon fusion process in semi-inclusive deep inelastic scattering. This process is studied using the p_T dependence of charged hadron asymmetries. The latest results indicate a positive gluon polarisation in the kinematic region of COMPASS

1 Introduction

One of the physics goals of the COMPASS experiment at CERN is the determination of the different contributions to the nucleon spin. Therefore data for deep inelastic scattering were recorded in 2002-2006 on a polarised ^{6}LiD target and in 2007/2011 on a polarised NH_3 target. In case of the 2011 data the beam energy has been increased from 160 GeV to 200 GeV to measure also lower values of the Bjorken scaling variable x_{Bj} and higher values of the photon virtuality Q^2. From these data the spin-dependent structure function g_1 can be determined. This structure function is of special interest as it can be used in a global QCD analysis to extract the helicity distributions of quarks and gluons. Using the deuteron and proton data from COMPASS only, it is also possible to calculate the non-singlet structure function, which allows an verification of the Bjorken sum rule. In case of the gluon polarisation a different approach can be used to extract this quantity directly in leading order.

2 Experiment

The COMPASS experiment is a fixed target experiment located at the M2 beamline of the CERN SPS. The complete setup is described in Reference [1]. In 2011 a polarised three cell target has been used. The outer most cells (30 cm long) were polarised in an opposite direction compared to the middle cell (60 cm long) allowing the measurement of both polarisation directions simultaneously. In order

[a]e-mail: mwilfert@cern.ch
[b]Supported by BMBF and GRK Symmetry Breaking

to cancel acceptance effects the polarisation of the cells is changed on a regular base by rotating the solenoid field of the target.

3 Results on A_1^{p} and g_1^{p}

The longitudinal photon-nucleon double-spin asymmetry is extracted from the measurement of DIS events from the oppositely polarised target cells. All details on this analysis can be found in Reference [2]. In order to obtain the asymmetry $A_1(x_{_{\mathrm{Bj}}}, Q^2)$ the raw asymmetry has to be corrected for the target and beam polarisation together with the dilution and depolarisation factor. This asymmetry is directly connected to the spin dependent structure function via $g_1 = F_2 A_1/(2x_{_{\mathrm{Bj}}}(1+R))$ using the ratio $R = \sigma_{\mathrm{L}}/\sigma_{\mathrm{T}}$ [3] of the absorption cross sections for longitudinal and transverse polarised virtual photons and the unpolarised structure function F_2 [4]. The results on the spin-dependent structure function from the 2011 is shown in Figure 1 as a function of Q^2 and $x_{_{\mathrm{Bj}}}$ together with the world data. In addition also the results from our NLO QCD fit to these data is shown.

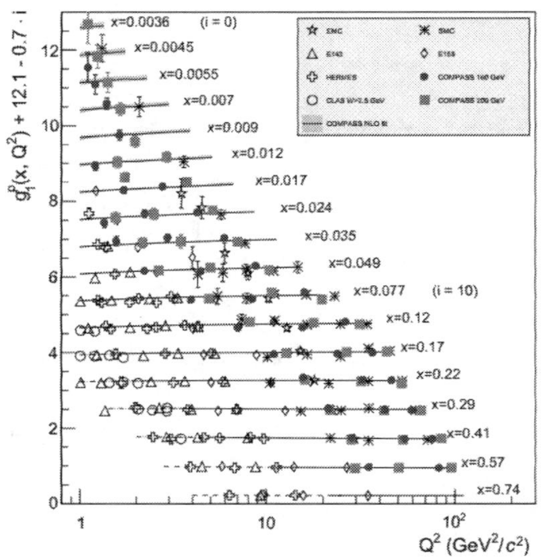

Figure 1. World data on g_1^{p} as a function of Q^2 for different $x_{_{\mathrm{Bj}}}$. The solid line represents the result of our NLO QCD fit. The dashed line indicates a extrapolation for $W^2 < 10\,\mathrm{GeV}^2$.

4 QCD fit

Performing a NLO QCD fit to the world data on g_1 in the DIS region allows for an extraction of the quark and gluon helicity. In addition to the new COMPASS data also the world data on proton, deuteron an ^{3}He targets from the CERN [4–7], HERA [8, 9], SLAC [10–14] and JLAB [15, 16] experiments are used. The QCD fit is performed in the $\overline{\mathrm{MS}}$ renormalisation and factorisation scheme and uses the DGLAP evolution equation for the Q^2 dependence. The $x_{_{\mathrm{Bj}}}$ dependence is given via

$$\Delta q_i(x) = \eta_i \frac{x^{\alpha_i}(1-x)^{\beta_i}(1+\gamma_i x)}{\int_0^1 x^{\alpha_i}(1-x)^{\beta_i}(1+\gamma_i x)dx}, \tag{1}$$

for the gluon, singlet and both non-singlet distributions. This functional form has been used at the reference scale of $Q_0^2 = 1\,(\mathrm{GeV}/c)^2$. In case of the gluon and non-singlet distributions γ_i is fixed to zero as the data is not sensitive to this parameter. The same is valid for β_g, which is fixed to the corresponding value from the unpolarised distribution taken from MSTW [17]. In addition the first moments η_3 and η_8 are fixed via the baryon decay constants. The unpolarised parton distributions are also used in each fit step to check the positivity constrain $|\Delta q(x_{\mathrm{Bj}})| < q(x_{\mathrm{Bj}})$ for all flavours (u, d, s, g). An extra term in the χ^2 penalises possible violations. Different systematic studies were performed to estimate the uncertainty on the fit. The largest contribution is the dependence on the reference scale at which the functional shape is initialised. From all studies two solutions are obtained with an equally good χ^2/NDF. One solution with γ_S fixed to zero results in a negative gluon distribution and one solution with γ_S as a free parameter results in a positive gluon distribution. The results are illustrated in Figure 2. From this fit the contribution from the quarks to the total spin of the nucleon is extracted to be in the range $0.26 < \Delta\Sigma < 0.36$, whereas the contribution from the gluon is not well constrained.

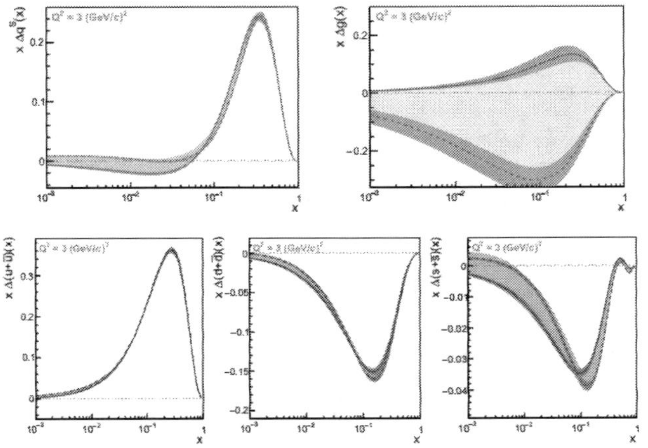

Figure 2. Result of the QCD fit to the world data. The solid lines indicate our two solutions with their statistical uncertainty indicated as a darker band. The lighter band corresponds to the systematic uncertainty of the fit.

5 Bjorken sum rule

The results from the QCD fit can also be used to extract first moments. The one of the non-singlet structure function

$$g_1^{\mathrm{NS}}(x_{\mathrm{Bj}}, Q^2) = g_1^{\mathrm{p}}(x_{\mathrm{Bj}}, Q^2) - g_1^{\mathrm{n}}(x_{\mathrm{Bj}}, Q^2) \tag{2}$$

is of special interest since it can be used to verify the Bjorken sum rule, which connects its first moment with the ratio of the weak coupling constants:

$$\Gamma_1^{\mathrm{NS}}(Q^2) = \int_0^1 g_1^{\mathrm{NS}}(x, Q^2)dx = \frac{1}{6}\left|\frac{g_{\mathrm{A}}}{g_{\mathrm{V}}}\right| C_1^{\mathrm{NS}}(Q^2). \tag{3}$$

Here C_1^{NS} is the non-singlet coefficient function, which has been calculated in perturbative QCD [18]. The first moment is calculated using only COMPASS data, which are evolved to a common Q^2 of the 2007 proton data set using the NLO QCD fit. The non-singlet structure function g_1^{NS} is calculated afterwards from the evolved data sets. These data are used in a second NLO QCD fit; only the non-singlet structure is fitted function via the non-singlet distribution Δq_3. The program used for the fit is the same as used to fit the world data on g_1. For the verification of the Bjorken sum rule it is necessary to calculate the first moment at a fixed scale. The evolution of the non-singlet structure function, obtained from the COMPASS data, to a common scale of $Q^2 = 3\,(\mathrm{GeV}/c)^2$ is performed using the results from the non-singlet fit. A comparison showing the results from the non-singlet fit and the COMPASS data is illustrated in Figure 3. The COMPASS data cover only the range of $0.0025 < x_{\mathrm{Bj}} < 0.7$. The results from the non-singlet fit are used to determine the contribution to the first moment of g_1^{NS} for the unmeasured region. The first moment at $Q^2 = 3\,(\mathrm{GeV}/c)^2$ is

$$\Gamma_1^{\mathrm{NS}} = 0.181 \pm 0.008(\mathrm{stat}) \pm 0.014(\mathrm{syst})\,, \tag{4}$$

from which 94% are within the measured x-range. Figure 4 shows the dependence of the integral of the non-singlet structure function as a function of the lower limit, also illustrating the size of the extrapolations. The ratio of the weak coupling constants is calculated using the non-singlet coefficient function in NLO

$$\left|\frac{g_{\mathrm{A}}}{g_{\mathrm{V}}}\right| = 1.22 \pm 0.05(\mathrm{stat}) \pm 0.10(\mathrm{syst})\,. \tag{5}$$

The systematic uncertainty is dominated by the uncertainty on the beam polarisation. Other contribution like the one from the target polarisation, the dilution factor, the depolarisation factor and F_2 are also taken into account. Comparing the result for $|g_{\mathrm{A}}/g_{\mathrm{V}}|$ with the results from the neutron β-decay ($|g_{\mathrm{A}}/g_{\mathrm{V}}| = 1.2701 \pm 0.002$ [19]) provides a validation of the Bjorken sum rule at the level of 9%.

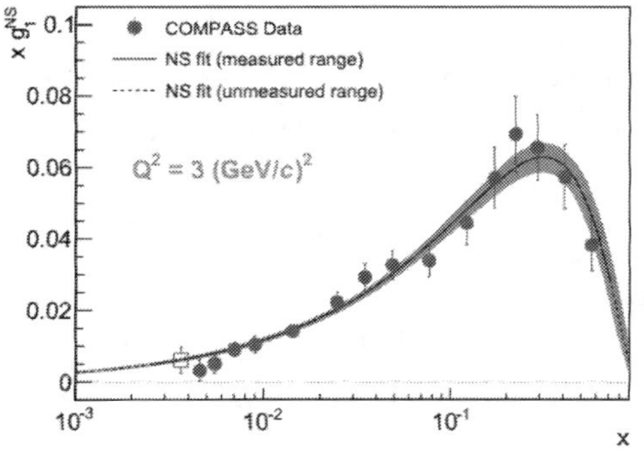

Figure 3. Values of g_1^{NS} at $Q^2 = 3\,(\mathrm{GeV}/c)^2$ compared to the NLO fit. The open square corresponds to the data point where the deuteron contribution is taken from the NLO QCD fit.

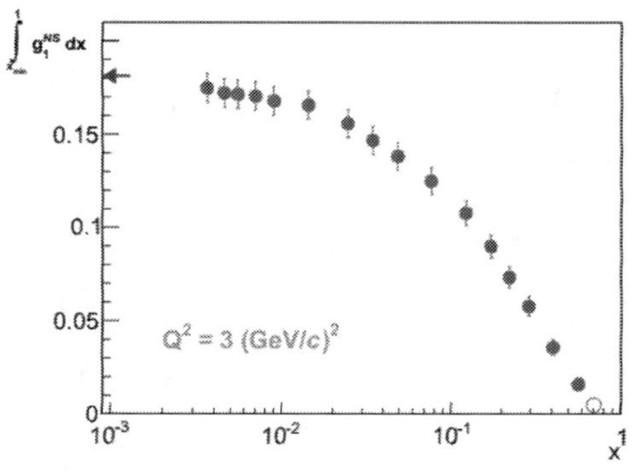

Figure 4. Dependence of the first moment as a function of x_{min}. The open circle at $x_{\mathrm{Bj}} = 0.7$ is obtained from the fit. The arrow on the left side indicates the value for the full range $0 < x_{\mathrm{Bj}} < 1$.

6 LO extraction of ΔG

An alternative access to the gluon polarisation uses the p_{T} dependence of the charged hadron asymmetry in semi-inclusive deep inelastic scattering, which can be split up in LO into contributions from different processes:

$$A_{\mathrm{LL}}^{h}(x_{\mathrm{Bj}}) = \alpha \cdot A_1^{LO}(x_{\mathrm{Bj}}) + \beta \cdot A_1^{LO}(x_{\mathrm{C}}) + \gamma \cdot \Delta g/g(x_{\mathrm{g}})\ , \tag{6}$$

the three processes are the leading process (LP), the QCD Compton process (QCDC) and the photon-gluon fusion (Figure 5). The factors α, β and γ depend on the partonic cross section asymmetries and the fraction to which each process contributes.

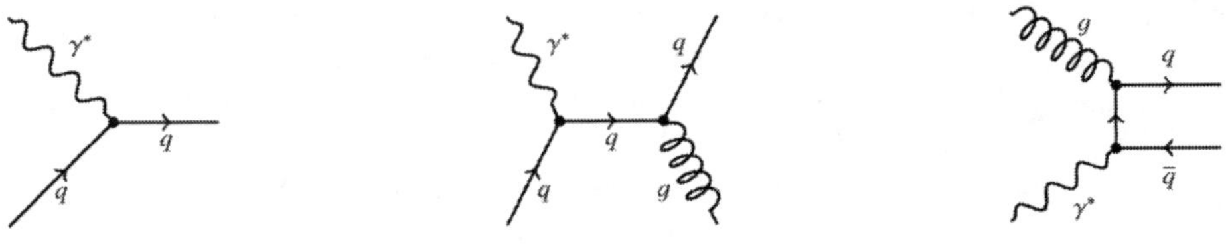

Figure 5. Feynman diagrams for the different processes. Leading-order process (left), QCD Compton (middle) and photon-gluon fussion (right).

The gluon polarisation is extracted from the 2002-2006 COMPASS deuteron data using a method developed for the extraction of the gluon polarisation from open charm events [20] distinguishing between signal and background events. A neural network (NN) is trained using Monte Carlo data to output the partonic cross section asymmetries $a^i_{\rm LL}$, the fractions R_i and x_i using $x_{\rm Bj}$, Q^2, $p_{\rm T}$ and $p_{\rm L}$ as input parameters. It assigns different weights to the events using the different kinematic dependence of the three processes. For systematic studies also different Monte Carlo simulations with different parameters for the hadronisation are used. The weights from the NN are used in a fit to obtain simultaneously the leading process asymmetries and the gluon polarisation. The final result is

$$\langle \Delta g/g \rangle = 0.113 \pm 0.038(\text{stat}) \pm 0.036(\text{syst}) \,, \tag{7}$$

at a scale of $\mu^2 = \langle Q^2 \rangle = 3\,(\text{GeV}/c)^2$ and $\langle x_g \rangle \approx 0.10$. This analysis is the first direct measurement of a positive value for the gluon polarisation. Since the statistics is good enough the result can be split up into three bins of x_g. These results are compared to the world data on $\Delta g/g$ extracted in LO in Figure 6. More informations can be found in Reference [21].

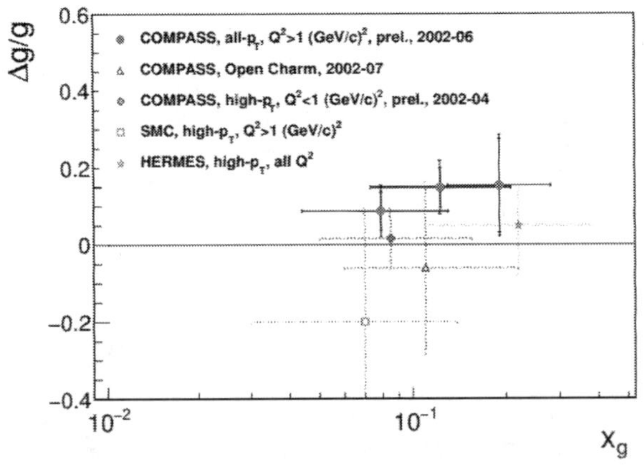

Figure 6. World data on $\Delta g/g$ [22–25] extracted in LO compared to our results in three bins of x_g.

7 Conclusion

The new data from the 2011 measurement complete the COMPASS results on A_1^{p} reaching lower values of x_{Bj} and higher values of Q^2. These data are used in a NLO QCD fit together with the world data on g_1 to extract the helicity distributions of the quarks and gluons. The contributions from quarks to the total spin of the nucleon are determined to be in the range of $0.26 < \Delta\Sigma < 0.36$. The COMPASS data taken on the proton and deuteron target are combined to extract the non-singlet structure function, which is used to verify the Bjorken sum rule. The obtained ratio of the weak coupling constants is $|g_{\mathrm{A}}/g_{\mathrm{V}}| = 1.22 \pm 0.05 \pm 0.10$ is obtained, which verifies the sum rule within an accuracy of 9%. A dedicated measurement of the gluon polarisation has been performed using p_{T} dependent muon-nucleon asymmetries from the DIS region. The value of the gluon polarisation obtained in LO QCD is $\langle \Delta g/g \rangle = 0.113 \pm 0.038 \pm 0.036$. This result is the first direct measurement of a positive gluon polarisation for the $x_g \approx 0.10$ region.

References

[1] COMPASS Collaboration, P. Abbon *et al.*, Nucl. Instr. Meth. A **577** (2007) 455.
[2] COMPASS Collaboration, C. Adolph *et al.*, Phys. Lett. B **753** (2016) 18.
[3] E143 Collaboration, K. Abe, *et al.*, Phys. Lett. B **452** (1999) 194.
[4] SMC, B. Adeva *et al.*, Phys. Rev. D **58** (1998) 112001.
[5] EMC, J. Ashman *et al.*, Phys. Lett. B **206** (1988) 364; Nucl. Phys. B **328** (1989) 1.
[6] COMPASS Collaboration, V. Yu. Alexakhin *et al.*, Phys. Lett. B **647** (2007) 8.
[7] COMPASS Collaboration, E. M.G. Alekseev *et al.*, Phys. Lett. B **690** (2010) 466.
[8] HERMES Collaboration, K. Ackerstaff *et al.*, Phys. Lett. B **404** (1997) 383.
[9] HERMES Collaboration, A. Airapetian *et al.*, Phys. Rev. D **75** (2007) 012007.
[10] E142 Collaboration, P. L. Anthony *et al.*, Phys. Rev. D **54** (1996) 6620.
[11] E143 Collaboration, K. Abe *et al.*, Phys. Rev. D **58** (1998) 112003.
[12] E154 Collaboration, K. Abe *et al.*, Phys. Lett. B **79** (1997) 26.
[13] E155 Collaboration, P. L. Anthony *et al.*, Phys. Lett. B **463** (1999) 339.
[14] E155 Collaboration, P. L. Anthony *et al.*, Phys. Lett. B **493** (2000) 19.
[15] JLab Hall A Collaboration, X. Zheng *et al.*, Phys. Rev. Lett. **92** (2004) 012004.
[16] CLAS Collaboration, K. V. Dharmawardane *et al.*, Phys. Lett. B **641** (2006) 11.
[17] A. D. Martin *et al.*, Eur. Phys. J. C **63** (2009) 189.
[18] S. A. Larin, T. van Ritbergen and J. A. M. Vermaseren, Phys. Lett. B **404** (1997) 153.
[19] J. Beringer *et al.* (Particle Data Group), Phys. Rev. D **86** (2012) 010001.
[20] J. Pretz and J.-M. Le Goff, Nucl. Instr. Meth. A **602** (2009) 594.
[21] COMPASS Collaboration, C. Adolph *et al.*, arXiv:1512.05053 [hep-ex].
[22] COMPASS Collaboration, E. S. Ageev *et al.*, Phys. Lett. B **633** (2006) 25.
[23] COMPASS Collaboration, C. Adolph *et al.*, Phys. Rev. D **87** (2013) 052018.
[24] SMC Collaboration, B. Adeva *et al.*, Phys. Rev. D **70** (2004) 012002.
[25] HERMES Collaboration, A. Airapetian *et al.*, Journal of High Energy Physics **1008** (2010) 130.

EPJ Web of Conferences **112**, 02001 (2016)
DOI: 10.1051/epjconf/201611202001

CGC beyond eikonal accuracy: finite width target effects

Tolga Altinoluk[1,2,a], Nestor Armesto[1], Guillaume Beuf[3], and Alexis Moscoso[1]

[1]*Departamento de Física de Partículas and IGFAE, Universidade de Santiago de Compostela, E-15706 Santiago de Compostela, Galicia-Spain*
[2]*CENTRA, Instituto Superior Técnico, Universidade de Lisboa, Av. Rovisco Pais, P-1049-001 Lisboa, Portugal*
[3]*European Centre for Theoretical Studies in Nuclear Physics and Related Areas (ECT*) and Fondazione Bruno Kessler, Strada delle Tabarelle 286, I-38123 Villazzano (TN), Italy*

Abstract. We present a method to systematically include the corrections to the eikonal approximation that are associated with the finite width of the target. The retarded gluon propagator in background field is calculated at next-to-next-to-eikonal (NNE) accuracy by using this method. The corrections to the strict eikonal limit of the gluon propagator are found to be Wilson lines decorated by gradients of the background field of the target. The result is then applied to single inclusive gluon production and to single transverse spin asymmetry for a polarized target in pA collisions.

1 Introduction

The high energy limit of the dilute-dense scattering processes adopts two key approximations within the Color Glass Condensate (CGC, see [1] and references therein) effective theory. The first approximation is called the semi-classical approximation and it amounts to replacing the target by an intense classical background field. The other one is referred to as eikonal approximation, in which the constituents of the projectile just experience color rotation upon scattering with the target, picking a Wilson line at a given transverse point but integrated along the light-cone direction of propagation of the projectile, for which the target appears as infinitely Lorentz contracted. Corrections suppressed by inverse powers of the energy of the collision are systematically neglected within the eikonal limit. We study such power suppressed corrections to the CGC, namely next-to-eikonal contributions due to finite width of the target [2, 3]. A similar analysis in the context of soft gluon resummation for hard scattering amplitudes is performed in [4, 5].

2 Eikonal expansion of the retarded gluon propagator in background field

One of the most important elements to describe the dilute-dense scattering processes at high energy is the retarded gluon propagator in a classical background field. A highly boosted left-moving target can be described by a classical background gluon field $\mathcal{A}_a^-(x^+, \mathbf{x})$. Only the $(-)$ component of this field is enhanced by a Lorentz gamma factor, so the other components are negligible. Moreover, one can also

[a]e-mail: tolga.altinoluk@usc.es

neglect the x^- dependence of this field due to time dilation. In the presence of such a field, it is natural to work in the light-cone gauge $\mathcal{A}_a^+ = 0$. Since the background field is independent of x^- component, it is convenient to introduce the one dimensional Fourier transform of the retarded gluon propagator

$$G_R^{\mu\nu}(x,y)_{ab} = \int \frac{dp^+}{2\pi}\, e^{-ik^+(x^- - y^-)}\, \frac{1}{2(k^+ + i\epsilon)}\, \mathcal{G}_{k^+}^{\mu\nu}(\underline{x};\underline{y})_{ab}\,. \tag{1}$$

For our purposes we only need the $(i-)$ component of the retarded gluon propagator which is written in terms of the background scale propagator as

$$\mathcal{G}_{k^+}^{i-}(\underline{x},\underline{y})^{ab} = \frac{i}{k^+ + i\epsilon}\partial_{\mathbf{y}^i}\mathcal{G}_{k^+}^{ab}(\underline{x},\underline{y}) \tag{2}$$

with the notation $\underline{x} \equiv (x^+, \mathbf{x})$, $\mathbf{x}$ being the transverse component. The scalar propagator $\mathcal{G}_{k^+}^{ab}(\underline{x},\underline{y})$ is the retarded solution of the Green's equation

$$\left[\delta^{ab}\left(i\partial_{x^+} + \frac{\partial_{\mathbf{x}}^2}{2(k^+ + i\epsilon)}\right) + g\big(\mathcal{A}^-(\underline{x})\cdot T\big)^{ab}\right]\mathcal{G}_{k^+}^{bc}(\underline{x};\underline{y}) = i\,\delta^{ac}\,\delta^{(3)}(\underline{x}-\underline{y})\,. \tag{3}$$

The scalar Green's equation (3) has a form of a Schrodinger equation in 2 + 1 dimensions with a matrix potential. Its solution can thus be written as a path integral [6–8]. For our purposes, we need the discretized form of the path integral, which reads

$$\mathcal{G}_{k^+}^{ab}(\underline{x},\underline{y}) = \lim_{N\to\infty}\int\left(\prod_{n=1}^{N-1} d^2\mathbf{z}_n\right)\left[\prod_{n=0}^{N-1}\mathcal{G}_{0,k^+}(z_{n+1}^+,\mathbf{z}_{n+1};z_n^+,\mathbf{z}_n)\right]\mathcal{U}^{ab}\left(x^+,y^+;\{\mathbf{z}_n\}\right), \tag{4}$$

with the boundary conditions $\mathbf{z}_0 \equiv \mathbf{y}$ and $\mathbf{z}_N \equiv \mathbf{x}$, and

$$z_n^+ = y^+ + \frac{n}{N}(x^+ - y^+)\,. \tag{5}$$

The free scalar propagator appearing in the path integral expression, given in equation (4) reads

$$\mathcal{G}_{0,k^+}(\underline{x},\underline{y}) = \theta(x^+ - y^+)\left(\frac{-ik^+}{2\pi(x^+ - y^+)}\right)\exp\left(\frac{ik^+}{2(x^+ - y^+)}(\mathbf{x}-\mathbf{y})^2\right). \tag{6}$$

The discretized Wilson line is defined as

$$\mathcal{U}^{ab}(x^+,y^+,\{\mathbf{z}_n\}) = \mathcal{P}_+\left\{\prod_{n=0}^{N-1}\exp\left[ig\frac{(x^+ - y^+)}{N}\big(\mathcal{A}^-(z_n^+,\mathbf{z}_n)\cdot T\big)\right]\right\}^{ab}, \tag{7}$$

with $\mathcal{P}_+$ denoting path ordering along the x^+ direction.

In the expressions of the observables related to the dilute-dense scattering at high energy, the background propagator appears typically through its Fourier transform $\int d^2\mathbf{x}\, e^{-i\mathbf{k}\cdot\mathbf{x}}\,\mathcal{G}_{k^+}^{ab}(\underline{x},\underline{y})$. In order to study the corrections to the eikonal limit, we define the medium modification factor $\widetilde{\mathcal{R}}_{\underline{k}}^{ab}(x^+,y^+;\mathbf{x})$ as

$$\int d^2\mathbf{x}\, e^{-i\mathbf{k}\cdot\mathbf{x}}\,\mathcal{G}_{k^+}^{ab}(\underline{x},\underline{y}) = \widetilde{\mathcal{R}}_{\underline{k}}^{ab}(x^+,y^+;\mathbf{y})\int d^2\mathbf{x}\, e^{-i\mathbf{k}\cdot\mathbf{x}}\,\mathcal{G}_{0,k^+}(\underline{x},\underline{y})\,. \tag{8}$$

In order to calculate the expansion of $\widetilde{\mathcal{R}}_{\underline{k}}^{ab}(x^+,y^+;\mathbf{x})$ beyond the eikonal approximation, it is more convenient to consider the large k^+ limit than the limit of the large boost of the target. The two limits of course equivalent due to the invariance of the whole scattering process under longitudinal boosts.

As it is discussed in details in [2, 3], the large k^+ expansion of the medium modification factor is performed in two steps. First, the k^+ expansion is calculated for fixed $\mathbf{k}/k^+$(semi-classical expansion) and then the result is re-expanded for small $\mathbf{k}/k^+$(small angle expansion). At the end, after performing both the semi-classical and the small angle expansion, the medium modification factor $\widetilde{\mathcal{R}}^{ab}_{\underline{k}}(x^+, y^+; \mathbf{x})$ reads

$$\widetilde{\mathcal{R}}_{\underline{k}}(x^+, y^+; \mathbf{y}) = \mathcal{U}(x^+, y^+; \mathbf{y}) + \frac{(x^+ - y^+)\mathbf{k}^i}{k^+} \mathcal{U}^i_{[0,1]}(x^+, y^+; \mathbf{y}) + i\frac{(x^+ - y^+)}{2k^+} \mathcal{U}_{[1,0]}(x^+, y^+; \mathbf{y}) \tag{9}$$
$$+ \frac{(x^+ - y^+)^2\mathbf{k}^i\mathbf{k}^j}{(k^+)^2} \mathcal{U}^{ij}_{[0,2]}(x^+, y^+; \mathbf{y}) + i\frac{(x^+ - y^+)^2\mathbf{k}^i}{2(k^+)^2} \mathcal{U}^i_{[1,1]}(x^+, y^+; \mathbf{y}) - \frac{(x^+ - y^+)^2}{4(k^+)^2} \mathcal{U}_{[2,0]}(x^+, y^+; \mathbf{y})$$
$$+ O\left(\left(\frac{(x^+ - y^+)|\mathbf{k}|}{k^+}\partial_\perp\right)^3\right) + O\left(\left(\frac{(x^+ - y^+)}{k^+}\partial^2_\perp\right)^3\right).$$

Here $\mathcal{U}(x^+, y^+; \mathbf{y})$ is the usual Wilson line operator and $\mathcal{U}_{[i,j]}(x^+, y^+; \mathbf{y})$ are the *decorated Wilson line* operators that appear beyond the strict eikonal limit. For the details of the calculation and the explicit expressions of the decorated Wilson lines, we refer to the reader [3].

3 Single inclusive gluon production in pA collisions

In the CGC formalism, a highly boosted left-moving nucleus is usually described by a classical gluon shockwave $\mathcal{A}^\mu_a(x) = \delta^{\mu -}\, \delta(x^+)\mathcal{A}^-_a(\mathbf{x})$ in the light-cone gauge $A^+_a = 0$. That field has indeed a vanishing longitudinal width and no x^- dependence in the limit of infinite boost.

Consider instead a background field

$$\mathcal{A}^\mu_a(x) = \delta^{\mu -}\ \mathcal{A}^-_a(x^+, \mathbf{x}) \tag{10}$$

with a finite support along the x^+ direction, from $x^+ = 0$ to $x^+ = L^+$. In the case of a large nucleus, this should be the dominant finite-boost correction with respect to the usual gluon shockwave. On the other hand, a highly boosted right-moving proton, considered as dilute, is described by a classical color current

$$j^\mu_a(x) = \delta^{\mu -}\ j^+_a(x) \tag{11}$$

with zero width along x^-: $j^+_a(x) \propto \delta(x^-)$. Let us consider a proton-nucleus collision with a particular impact parameter $\mathbf{B}$ and choose the center of the nucleus as the reference point for the transverse plane, so that a generic point $\mathbf{x}$ in the transverse plane is at a distance $|\mathbf{x} - \mathbf{B}|$ from the center of the proton and at a distance $|\mathbf{x}|$ from the center of the nucleus.Then, the color current $j^+_a(x)$ can be written as $j^+_a(x) = \delta(x^-)\ \mathcal{U}^{ab}(x^+, -\infty, \mathbf{x})\rho^b(\mathbf{x}-\mathbf{B})$ where ρ^b is the transverse density of color charges inside the proton before it reaches the nucleus, and $\mathcal{U}^{ab}(x^+, -\infty, \mathbf{x})$ is the Wilson line implementing the color precession of these color charges in the background field $\mathcal{A}^-_a(x^+, \mathbf{x})$ of the nucleus. One can define the Fourier transform of the color charge density

$$\rho^a(\mathbf{y} - \mathbf{B}) = \int \frac{d^2\mathbf{q}}{(2\pi)^2} e^{i\mathbf{q}\cdot(\mathbf{y}-\mathbf{B})}\tilde{\rho}(\mathbf{q}) \tag{12}$$

and the gluon-nucleus reduced amplitude, $\overline{\mathcal{M}}^{ab}_\lambda(\underline{k}, \mathbf{q})$, as

$$\mathcal{M}^a_\lambda(\underline{k}, \mathbf{B}) = \int \frac{d^2\mathbf{q}}{(2\pi)^2}\ e^{-i\mathbf{q}\cdot\mathbf{B}}\ \overline{\mathcal{M}}^{ab}_\lambda(\underline{k}, \mathbf{q})\ \tilde{\rho}^b(\mathbf{q})\,, \tag{13}$$

where the gluon production amplitude is $\mathcal{M}^a_\lambda(\underline{k}, \mathbf{B})$ is given as

$$\mathcal{M}^a_\lambda(\underline{k}, \mathbf{B}) = \epsilon^{i*}(2k^+) \lim_{x^+\to\infty} \int d^2\mathbf{x} \int dx^- e^{ik\cdot x} \int d^4y\, G^{i-}_R(x,y)_{ab}\, j^+_b(y). \tag{14}$$

The cross section for single inclusive gluon production can be written as

$$k^+ \frac{d\sigma}{dk^+\, d^2\mathbf{k}} = \int \frac{d^2\mathbf{q}}{(2\pi)^2}\, \varphi_p(\mathbf{q})\, \frac{\mathbf{q}^2}{4}\, \frac{1}{N_c^2-1} \sum_{\lambda\,\text{phys.}} \left\langle \overline{\mathcal{M}}^{ab}_\lambda(\underline{k},\mathbf{q})^\dagger\, \overline{\mathcal{M}}^{ab}_\lambda(\underline{k},\mathbf{q}) \right\rangle_A , \tag{15}$$

where $\varphi_p(\mathbf{q})$ is the unintegrated gluon distribution. By using the expanded expression of the medium modification factor one can write the reduced gluon-nucleus amplitude at next-to-next-to-eikonal accuracy as

$$\begin{aligned}
\overline{\mathcal{M}}^{ab}_\lambda\left(\underline{k};\mathbf{q}\right) &= i\,\varepsilon^{i*}_\lambda \int d^2\mathbf{y}\, e^{i\mathbf{y}\cdot(\mathbf{q}-\mathbf{k})} \Bigg\{ 2\left(\frac{\mathbf{k}^i}{\mathbf{k}^2} - \frac{\mathbf{q}^i}{\mathbf{q}^2}\right) \mathcal{U}(L^+,0;\mathbf{y}) \\
&+ \left(\frac{L^+}{k^+}\right)\left[\left(\delta^{ij} - 2\,\frac{\mathbf{q}^i\mathbf{k}^j}{\mathbf{q}^2}\right)\mathcal{U}^j_{[0,1]}(L^+,0;\mathbf{y}) - i\frac{\mathbf{q}^i}{\mathbf{q}^2}\,\mathcal{U}_{[1,0]}(L^+,0;\mathbf{y})\right] \\
&+ \left(\frac{L^+}{k^+}\right)^2 \Bigg[-2\frac{\mathbf{q}^i}{\mathbf{q}^2}\mathbf{k}^j\mathbf{k}^l\,\mathcal{U}^{jl}_{[0,2]}(L^+,0;\mathbf{y}) - i\frac{\mathbf{q}^i\mathbf{k}^j}{\mathbf{q}^2}\,\mathcal{U}^j_{[1,1]}(L^+,0;\mathbf{y}) + \frac{1}{2}\frac{\mathbf{q}^i}{\mathbf{q}^2}\,\mathcal{U}_{[2,0]}(L^+,0;\mathbf{y}) \\
&+ \frac{i}{4}\left(\mathbf{k}^2\delta^{ij} - 2\mathbf{k}^i\mathbf{k}^j\right)\mathcal{U}^j_{(A)}(L^+,0;\mathbf{y}) + \frac{\mathbf{k}^j}{4}\,\mathcal{U}^{ij}_{(B)}(L^+,0;\mathbf{y}) + \frac{i}{4}\,\mathcal{U}^i_{(C)}(L^+,0;\mathbf{y})\Bigg]\Bigg\}^{ab} .
\end{aligned} \tag{16}$$

After squaring the reduced gluon-nucleus amplitude without summing over the helicity λ, one can realise that the transverse momentum structure of the strict eikonal and next-to-next-to-eikonal terms are symmetric under the exchange of $i \leftrightarrow j$ where i and j are the indices of the transverse components of the physical polarisation vectors in the amplitude and complex conjugate amplitude. On the other hand, the transverse momentum structure of the next-to-eikonal terms is anti-symmetric under the same exchange.

In the calculation of the single inclusive gluon production cross section one should sum over the gluon polarisations, which leads to

$$\sum_\lambda \epsilon^{i*}_\lambda \epsilon^j_\lambda = \delta^{ij} . \tag{17}$$

Since the transverse momentum structure of the next-to-eikonal terms are anti-symmetric under the exchange of $i \leftrightarrow j$, this contribution to the single inclusive gluon production cross section vanishes. The final expression for the single inclusive gluon cross section reads at next-to-next-to-eikonal accuracy

$$\begin{aligned}
k^+\frac{d\sigma}{dk^+d^2\mathbf{k}} &= \int \frac{d^2\mathbf{q}}{(2\pi)^2}\varphi(\mathbf{q})\frac{\mathbf{q}^2}{4} \int d^2\mathbf{r}e^{i\mathbf{r}\cdot(\mathbf{q}-\mathbf{k})}\Bigg\{4C^i(\mathbf{k},\mathbf{q})C^j(\mathbf{k},\mathbf{q})O(\mathbf{r}) + \left(\frac{L^+}{k^+}\right)^2\Bigg[2\frac{\mathbf{q}^i}{\mathbf{q}^2}C^i(\mathbf{k},\mathbf{q}) \\
&\times\Big[-2\mathbf{k}^l\mathbf{k}^m O^{l,m}_{[0,2]}(\mathbf{r}) - 2i\mathbf{k}^l O^l_{[1,1]}(\mathbf{r}) + O_{[2,0]}(\mathbf{r})\Big] + C^i(\mathbf{k},\mathbf{q})\Big[i(\mathbf{k}^2\delta^{il} - 2\mathbf{k}^i\mathbf{k}^l)O^l_{(A)}(\mathbf{r}) + \mathbf{k}^m O^{im}_{(B)}(\mathbf{r}) \\
&+iO^i_{(C)}(\mathbf{r})\Big] + \tilde{C}^{li}(\mathbf{k},\mathbf{q})\Big[\tilde{C}^{li}(\mathbf{k},\mathbf{q})O^{l;m}_{[0,1];[0,1]}(\mathbf{r}) + 2i\frac{\mathbf{q}^i}{\mathbf{q}^2}O^l_{[0,1];[1,0]}(\mathbf{r})\Big] + \frac{1}{\mathbf{q}^2}O_{[1,0];[1,0]}(\mathbf{r})\Bigg\}
\end{aligned} \tag{18}$$

where $C^i(\mathbf{k},\mathbf{q}) = \left(\frac{\mathbf{k}^i}{\mathbf{k}^2} - \frac{\mathbf{q}^i}{\mathbf{q}^2}\right)$, $\tilde{C}^{ij}(\mathbf{k},\mathbf{q}) = \left(\delta^{ij} - 2\mathbf{k}^i\frac{\mathbf{q}^j}{\mathbf{q}^2}\right)$ and the dipole operators are defined as

$$O(\mathbf{r}) = \int d^2\mathbf{b}\frac{1}{N_c^2-1}\left\langle \mathrm{tr}\left[\mathcal{U}\left(\mathbf{b}+\frac{\mathbf{r}}{2}\right)\mathcal{U}^\dagger\left(\mathbf{b}-\frac{\mathbf{r}}{2}\right)\right]\right\rangle_A , \tag{19}$$

$$O^{i\cdots j}_{[\alpha,\beta]}(\mathbf{r}) = \int d^2\mathbf{b}\frac{1}{N_c^2-1}\left\langle \mathrm{tr}\left[\mathcal{U}^{i\cdots j}_{[\alpha,\beta]}\left(\mathbf{b}+\frac{\mathbf{r}}{2}\right)\mathcal{U}^\dagger\left(\mathbf{b}-\frac{\mathbf{r}}{2}\right)\right]\right\rangle_A , \tag{20}$$

$$O^{i\cdots j;l\cdots m}_{[\alpha,\beta];[\gamma,\delta]}(\mathbf{r}) = \int d^2\mathbf{b}\frac{1}{N_c^2-1}\left\langle \mathrm{tr}\left[\mathcal{U}^{i\cdots j}_{[\alpha,\beta]}\left(\mathbf{b}+\frac{\mathbf{r}}{2}\right)\mathcal{U}^{l\cdots m\dagger}_{[\gamma,\delta]}\left(\mathbf{b}-\frac{\mathbf{r}}{2}\right)\right]\right\rangle_A . \tag{21}$$

Recently, the decorated Wilson lines that appear beyond the eikonal limit have been calculated explicitly to leading order in the field of the target [9]. These corrections, indeed, determine the next-to-eikonal and next-to-next-to-eikonal corrections to the Lipatov vertex.

4 SIngle transverse spin asymmetry: polarized target

Another interesting observable that one can discuss beyond the eikonal limit is the single transverse spin asymmetry (SSA) for the process $p + A^\uparrow \to g + X$ i.e. with a transversely polarized target. The SSA is defined as

$$A_N = \frac{k^+\frac{d\sigma^\uparrow}{dk^+d^2\mathbf{k}} - k^+\frac{d\sigma^\downarrow}{dk^+d^2\mathbf{k}}}{k^+\frac{d\sigma^\uparrow}{dk^+d^2\mathbf{k}} + k^+\frac{d\sigma^\downarrow}{dk^+d^2\mathbf{k}}} = \frac{k^+\frac{d\sigma^\uparrow}{dk^+d^2\mathbf{k}} - k^+\frac{d\sigma^\downarrow}{dk^+d^2\mathbf{k}}}{2k^+\frac{d\sigma}{dk^+d^2\mathbf{k}}} , \tag{22}$$

where the denominator corresponds to the unpolarized cross section.

Within the CGC framework, the dependence on the transverse spin vector $\mathbf{s}$ should appear through the probability distribution for the background field that leads to $\mathbf{s}$-dependent target averaged quantities, i.e. $\mathbf{s}$-dependent standard and decorated adjoint dipoles. The transverse rotational symmetry around the center of the target still holds for the $\mathbf{s}$-dependent decorated dipoles. Hence, under the transformation $\mathbf{r} \to -\mathbf{r}$ and $\mathbf{s} \to -\mathbf{s}$, the decorated dipole operators with even number of transverse indices are invariant and the ones with odd number of indices flip sign.

Due to transverse rotational symmetry around the center of the target, one gets

$$O(-\mathbf{r},-\mathbf{s}) = O(\mathbf{r},\mathbf{s}) . \tag{23}$$

Moreover, the dipole operator satisfies

$$O(-\mathbf{r},\mathbf{s}) = O(\mathbf{r},\mathbf{s}) . \tag{24}$$

As discussed above, under the transformations $\mathbf{r} \to -\mathbf{r}$ and $\mathbf{s} \to -\mathbf{s}$, the decorated dipole operators with even number of transverse indices are invariant and the ones with odd number of indices flip sign. Using these symmetry properties, it is straight forward to show that the strict eikonal and next-to-next-to-eikonal contributions to the numerator of SSA vanish whereas the next-to-eikonal contributions do not vanish. After summing over the gluon polarizations, it can be written as

$$k^+\left(\frac{d\sigma^\uparrow}{dk^+d^2\mathbf{k}} - \frac{d\sigma^\downarrow}{dk^+d^2\mathbf{k}}\right)\Bigg|_{\mathrm{NE}} = 2\frac{L^+}{k^+}\int\frac{d^2\mathbf{q}}{(2\pi)^2}\,\varphi_p(\mathbf{q})\,\mathbf{q}^2\int d^2\mathbf{r}\; e^{i\mathbf{r}\cdot(\mathbf{q}-\mathbf{k})}$$
$$\times C^i(\mathbf{k},\mathbf{q})\left\{\tilde{C}^{ji}(\mathbf{k},\mathbf{q})\,O^j_{[0,1]}(\mathbf{r},\mathbf{s}) - i\frac{\mathbf{q}^i}{\mathbf{q}^2}O_{[1,0]}(\mathbf{r},\mathbf{s})\right\}. \tag{25}$$

5 Conclusions

We have presented a systematic expansion of the retarded gluon propagator in a background field. The corrections associated with the finite width of the target were calculated at next-to-next-to-eikonal accuracy. These corrections involve new operators referred to as decorated Wilson lines since they include the gradients of the background field along the longitudinal extend of the target. The resulting gluon background propagator is then used to calculate two different observables within the CGC framework: the single inclusive gluon production cross section and the single transverse spin asymmetry for a transversely polarized target. For the single inclusive gluon cross section, it has been shown that the next-to-eikonal corrections vanish and the first non-vanishing contribution to the strict eikonal limit that appear at next-to-next-to-eikonal order have been calculated. On the other hand, for the single transverse spin asymmetry for a transversely polarized target, it has been shown that both strict eikonal terms and next-to-next-to-eikonal terms vanish and the leading contribution to this observable turns out to be next-to-eikonal ones.

References

[1] F. Gelis, E. Iancu, J. Jalilian-Marian and R. Venugopalan, Ann. Rev. Nucl. Part. Sci. **60**, 463 (2010) [arXiv:1002.0333 [hep-ph]].
[2] T. Altinoluk, N. Armesto, G. Beuf, M. Martinez and C. A. Salgado, JHEP **1407**, 068 (2014) [arXiv:1404.2219 [hep-ph]].
[3] T. Altinoluk, N. Armesto, G. Beuf and A. Moscoso, arXiv:1505.01400 [hep-ph].
[4] E. Laenen, G. Stavenga and C. D. White, JHEP **0903**, 054 (2009) [arXiv:0811.2067 [hep-ph]].
[5] E. Laenen, L. Magnea, G. Stavenga and C. D. White, JHEP **1101**, 141 (2011) [arXiv:1010.1860 [hep-ph]].
[6] R. Baier, Y. L. Dokshitzer, A. H. Mueller, S. Peigne and D. Schiff, Nucl. Phys. B **483**, 291 (1997) [hep-ph/9607355].
[7] B. G. Zakharov, JETP Lett. **63**, 952 (1996) [hep-ph/9607440].
[8] B. G. Zakharov, Phys. Atom. Nucl. **61**, 838 (1998) [Yad. Fiz. **61**, 924 (1998)] [hep-ph/9807540].
[9] T. Altinoluk and A. Dumitru, arXiv:1512.00279 [hep-ph].

EPJ Web of Conferences **112**, 02002 (2016)
DOI: 10.1051/epjconf/201611202002

Rapidity evolution of gluon TMD from low to moderate x

Ian Balitsky[1,2,a]

[1] JLab, 12000 Jefferson Ave, Newport News, VA 23606, USA
[2] Phys. Dept., Old Dominion Univ., 1600 Elkhorn Ave, Norfolk, VA 23529, USA

Abstract. I discuss how the rapidity evolution of gluon transverse momentum dependent distribution changes from nonlinear evolution at small $x \ll 1$ to linear evolution at moderate $x \sim 1$.

1 Introduction

The TMDs [1–3] (also called unintegrated parton distributions) are widely used in the analysis various scattering processes in JLab and elsewhere on low-energy accelerators. However, since at the future EIC accelerator the majority of the produced particles will be gluons (turning to hadrons due to confinement) one needs to study also the evolution of gluon TMDs. The TMD generalizes the usual concept of parton density by allowing PDFs to depend on intrinsic transverse momenta in addition to the usual longitudinal momentum fraction variable. The unintegrated gluon distribution is defined as [4]

$$\mathcal{D}(x_B, k_\perp, \eta) = \int d^2 z_\perp \; e^{i(k,z)_\perp} \mathcal{D}(x_B, z_\perp, \eta), \tag{1}$$

$$\alpha_s \mathcal{D}(x_B, z_\perp, \eta) = \frac{-x_B^{-1}}{8\pi^2 (p \cdot n)} \int du \; e^{-ix_B u(pn)} \langle p | \mathcal{F}_\xi^a (z_\perp + un) [z_\perp + \infty n, \infty n]^{ab} \mathcal{F}^{b\xi}(0) | p \rangle$$

where $|p\rangle$ is an unpolarized target with momentum p (typically proton) and ∞ can be plus or minus infinity. Hereafter we use the notation

$$\mathcal{F}_\xi^a (z_\perp + un) \equiv [\infty n + z_\perp, un + z_\perp]^{am} n^\mu g F_{\mu\xi}^m (un + z_\perp) \tag{2}$$

where $[x, y]$ denotes straight-line gauge link connecting points x and y:

$$[x, y] \equiv \mathrm{P} e^{ig \int du \; (x-y)^\mu A_\mu (ux + (1-u)y)} \tag{3}$$

There are more involved definitions with Eq. (1) multiplied by some Wilson-line factors [3, 5] following from CSS factorization [6] but we will discuss the "primordial" TMD (1).

---

[a] e-mail: balitsky@jlab.org

2 Rapidity factorization and evolution

In the spirit of high-energy OPE, the rapidity of the gluons in the operator (2) is restricted from above by the "rapidity divide" η separating the impact factor and the matrix element

$$A^{\eta}_{\mu}(x) = \int \frac{d^4k}{16\pi^4}\,\theta(e^{\eta} - |\alpha|)e^{-ik\cdot x}A_{\mu}(k) \tag{4}$$

where the Sudakov variable α is defined as usual, $k = \alpha p_1 + \beta p_2 + k_\perp$. We define the light-like vectors p_1 and p_2 such that $p_1 = n$ and $p_2 = p - \frac{m^2}{s}n$, where p is the momentum of the target particle of mass m. We use metric $g^{\mu\nu} = (1,-1,-1,-1)$ so $p\cdot q = (\alpha_p\beta_q + \alpha_q\beta_p)\frac{s}{2} - (p,q)_\perp$. For the coordinates we use the notations $x_\bullet \equiv x_\mu p_1^\mu$ and $x_* \equiv x_\mu p_2^\mu$ related to the light-cone coordinates by $x_* = \sqrt{\frac{s}{2}}x_+$ and $x_\bullet = \sqrt{\frac{s}{2}}x_-$. Also, hereafter we use the notation $[\infty, z_*]_z \equiv [\infty_* p_1 + z_\perp, \frac{2}{s}z_* p_1 + z_\perp]$ where $[x,y]$ stands for the straight-line gauge link connecting points x and y as defined in Eq. (3). Our convention is that the Latin Lorentz indices always correspond to transverse coordinates while Greek Lorentz indices are four-dimensional.

Similarly, we define

$$\tilde{\mathcal{F}}^{a\eta}_i(x_B, z_\perp) \equiv \frac{2}{s}\int dz_*\; e^{-ix_B z_*} g(\tilde{F}^m_{\bullet i}(z_*, z_\perp)[z_*,\infty]^{ma}_z)^{\eta} \tag{5}$$

in the complex-conjugate part of the amplitude.

In this notations the unintegrated gluon TMD (1) can be represented as

$$\begin{aligned}
\langle p|\tilde{\mathcal{F}}^{a\eta}_i(x_B, z_\perp)\mathcal{F}^{ai\eta}(x_B, 0_\perp)|p + \xi p_2\rangle &\equiv \sum_X \langle p|\tilde{\mathcal{F}}^{a\eta}_i(x_B, z_\perp)|X\rangle\langle X|\mathcal{F}^{ai\eta}(x_B, 0_\perp)|p + \xi p_2\rangle \\
&= -4\pi^2\delta(\xi)x_B g^2\mathcal{D}(x_B, z_\perp, \eta)
\end{aligned} \tag{6}$$

Hereafter we use a short-hand notation

$$\langle p|\tilde{O}_1...\tilde{O}_m O_1...O_n|p'\rangle \equiv \sum_X \langle p|\tilde{T}\{\tilde{O}_1...\tilde{O}_m\}|X\rangle\langle X|T\{O_1...O_n\}|p'\rangle \tag{7}$$

where tilde on the operators in the l.h.s. of this formula stands as a reminder that they should be inverse time ordered as indicated by inverse-time ordering $\tilde{T}$ in the r.h.s. of the above equation.

As discussed e.g. in Ref. [7], such martix element can be represented by a double functional integral

$$\begin{aligned}
&\langle \tilde{O}_1...\tilde{O}_m O_1...O_n\rangle \\
&= \int D\tilde{A}D\tilde{\bar{\psi}}D\tilde{\psi}\; e^{-iS_{\rm QCD}(\tilde{A},\tilde{\psi})}\int DAD\bar{\psi}D\psi\; e^{iS_{\rm QCD}(A,\psi)}\tilde{O}_1...\tilde{O}_m O_1...O_n
\end{aligned} \tag{8}$$

with the boundary condition $\tilde{A}(\vec{x}, t = \infty) = A(\vec{x}, t = \infty)$ (and similarly for quark fields) reflecting the sum over all intermediate states X. Due to this condition, the matrix element (6) can be made gauge-invariant by connecting the endpoints of Wilson lines at infinity with the gauge link

$$\langle p|\tilde{\mathcal{F}}^a_i(x_B, x_\perp)\mathcal{F}^{ai}(x'_B, y_\perp)|p'\rangle \to \langle p|\tilde{\mathcal{F}}^a_i(x_B, x_\perp)[x_\perp + \infty p_1, y_\perp + \infty p_1]\mathcal{F}^{ai}(\beta'_B, y_\perp)|p'\rangle \tag{9}$$

This gauge link is important if we use the light-like gauge $p_1^\mu A_\mu = 0$ for calculations but in all other gauges it can be neglected. We will not write it down explicitly but will always assume it in our formulas.

2.1 Evolution kernel in the leading order

We will study the rapidity evolution of the operator

$$\tilde{\mathcal{F}}_i^{a\eta}(x_B, x_\perp)\mathcal{F}_j^{a\eta}(x_B, y_\perp) \tag{10}$$

Our aim is to construct an evolution of the TMD operator with $\eta = \ln\sigma$, where σ is a corresponding cutoff in α. Tofindthe evolution kernel, we introduce a new cutoff η' and integrate over "slow" fields with rapidity $\eta' < y < \eta$, where $y = \ln\alpha$. We suppose that the "fast" fields with $y < \eta'$ are fixed and serve as a background field. The typical diagrams are given in Fig. 1.

Figure 1. Typical diagrams for production (a) and virtual (b) contributions to the evolution kernel. The dashed lines denote gauge links.

We aim to obtain evolution at the one loop order so there is only one "quantum" (blue) line of the "slow" fields propagating in the "classical" (red) background of the "fast" fields. In calculation of this diagrams we distinguish between two limits: $k_\perp^2 \sim l_\perp^2$ and $k_\perp^2 \gg l_\perp^2$, where $k_\perp$ is a transverse momentum of the "quantum" fields, while $l_\perp$ is the characteristic transverse scale of the external fields.

It is easy to understand that the typical longitudinal distance traveled by the "slow" fields is $\alpha s/k_\perp^2 \sim \sigma s/k_\perp^2$. At the same time the longitudinal scale of the "fast" fields is $\sigma_* \sim \sigma' s/l_\perp^2$. We suppose that $\sigma \gg \sigma'$, so in the limit $k_\perp^2 \sim l_\perp^2$ the "quantum" fields effectively propagate through a thin area of the "fast" fields of the background. This is the shock-wave approximation: the red (external) fields in Fig. 1 are concentrated within a segment $(-\sigma_*, \sigma)$. Of course this segment is very short, but in our calculation it is nevertheless finite. We consider propagation of the "quantum" (blue) fields through this segment and expand the corresponding propagator in powers of $\frac{k_\perp^2}{\alpha s}\sigma * \sim \frac{\sigma'}{\sigma} \ll 1$.

However, in the limit $k_\perp \gg l_\perp$ this scheme is not valid: there is no separation in distance between the "slow" and "fast" fields. Insteed of the shock-wave approximation, in this case we apply the light-cone expansion approach [8]. We take a ration $l_\perp/k_\perp \ll 1$ as an expansion parameter and expand the gluon propagator in Fig. 1 around the light ray $y_\perp + \frac{2}{s}y_* p_1$, which is the direction of the Wilson line in definition (1).

2.2 The Lipatov vertex

As usual, there are two contributions into the evolution kernel: the real gluon emission (Fig. 1a) and the virtual diagrams (Fig. 1b). The "real" correction to the TMD operator (10) is

$$\langle\tilde{\mathcal{F}}_i^a(x_\perp, x_B)\mathcal{F}_j^a(y_\perp, x_B)\rangle^{\ln\sigma} = -\int_{\sigma'}^{\sigma}\frac{d\alpha}{2\alpha}d^2k_\perp\,\left(\tilde{L}_i^{ba;\mu}(k, x_\perp, x_B)L_{\mu j}^{ab}(k, y_\perp, x_B)\right)^{\ln\sigma'}, \tag{11}$$

where the Lipatov vertex $L^{ab}_{\mu i}(k, y_\perp, x_B) = i\lim_{k^2\to 0} k^2\langle A^a_\mu(k)\mathcal{F}^b_i(y_\perp, x_B)\rangle$ is an amplitude of the single-gluon production by the operator $\mathcal{F}^b_i$. There are several possible types of emission. The corresponding diagrams can be found in Fig. 2. We start with the calculation of the Lipatov vertex in the light-cone

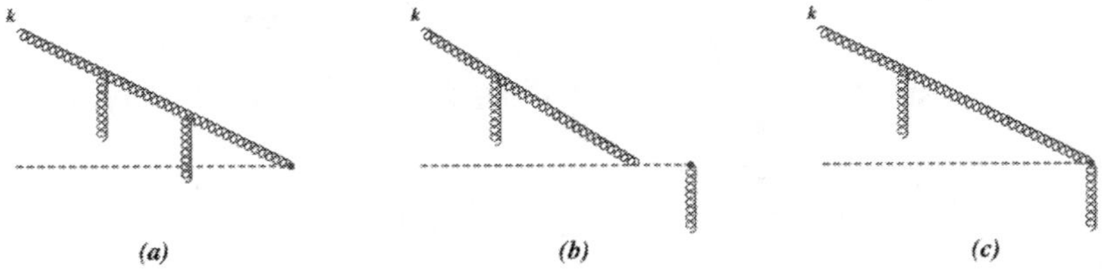

Figure 2. Diagrams contributing to Lipatov vertex

approach. We expand the gluon propagator around direction of the Wilson line and neglect terms suppressed by $l_\perp/k_\perp \ll 1$. One can show that the leading contribution comes from the terms linear in the gluon strength tensor $F_{\bullet i}$. The external (red) field has no specific longitudinal properties and we take into account all possible emissions from the Wilson line and the strength tensor in the definition (2) as well. The details of the calculation can be found in [9]. Here we show only final result:

$$L^{ab}_{\mu i}(k, y_\perp, x_B) = i\lim_{k^2\to 0} k^2\langle A^{aq}_\mu(k)(\mathcal{F}^b_i(x_B, y_\perp))^{1\text{st}}\rangle = \frac{2ge^{-i(k,y)_\perp}}{\alpha x_B s + k_\perp^2} \quad (12)$$
$$\times\Big[\frac{\alpha x_B s}{k_\perp^2}\Big(\frac{k_\perp^2}{\alpha s}p_{2\mu} - \alpha p_{1\mu}\Big)\delta^l_i - \delta^l_\mu k_i + \frac{\alpha x_B s g_{\mu i}k^l}{k_\perp^2 + \alpha x_B s} + \frac{2\alpha k_i k^l}{k_\perp^2 + \alpha x_B s}p_{1\mu}\Big]\mathcal{F}^{ab}_l\Big(x_B + \frac{k_\perp^2}{\alpha s}, y_\perp\Big).$$

On the other hand, in the limit $k_\perp \sim l_\perp$ we apply the shock-wave approximation. In this case the external field in Fig. 2 shrinks to fill a small area from $-\sigma_*$ to σ_* in longitudinal direction. Effectively the diagrams in Fig. 2 are the diagrams in the shock-wave background, see Fig. 3. Now we take

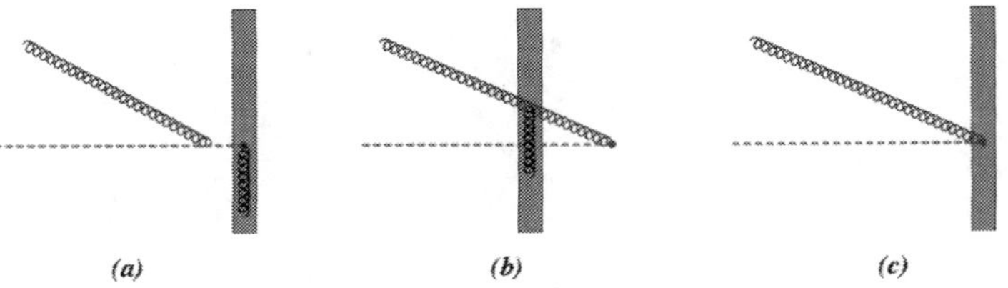

Figure 3. Lipatov vertex in the shock-wave approximation

into account not only different types of emission, but the position of the emission point as well: we separate the emission from the left, right sides and from inside of the shock-wave. The sum of all diagrams in Fig. 3 is

$$L^{ab}_{\mu i}(k, y_\perp, x_B) = 2ge^{-i(k,y)_\perp}\Big(\frac{p_{2\mu}}{\alpha s} - \frac{\alpha p_{1\mu}}{k_\perp^2}\Big)[\mathcal{F}_i(x_B, y_\perp) - U_i(y_\perp)]^{ab} \quad (13)$$
$$+ g(k_\perp|g_{\mu i}\Big(\frac{\alpha x_B s}{\alpha x_B s + p_\perp^2} - U\frac{\alpha x_B s}{\alpha x_B s + p_\perp^2}U^\dagger\Big) + 2\alpha p_{1\mu}\Big(\frac{p_i}{\alpha x_B s + p_\perp^2} - U\frac{p_i}{\alpha x_B s + p_\perp^2}U^\dagger\Big)$$
$$+ \Big[2ix_B p_{2\mu}\partial_i U - 2i\partial^\perp_\mu U p_i + \frac{2p_{2\mu}}{\alpha s}\partial^2_\perp U p_i\Big]\frac{1}{\alpha x_B s + p_\perp^2}U^\dagger - \frac{2\alpha p_{1\mu}}{p_\perp^2}U_i|y_\perp)^{ab}.$$

In this calculation we expand the "slow" gluon propagator in powers of the size of the shock-wave and neglect terms suppressed by $\frac{p_\perp^2}{\alpha s}\sigma_* \ll 1$.

Next, we combine the two results (12 and (13) in an interpolation formula:

$$L^{ab}_{\mu i}(k, y_\perp, x_B)$$
$$= g(k_\perp|\frac{\alpha x_B s g_{\mu i} + 2\alpha p_{1\mu}k_i}{\alpha x_B s + k_\perp^2}(2ik^j\partial_j U - \partial_\perp^2 U)\frac{1}{\alpha x_B s + p_\perp^2}U^\dagger + 2i\alpha p_{1\mu}\partial_i U\frac{1}{\alpha x_B s + p_\perp^2}U^\dagger$$
$$+ \frac{2i}{\alpha s}p_{2\mu}\partial_i U\frac{\alpha x_B s}{\alpha x_B s + p_\perp^2}U^\dagger - [2i\partial_\mu U - \frac{2p_{2\mu}}{\alpha s}\partial_\perp^2 U]\frac{p_i}{\alpha x_B s + p_\perp^2}U^\dagger - \frac{2\alpha p_{1\mu}}{p_\perp^2}i(\partial_i U)U^\dagger|y_\perp)^{ab}$$
$$+ \frac{2ge^{-i(k,y)_\perp}}{\alpha x_B s + k_\perp^2}[-\delta^j_\mu k_i + \frac{2\alpha k_i k^j p_{1\mu}}{\alpha x_B s + k_\perp^2} + \frac{\alpha x_B s g_{\mu i}k^j}{\alpha x_B s + k_\perp^2} + x_B p_{2\mu}\delta^j_i - \alpha p_{1\mu}\frac{\alpha x_B s}{k_\perp^2}\delta^j_i]$$
$$\times [\mathcal{F}_j(x_B + \frac{k_\perp^2}{\alpha s}, y_\perp) - U_j(y_\perp)]^{ab} \quad (14)$$

In Ref. [9] we prove that it is correct in the whole region of values of x_B. It is also important that (14) doesn't break the gauge invariance: $k^\mu L^{ab}_{\mu i}(k, y_\perp, x_B) = 0$. Finally, we use the Lipatov vertex (14) to calculate the "real" part of the evolution kernel according to (11).

2.3 The virtual emission

The second part of the evolution kernel is the virtual correction, see Fig. 1b. For calculation of this diagrams we apply the same strategy we used in derivation of the Lipatov vertex. In the limit $k_\perp \gg l_\perp$ we expand the gluon propagator around $y_\perp + \frac{2}{s}y_* p_1$ direction and neglect terms suppressed by $l_\perp/k_\perp \ll 1$. What is left is used in the explicit calculation of the diagrams if Fig. 1b. (There are actually several of them, see [9]). In the limit $k_\perp \sim l_\perp$ we use the shock-wave approximation so the

Figure 4. The virtual correction in the shock-wave approximation

virtual emission diagrams take a form given in Fig. 4. We show that the sum of this diagrams can be combined with the result obtained at $k_\perp \gg l_\perp$ and one can write an expression which describes both limits and is valid for all values of x_B:

$$\langle \mathcal{F}^n_i(y_\perp, x_B)\rangle^{\text{virt}} = \quad (15)$$
$$= -ig^2 f^{nkl}\int_{\sigma'}^{\sigma}\frac{d\alpha}{\alpha}(y_\perp|\frac{p^j}{p_\perp^2}(2\partial_i\partial_j U + g_{ij}\partial_\perp^2 U)\frac{1}{\alpha x_B s + p_\perp^2}U^\dagger + \frac{\alpha x_B s p_\perp^{-2}}{\alpha x_B s + p_\perp^2}\mathcal{F}_i(x_B)|y_\perp)^{kl}.$$

3 Evolution equation for gluon TMD

Now we can write the one-loop correction to the operator (1). It is a sum of the product of two Lipatov vertexes (14) and the virtual contribution (15):

$$\langle \tilde{\mathcal{F}}^a_i(x_\perp, x_B)\mathcal{F}^a_j(y_\perp, x_B)\rangle^{\ln\sigma}_{\text{one-loop}} = -\int_{\sigma'}^{\sigma}\frac{d\alpha}{2\alpha}d^2k_\perp\,(\tilde{L}^{ba;\mu}_i(k, x_\perp, x_B)L^{ab}_{\mu j}(k, y_\perp, x_B))^{\ln\sigma'}$$
$$+\tilde{\mathcal{F}}^a_i(x_\perp, x_B)\langle\mathcal{F}^a_j(y_\perp, x_B)\rangle^{\text{virt}} + \langle\tilde{\mathcal{F}}^a_i(x_\perp, x_B)\rangle^{\text{virt}}\mathcal{F}^a_j(y_\perp, x_B). \quad (16)$$

To get the evolution equation we differentiate this expression over σ. Combining all results together we get

$$
\begin{aligned}
&\frac{d}{d\log\sigma}\langle p|\tilde{\mathcal{F}}^a_i(x_\perp,x_B)\mathcal{F}^a_j(y_\perp,x_B)|p\rangle \qquad (17)\\
&= -\alpha_s\langle p|\mathrm{Tr}\Big\{\int d^2k_\perp\theta\Big(1-x_B-\frac{k_\perp^2}{\sigma s}\Big)\Big[(x_\perp|\Big(\tilde U\frac{1}{\sigma x_B s+p_\perp^2}(\tilde U^\dagger k_k+p_k\tilde U^\dagger)\\
&\times\frac{\sigma x_B s g_{\mu i}-2k^\perp_\mu k_i}{\sigma x_B s+k_\perp^2} - 2k^\perp_\mu g_{ik}\tilde U\frac{1}{\sigma x_B s+p_\perp^2}\tilde U^\dagger - 2g_{\mu k}\tilde U\frac{p_i}{\sigma x_B s+p_\perp^2}\tilde U^\dagger\Big)\tilde{\mathcal{F}}^k\Big(x_B+\frac{k_\perp^2}{\sigma s}\Big)|k_\perp)\\
&\times(k_\perp|\mathcal{F}^l\Big(x_B+\frac{k_\perp^2}{\sigma s}\Big)\Big(\frac{\sigma x_B s\delta^\mu_j-2k^\mu_\perp k_j}{\sigma x_B s+k_\perp^2}(k_lU+Up_l)\frac{1}{\sigma x_B s+p_\perp^2}U^\dagger\\
&-2k^\mu_\perp g_{jl}U\frac{1}{\sigma x_B s+p_\perp^2}U^\dagger - 2\delta^\mu_l U\frac{p_j}{\sigma x_B s+p_\perp^2}U^\dagger\Big)|y_\perp)\\
&+2(x_\perp|\tilde{\mathcal{F}}_i\Big(x_B+\frac{k_\perp^2}{\sigma s}\Big)|k_\perp)(k_\perp|\mathcal{F}^l\Big(x_B+\frac{k_\perp^2}{\sigma s}\Big)\Big(\frac{k_j}{k_\perp^2}\frac{\sigma x_B s+2k_\perp^2}{\sigma x_B s+k_\perp^2}(k_lU+Up_l)\frac{1}{\sigma x_B s+p_\perp^2}U^\dagger\\
&+2U\frac{g_{jl}}{\sigma x_B s+p_\perp^2}U^\dagger-2\frac{k_l}{k_\perp^2}U\frac{p_j}{\sigma x_B s+p_\perp^2}U^\dagger\Big)|y_\perp)\\
&+2(x_\perp|\Big(\tilde U\frac{1}{\sigma x_B s+p_\perp^2}(\tilde U^\dagger k_k+p_k\tilde U^\dagger)\frac{k_i}{k_\perp^2}\frac{\sigma x_B s+2k_\perp^2}{\sigma x_B s+k_\perp^2}+2\tilde U\frac{g_{ik}}{\sigma x_B s+p_\perp^2}\tilde U^\dagger\\
&-2\tilde U\frac{p_i}{\sigma x_B s+p_\perp^2}\tilde U^\dagger\frac{k_k}{k_\perp^2}\Big)\tilde{\mathcal{F}}^k\Big(x_B+\frac{k_\perp^2}{\sigma s}\Big)|k_\perp)(k_\perp|\mathcal{F}_j\Big(x_B+\frac{k_\perp^2}{\sigma s}\Big)|y_\perp)\Big]\\
&+2\tilde{\mathcal{F}}_i(x_\perp,x_B)(y_\perp|-\frac{p^m}{p_\perp^2}\mathcal{F}_k(x_B)(i\overleftarrow{\partial}_l+U_l)(2\delta^k_m\delta^l_j-g_{jm}g^{kl})U\frac{1}{\sigma x_B s+p_\perp^2}U^\dagger|y_\perp)\\
&+2(x_\perp|\tilde U\frac{1}{\sigma x_B s+p_\perp^2}\tilde U^\dagger(2\delta^k_i\delta^l_m-g_{im}g^{kl})(i\partial_k-\tilde U_k)\tilde{\mathcal{F}}_l(x_B)\frac{p^m}{p_\perp^2}|x_\perp)\mathcal{F}_j(y_\perp,x_B)\\
&-4\int\frac{d^2k_\perp}{k_\perp^2}\Big[\theta\Big(1-x_B-\frac{k_\perp^2}{\sigma s}\Big)\tilde{\mathcal{F}}_i\Big(x_\perp,x_B+\frac{k_\perp^2}{\sigma s}\Big)\mathcal{F}_j\Big(y_\perp,x_B+\frac{k_\perp^2}{\sigma s}\Big)e^{i(k,x-y)_\perp}\\
&-\frac{\sigma x_B s}{\sigma x_B s+k_\perp^2}\tilde{\mathcal{F}}_i(x_\perp,x_B)\mathcal{F}_j(y_\perp,x_B)\Big]\Big\}|p\rangle\ +\ O(\alpha_s^2).
\end{aligned}
$$

The kinematical constraint $k_\perp^2\leq\alpha(1-\beta_B)s$ in the production part of the amplitude reflects the fact that so the sum of the fraction $\beta_B p_2$ and the fraction $\frac{k_\perp^2}{\alpha s}p_2$ carried by the emitted gluon should be smaller than the initial hadron's momentum is $p\simeq p_2$.

The formula (17) is the main result of our study. It is an evolution equation of the gluon TMD operator (1) and is valid for all (except too small) values of transverse momentum $k_\perp\sim(x_\perp-y_\perp)^{-1}$ and longitudinal momentum fraction variable x_B.

Let us discuss the gauge invariance of this equation. The l.h.s. is gauge invariant after taking into account gauge link at $+\infty$ as shown in Eq. (9). As to the right side, it was obtained by calculation in the background field and promoting the background fields to operators in a usual way. However, we performed our calculations in a specific background field $A_\bullet(x_*,x_\perp)$ with a finite support in $x_\perp$ and we need to address the question how can we restore the r.h.s. of Eq. (17) in a generic field A_μ. It is easy to see how one can restore the gauge-invariant form: just add gauge link at $+\infty p_1$ or $-\infty p_1$ appropriately. For example, the terms $U_z(z|\frac{1}{\sigma\beta s+p_\perp^2}|z')U^\dagger_{z'}$ in r.h.s. of should be replaced by

$U_z[z_\perp - \infty p_1, z'_\perp - \infty p_1](z|\frac{1}{\sigma\beta s + p_\perp^2}|z')U^\dagger_{z'}$. After performing these insertions we will have the result which is (i) gauge invariant and (ii) coincides with Eq. (17) for our choice of background field. At this step, the background fields in the r.h.s. of Eq. (17) can be promoted to operators. However, the explicit display of these gauge links at $\pm\infty$ will make the evolution equation much less readable so we assume they are always in place rather than written explicitly.

It is demonstrated in Ref. [9] that the equation (17) has three correct limits: at $x \sim 1$ and large transverse momentum it yields the linear DGLAP equation, at $x \sim 1$ and small $k_\perp$ it gives the Sudakov double logarithm evolution, and at small x it reduces to the non-linear equation of the Balitsky-Kovchegov (BK) type.

4 Conclusions

I have described the rapidity evolution of gluon TMD (1) in the whole range of Bjorken x_B and the whole range of transverse momentum $k_\perp$. It should be emphasized that with our definition of rapidity cutoff (4) the leading-order matrix elements of TMD operators are UV-finite so the rapidity evolution is the only evolution and it describes all the dynamics of gluon TMDs (1) in the leading-log approximation.

As an outlook, it would be very interesting to obtain the NLO correction to the evolution equation (17). The NLO corrections to the BFKL and BK equation are available but they suffer from the well-known problem that they lead to negative cross sections. This difficulty can be overcome by the "collinear resummation" of double-logarithmic contributions for the BFKL [10] and BK [11] equations and we hope that our Eq. (17) and especially its future NLO version will help to solve the problem of negative cross sections of NLO amplitudes at high energies.

This work on TMD evolution was done in collaboration with A. Tarasov. In addition, the author is grateful G.A. Chirilli and A.V. Radyushkin for valuable discussions. The work was supported by contract DE-AC05-06OR23177 under which the Jefferson Science Associates, LLC operates the Thomas Jefferson National Accelerator Facility, and by the grant DE-FG02-97ER41028.

References

[1] J. C. Collins and D. E. Soper, Nucl. Phys. **B194**, 445 (1982).

[2] X. Ji, Jian-Ping Ma, and F. Yuan, Phys. Rev. **D71**, 034005 (2005).

[3] J.C. Collins , *Foundations of Perturbative QCD* (Cambridge University Press, Cambridge, 2011)

[4] P. J. Mulders and J. Rodrigues, Phys. Rev. **D63**, 094021 (2001).

[5] M.G. Echevarria, A. Idilbi, and I. Scimeni, JHEP bf 07 002 (2012).

[6] J.C. Collins, D. E. Soper and G. Sterman, Nucl. Phys. **B250**, 199 (1985).

[7] I. Balitsky and V.M. Braun, Phys. Lett.**B222**, 121 (1989); Nucl. Phys. **B361**, 93 (1991); Nucl. Phys. **B380**, 51 (1992).

[8] I. Balitsky, Phys. Lett. **B124**, 230 (1983); I. Balitsky and V.M. Braun, Nucl. Phys. **B311**, 541 (1989).

[9] I. Balitsky and A. Tarasov, JHEP. **1510**, 017 (2015)

[10] G.P. Salam, JHEP**9807** 019 (1998); M. Ciafaloni, D. Colferai, G.P. Salam, Phys.Rev. **D60** 114036 (1999); M. Ciafaloni, D. Colferai, G.P. Salam, and A.M. Stasto, Phys.Rev. **D68** 114003 (2003);

[11] L. Motyka and A. M. Stasto Phys.Rev. **D79** 085016 (2009); G. Beuf, Phys.Rev. **D89** 074039 (2014); E. Iancu, J.D. Madrigal, A.H. Mueller, G. Soyez, D.N. Triantafyllopoulos, Phys.Lett. **B744** 293 (2015).

EPJ Web of Conferences **112**, 02003 (2016)
DOI: 10.1051/epjconf/201611202003

Gluon TMD studies at EIC

Daniël Boer[1,a]

[1] *Van Swinderen Institute for Particle Physics and Gravity, University of Groningen Nijenborgh 4, NL-9747 AG Groningen, The Netherlands*

Abstract. A high-energy Electron-Ion Collider (EIC) would offer a most promising tool to study in detail the transverse momentum distributions of gluons inside hadrons. This applies to unpolarized as well as linearly polarized gluons inside unpolarized protons, and to left-right asymmetric distributions of gluons inside transversely polarized protons, the so-called gluon Sivers effect. The inherent process dependence of these distributions can be studied by comparing to similar, but often complementary observables at LHC.

1 Introduction

Transverse momentum dependent parton distributions (TMDs) are currently under active investigation, both theoretically and experimentally. Typical TMD processes are semi-inclusive Deep Inelastic Scattering (SIDIS) or the Drell-Yan process. The SIDIS process ($e\,p \to e'\,h\,X$) is sensitive to the transverse momentum of quarks, while for instance D-meson pair production ($e\,p \to e' D\,\overline{D}\,X$) is sensitive to the transverse momentum of gluons in the back-to-back correlation limit. For studies of the gluon TMDs, higher energy ($\sqrt{s}$) or smaller x is required. A high-energy Electron-Ion Collider (EIC) can offer clean probes of the distributions of unpolarized and linearly polarized gluons inside unpolarized protons, and of the gluon Sivers effect for transversely polarized protons. These distributions and what we can learn about them at an EIC will be reviewed here, with emphasis on the most promising observables, the process dependence, and the expected small-x behavior of the distributions.

Describing the transverse momentum of partons in a process is not just a matter of adding a transverse momentum dependence in collinear distributions, i.e. $f_1(x) \to f_1(x, k_T^2)$, that appear in collinear factorization expressions. Rather one has to deal with TMD factorization, in which new factors and new distributions appear, such as the Sivers effect TMD that describes a correlation between the transverse momentum and the proton spin.

For gluons there are eight leading twist TMDs [1] that parametrize the gluon correlator

$$\Gamma_g^{\mu\nu\,[\mathcal{U},\mathcal{U}']}(x, k_T) \equiv \int \frac{d(\xi \cdot P)\, d^2\xi_T}{(xP \cdot n)^2 (2\pi)^3} e^{i(xP+k_T)\cdot\xi} \langle P|\mathrm{Tr}_c\big[F^{n\nu}(0)\, \mathcal{U}_{[0,\xi]}\, F^{n\mu}(\xi)\, \mathcal{U}'_{[\xi,0]}\big]|P\rangle_{\xi\cdot P'=0}. \quad (1)$$

The dependence on the gauge links $\mathcal{U}$ and $\mathcal{U}'$ will be discussed later on. For unpolarized hadrons the correlator Γ_g is parametrized by two gluon TMDs [1] (here $\boldsymbol{k}_T^2 = -k_T^2$):

$$\Gamma_g^{\mu\nu}(x, \boldsymbol{k}_T) = -\frac{1}{2x}\left\{ g_T^{\mu\nu} f_1^g - \left(\frac{k_T^\mu k_T^\nu}{M^2} + g_T^{\mu\nu} \frac{\boldsymbol{k}_T^2}{2M^2} \right) h_1^{\perp g} \right\}. \quad (2)$$

[a] e-mail: d.boer@rug.nl

The unpolarized gluon TMD f_1^g and linearly polarized gluon TMD $h_1^{\perp g}$ are both functions of x and $\boldsymbol{k}_T^2$. Nonzero $h_1^{\perp g}$ requires nonzero transverse momentum and stems from an interference between ± 1 gluon helicities. For positive $h_1^{\perp g}$ the gluon polarization $\boldsymbol{\epsilon}_T$ is distributed around $\boldsymbol{k}_T$ with a $\cos 2\phi$ distribution ($\phi = \angle(\boldsymbol{k}_T, \boldsymbol{\epsilon}_T)$). Linear gluon polarization modifies among others the transverse momentum distribution of Higgs production (perturbatively at NNLO [2, 3] and nonperturbatively at LO in pQCD [4, 5]), which can be studied at LHC. In TMD factorization the cross section takes the form [5]

$$\frac{E\, d\sigma^{pp\to HX}}{d^3\vec{q}}\bigg|_{q_T \ll m_H} \propto \left(\mathcal{C}\left[f_1^g\, f_1^g\right] + \mathcal{C}\left[w_H\, h_1^{\perp g}\, h_1^{\perp g}\right]\right) + O\left(\frac{q_T}{m_H}\right), \tag{3}$$

where $\mathcal{C}$ denotes a convolution of TMDs and $w_H = \left((\boldsymbol{k}_{1T}\cdot \boldsymbol{k}_{2T})^2 - \frac{1}{2}\boldsymbol{k}_{1T}^2\boldsymbol{k}_{2T}^2\right)/2M^4$. Including resummation of large logarithms the contribution of linearly polarized gluons relative to unpolarized gluons is given by (for explanations cf. [5]):

$$\mathcal{R}(Q_T) \equiv \frac{\mathcal{C}[w_H\, h_1^{\perp g}\, h_1^{\perp g}]}{\mathcal{C}[f_1^g\, f_1^g]} = \frac{\int d^2\boldsymbol{b}\, e^{i\boldsymbol{b}\cdot\boldsymbol{q}_T}\, e^{-S_A(b_*,Q)-S_{NP}(b,Q)}\, \widetilde{h}_1^{\perp g}(x_A, b_*^2; \mu_{b_*})\, \widetilde{h}_1^{\perp g}(x_B, b_*^2; \mu_{b_*})}{\int d^2\boldsymbol{b}\, e^{i\boldsymbol{b}\cdot\boldsymbol{q}_T}\, e^{-S_A(b_*,Q)-S_{NP}(b,Q)}\, \widetilde{f}_1^g(x_A, b_*^2; \mu_{b_*})\, \widetilde{f}_1^g(x_B, b_*^2; \mu_{b_*})}, \tag{4}$$

where $\widetilde{f}_1^g$ denotes the Fourier transform of f_1^g and

$$\widetilde{h}_1^{\perp g}(x, b^2) \equiv \int d^2\boldsymbol{k}_T\, \frac{(\boldsymbol{b}\cdot\boldsymbol{k}_T)^2 - \frac{1}{2}\boldsymbol{b}^2\boldsymbol{k}_T^2}{b^2M^2}\, e^{-i\boldsymbol{b}\cdot\boldsymbol{k}_T}\, h_1^{\perp g}(x, \boldsymbol{k}_T^2) = -\pi \int dk_T^2 \frac{k_T^2}{2M^2} J_2(bk_T) h_1^{\perp g}(x, k_T^2). \tag{5}$$

The integrand in b space has been split into a calculable perturbative part and a nonperturbative (NP) part that should be obtained from fits to data. Although the nonperturbative Sudakov factor S_{NP} for $gg \to H$ is unknown, at the Higgs scale it does not matter too much. What matters most is the small-b part of the TMDs, which is perturbatively calculable [2, 4, 6]: $\widetilde{f}_1^g(x, b^2; \mu_b) = f_{g/P}(x; \mu_b) + O(\alpha_s)$, while

$$\widetilde{h}_1^{\perp g}(x, b^2; \mu_b) = \frac{\alpha_s(\mu_b) C_A}{2\pi} \int_x^1 \frac{d\hat{x}}{\hat{x}} \left(\frac{\hat{x}}{x} - 1\right) f_{g/P}(\hat{x}; \mu_b) + O(\alpha_s^2). \tag{6}$$

Note that the perturbative tail of $h_1^{\perp g}$ is driven by the unpolarized collinear gluon distribution $f_{g/P}(x; \mu)$.

In [7] and [8] the above expressions were studied numerically (cf. [9] for the ranges of the predictions). The conclusion from those studies is that $\mathcal{R}(Q_T)$ is on the order of 2-5% in Higgs production at low Q_T. This probably means that the extraction of $h_1^{\perp g}$ from Higgs production will be too challenging. In [7] and [8] also heavy (pseudo-)scalar $C = +$ quarkonium production, $p\,p \to [\overline{Q}Q]\,X$, has been studied. Much larger effects from linear gluon polarization are possible in this case, but there are very large uncertainties (cf. [9]). It is much more sensitive to the unknown NP part than Higgs production. From this perspective the heavier bottomonium states are probably best to consider. Employing the color singlet model [10] and LO NRQCD results [11, 12], the differential cross sections for η_b, χ_{b0} and χ_{b2} production have been obtained in [13]. By forming ratios of ratios, in which the hadronic uncertainties cancel, it becomes in principle possible to probe $\mathcal{R}(Q_T)$ directly:

$$\frac{\sigma(\chi_{b2})}{\sigma(\chi_{b0})}\frac{d\sigma(\chi_{b0})/d^2\boldsymbol{q}_T}{d\sigma(\chi_{b2})/d^2\boldsymbol{q}_T} \approx 1 + \mathcal{R}(Q_T), \qquad \frac{\sigma(\chi_{b0})}{\sigma(\eta_b)}\frac{d\sigma(\eta_b)/d^2\boldsymbol{q}_T}{d\sigma(\chi_{b0})/d^2\boldsymbol{q}_T} \approx \frac{1 - \mathcal{R}(Q_T)}{1 + \mathcal{R}(Q_T)}. \tag{7}$$

These are color singlet model expressions, which may be justified for $C = +$ bottomonium states from NRQCD considerations [12, 14] and by several numerical studies of color octet contributions [14–16]. TMD factorization for the p-wave states χ_{bJ} has been called into question though [17]. Consistency between the experimental results for $\mathcal{R}(Q_T)$ from (7), e.g. at LHCb, can be used to assess the possible factorization breaking contributions. Because of the small energy scale differences ($m_{\eta_b} = 9.4$ GeV, $m_{\chi_{b0}} = 9.9$ GeV, $m_{\chi_{b2}} = 10.3$ GeV), evolution effects should be negligible in this comparison.

2 EIC probes

At EIC $h_1^{\perp g}$ can be probed in open charm and bottom quark pair electro-production, $e\,p \to e'\,Q\,\overline{Q}\,X$, where Q and $\overline{Q}$ are almost back-to-back in the transverse plane. Unlike Higgs production one needs to study angular distributions now, e.g. a $\cos 2\phi$ asymmetry where $\phi = \phi_T - \phi_\perp$ and $\phi_{T/\perp}$ are the angles of $K_\perp^Q \pm K_\perp^{\overline{Q}}$ [18], under the restriction $q_T \equiv (K_\perp^Q + K_\perp^{\overline{Q}}) \ll K_\perp \equiv (K_\perp^Q - K_\perp^{\overline{Q}})/2$. In the asymmetry expression, $h_1^{\perp g}$ appears by itself, as opposed to in a product of two. Therefore, larger effects are expected and the sign of $h_1^{\perp g}$ can be determined. The asymmetry depends on Q^2, $K_\perp^2$ and M_Q^2, but the maximum of the asymmetry is to a large extent independent of these scales, and around 15% [19]. There are also angular asymmetries w.r.t. the lepton scattering plane that probe $h_1^{\perp g}$. These are mostly relevant at smaller $|K_\perp|$ [19]. Dijet DIS, $e\,p \to e'\,\text{jet}\,\text{jet}\,X$, is similar except that also quark TMDs enter. The analogous processes $p\,p \to Q\,\overline{Q}\,X$ and $p\,p \to \text{jet}\,\text{jet}\,X$ at RHIC or LHC are not expected to be TMD factorizing [20].

At an EIC one also can consider transversely polarized protons, where $e\,p^\uparrow \to e'\,Q\,\overline{Q}\,X$ is a very promising process for probing the gluon Sivers effect. For a review of the status and prospects of the gluon Sivers distribution, cf. [21], and for specific model studies, cf. [22]. There are also suggestions to measure the gluon Sivers effect in proton-proton/ion collisions (RHIC, AFTER@LHC), in processes for which TMD factorization may hold: $p^\uparrow\,p \to \gamma\,\text{jet}\,X$ [23, 24], $p^\uparrow\,p \to J/\psi\,\gamma\,X$ [25], $p^\uparrow\,p \to J/\psi\,J/\psi\,X$ [26, 27]. According to [25, 26], the color singlet contribution to a large extent dominates over the color octet one in the J/ψ production processes.

Such gluon Sivers effect measurements in $p\,p$ collisions are complementary to EIC studies, because TMDs are actually process dependent, as will be discussed next. Although this process dependence can be calculated, not all Sivers functions from all processes can be related to each other!

3 Process dependence

It has been realized that TMDs in general are not universal [28–30]. Gluon rescattering corrections can be summed into path-ordered exponentials $\mathcal{U}_C$ in TMD correlators [31], where the gauge link or Wilson line is along a path C. The path in this Wilson line depends on whether color charges are coming from the initial state or going into the final state [29, 30, 32–35]. Surprisingly, it has turned out that in certain cases the shape of the Wilson lines affects observables, such as the Sivers effect asymmetries. In SIDIS the quark TMD correlator has a future pointing staple-like Wilson line arising from final state interactions (FSI), referred to as a + link. In the Drell-Yan (DY) process it is past pointing from initial state interactions (ISI), a − link. The quark Sivers functions with + and − links are related by parity and time reversal invariance by an overall minus sign: $f_{1T}^{\perp[\text{SIDIS}]} = -f_{1T}^{\perp[\text{DY}]}$ [29]. In general, the more hadrons observed in a process, the more complicated the resulting Wilson lines and the possible relations among TMDs of various processes [36–38]. Wilson lines may even become entangled or trapped, leading to factorization breaking [20, 39].

The processes that allow access to the linearly polarized gluon distribution and the gluon Sivers distribution depend on two gauge links as in Eq. (1). The subprocess $\gamma^*\,g \to Q\,\overline{Q}$ for $e\,p \to e'\,Q\,\overline{Q}\,X$ probes a gluon correlator with two + links, i.e. both are future pointing. In the kinematic regime where gluons in one proton in $p\,p \to \gamma\,\text{jet}\,X$ dominate, one effectively selects the subprocess $q\,g \to \gamma\,q$. The latter subprocess probes a gluon correlator with a + and − link (future and past pointing), enclosing an area. As a consequence, these two processes probe two distinct, independent gluon Sivers functions. They correspond to antisymmetric (f_{abc}) and symmetric (d_{abc}) color structures as discussed in [40].

At LHC $gg \to H$ and $gg \to [\overline{Q}Q]$ both probe a gluon correlator with two − links. As $h_1^{\perp g\,[+,+]} = h_1^{\perp g\,[-,-]}$ (and $h_1^{\perp g\,[+,-]} = h_1^{\perp g\,[-,+]}$), one concludes that EIC and LHC can probe the same $h_1^{\perp g}$ function.

Table 1. List of processes that probe the WW and/or DP unpolarized gluon TMD at small x [41].

	DIS & DY	SIDIS	$pA \to hX$	$pA \to \gamma\,\mathrm{jet}\,X$	Dijet in DIS	Dijet in pA
$f_1^{g\,[+,+]}$ (WW)	×	×	×	×	√	√
$f_1^{g\,[+,-]}$ (DP)	√	√	√	√	×	√

But e.g. $gg \to Hg$ probes a more complicated link structure. On the other hand, for the gluon Sivers function it holds that $f_{1T}^{\perp g\,[+,+]} = -f_{1T}^{\perp g\,[-,-]}$ and $f_{1T}^{\perp g\,[+,-]} = -f_{1T}^{\perp g\,[-,+]}$. One thus concludes that the proposed gluon Sivers TMD studies at EIC and at RHIC or AFTER@LHC are complementary [21].

This TMD nonuniversality is not just a polarization issue. It was first realized in a small-x context that this process dependence also applies to the unpolarized gluon TMD f_1^g [41].

4 Small-x: a tale of two gluon distributions

At small x (and large N_c) there are two unpolarized gluon distributions that matter [41], the gluon correlator with two + links (for which $f_1^{g\,[+,+]} = f_1^{g\,[-,-]}$) and the one with a + and a − link (for which $f_1^{g\,[+,-]} = f_1^{g\,[-,+]}$). In [41] these were denoted by $G^{(1)}$ and $G^{(2)}$, respectively. At small x they correspond to the Weizsäcker-Williams (WW) and dipole (DP) distributions, which are in general different. The fact that there are two distinct but equally valid definitions for the gluon distribution was noted first in "A tale of two gluon distributions" by Kharzeev, Kovchegov & Tuchin (KKT) in [42], where the authors say that they "cannot offer any simple physical explanation of this paradox". The explanation turns out to be the process dependence of the gluon distribution, in other words, its sensitivity to the ISI/FSI in a process. Here it is not so much the direction, but rather whether a process is only sensitive to *either* ISI or FSI or to *both* ISI and FSI. The difference between the WW and DP distributions would disappear without ISI/FSI. In the MV model considered by KKT, one may not notice the origin for the difference, because the two gluon distributions become related: $xG^{(2)}(x,q_\perp) \overset{MV}{\propto} q_\perp^2 \nabla^2_{q_\perp} xG^{(1)}(x,q_\perp)$ [41, 42]. For instance, the process $\gamma A \to Q\overline{Q}X$ has been studied in [43] in the MV model, where the cross section $d\sigma_T/dy dk_\perp$ is expressed in terms of $C(k_\perp) = \int d^2x_\perp e^{ik_\perp \cdot x_\perp}\langle U(0)U^\dagger(x_\perp)\rangle \sim G^{(2)}$ which is the DP distribution, whereas the process rather probes the WW distribution $G^{(1)}$ ([+, +]).

Different processes probe $G^{(1)}$ or $G^{(2)}$ or a mixture, as listed in Table 1. For dijet production in pA collisions, the result requires large N_c, otherwise (four) additional functions appear (cf. [44]).

This process dependence of TMDs implies that also their p_T-widths are process dependent, and as a consequence, it gives an additional process dependence to p_T-broadening [45].

The WW and DP $h_1^{\perp g}$ distributions will be different too. Within the MV model [46] the DP $h_1^{\perp g}$ distribution is found to be maximal for all transverse momenta, while the WW $h_1^{\perp g}$ distribution is maximal only at large k_T ($\gg Q_s$) and suppressed w.r.t. f_1^g in the saturation region ($k_T \ll Q_s$). Maximal (positive) linear polarization also arises in the small-x "k_T-factorization" approach [47]:

$$\Gamma_g^{\mu\nu}(x, \boldsymbol{k}_T)_{\text{max pol}} = \frac{1}{x}\,\frac{k_T^\mu k_T^\nu}{\boldsymbol{k}_T^2}\, f_1^g(x, \boldsymbol{k}_T^2). \tag{8}$$

Finally, the perturbative tail of $h_1^{\perp g}$ in Eq. (6) has a $1/x$ growth, which keeps up with f_1^g towards small x. Clearly there is no theoretical reason why $h_1^{\perp g}$ should be small, especially at small x. In analogy to f_1^g, in Table 2 we list processes where the WW and DP $h_1^{\perp g}$ distributions enter (or not). It turns out that the processes DIS, DY, SIDIS, hadron and γ + jet production in pA collisions do

Table 2. List of processes that probe the WW and/or DP linearly polarized gluon TMD at small x.

	DIS & DY	SIDIS	$pA \to hX$	$pA \to \gamma\,\text{jet}\,X$	Dijet in DIS	Dijet in pA
$h_1^{\perp g\,[+,+]}$ (WW)	×	×	×	×	√	√
$h_1^{\perp g\,[+,-]}$ (DP)	×	×	×	×	×	√

not probe $h_1^{\perp g}$ in leading power [48]. Dijet production in ep and eA collisions at small x probes the WW distribution. Since there are different expectations inside and outside the saturation region, it would thus be very interesting to study $h_1^{\perp g}$ (via the $\cos 2\phi$ asymmetries discussed earlier) in dijet DIS at a high-energy EIC. The relevant expressions for general x can be found in [19] and small-x expressions in [46, 49]. As said, these expressions involve only the WW-type distributions (at any N_c). In contrast, dijet and open heavy quark pair production in pp and pA collisions suffer from factorization breaking [20]. Although at small x the factorization breaking contributions may become suppressed, effectively restoring TMD factorization [44, 50], still a combination of six distinct distributions is probed, complicating the analysis considerably, probably too much.

5 Summary

Production of (pseudo-)scalar particles at LHC is a good way to probe gluon distributions, but unfortunately the effect of linear gluon polarization on Higgs production is small (2-5% level), smaller than the current theoretical uncertainty in the perturbative treatment (NNLL+NNLO). $C = +$ quarkonium states may offer alternative probes, but in this case the predictions have large theoretical uncertainties. Future LHC data on bottomonium states $\chi_{b0/2}$ and η_b are most promising. Linear gluon polarization is expected to lead to large differences between these three states.

Heavy quark pair and dijet production in DIS at a high-energy EIC offer clean channels for probing linearly polarized gluons and the gluon Sivers effect. Specific $\cos 2\phi$ asymmetries may exhibit large $h_1^{\perp g}$ effects, allowing to study its sign, its small-x behavior. It may even show saturation effects, as it probes the WW or f-type ([+, +]) distribution, which is expected to show a significant change in behavior around $k_\perp \sim Q_s$. This same distribution happens to appear in Higgs or $J^{PC} = 0^{\pm +}$ quarkonium production at LHC. In contrast, for the gluon Sivers TMD the cleanest probes at EIC and at RHIC and/or AFTER@LHC are actually entirely complementary.

Acknowledgements

I would like to thank Maarten Buffing, Wilco den Dunnen, Jean-Philippe Lansberg, Piet Mulders, Elena Petreska, Cristian Pisano, and Jian Zhou, for fruitful discussions and/or collaborations.

References

[1] P. J. Mulders and J. Rodrigues, Phys. Rev. D **63**, 094021 (2001).
[2] S. Catani and M. Grazzini, Nucl. Phys. B **845**, 297 (2011).
[3] J. Wang, C. S. Li, Z. Li, C. P. Yuan and H. T. Li, Phys. Rev. D **86**, 094026 (2012).
[4] P. Sun, B.-W. Xiao and F. Yuan, Phys. Rev. D **84**, 094005 (2011).
[5] D. Boer *et al.* Phys. Rev. Lett. **108**, 032002 (2012).
[6] P. M. Nadolsky, C. Balazs, E. L. Berger and C.-P. Yuan, Phys. Rev. D **76**, 013008 (2007).

[7] D. Boer and W. J. den Dunnen, Nucl. Phys. B **886**, 421 (2014).
[8] M. G. Echevarría, T. Kasemets, P. J. Mulders and C. Pisano, JHEP **1507**, 158 (2015).
[9] D. Boer, arXiv:1510.05915.
[10] R. Baier and R. Rückl, *Z. Phys. C* **19** (1983) 251.
[11] G. T. Bodwin, E. Braaten and G. P. Lepage, Phys. Rev. D **51**, 1125 (1995).
[12] G. T. Bodwin, E. Braaten and J. Lee, Phys. Rev. D **72**, 014004 (2005).
[13] D. Boer and C. Pisano, Phys. Rev. D **86**, 094007 (2012).
[14] P. Hägler *et al.* Phys. Rev. Lett. **86**, 1446 (2001).
[15] F. Maltoni and A. D. Polosa, Phys. Rev. D **70**, 054014 (2004).
[16] J. P. Lansberg, S. J. Brodsky, F. Fleuret and C. Hadjidakis, Few Body Syst. **53**, 11 (2012).
[17] J. P. Ma, J. X. Wang and S. Zhao, Phys. Lett. B **737** (2014) 103.
[18] D. Boer, S. J. Brodsky, P. J. Mulders and C. Pisano, Phys. Rev. Lett. **106**, 132001 (2011).
[19] C. Pisano, D. Boer, S. J. Brodsky, M. G. A. Buffing and P. J. Mulders, JHEP **1310**, 024 (2013).
[20] T.C. Rogers and P.J. Mulders, Phys. Rev. D **81**, 094006 (2010).
[21] D. Boer, C. Lorcé, C. Pisano and J. Zhou, Adv. High Energy Phys. **2015**, 371396 (2015).
[22] D. Boer *et al.*, arXiv:1108.1713.
[23] I. Schmidt, J. Soffer and J. J. Yang, Phys. Lett. B **612**, 258 (2005).
[24] A. Bacchetta *et al.*, Phys. Rev. Lett. **99**, 212002 (2007).
[25] W. J. den Dunnen *et al.*, Phys. Rev. Lett. **112**, 212001 (2014).
[26] J. P. Lansberg *et al.*, EPJ Web Conf. **85**, 02038 (2015).
[27] J. P. Lansberg and H. S. Shao, Nucl. Phys. B **900**, 273 (2015).
[28] S. J. Brodsky, D. S. Hwang and I. Schmidt, Phys. Lett. B **530**, 99 (2002).
[29] J. C. Collins, Phys. Lett. B **536**, 43 (2002).
[30] A. V. Belitsky, X. Ji and F. Yuan, Nucl. Phys. B **656**, 165 (2003).
[31] A. V. Efremov and A. V. Radyushkin, Theor. Math. Phys. **44**, 774 (1981).
[32] J. C. Collins, D. E. Soper and G. F. Sterman, Phys. Lett. B **126**, 275 (1983).
[33] D. Boer and P. J. Mulders, Nucl. Phys. B **569**, 505 (2000).
[34] S. J. Brodsky, D. S. Hwang and I. Schmidt, Nucl. Phys. B **642**, 344 (2002).
[35] D. Boer, P. J. Mulders and F. Pijlman, Nucl. Phys. B **667**, 201 (2003).
[36] C. J. Bomhof, P. J. Mulders and F. Pijlman, Phys. Lett. B **596**, 277 (2004).
[37] M. G. A. Buffing and P. J. Mulders, Phys. Rev. Lett. **112**, 092002 (2014).
[38] M. G. A. Buffing, A. Mukherjee and P. J. Mulders, Phys. Rev. D **86**, 074030 (2012).
[39] J. Collins and J. W. Qiu, Phys. Rev. D **75**, 114014 (2007).
[40] M. G. A. Buffing, A. Mukherjee and P. J. Mulders, Phys. Rev. D **88**, 054027 (2013).
[41] F. Dominguez, C. Marquet, B. W. Xiao and F. Yuan, Phys. Rev. D **83**, 105005 (2011).
[42] D. Kharzeev, Y. V. Kovchegov and K. Tuchin, Phys. Rev. D **68**, 094013 (2003).
[43] F. Gelis and A. Peshier, Nucl. Phys. A **697**, 879 (2002).
[44] P. Kotko *et al.*, JHEP **1509**, 106 (2015).
[45] D. Boer, M. G. A. Buffing and P. J. Mulders, JHEP **1508**, 053 (2015).
[46] A. Metz and J. Zhou, Phys. Rev. D **84**, 051503 (2011).
[47] S. Catani, M. Ciafaloni and F. Hautmann, Nucl. Phys. B **366**, 135 (1991).
[48] D. Boer, P. J. Mulders and C. Pisano, Phys. Rev. D **80**, 094017 (2009).
[49] A. Dumitru, T. Lappi and V. Skokov, Phys. Rev. Lett. **115**, 252301 (2015).
[50] G. A. Chirilli, B. W. Xiao and F. Yuan, Phys. Rev. Lett. **108**, 122301 (2012).

EPJ Web of Conferences **112**, 02004 (2016)
DOI: 10.1051/epjconf/201611202004

Classical gluon production amplitude in heavy-ion collisions

Giovanni Antonio Chirilli[1,2,a]

[1]*Department of Physics, The Ohio State University, Columbus, OH 43210, USA*
[2]*Institut für Theoretische Physik, University of Regensburg, D-93040 Regensburg, Germany*

Abstract. The distribution of quarks and gluons produced in the initial stages of nuclear collisions, known as the initial condition of the Quark-Gluon Plasma formation, is the fundamental building block of heavy-ion theory. I will present the scattering amplitude, beyond the leading order, of the classical gluon produced in heavy-ion collisions. The result is obtained in the framework of saturation physics and Wilson lines formalism.

1 Introduction

To understand the initial stages of nuclear collisions and the formation of quark gluon plasma, one has to study the dynamics of collisions from first principle in Quantum Chromodynamic (QCD). To this end, we calculate the scattering amplitude, beyond the leading order, of the classical gluon produced in Nucleus-Nucleus collisions.

As a first approximation, we study the collision of two large nuclei with atomic number $1 \ll A_1 \ll A_2$. The smaller nucleus can be treated as a diluted system of constituents when compared to the large nucleus. In particular, the saturation scale of nucleus A_1, which is proportional to its atomic number, $Q^2_{s1} \sim \alpha_s^2 A_{s1}^{1/3}$ is much smaller then the saturation scale of the large nucleus, $Q^2_{2s} \sim \alpha_s^2 A_2^{1/3}$. The atomic number of the nucleus enters in the resummation parameter. Indeed, if we consider a particle propagation through a large nucleus with atomic number A, with good probability, will encounter, along its path, $A^{1/3}$ nucleons. In the two-gluon approximation, the particle propagating through the nucleus, interacts with each nucleon with the exchange of two gluons. Hence, for large nuclei, there are contributions to the scattering amplitude that are proportional to $\alpha_s^2 A^{1/3}$ and since A is a large number, these contributions are the leading ones. In our case we have $\alpha_s^2 A_2^{1/3} \sim 1$ for the larger nucleus (the target) which in the boosted Lorentz frame reduces to a shock-wave (represented as a red stripe in all diagrams below), and $\alpha_s^2 A_1^{1/3} \leq 1$ for the smaller one (the projectile).

2 From proton-Nucleus to Heavy-light Ion Collisions

The leading-order (LO) contribution to the classical gluon production amplitude in heavy-light ions collisions coincides with proton-Nucleus (pA) collisions. The diagrams for pA collisions are shown in Fig. 1. Using light-cone coordinates $x^{\pm} = \frac{1}{\sqrt{2}}(x^0 \pm x^3)$, the proton propagates along x^+ direction and the target (large nucleus) in x^- direction. Transverse coordinate are $\vec{x}_\perp = (x^1, x^2)$. We will perform

[a]e-mail: giovanni.chirilli@physik.uni-regensburg.de

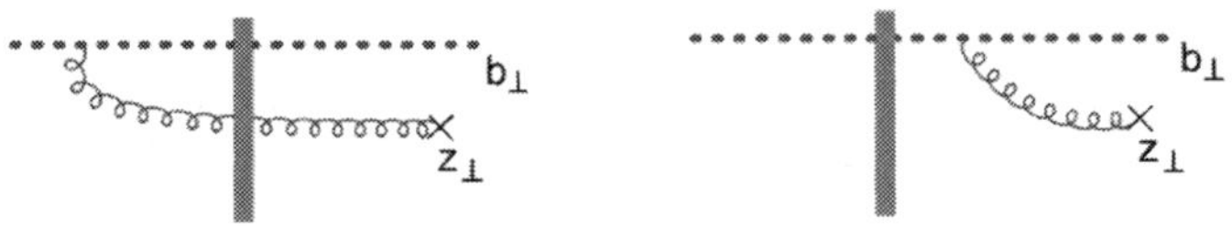

Figure 1. Feynman diagrams for pA collisions. Dashed lines represent Wilson lines.

the calculation in the $A^+ = 0$ light-cone gauge. Note that the blue dashed lines represent Wilson lines. The scattering amplitude for pA collision is [1]

$$A(\vec{z}_\perp, \vec{b}_\perp) = \frac{i\,g}{\pi}\, \frac{\vec{\epsilon}_\perp^{\,\lambda *}\cdot(\vec{z}_\perp - \vec{b}_\perp)}{|\vec{z}_\perp - \vec{b}_\perp|^2} \left[U^{ab}_{\vec{z}_\perp} - U^{ab}_{\vec{b}_\perp}\right]\left(V_{\vec{b}_\perp} t^b\right) \tag{1}$$

where $\vec{\epsilon}_\perp^{\,*}$ is the polarization vector of the emitted gluon, $U^{ab}_{\vec{z}_\perp}$ is the Wilson line in the adjoint representation at transverse position $\vec{z}_\perp$, while $V_{\vec{b}_\perp}$ is the Wilson line in the fundamental representation. The gluon production cross section is proportional to the amplitude square and is given by

$$\frac{d\sigma}{d^2k_T\, dy} = \frac{1}{2\,(2\pi)^3} \int d^2z\, d^2z'\, d^2b\, e^{-i\vec{k}_\perp\cdot(\vec{z}_\perp - \vec{z}'_\perp)} \left\langle A(\vec{z}_\perp, \vec{b}_\perp)\, A^*(\vec{z}'_\perp, \vec{b}_\perp)\right\rangle . \tag{2}$$

We are interested in calculating the next-to-leading order (NLO) contribution to the gluon production cross-section which is proportional to $\alpha_s^3 A_1^{2/3}$. The diagrams that contribute at this order may be divided in three classes as shown in Fig. 2 [2]. It is possible to reduce dramatically the number of

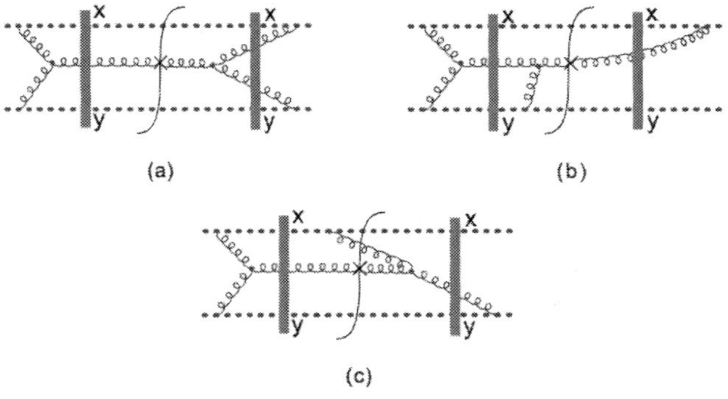

Figure 2. Here we show the three different classes of diagrams at order g^6: (a) Square of order g^3 amplitudes; (b) Interference between order g^5 and order g amplitudes; (c) Interference between order g^4 and order g^2 amplitudes.

diagrams if we sum up diagrams having gluons on one side of the cut with the corresponding diagrams having gluons crossing the cut, thus reducing the two diagrams in a single diagram with a retarded gluon propagator (see Fig.3)

$$\frac{-id^{\mu\nu}(k)}{k^2 + i\epsilon} + 2\pi\theta(-k^+)\delta(k^2)d^{\mu\nu}(k) = \frac{-id^{\mu\nu}(k)}{k^2 + i\epsilon k^+} \tag{3}$$

where $d^{\mu\nu}(k) = g^{\mu\nu} - \frac{\eta^\mu k^\nu + \eta^\nu k^\mu}{\eta\cdot k}$. A consequence of this observation is that, the only diagrams we will have to consider are those at order g^3 and those at order g^5 (at amplitude level), but with each gluon propagator being retarded propagator.

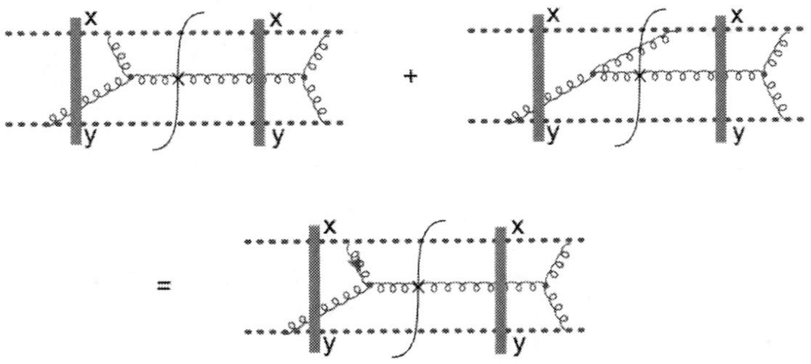

Figure 3. Summing up diagrams with gluons on one side of the cut with the corresponding diagram with gluons crossing the cut, we get a single diagram with a retarded gluon propagator.

A further reduction of the number of diagrams to be calculated is given by the following two observations. First, we notice that the sum of diagrams of the type in Fig. 5 cancel. Second, using the

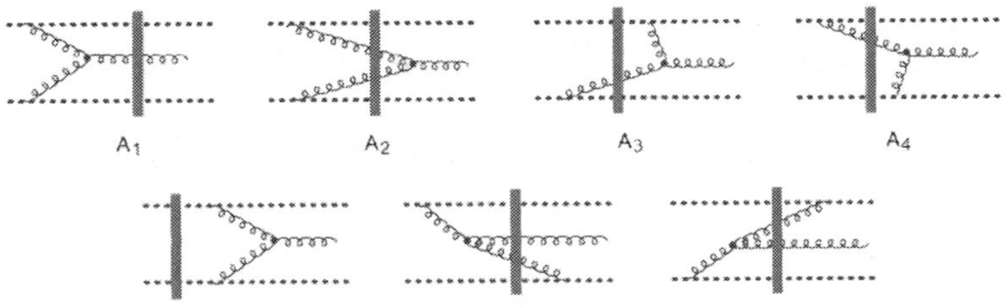

Figure 4. Diagrams A have single 3-gluon vertex with gluons connecting to both quarks. All gluons are retarded and the time flows to the right.

cancellation described in Fig. 5 we can combine diagrams as shown in Fig. 6 thus reducing further the number of diagrams (see Fig. 7).

The result for the calculation of diagrams in Fig. 4 and in Fig. 7 is [2]

$$\sum_{i=1}^{7} A_i + {\sum_{i=1}^{12}}' B_i + {\sum_{i=1}^{12}}' C_i = -\frac{g^3}{4\pi^4} \int d^2x_1\, d^2x_2\, \delta[(\vec{z}_\perp - \vec{x}_{1\perp}) \times (\vec{z}_\perp - \vec{x}_{2\perp})]$$
$$\times \left[\frac{\vec{\epsilon}_\perp^{\lambda *} \cdot (\vec{x}_{2\perp} - \vec{x}_{1\perp})}{|\vec{x}_{2\perp} - \vec{x}_{1\perp}|^2} \frac{\vec{x}_{1\perp} - \vec{b}_{1\perp}}{|\vec{x}_{1\perp} - \vec{b}_{1\perp}|^2} \cdot \frac{\vec{x}_{2\perp} - \vec{b}_{2\perp}}{|\vec{x}_{2\perp} - \vec{b}_{2\perp}|^2} - \frac{\vec{\epsilon}_\perp^{\lambda *} \cdot (\vec{x}_{1\perp} - \vec{b}_{1\perp})}{|\vec{x}_{1\perp} - \vec{b}_{1\perp}|^2} \frac{\vec{z}_\perp - \vec{x}_{1\perp}}{|\vec{z}_\perp - \vec{x}_{1\perp}|^2} \cdot \frac{\vec{x}_{2\perp} - \vec{b}_{2\perp}}{|\vec{x}_{2\perp} - \vec{b}_{2\perp}|^2} \right.$$
$$\left. + \frac{\vec{\epsilon}_\perp^{\lambda *} \cdot (\vec{x}_{2\perp} - \vec{b}_{2\perp})}{|\vec{x}_{2\perp} - \vec{b}_{2\perp}|^2} \frac{\vec{x}_{1\perp} - \vec{b}_{1\perp}}{|\vec{x}_{1\perp} - \vec{b}_{1\perp}|^2} \cdot \frac{\vec{z}_\perp - \vec{x}_{2\perp}}{|\vec{z}_\perp - \vec{x}_{2\perp}|^2} \right] \times$$

$$
\times f^{abc}\left[U^{bd}_{\vec{x}_{1\perp}}-U^{bd}_{\vec{b}_{1\perp}}\right]\left[U^{ce}_{\vec{x}_{2\perp}}-U^{ce}_{\vec{b}_{2\perp}}\right]\left(V_{\vec{b}_{1\perp}}t^{d}\right)_1\left(V_{\vec{b}_{2\perp}}t^{e}\right)_2+\frac{i g^3}{4\pi^3}f^{abc}\left(V_{\vec{b}_{1\perp}}t^{d}\right)_1\left(V_{\vec{b}_{2\perp}}t^{e}\right)_2
$$

$$
\times\int d^2x\left[U^{bd}_{\vec{b}_{1\perp}}\left(U^{ce}_{\vec{x}_\perp}-U^{ce}_{\vec{b}_{2\perp}}\right)\left(\frac{\vec{\epsilon}^{\lambda *}_{\perp}\cdot(\vec{z}_\perp-\vec{x}_\perp)}{|\vec{z}_\perp-\vec{x}_\perp|^2}\frac{\vec{x}_\perp-\vec{b}_{1\perp}}{|\vec{x}_\perp-\vec{b}_{1\perp}|^2}\cdot\frac{\vec{x}_\perp-\vec{b}_{2\perp}}{|\vec{x}_\perp-\vec{b}_{2\perp}|^2}\right.\right.
$$

$$
\left.-\frac{\vec{\epsilon}^{\lambda *}_{\perp}\cdot(\vec{z}_\perp-\vec{b}_{1\perp})}{|\vec{z}_\perp-\vec{b}_{1\perp}|^2}\frac{\vec{z}_\perp-\vec{x}_\perp}{|\vec{z}_\perp-\vec{x}_\perp|^2}\cdot\frac{\vec{x}_\perp-\vec{b}_{2\perp}}{|\vec{x}_\perp-\vec{b}_{2\perp}|^2}-\frac{\vec{\epsilon}^{\lambda *}_{\perp}\cdot(\vec{z}_\perp-\vec{b}_{1\perp})}{|\vec{z}_\perp-\vec{b}_{1\perp}|^2}\frac{\vec{x}_\perp-\vec{b}_{1\perp}}{|\vec{x}_\perp-\vec{b}_{1\perp}|^2}\cdot\frac{\vec{x}_\perp-\vec{b}_{2\perp}}{|\vec{x}_\perp-\vec{b}_{2\perp}|^2}\right)
$$

$$
-\left(U^{bd}_{\vec{x}_\perp}-U^{bd}_{\vec{b}_{1\perp}}\right)U^{ce}_{\vec{b}_{2\perp}}\left(\frac{\vec{\epsilon}^{\lambda *}_{\perp}\cdot(\vec{z}_\perp-\vec{x}_\perp)}{|\vec{z}_\perp-\vec{x}_\perp|^2}\frac{\vec{x}_\perp-\vec{b}_{1\perp}}{|\vec{x}_\perp-\vec{b}_{1\perp}|^2}\cdot\frac{\vec{x}_\perp-\vec{b}_{2\perp}}{|\vec{x}_\perp-\vec{b}_{2\perp}|^2}\right.
$$

$$
\left.\left.-\frac{\vec{\epsilon}^{\lambda *}_{\perp}\cdot(\vec{z}_\perp-\vec{b}_{2\perp})}{|\vec{z}_\perp-\vec{b}_{2\perp}|^2}\frac{\vec{z}_\perp-\vec{x}_\perp}{|\vec{z}_\perp-\vec{x}_\perp|^2}\cdot\frac{\vec{x}_\perp-\vec{b}_{1\perp}}{|\vec{x}_\perp-\vec{b}_{1\perp}|^2}-\frac{\vec{\epsilon}^{\lambda *}_{\perp}\cdot(\vec{z}_\perp-\vec{b}_{2\perp})}{|\vec{z}_\perp-\vec{b}_{2\perp}|^2}\frac{\vec{x}_\perp-\vec{b}_{1\perp}}{|\vec{x}_\perp-\vec{b}_{1\perp}|^2}\cdot\frac{\vec{x}_\perp-\vec{b}_{2\perp}}{|\vec{x}_\perp-\vec{b}_{2\perp}|^2}\right)\right]
$$

$$
-\frac{i g^3}{4\pi^2}f^{abc}\left(V_{\vec{b}_{1\perp}}t^{d}\right)_1\left(V_{\vec{b}_{2\perp}}t^{e}\right)_2\left[(U^{bd}_{\vec{z}_\perp}-U^{bd}_{\vec{b}_{1\perp}})\,U^{ce}_{\vec{b}_{2\perp}}\frac{\vec{\epsilon}^{\lambda *}_{\perp}\cdot(\vec{z}_\perp-\vec{b}_{1\perp})}{|\vec{z}_\perp-\vec{b}_{1\perp}|^2}\ln\frac{1}{|\vec{z}_\perp-\vec{b}_{2\perp}|\Lambda}-U^{bd}_{\vec{b}_{1\perp}}(U^{ce}_{\vec{z}_\perp}\right.
$$

$$
\left.-U^{ce}_{\vec{b}_{2\perp}})\frac{\vec{\epsilon}^{\lambda *}_{\perp}\cdot(\vec{z}_\perp-\vec{b}_{2\perp})}{|\vec{z}_\perp-\vec{b}_{2\perp}|^2}\ln\frac{1}{|\vec{z}_\perp-\vec{b}_{1\perp}|\Lambda}\right]-\frac{i g^3}{4\pi^3}\int d^2x\left[U^{ab}_{x_\perp}-U^{ab}_{z_\perp}\right]f^{bde}\left(V_{\vec{b}_{1\perp}}t^{d}\right)_1\left(V_{\vec{b}_{2\perp}}t^{e}\right)_2
$$

$$
\times\frac{\vec{\epsilon}^{\lambda *}_{\perp}\cdot(\vec{z}_\perp-\vec{x}_\perp)}{|\vec{z}_\perp-\vec{x}_\perp|^2}\frac{\vec{x}_\perp-\vec{b}_{1\perp}}{|\vec{x}_\perp-\vec{b}_{1\perp}|^2}\cdot\frac{\vec{x}_\perp-\vec{b}_{2\perp}}{|\vec{x}_\perp-\vec{b}_{2\perp}|^2}\,\mathrm{Sign}(b_2^--b_1^-)\,. \tag{4}
$$

Another class of diagrams that contribute at g^5 order is shown in Fig. 8 and the result is [2]

$$
\sum_{i=1}^{6}{}'\,D_i
$$

$$
=-\frac{g^3}{8\pi^4}\int d^2x_1\,d^2x_2\,\delta[(\vec{z}_\perp-\vec{x}_{1\perp})\times(\vec{z}_\perp-\vec{x}_{2\perp})]\left[\frac{\vec{\epsilon}^{\lambda *}_{\perp}\cdot(\vec{x}_{2\perp}-\vec{x}_{1\perp})}{|\vec{x}_{2\perp}-\vec{x}_{1\perp}|^2}\frac{\vec{x}_{1\perp}-\vec{b}_{2\perp}}{|\vec{x}_{1\perp}-\vec{b}_{2\perp}|^2}\cdot\frac{\vec{x}_{2\perp}-\vec{b}_{2\perp}}{|\vec{x}_{2\perp}-\vec{b}_{2\perp}|^2}\right.
$$

$$
\left.-\frac{\vec{\epsilon}^{\lambda *}_{\perp}\cdot(\vec{x}_{1\perp}-\vec{b}_{2\perp})}{|\vec{x}_{1\perp}-\vec{b}_{2\perp}|^2}\frac{\vec{z}_\perp-\vec{x}_{1\perp}}{|\vec{z}_\perp-\vec{x}_{1\perp}|^2}\cdot\frac{\vec{x}_{2\perp}-\vec{b}_{2\perp}}{|\vec{x}_{2\perp}-\vec{b}_{2\perp}|^2}+\frac{\vec{\epsilon}^{\lambda *}_{\perp}\cdot(\vec{x}_{2\perp}-\vec{b}_{2\perp})}{|\vec{x}_{2\perp}-\vec{b}_{2\perp}|^2}\frac{\vec{x}_{1\perp}-\vec{b}_{2\perp}}{|\vec{x}_{1\perp}-\vec{b}_{2\perp}|^2}\cdot\frac{\vec{z}_\perp-\vec{x}_{2\perp}}{|\vec{z}_\perp-\vec{x}_{2\perp}|^2}\right]
$$

$$
\times f^{abc}\left[U^{bd}_{\vec{x}_{1\perp}}-U^{bd}_{\vec{b}_{2\perp}}\right]\left[U^{ce}_{\vec{x}_{2\perp}}-U^{ce}_{\vec{b}_{2\perp}}\right]\left(V_{\vec{b}_{1\perp}}\right)_1\left(V_{\vec{b}_{2\perp}}t^{e}t^{d}\right)_2
$$

$$
+\frac{i g^3}{4\pi^3}\int d^2x\,f^{abc}\,U^{bd}_{\vec{b}_{2\perp}}\left[U^{ce}_{\vec{x}_\perp}-U^{ce}_{\vec{b}_{2\perp}}\right]\left(V_{\vec{b}_{1\perp}}\right)_1\left(V_{\vec{b}_{2\perp}}t^{e}t^{d}\right)_2\left(\frac{\vec{\epsilon}^{\lambda *}_{\perp}\cdot(\vec{z}_\perp-\vec{x}_\perp)}{|\vec{z}_\perp-\vec{x}_\perp|^2}\frac{1}{|\vec{x}_\perp-\vec{b}_{2\perp}|^2}\right.
$$

$$
\left.-\frac{\vec{\epsilon}^{\lambda *}_{\perp}\cdot(\vec{z}_\perp-\vec{b}_{2\perp})}{|\vec{z}_\perp-\vec{b}_{2\perp}|^2}\frac{\vec{z}_\perp-\vec{x}_\perp}{|\vec{z}_\perp-\vec{x}_\perp|^2}\cdot\frac{\vec{x}_\perp-\vec{b}_{2\perp}}{|\vec{x}_\perp-\vec{b}_{2\perp}|^2}-\frac{\vec{\epsilon}^{\lambda *}_{\perp}\cdot(\vec{z}_\perp-\vec{b}_{2\perp})}{|\vec{z}_\perp-\vec{b}_{2\perp}|^2}\frac{1}{|\vec{x}_\perp-\vec{b}_{2\perp}|^2}\right)
$$

$$
+\frac{i g^3}{4\pi^2}f^{abc}\,U^{bd}_{\vec{b}_{2\perp}}\left[U^{ce}_{\vec{z}_\perp}-U^{ce}_{\vec{b}_{2\perp}}\right]\left(V_{\vec{b}_{1\perp}}\right)_1\left(V_{\vec{b}_{2\perp}}t^{e}t^{d}\right)_2\frac{\vec{\epsilon}^{\lambda *}_{\perp}\cdot(\vec{z}_\perp-\vec{b}_{2\perp})}{|\vec{z}_\perp-\vec{b}_{2\perp}|^2}\ln\frac{1}{|\vec{z}_\perp-\vec{b}_{2\perp}|\Lambda}\,. \tag{5}
$$

3 Gauge invariance

As we stated above, the calculation was carried out in the light-cone gauge. It is well known that the light-cone gauge posses an ambiguity about the light-cone singularity $k^+ = 0$. In Ref. [3] we showed that this ambiguity is related to the residual sub-gauge choice, and that different prescriptions

Figure 5. The shaded region represents a general interaction. The two diagrams, with retarded gluon propagators with time flow in the direction of the arrow, cancel.

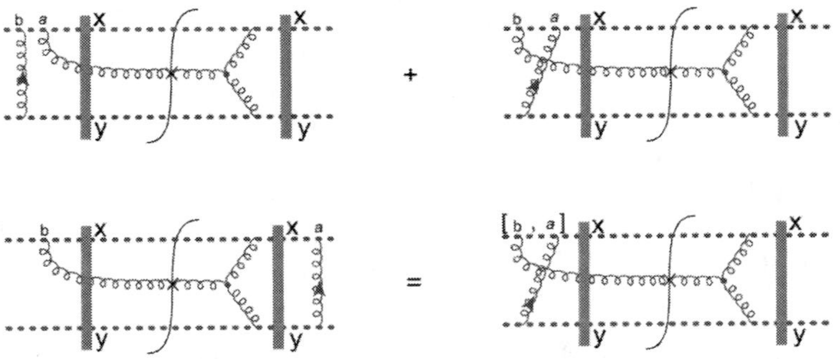

Figure 6. The sum these three diagrams is equal to the diagram with the color factor associated with the top quark replaced by a commutator. The commutator is represented by a bracket, [,].

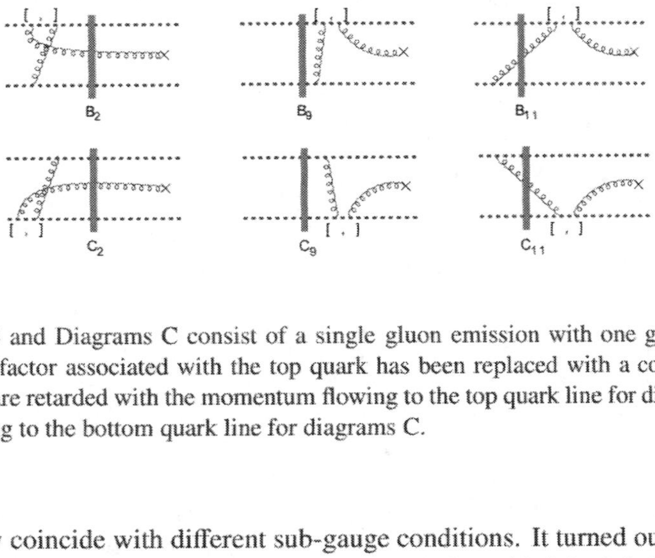

Figure 7. Diagrams B and Diagrams C consist of a single gluon emission with one gluon traveling between the quarks. The color factor associated with the top quark has been replaced with a commutator. The gluons connecting the quarks are retarded with the momentum flowing to the top quark line for diagrams B, and retarded with momentum flowing to the bottom quark line for diagrams C.

for the k^+ singularity coincide with different sub-gauge conditions. It turned out that the quark lines reduce to Wilson lines if we regulate the light-cone singularity with Principle-Value (PV). The sub-gauge condition that leads to PV-prescription is $\partial_\perp \cdot A_\perp(x^- = +\infty) + \partial_\perp \cdot A_\perp(x^- = -\infty) = 0$ and the light-cone propagator is

$$D^{\mu\nu}_{PV}(x,y) \equiv \int \frac{d^4k}{(2\pi)^4} e^{-ik\cdot(x-y)} \frac{-i}{k^2+i\epsilon}\left[g^{\mu\nu}_\perp - \left(k^\mu_\perp \eta^\nu + k^\nu_\perp \eta^\mu\right)\mathrm{PV}\left\{\frac{1}{k^+}\right\} - 2\eta^\mu\eta^\nu k^- \mathrm{PV}\left\{\frac{1}{k^+}\right\}\right] \quad (6)$$

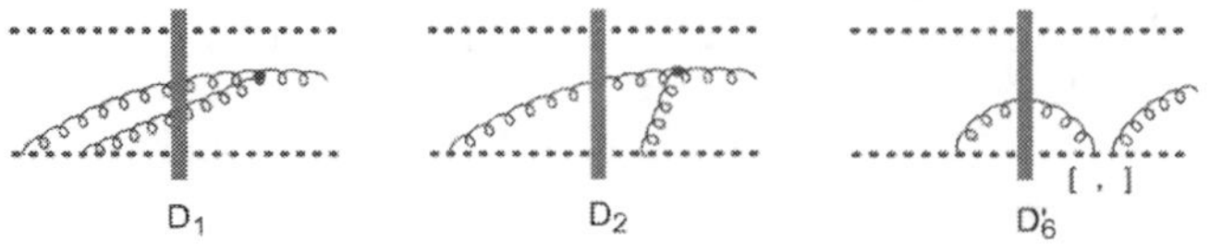

Figure 8. Diagrams involving only one quark at the amplitude level. All gluons are retarded with momentum flowing to the right.

where $\mathrm{PV}\left\{\frac{1}{k^+}\right\} \equiv \frac{1}{2}\left(\frac{1}{k^++i\epsilon} + \frac{1}{k^+-i\epsilon}\right)$. Alternatively, one may choose sub-gauge condition $\vec{\partial}_\perp \cdot \vec{A}_\perp(x^- = +\infty) = 0$ and the propagator is

$$D_1^{\mu\nu}(x,y) \equiv \int \frac{d^4k}{(2\pi)^4} e^{-ik\cdot(x-y)} \frac{-i}{k^2+i\epsilon}\left[g^{\mu\nu} - \frac{k^\mu\eta^\nu}{k^+-i\epsilon} - \frac{k^\nu\eta^\mu}{k^++i\epsilon}\right]. \tag{7}$$

For sub-gauge condition at $x^- = -\infty$ *i.e* $\vec{\partial}_\perp \cdot \vec{A}_\perp(x^- = -\infty) = 0$ the propagator is

$$D_2^{\mu\nu}(x,y) \equiv \int \frac{d^4k}{(2\pi)^4} e^{-ik\cdot(x-y)} \frac{-i}{k^2+i\epsilon}\left[g^{\mu\nu} - \frac{k^\mu\eta^\nu}{k^++i\epsilon} - \frac{k^\nu\eta^\mu}{k^+-i\epsilon}\right]. \tag{8}$$

The results we obtained in Ref. [2] is shown to be invariant under the above sub-gauge choices.

4 Conclusion and Outlook

The results presented here represent a first step toward obtaining the single inclusive gluon production cross-section for heavy-light ions collisions at the classical level beyond the LO. The results we obtained are reasonably compact considering the number of diagrams that contribute at this order. For the success of this calculation it was important to organize the diagrams in different classes and to combine many of them to fewer *effective* diagrams.

It is certainly desirable to compare our result to the similar one obtained in Ref. [4], but such comparison will not be possible until all the diagrams of order g^5 (at amplitude level) will be calculated. Moreover, since our result is at the classical level and there are no quantum fluctuations, it also desirable to check its conformal invariance in the transverse coordinate space. However, at this point this check cannot be performed since diagrams at order g^5 at the amplitude level have still to be calculated.

The author is grateful to the organizers of POETIC06 for warm hospitality.

References

[1] Y. V. Kovchegov and A. H. Mueller, *Gluon production in current nucleus and nucleon nucleus collisions in a quasi-classical approximation, Nucl. Phys.* **B529** (1998) 451-479 [hep-ph/9802440].

[2] G. A. Chirilli, Y. V. Kovchegov, D. E. Wertepny, *Classical Gluon Production Amplitude for Nucleus-Nucleus Collisions: First Saturation Correction in the Projectile, JHEP* **1513** (2015) 015 [arxiv:1501.03106].

[3] G. A. Chirilli, Y. V. Kovchegov and D. E. Wertepny, JHEP **1512**, 138 (2015) doi:10.1007/JHEP12(2015)138 [arXiv:1508.07962 [hep-ph]].

[4] I. Balitsky, *Scattering of shock waves in QCD, Phys. Rev.* **D70** (2004) 114030 [hep-ph/0409314].

EPJ Web of Conferences **112**, 02005 (2016)
DOI: 10.1051/epjconf/201611202005

Quarkonium photoproduction in pp and AA collisions at the LHC

M.B. Gay Ducati[1,a], M.T. Griep[1,b], and M.V.T. Machado[1,c]

[1]*High Energy Physics Phenomenology Group, GFPAE IF-UFRGS. Caixa Postal 15051, CEP 91501-970, Porto Alegre, RS, Brazil*

Abstract. In this contribution we report on the investigations of the exclusive production of quarkonium states in proton-proton and nucleus-nucleus reactions at the LHC using the theoretical framework of the light-cone dipole formalism. In particular, we discuss the rapidity distribution for J/ψ, $\psi(2S)$ and $\Upsilon(1S)$ states and the transverse momentum distribution for the J/ψ meson. The numerical calculations are compared to recent experimental results and prediction are done for future LHC runs in the ultraperipheral channel.

1 Introduction

The exclusive vector meson photoproduction has being investigated both experimentally and theoretically in recent years as it allows to test perturbative Quantum Chromodynamics. The quarkonium masses, m_V, give a perturbative scale for the problem even in the photoproduction limit. An important feature of these processes at the high energy regime is the possibility to investigate the Pomeron exchange. For this energy domain hadrons and photons can be considered as color dipoles in the mixed light cone representation, where their transverse size can be considered frozen during the interaction. Therefore, the scattering process is characterized by the color dipole cross section describing the interaction of those color dipoles with the nucleon/nucleus target. Here, we summarize the main results published in Refs. [1, 2] where the exclusive production of J/ψ and the radially excited $\psi(2S)$ mesons were studied in pp and AA collisions in the LHC energy range. In addition, in this contribution we address the case for $\Upsilon(1S)$ state and the transverse momentum distribution for the J/ψ meson. The theoretical framework considered is the light-cone dipole formalism, where the $Q\bar{Q}$ fluctuation (color dipole) of the incoming quasi-real photon interacts with the nucleon or nucleus target via the dipole cross section and the result is projected in the wavefunction of the observed hadron. At high energies, the transition of the regime described by the linear dynamics of emissions chain to a new regime where the physical process of recombination of partons becomes important is expected. It is characterized by the limitation on the maximum phase-space parton density that can be reached in the hadron wavefunction, the so-called parton saturation phenomenon. The transition is set by saturation scale $Q_{\rm sat} \propto x^{-\lambda}$ (with $\lambda \simeq 0.3$), which is enhanced in the nuclear case. The Bjorken-x variable is connected in the photoproduction case to the center-of-mass-energy of the photon-nucleon (or nucleus) system, W, by the relation $x = M_V^2/W^2$.

[a]e-mail: beatriz.gay@ufrgs.br
[b]e-mail: mirian.fis@gmail.com
[c]e-mail: magnus@if.ufrgs.br

2 Theoretical framework

The exclusive meson photoproduction in hadron-hadron collisions can be factorized in terms of the equivalent flux of photons of the hadron projectile and photon-target production cross section [3]. The photon energy spectrum, $dN_\gamma^p/d\omega$, which depends on the photon energy ω, is well known [3]. The rapidity distribution y for quarkonium photoproduction in pp collisions can be written down as,

$$\frac{d\sigma}{dy}(pp \to p \otimes V \otimes p) \;=\; S^2_{\text{gap}} \left[\omega \frac{dN_\gamma^p}{d\omega} \sigma(\gamma p \to V + p) + (y \to -y) \right]. \tag{1}$$

The produced state with mass m_V has rapidity $y \simeq \ln(2\omega/m_V)$ and the square of the γp centre-of-mass energy is given by $W^2_{\gamma p} \simeq 2\omega\sqrt{s}$. The symbol $\otimes$ denotes a large rapidity gap between the proton and the meson. The absorptive corrections due to spectator interactions between the two hadrons are represented by the factor S_{gap}. The photon-Pomeron interaction will be described within the light-cone dipole frame, where the probing projectile fluctuates into a quark-antiquark pair with transverse separation r (and momentum fraction z) long after the interaction, which then scatters off the hadron. The cross section for exclusive photoproduction of quarkonia off a nucleon target is given by,

$$\sigma(\gamma p \to V + p) = \frac{1}{16\pi B_V} \left| \sum_{h,\bar{h}} \int dz\, d^2r\, \Psi^\gamma_{h,\bar{h}}(z,r)\, \sigma_{dip}(x,r)\, \Psi^{V*}_{h,\bar{h}}(z,r,M_V) \right|^2 , \tag{2}$$

where Ψ^γ and Ψ^V are the light-cone wavefunction of the photon and of the vector meson, respectively. In the photoproduction limit, only the transverse polarization contribution is relevant. The Bjorken variable is denoted by x, the dipole cross section by $\sigma_{dip}(x,r)$ and the diffractive slope parameter by B_V. Here, we consider the energy dependence of the slope using the Regge motivated expression (see Refs. [1, 2] for details). Similarly, the rapidity distribution y in nucleus-nucleus collisions has the same factorized form,

$$\frac{d\sigma}{dy}(AA \to A \otimes V \otimes Y) = \left[\omega \frac{dN_\gamma^A}{d\omega}\, \sigma(\gamma A \to V + Y) + (y \to -y) \right], \tag{3}$$

where the photon flux in nucleus is denoted by $dN_\gamma^A/d\omega$ and $Y = A$ (coherent case) or $Y = A^*$ (incoherent case). The exclusive photoproduction off nuclei for coherent and incoherent processes can be simply computed in high energies where the large coherence length $l_c \gg R_A$ is fairly valid. the expressions for both cases are given by [4],

$$\frac{d^2\sigma_{coh}}{d^2b} \;=\; \left| \sum_{h,\bar{h}} \int dz\, d^2r\, \Psi^{V*}(z,r,M_V) \left[1 - \exp\left(-\frac{1}{2} R_G \sigma_{dip}(x,r) T_A(b) \right) \right] \Psi^\gamma(z,r) \right|^2 , \tag{4}$$

$$\frac{d^2\sigma_{incoh}}{d^2b} \;=\; \frac{T_A(b)}{16\pi\, B_V} \left| \sum_{h,\bar{h}} \int dz\, d^2r\, \Psi^{V*} \left[R_G \sigma_{dip}(x,r) \exp\left(-\frac{1}{2} R_G \sigma_{dip}(x,r) T_A(b) \right) \right] \Psi^\gamma \right|^2 , \tag{5}$$

where $T_A(b) = \int dz \rho_A(b,z)$ is the nuclear thickness function. In the numerical evaluations, we have considered the boosted Gaussian (BG) wavefunction and the phenomenological saturation model proposed in Ref. [5] (IIM model) which encodes the main properties of the saturation approaches. For that model we used the fitted parameters from DESY-HERA data including the contribution of charm quark, with $m_c = 1.4$ GeV [6] (for the bottom mass we will use the fixed value $m_b = 4.5$ GeV). The nuclear ratio for the gluon density is denoted by $R_G(x, Q^2 = m_V^2/4)$. In the present investigation we will use the nuclear ratio from the leading twist theory of nuclear shadowing based on generalization of the Gribov-Glauber multiple scattering formalism as investigated in Ref. [7]. We used the two models available for $R_G(x, Q^2)$ in [7], Models 1 and 2, which correspond to higher nuclear shadowing and lower nuclear shadowing, respectively. Such a choice is completely arbitrary and other nuclear gluon ratios available in literature could be considered.

In the numerical calculation shown in next section, the corrections due to skewedness effect (off-diagonal gluon exchange) and real part of amplitude are also taken into account. Detail on the model dependence on these corrections can be found for instance in Ref. [1, 2]. In order to give details on the meson wavefunction, the boosted Gaussian is given by the expression:

$$\psi^{nS}_{\lambda,h\bar{h}} = \sqrt{\frac{N_c}{4\pi}}\frac{\sqrt{2}}{z(1-z)}\left\{\delta_{h,\bar{h}}\delta_{\lambda,2h}m_c + i(2h)\delta_{h,-\bar{h}}e^{i\lambda\phi}\left[(1-z)\delta_{\lambda,-2h} + z\delta_{\lambda,2h}\right]\partial_r\right\}\phi_{nS}(z,r). \tag{6}$$

Here, $\phi(z,r)$ in the mixed (r,z) representation is obtained by boosting a Schrödinger Gaussian wavefunction in momentum representation, $\Psi(z,\mathbf{k})$. In this case, one obtains the following expression for the $1S$ state:

$$\phi_{1S}(r,z) = N_T^{(1S)}\left\{4z(1-z)\sqrt{2\pi R_{1S}^2}\exp\left[-\frac{m_q^2R_{1S}^2}{8z(1-z)}\right]\exp\left[-\frac{2z(1-z)r^2}{R_{1S}^2}\right]\exp\left[\frac{m_q^2R_{1S}^2}{2}\right]\right\}, \tag{7}$$

where for the $1S$ ground state vector mesons we determine the parameters R_{1S}^2 and N_T by considering the normalization property of wavefunctions and the predicted decay widths. The radial wavefunction of the $2S$ state is obtained by the following modification of the $1S$ state:

$$\begin{aligned}\phi_{2S}(r,z) &= N_T^{(2S)}\left\{4z(1-z)\sqrt{2\pi R_{2S}^2}\exp\left[-\frac{m_q^2R_{2S}^2}{8z(1-z)}\right]\exp\left[-\frac{2z(1-z)r^2}{R_{2S}^2}\right]\exp\left[\frac{m_q^2R_{2S}^2}{2}\right]\right.\\ &\times \left.\left[1-\alpha\left(1+m_q^2R_{2S}^2-\frac{m_q^2R_{2S}^2}{4z(1-z)}+\frac{4z(1-z)}{R_{2S}^2}r^2\right)\right]\right\},\end{aligned} \tag{8}$$

with the new parameter α controlling the position of the node. In addition, the two parameters α and R_{2S} are determined from the orthogonality conditions for the meson wavefunction.

Finally, as we mentioned above the phenomenological parton saturation model proposed in Ref. [5] (hereafter IIM-old or CGC model) was taken into account which encodes the main properties of the saturation approaches. Thus, the dipole cross section is parameterized as follows

$$\sigma_{dip}(x,r) = \sigma_0 \begin{cases} \mathcal{N}_0\left(\frac{\bar{\tau}^2}{4}\right)^{\gamma_{\text{eff}}(x,r)}, & \text{for } \bar{\tau} \le 2, \\ 1-\exp\left[-a\,\ln^2(b\,\bar{\tau})\right], & \text{for } \bar{\tau} > 2, \end{cases}$$

where $\bar{\tau} = rQ_{\text{sat}}(x)$ and the expression for $\bar{\tau} > 2$ (saturation region) has the correct functional form, as obtained from the theory of the Color Glass Condensate (CGC). For the color transparency region near saturation border ($\bar{\tau} \le 2$), the behavior is driven by the effective anomalous dimension $\gamma_{\text{eff}}(x,r) = \gamma_{\text{sat}} + \frac{\ln(2/\bar{\tau})}{\kappa\lambda y}$, where $\gamma_{\text{sat}} = 0.63$ is the LO BFKL anomalous dimension at saturation limit.

3 Numerical results and data comparison

Let us start by the proton-proton mode at the LHC. First, we compare our study to the data measured by LHCb Collaboration at 7 TeV in pp collisions at the forward region $2.0 < \eta_{\pm} < 4.5$ [8], which cover values of Bjorken-x variable down to $x \approx 5 \times 10^{-6}$. We assume for the absorption factor the

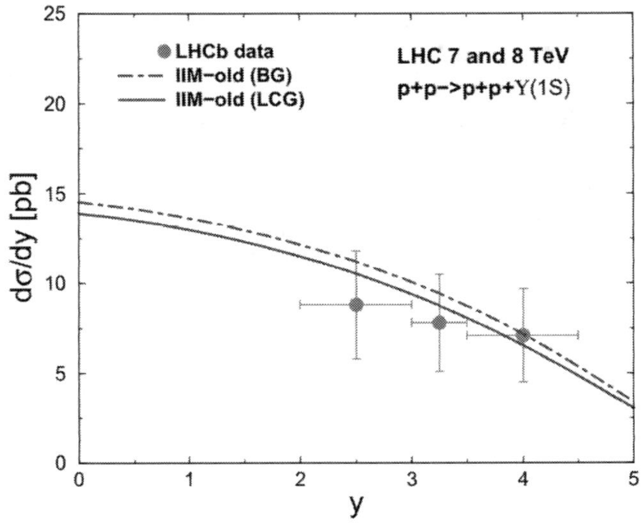

Figure 1. The rapidity distribution for $\Upsilon(1S)$ exclusive production in proton-proton collisions at 7 and 8 TeV. See discussion in text. Data from LHCb Collaboration [10].

average value $S^2_{\text{gap}} = 0.8$. We obtain $\sigma(pp \to p + J/\psi + p) \times \text{Br} = 698$ pb for rapidity between 2 and 4.5. In terms of muon pseudorapidities we get $\sigma_{pp\to J/\psi(\to\mu^+\mu^-)}(2.0 < \eta_{\mu^\pm} < 4.5) = 298$ pb. This is in good agreement to the experimental result 307 ± 42 pb [8]. For the $\psi(2S)$ mesons it is obtained $\sigma(pp \to p + \psi(2S) + p) \times \text{Br} = 18$ pb for rapidities $2.0 < y < 4.5$. Accordingly, we now predict $\sigma_{pp\to\psi(2S)(\to\mu^+\mu^-)}(2.0 < \eta_{\mu^\pm} < 4.5) = 7.7$ pb compared to 7.8 ± 1.6 pb measured value [8]. We performed predictions for the next LHC runs in pp mode. We have found $\frac{d\sigma_{J/\psi}}{dy} = 6.2$ nb and 7.9 nb for central rapidities at energies of 8 and 14 TeV, respectively. For the $\psi(2S)$ state the extrapolation gives $\frac{d\sigma_{\psi(2S)}}{dy} = 1.0$ nb and 1.4 nb for the same energies at central rapidity. In addition, the rapidity distributions for J/ψ and $\psi(2S)$ are completely consistent within the errors with the more precise LHCb data published in Ref. [9] (we quote Fig. 7 of that paper, where our predictions are compared to experimental results). Still in the proton-proton mode, we have predicted the rapidity distribution for the $\Upsilon(1S)$ state and compare the analysis to the recent measurement of exclusive Υ production at 7 and 8 TeV by LHCb Collaboration [10]. The results is presented in Fig. 1, where the uncertainty on the choice for the meson wavefunction is presented using the IIM model (labeled here as IIM-old) for the dipole cross section. The solid line corresponds to the results for the light cone Gaussian wavefunction (LCG) and the dot-dashed curve stands for the boosted Gaussian (BG) wavefunction. Both predictions are somewhat consistent with the present data accuracy.

For sake of completeness, in Fig. 2 we present the transverse momentum distribution for the J/ψ exclusive production in the fixed energies of 7 TeV(solid line), 8 TeV (dashed line) and 14 TeV (dot-dashed line). We consider the CGC model (IIM model) and BG wavefunction and central rapidity, $y = 0$. The p_T-distribution is directly connected to the t-dependence of the dipole cross section. In our case, we use the hypothesis of exponential behavior on t for the differential cross section. Distinct distributions will be obtained using more sophisticated models for the b-dependence on the dipole cross section.

In Fig. 3 (left panel) we present the calculations for the rapidity distribution of coherent J/ψ state, using distinct scenarios for the nuclear gluon shadowing. The dot-dashed curve represents the

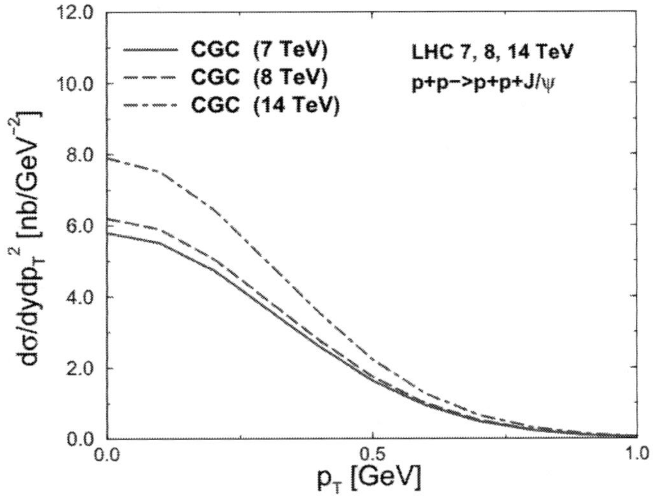

Figure 2. The transverse momentum distribution for J/ψ exclusive production in proton-proton collisions at 7, 8 and 14 TeV. See discussion in text.

result using $R_G = 1$. It overestimates the ALICE data [11, 12] on the backward (forward) and mainly in central rapidities. The situation is improved if we consider nuclear shadowing renormalising the dipole cross section. The reason is that the gluon density in nuclei at small Bjorken x is expected to be suppressed compared to a free nucleon due to interferences. For R_G, we have considered the theoretical evaluation of Ref. [7]. As a prediction at central rapidity, one obtains $\frac{d\sigma}{dy}(y = 0) =$ 4.95, 1.68 and 2.27 mb for calculation using $R_G = 1$, Model 1 (strong shadowing) and Model 2 (weak shadowing), respectively. In Fig. 3 (right panel) is presented the incoherent cross section for both J/ψ and $\psi(2S)$ states with $R_G = 1$. In this case, the data description is very good compared to the ALICE data. Here some comments are in order. It was recently shown [13] that the color dipole approach with $R_G = 1$ is not able to describe the ALICE data for coherent production of J/ψ at 2.76 GeV [11, 12] at midrapidity even including the theoretical uncertainties related to the meson wavefunction and models for the dipole cross section (and distinct choices for the charm quark mass). The mid-rapidity cross section is overestimated by a factor two in the case presented here. It is advocated in Ref. [14] the reason is that for $R_G = 1$ in the photon-nucleus cross section the nuclear effect included via eikonalization corresponds to lowest $Q\bar{Q}$ Fock component, $|Q\bar{Q}\rangle$. It does not include any correction for gluon shadowing, but rather correspond to shadowing of sea quarks in nuclei. Although σ_{dip} includes all possibles effects of gluon radiation, the eikonal assumes that none of radiated gluons take part in multiple interactions in the nucleus. The leading order correction corresponding to gluon shadowing comes from the eikonalization of the next Fock component $|Q\bar{Q}G\rangle$. Explicitly, in the large coherence length limit $\ell_c \to \infty$ the general formula for the γA cross section is given by,

$$\begin{aligned} \frac{d^2\sigma_{\gamma A}}{d^2b} &= 2\int dz \int d^2r \left|\psi_{q\bar{q}}\right|^2 \left[1 - \exp\left(-\frac{1}{2}\sigma_{q\bar{q}} T_A(b)\right)\right] \\ &+ 2\int dz \int \frac{dz_G}{z_G} \int d^2r_1 \int d^2r_2 \left|\psi_{q\bar{q}G}\right|^2 \left[1 - \exp\left(-\frac{1}{2}\sigma_{q\bar{q}G} T_A(b)\right)\right], \end{aligned} \quad (9)$$

where $\psi_{q\bar{q}G}$ is the wavefunction for the gluonic component and the cross section for the three body system $\sigma_{q\bar{q}G}$ can be expressed in terms of the dipole cross section, σ_{dip}. The complete calculation

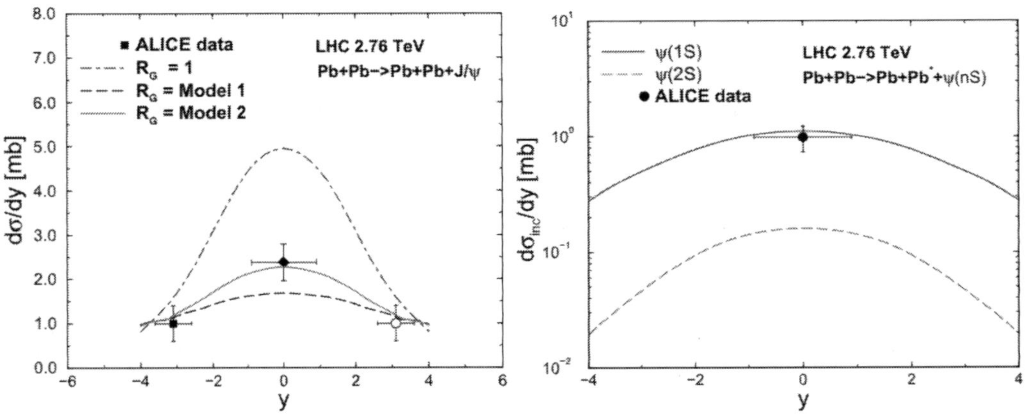

Figure 3. The rapidity distribution of coherent (left panel) J/ψ production and the incoherent (right panel) J/ψ and $\psi(2S)$ exclusive photoproduction at $\sqrt{s} = 2.76$ TeV in PbPb collisions at the LHC (see text).

has been done in Ref. [14], and it was shown that the $|Q\bar{Q}G\rangle$ contribution can be absorbed in a factor $R_G(x, Q^2, b)$ multiplying the original dipole cross section.

As a summary, we calculate the results for the exclusive production of quarkonium states in proton-proton and nucleus-nucleus reactions at the LHC using the theoretical framework of light-cone dipole formalism. The exclusive production in proton-proton case is consistently described by our approach being in good agreement wit the rapidity distribution for J/ψ, $\psi(2S)$ and $\Upsilon(1S)$ states. In the nucleus-nucleus case, the approach fairly describes the incoherent cross section and we discuss the role played by the gluon nuclear shadowing in order to evaluate the coherent cross section at midrapidity.

References

[1] M.B. Gay Ducati, M.T. Griep, M.V.T. Machado, Phys. Rev. D **88**, 017504 (2013).
[2] M.B. Gay Ducati, M.T. Griep, M.V.T. Machado, Phys. Rev. C **88**, 014910 (2013).
[3] C. A. Bertulani, S. R. Klein and J. Nystrand, Ann. Rev. Nucl. Part. Sci. **55**, 271 (2005).
[4] B. Z. Kopeliovich and B. G. Zakharov, Phys. Rev. D **44**, 3466 (1991).
[5] E. Iancu, K. Itakura and S. Munier, Phys. Lett. B **590**, 199 (2004).
[6] G. Soyez, Phys. Lett. B **655**, 32 (2007).
[7] L. Frankfurt, V. Guzey and M. Strikman, Phys. Rept. **512**, 255 (2012).
[8] R. Aaij *et al.* [LHCb Collaboration], J. Phys. G **40**, 045001 (2013).
[9] R. Aaij *et al.* [LHCb Collaboration], J. Phys. G **41**, 055002 (2014).
[10] R. Aaij *et al.* [LHCb Collaboration], JHEP **1509**, 084 (2015).
[11] B. Abelev *et al.* [ALICE Collaboration], Phys. Lett. B **718**, 1273 (2013).
[12] E. Abbas *et al.* [ALICE Collaboration], Eur. Phys. J. C **73**, 2617 (2013).
[13] G. Sampaio dos Santos and M.V.T. Machado, J. Phys. G **42**, 105001 (2015).
[14] B. Z. Kopeliovich, A. Schaefer and A. V. Tarasov, Phys. Rev. C **62**, 054022 (2000).

EPJ Web of Conferences **112**, 02006 (2016)
DOI: 10.1051/epjconf/201611202006

Searching saturation effects in inclusive and exclusive eA processes

V. P. Goncalves[1,2,a]

[1] Department of Astronomy and Theoretical Physics, Lund University, SE-223 62 Lund, Sweden
[2] High and Medium Energy Group, Instituto de Física e Matemática, Universidade Federal de Pelotas Caixa Postal 354, 96010-900, Pelotas, RS, Brazil.

Abstract. The searching of saturation effects in inclusive and exclusive eA collisions is reviewed. In particular, we present a comparison between the linear and non-linear predictions for the nuclear structure functions as well as for the Deep Virtual Compton Scattering and vector meson production in future eA colliders.

1 Introduction

The study of the hadronic structure in the non-linear regime of the Quantum Chromodynamics (QCD) is one of the main goals of the future Electron - Ion Colliders (EIC) [1]. It is expected that the resulting experimental data will be able to determine the presence of gluon saturation effects, the magnitude of the associated non-linear corrections and what is the correct theoretical framework for their description [2]. This expectation is easily understood, if we analyse the Bjorken-x and nuclear mass number A behaviours of the nuclear saturation scale, $Q_{s,A}$, which determines the onset of non-linear effects in the QCD dynamics [2]. Assuming that the saturation scale of the nucleus is enhanced with respect to the nucleon one by the *oomph* factor $A^{\frac{1}{3}}$, it can be expressed by the parametrization $Q^2_{s,A} = A^{\frac{1}{3}} \times Q_0^2 (\frac{x_0}{x})^\lambda$. As a consequence nuclei are an efficient amplifier of the non-linear effects. In Fig. 1 we present the theoretical expectations for the saturation scale as a function of x and A, considering the parameters $Q_0^2 = 1.0$ GeV^2, $x_0 = 1.632 \times 10^{-5}$ and $\lambda = 0.2197$. We can observe that, while in the proton case we need very small values of x to obtain large values of Q_s^2, in the nuclear case a similar value can be obtained for values of x approximately two orders of magnitude greater. Consequently, the parton density that is accessed in electron - nucleus (eA) collisions would be equivalent to that obtained in an electron - proton collider at energies that are at least one order of magnitude higher than at HERA.

The enhancement of the non-linear effects with the nuclear mass number has motivated the development of an intense phenomenology about the implications of these effects in inclusive and diffractive observables which could be measured in the Electron - Ion Colliders [3–22]. In this contribution we review the results obtained in Refs. [11, 18] which have indicated that the analysis of inclusive observables, as e.g. the nuclear structure functions F_2^A and F_L^A, probably will not be the best way to obtain a signature of the non-linear effects due to the large uncertainty present in the collinear predictions at small x. As a alternative we discuss the possibility to constrain these effects by the analysis

[a] e-mail: barros@ufpel.edu.br

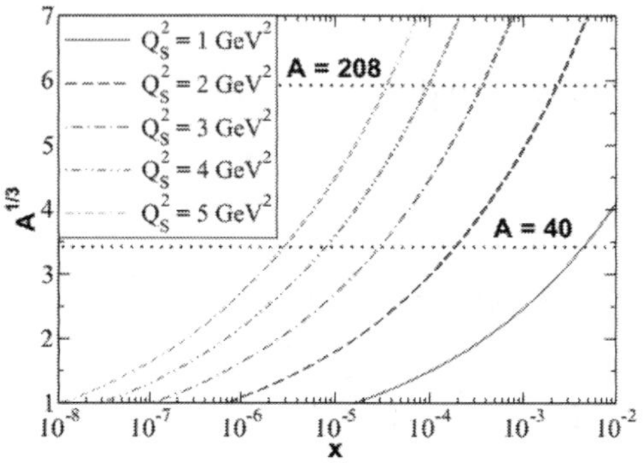

Figure 1. Nuclear saturation scale as a function of the Bjorken x and nuclear mass number A.

of the logarithmic Q^2 slope of the nuclear structure functions, proposed originally in Refs. [3, 4]. Moreover, we will discuss the results from Refs. [12, 15, 21, 22] which demonstrated that, in contrast, diffractive events in eA collisions can be considered a smoking gun for the gluon saturation physics. In particular, we will present our recent results for the nuclear deeply virtual Compton scattering (DVCS) and vector meson production in eA processes.

2 Saturation physics in inclusive eA processes

In the color dipole formalism the nuclear F_2 structure function is given by $F_2^A(x,Q^2) = \frac{Q^2}{4\pi^2\alpha_{em}}\,\sigma_{tot}$, with

$$\sigma_{tot} = \sigma_T + \sigma_L \text{ and } \sigma_{T,L} = \int d^2r\, dz\, |\Psi_{T,L}(r,z,Q^2)|^2\, \sigma_{dip}^A(x,r), \tag{1}$$

where $\Psi_{T,L}$ is the light-cone wave function of the virtual photon and σ_{dip}^A is the dipole nucleus cross section describing the interaction of the $q\bar{q}$ dipole with the nucleus target. In equation (1) r is the transverse separation of the $q\bar{q}$ pair and z is the photon momentum fraction carried by the quark . Moreover, the nuclear longitudinal structure function is given by $F_L^A(x,Q^2) = (Q^2/4\pi^2\alpha_{em})\,\sigma_L$.

The main input for the calculations of inclusive observables in the dipole picture is $\sigma_{dip}^A(x,r)$ which is determined by the QCD dynamics at small x. In the eikonal approximation it is given by:

$$\sigma_{dip}^A(x,r) = 2\int d^2b\, \mathcal{N}^A(x,r,b) \tag{2}$$

where $\mathcal{N}^A(x,r,b)$ is the forward scattering amplitude for a dipole with size r and impact parameter b which encodes all the information about the hadronic scattering, and thus about the non-linear and quantum effects in the hadron wave function. In this contribution we assume that the forward dipole-nucleus amplitude was parametrized as follows

$$\mathcal{N}^A(x,r,b) = 1 - \exp\left[-\frac{1}{2}A\,T_A(b)\,\sigma_{dip}^p(x,r^2)\right], \tag{3}$$

where $T_A(b)$ is the nuclear profile function, which is obtained from a 3-parameter Fermi distribution for the nuclear density normalized to unity. The above equation sums up all the multiple elastic rescattering diagrams of the $q\bar{q}$ pair and is justified for large coherence length, where the transverse

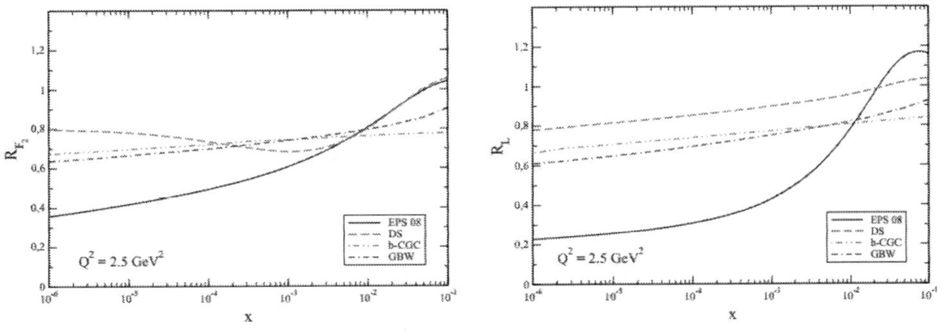

Figure 2. Comparison between the predictions for $R_{F_2} \equiv 2F_2^A/AF_2^D$ and $R_L \equiv 2F_L^A/AF_L^D$ of the saturation models (b-CGC and GBW) and the collinear ones (DS and EPS08) for $Q^2 = 2.5$ GeV^2 and $A = Pb$.

separation r of partons in the multiparton Fock state of the photon becomes a conserved quantity, *i. e.* the size of the pair r becomes eigenvalue of the scattering matrix. In [11] we have considered that σ^p_{dip} is given by the bCGC model [23], which describes the *ep* HERA data.

In Fig. 2 we present our predictions for the x dependence of the ratios R_{F_2} and R_{F_L} for $Q^2 = 2.5$ GeV^2 and $A = Pb$. The saturation models predict a similar behaviour in the small x region. For comparison we also show the predictions obtained using the collinear factorization and nuclear parton distributions resulting from the global analysis of the nuclear experimental data using the DGLAP evolution equation. Here we consider two different sets of nuclear parton distributions: the DS and EPS 08 nuclear parametrizations. It is important to emphasize that the collinear predictions *do not* include dynamical effects associated to non-linear (saturation) physics, since they are based on the linear DGLAP dynamics. Therefore, the comparison between the saturation and collinear predictions could, in principle, tell us if the observable considered can be used to discriminate between linear and non-linear dynamics. The results shown in Fig. 2 (left panel) demonstrate that this is not the case of the nuclear structure function. Although the predictions of the two collinear models are similar for large x ($\geq 10^{-2}$), where there exist experimental data, the predicted behaviour at small x is very distinct. This difference is a consequence of the choice of different experimental data used in the global analysis and of the assumptions related to the behaviour of the nuclear gluon density. The difference between the collinear predictions is so large at small x that it is not possible to extract any information about the presence or not of new QCD dynamical effects from the study of the nuclear structure function. Let us now consider the behaviour of the nuclear longitudinal structure function. Our results are show in Fig. 2 (right panel). The predictions of saturation models are, as in the case of R_{F_2}, contained within the uncertainty range of the collinear models. On the other hand, the difference between the two collinear models is bigger. In this case the results for large x are distinct too. This is due to the strong dependence of F_L^A on the nuclear gluon distribution, which is very different in the two models considered. The conclusion that we can draw in Ref. [11] is that the two dipole models yield similar predictions for the structure functions and they fall inside the range of predictions of the EPS and DS parametrizations. The latter differ among each other because of the large freedom inherent to global data analysis. Although F_2 and F_L are sensitive to saturation effects, it is not yet possible to draw any firm conclusion concerning the QCD dynamics from inclusive quantities. It implies that in order to discriminate the saturation effects we should consider less inclusive observables.

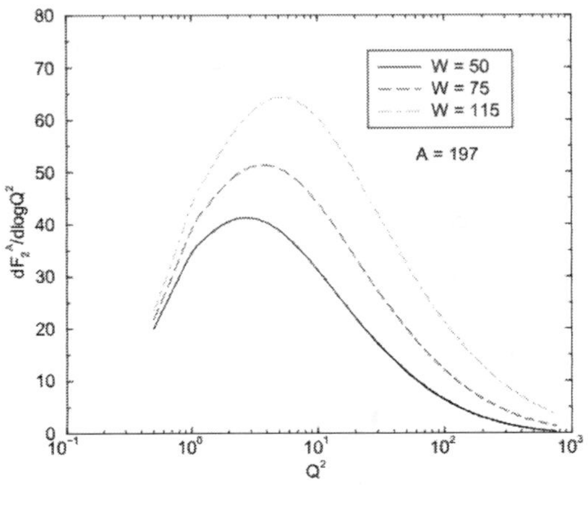

Figure 3. The logarithmic Q^2 slope of F_2^A as function of the variable Q^2 at different values of W and $A = 197$.

An alternative to search saturation effects in eA processes is the analysis of the logarithmic slope of the nuclear structure function proposed in Refs. [3, 4]. In Refs. [3, 4] we considered that the Bjorken variable x and the photon virtuality Q^2 are related by the expression $x = Q^2/W^2$, where W is the $\gamma^* h$ c.m. energy, and calculate the x and Q^2 dependences for W and A fixed. The Fig. 3 shows the logarithmic Q^2 slope of F_2^A plotted for different values of W^2 as a function of Q^2. The remarkable property of the plot is the presence of a distinct maximum for slope which depends of the center of mass energy. Moreover, as shown in Ref. [4], the maximum also is dependent on the atomic number. At fixed A, if we increase the value of the energy, the maximum value of the slope occur at larger values of Q^2. The remarkable properties of the collisions with nuclei targets are the large values of Q^2 for the turnover of F_2^A slope and the large shift of the turnover at large values of Q^2 with the growth of the energy. In Ref. [4] we verified that, at fixed W, the turnover is displaced at larger values of x and Q^2 if we increase the number of nucleons A. These behaviours can be understood intuitively. The turnover is associated with the regime in which the partons in the nucleus form a dense system with mutual interactions and recombinations, with a transition between the linear and non linear regime at the saturation scale. As the partonic density growth at larger values of the number of nucleons A and smaller values of x we have that, at fixed A, the saturation scale Q_s^2 will increase at small values of x, since this is directly proportional to the gluon distribution. Moreover, at fixed energy W, the same density at $A = 12, 32, 197$ will be obtained at larger values of x and Q^2. These properties of the high density effects are verified in the F_2^A slope. The main result from the analysis presented in Ref. [4] is that the saturation should occur already at rather small distances (large Q_s^2) well below where soft dynamics is supposed to set in, justifying the use of perturbative QCD to approach a highly dense system.

3 Saturation effects in exclusive eA processes

An alternative to search the saturation effects can be the study of observables measured in diffractive deep inelastic scattering (DDIS), since the total diffractive cross section is much more sensitive to large-size dipoles than the inclusive one. Basically, the saturation effects screen large-size dipole (soft) contributions, so that a fairly large fraction of the cross section is hard and hence eligible for

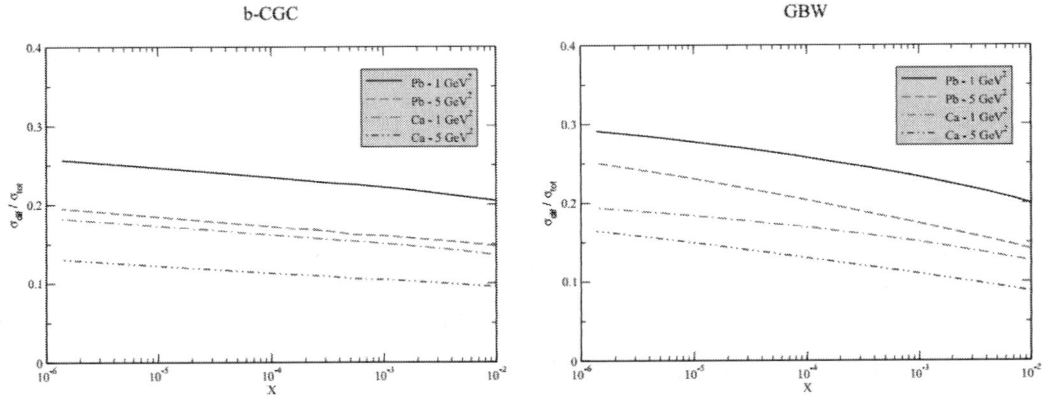

Figure 4. Ratio of diffractive to total cross sections, R_σ, as a function of x with the b-CGC and GBW models. Long dashed and solid lines are for Pb targets at $Q^2 = 5.0$ and $1.0\ \mathrm{GeV}^2$, respectively. Dot-dot-dash and dot-dash lines are the same for Ca targets.

a perturbative treatment. Here we estimate the ratio $R_\sigma = \sigma_{diff}/\sigma_{tot}$, with the total diffractive cross section given by

$$\sigma_{diff} = \sigma_L^D + \sigma_T^D \text{ and } \sigma_{L,T}^D = \frac{1}{4}\int d^2r\,dz\,|\Psi_{T,L}(r,z,Q^2)|^2 \int d^2b \left(\frac{d\sigma_{dip}^A}{d^2b}\right)^2 \tag{4}$$

where $d\sigma_{dip}^A/d^2b = 2\,\mathcal{N}^A(x,r,b)$, with $\mathcal{N}^A$ given by Eq. (3). In Fig. 4 we present the predictions obtained in Ref. [11] for the ratio R_σ as a function of x for two values of Q^2 and the atomic number considering the b-CGC and GBW models. These results indicated that the contribution of diffractive processes to the total cross section is about 20 % at large A and small Q^2, being factor two larger than observed in ep collisions at HERA. Consequently, the study of diffractive processes should be an easy task in an eA collider, allowing for a detailed analysis of the QCD dynamics. In particular, this observation have motivated more detailed studies of the diffractive interactions, with special attention to exclusive diffractive vector meson production and the deeply virtual Compton scattering (DVCS), which are experimentally clean and can be unambiguously identified by the presence of a rapidity gap in final state. As these exclusive processes are driven by the gluon content of the target, with the cross sections being proportional to the square of the scattering amplitude, they are strongly sensitive to the underlying QCD dynamics.

In the color dipole approach the exclusive production $\gamma^* A \to EY$ ($E = \gamma, \rho, J/\Psi, ...$) in electron-nucleus interactions at high energies can be factorized in terms of the fluctuation of the virtual photon into a $q\bar{q}$ color dipole, the dipole-nucleus scattering by a color singlet exchange and the recombination into the exclusive final state E. This process is characterized by a rapidity gap in the final state. If the nucleus scatters elastically, $Y = A$, the process is called coherent production. On the other hand, if the nucleus scatters inelastically, i.e. breaks up ($Y = A'$), the process is denoted incoherent production. In Refs. [21, 22] we have presented a detailed analysis of the impact of the saturation effects in the nuclear DVCS processes and in the ρ meson production considering coherent and incoherent eA collisions. In particular, results for the distributions on the square of the momentum transfer (t) were presented, which are an important source of information about the spatial distribution of the gluons in a nucleus and about fluctuations of the nuclear color fields. In Fig. 5 we present a comparison

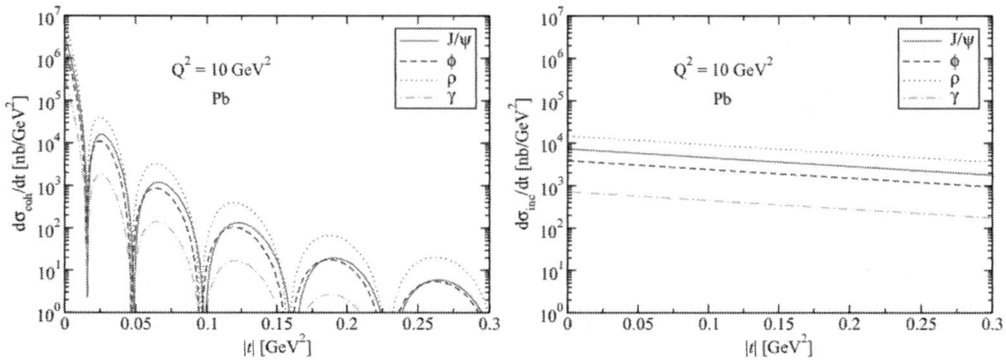

Figure 5. Comparison between the differential cross sections for coherent and incoherent interactions considering $Q^2 = 10$ GeV^2.

between our predictions for the ρ production with those for other exclusive final states: J/Ψ, ϕ and γ. The results are presented for $Q^2 = 10$ GeV^2 and $A = Pb$. In Ref. [22] we have demonstrated that for $Q^2 = 0$ the differential cross sections decrease at heavier vector mesons and the position of the dips for the J/Ψ production is slightly different of the other mesons, which is associated to the dominance of small dipoles in this final state. In contrast, the results for $Q^2 = 10$ GeV^2 presented in Fig. 5, indicate that the position of the dips becomes almost identical for all exclusive final states.

Finally, in order to estimate the magnitude of the non linear effects, we present in Fig. 6 a comparison between our predictions and those obtained disregarding the multiple scatterings of the dipole with the nuclei and the presence of the non linear effects in the nucleon. Basically, we will assume as input in our calculations the following models for the scattering amplitudes:

$$\mathcal{N}^A(x, r, b) = \frac{1}{2}\sigma_{dip}(x, r) A T_A(b) \tag{5}$$

with σ_{dip} given by Eq. (2) and $\mathcal{N}^p(x, \boldsymbol{r}, \boldsymbol{b})$ given by the linear part of the bCGC model

$$\mathcal{N}^p(x, \boldsymbol{r}, \boldsymbol{b}) = \mathcal{N}_0 \left(\frac{r\, Q_s(b)}{2}\right)^{2\left(\gamma_s + \frac{\ln(2/rQ_s(b))}{\kappa \lambda Y}\right)} . \tag{6}$$

We can observe that the incoherent cross sections are not strongly modified by the non linear effects. In contrast, in the case of coherent interactions, the magnitude of the cross section and position of the dips are distinct in the non linear and linear predictions, which makes the analysis of this observable a sensitive probe of the non linear QCD dynamics.

4 Conclusions

The study of exclusive processes in deep inelastic scattering (DIS) processes have been one of the main focuses of hadronic physics in the last years. It is expected that analysis of these processes allow us to probe the QCD dynamics at high energies, driven by the gluon content of the target (proton or nucleus) which is strongly subject to non linear effects (parton saturation) effects. In particular, electron - nucleus collisions the gluon density probed is amplified due to the coherent contributions from many nucleons. In this contribution we have discussed the search of saturation effects in inclusive and exclusive eA processes. These results demonstrated that although the inclusive observables

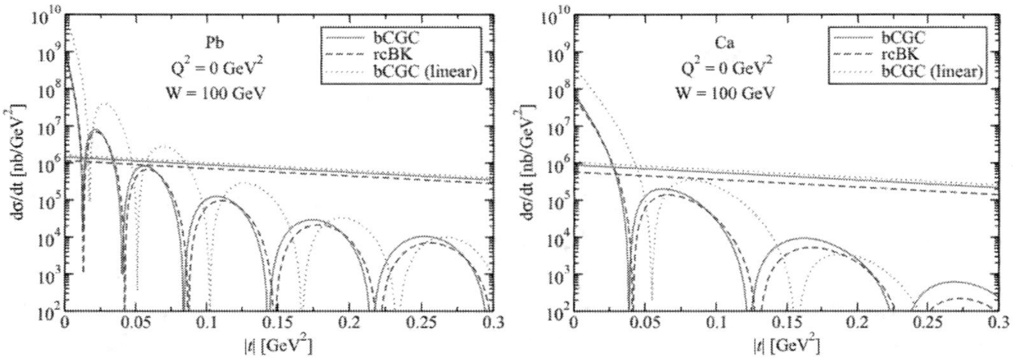

Figure 6. Comparison between non linear and linear predictions for the differential cross sections in coherent and incoherent interactions.

are sensitive to saturation effects, it is not yet possible to draw any firm conclusion concerning the QCD dynamics from inclusive quantities due to the large uncertainty present in the collinear predictions. In contrast, exclusive processes are promising observables to search saturation effects, due to the quadratic dependence on the forward scattering amplitude. In particular, the analysis of the nuclear DVCS and vector meson production demonstrated that the energy dependence of the differential cross sections are strongly modified with the increasing of the atomic mass number and that coherent cross section dominates at small t and the incoherent one at large t. Moreover, the number of dips at small t increases with the atomic number, with the position of the dips being almost independent of the model used to treat the dipole - proton interaction. Furthermore, the non linear effects implies different positions for the dips in comparison to the linear one. These results are robust predictions from the saturation physics, which can be used to challenge the treatment of the non linear QCD dynamics in the kinematical range of future electron - ion experiments.

Acknowledgements

The author thank E.R. Cazaroto, F. Carvalho, M. S. Kugeratski, E. G. de Oliveira, D. Spiering and F.S. Navarra by the collaboration in the subjects discussed in this contribution. This work was partially financed by the Brazilian funding agencies CNPq, CAPES and FAPERGS.

References

[1] A. Deshpande, R. Milner, R. Venugopalan and W. Vogelsang, Ann. Rev. Nucl. Part. Sci. **55**, 165 (2005); D. Boer, M. Diehl, R. Milner, R. Venugopalan, W. Vogelsang, D. Kaplan, H. Montgomery and S. Vigdor *et al.*, arXiv:1108.1713 [nucl-th]; A. Accardi, J. L. Albacete, M. Anselmino, N. Armesto, E. C. Aschenauer, A. Bacchetta, D. Boer and W. Brooks *et al.*, arXiv:1212.1701 [nucl-ex]; J. L. Abelleira Fernandez *et al.* [LHeC Study Group Collaboration], J. Phys. G **39**, 075001 (2012)

[2] F. Gelis, E. Iancu, J. Jalilian-Marian and R. Venugopalan, Ann. Rev. Nucl. Part. Sci. **60**, 463 (2010);E. Iancu and R. Venugopalan, arXiv:hep-ph/0303204; H. Weigert, Prog. Part. Nucl. Phys. **55**, 461 (2005); J. Jalilian-Marian and Y. V. Kovchegov, Prog. Part. Nucl. Phys. **56**, 104 (2006); J. L. Albacete and C. Marquet, Prog. Part. Nucl. Phys. **76**, 1 (2014).

[3] M. B. Gay Ducati and V. P. Goncalves, Phys. Lett. B **466**, 375 (1999)
[4] V. P. Goncalves, Phys. Lett. B **495**, 303 (2000)
[5] M. S. Kugeratski, V. P. Goncalves and F. S. Navarra, Eur. Phys. J. C **46**, 465 (2006)
[6] M. S. Kugeratski, V. P. Goncalves and F. S. Navarra, Eur. Phys. J. C **46**, 413 (2006)
[7] N. N. Nikolaev, W. Schafer, B. G. Zakharov and V. R. Zoller, JETP Lett. **84**, 537 (2007)
[8] H. Kowalski, T. Lappi and R. Venugopalan, Phys. Rev. Lett. **100**, 022303 (2008)
[9] H. Kowalski, T. Lappi, C. Marquet and R. Venugopalan, Phys. Rev. C **78**, 045201 (2008)
[10] E. R. Cazaroto, F. Carvalho, V. P. Goncalves and F. S. Navarra, Phys. Lett. B **669**, 331 (2008)
[11] E. R. Cazaroto, F. Carvalho, V. P. Goncalves and F. S. Navarra, Phys. Lett. B **671**, 233 (2009)
[12] V. P. Goncalves, M. S. Kugeratski, M. V. T. Machado and F. S. Navarra, Phys. Rev. C **80**, 025202 (2009).
[13] V. P. Goncalves, M. S. Kugeratski and F. S. Navarra, Phys. Rev. C **81**, 065209 (2010)
[14] A. Caldwell and H. Kowalski, Phys. Rev. C **81**, 025203 (2010).
[15] E. R. Cazaroto, F. Carvalho, V. P. Goncalves, M. S. Kugeratski and F. S. Navarra, Phys. Lett. B **696**, 473 (2011)
[16] T. Lappi and H. Mantysaari, Phys. Rev. C **83**, 065202 (2011)
[17] T. Toll and T. Ullrich, Phys. Rev. C **87**, 024913 (2013)
[18] F. Carvalho, V. P. Gonçalves, F. S. Navarra and E. G. de Oliveira, Phys. Rev. C **87**, no. 6, 065205 (2013)
[19] J. T. Amaral, V. P. Goncalves and M. S. Kugeratski, Nucl. Phys. A **930**, 104 (2014)
[20] T. Lappi, H. Mantysaari and R. Venugopalan, Phys. Rev. Lett. **114**, 082301 (2015)
[21] V. P. Gonçalves and D. S. Pires, Phys. Rev. C **91**, 055207 (2015)
[22] V. P. Goncalves, F. S. Navarra and D. Spiering, arXiv:1510.01512 [hep-ph].
[23] H. Kowalski, L. Motyka and G. Watt, Phys. Rev. D **74**, 074016 (2006); G. Watt and H. Kowalski, Phys. Rev. D **78**, 014016 (2008)

EPJ Web of Conferences **112**, 02007 (2016)
DOI: 10.1051/epjconf/201611202007

Di-jet production and angular correlations in DIS at NLO

Jamal Jalilian-Marian[1,a]

[1] *Department of Natural Sciences, Baruch College, 17 Lexington Ave, New York NY 10010 and CUNY Graduate Center, 365 Fifth Ave., New York NY 10016*

Abstract. Angular correlations are a sensitive probe of the dynamics of QCD at high energy. In particular azimuthal angular correlations between two hadrons produced in Deeply Inelastic Scattering (DIS) of a virtual photon on a hadron or nucleus offer the best environment in which to investigate high gluon density (gluon saturation) effects expected to arise at small x. Here we give a progress report on our derivation of Next to Leading Order (NLO) corrections to di-jet (di-hadron) production in DIS.

1 Introduction

Behavior of hadronic cross sections at high energy is one of the most interesting aspects of QCD where significant progress has been made in the last few years. Perturbative QCD calculations show a growth of hadronic cross sections with energy which would eventually violate unitarity bounds due to the untamed growth of parton distribution functions of a hadron with $1/x$. This fast rise of parton (specially gluon) distribution functions with decreasing x was one of the two surprises seen in DIS experiments at HERA [1]. Even before this experimental observation, theoretical considerations of gluon radiation in pQCD had predicted a power-like growth of gluon distribution function $xG(x, Q^2) \sim x^{-\delta}$ with $\delta \sim 0.3-0.5$ [2]. It was suggested that due to the large number of radiated gluons into a small area they would eventually overlap in the transverse plane and "recombine" [3, 4]. This lead to the idea of gluon saturation and its subsequent evolution into a Color Glass Condensate (CGC) [5]. It has been applied to high energy collisions involving at least one hadron. The two main physics effects encoded in the CGC formalism are *i*) multiple scattering (in the target frame) and *ii*) non-linear evolution of a hadronic or nuclear wave function with x (equivalently with energy or rapidity) [6, 7]. The first effect leads to a broadening of the transverse momentum spectra which appears as the so-called Cronin peak (enhancement) in pA collisions. The second effect due to non-linear evolution leads to a suppression of the transverse momentum spectra in pA as compared to a pp collisions, when properly normalized. Depending on the center of mass energy of the collision and the rapidity of the produced particles one or the the other effect will dominate. These two effects distinguish the CGC approach from that of collinear factorization based pQCD approaches and lead to perturbative unitarization of hadronic cross sections (at fixed impact parameter).

The CGC formalism, when applied to single inclusive particle production [8] in a high energy proton-nucleus (pA) collision, predicts an enhancement of the transverse momentum spectra in pA

[a] e-mail: jamal.jalilian-marian@baruch.cuny.edu

collisions as compared to that in a proton-proton collision (normalized) in the mid rapidity region. This is knows as the Cronin effect. More significantly the CGC formalism successfully predicted a suppression of the transverse momentum spectra in pA collisions as one moved from the mid rapidity to the forward rapidity region [9]. This is due to non-linear evolution of the nucleus wave function with x which reduces the number of partons in the nucleus. This theoretical prediction was subsequently verified experimentally and was considered a major indication of presence saturation effects in the target nucleus wave function.

More significantly, a disappearance of the away side peak (at 180 degrees) in di-hadron production in the forward rapidity region of proton-nucleus collisions was predicted [10]. The effect is two-fold, first there is broadening of the away side peak due to multiple scatterings and second, there is the reduction in the number of partons in the target wave function due to the non-linear evolution with x [11]. The disappearance of the away side was subsequently observed in the forward proton (deuteron)- gold collisions at RHIC and was a significant indication of dominance of saturation dynamics in the forward rapidity region at RHIC. It should however be noted that approaches based on collinear factorization in pQCD which include a modification of nuclear parton distribution unctions (nuclear shadowing) and some cold matter energy loss have also been able to fit the forward rapidity data at RHIC [12].

Despite the many successes of CGC formalism in describing particle production at RHIC, there are still some aspects that need improvement. Often particle production in the CGC formalism is know only at the Leading Order (LO) level. To be able to make truly quantitative predictions one need a NLO description [13]. In this work we take the first step in this direction by considering three parton production in DIS using the CGC formalism [14]. This will be useful in two aspects; first it gives us another (and less inclusive) particle production channel in which the CGC formalism can be applied and tested. While in di-hadron angular correlation studies one fixed the kinematics of the leading hadron and varies the momenta of the away side hadron, here one can vary the momenta of two away side hadrons which provides an additional handle on probing the dynamics of gluon saturation. Second, upon integrating out one of the three final state particles it will give us the real part of the NLO di-hadron production cross section. Including the virtual corrections then will give us the complete NLO di-hadron production cross section which can then be used to investigate azimuthal angular correlations between the produced hadrons.

2 Three parton production in DIS at small x

We start by considering $\gamma^\star \to q\,\bar{q}\,g$ at high energy. There are two diagrams as shown in figure 1 which can be understood as a two step process. First the virtual photon splits into a quark anti-quark pair that interacts with the target and then either one of the quark or anti-quark radiates a gluon. This is shown in the first diagram. The second diagram corresponds to the virtual photon splitting into a quark anti-quark pair, one of which radiates a gluon and the three final state partons interact with the target. Alternatively one can think of the gluon radiation happening either before or after the scattering from the target. Finally, only radiation from a quark (top line) is shown so that there are two additional diagrams where the gluon is radiated from the anti-quark (bottom line). These additional diagram will give results which can be obtained from the diagrams shown by a trivial change of momenta.

It should be noted that there is no radiation during the passage of the quark anti-quark pair through the target due to the eikonal kinematics (target moves with the speed of light). Furthermore it can be shown that all diagrams where only one (either one or two) of the lines in the first (second) diagram interacts with the target cancel amongst each other.

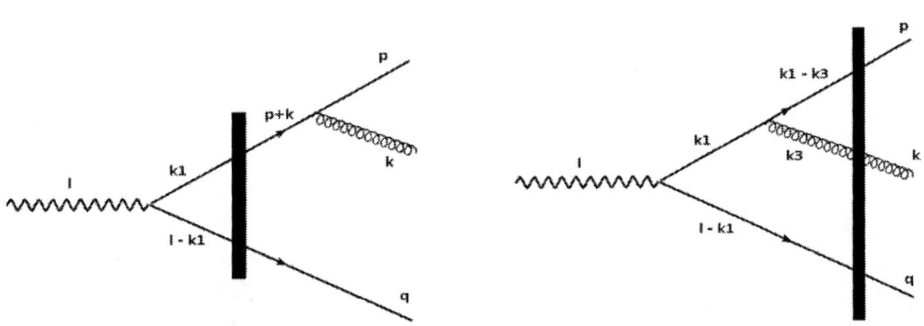

Figure 1. Leading Order 3-parton production diagrams corresponding to $\mathcal{A}_1$ (left) and $\mathcal{A}_2$ (right). The solid thick line represents interactions with the target (shock wave). The diagrams where the anti-quark radiates the gluon ($\mathcal{A}_3$ and $\mathcal{A}_4$) are not shown.

The Leading Order (LO) amplitude can be written as

$$\mathcal{A}^a = \mathcal{A}_1^a + \mathcal{A}_2^a + \mathcal{A}_3^a + \mathcal{A}_4^a \tag{1}$$

with

$$\begin{aligned}
\mathcal{A}_1^a &= (ie)(ig)\bar{u}(p)\,\gamma^\mu\, t^a\, S_F(p+k,k_1)\gamma^\nu\, S_F(l-k_1,q)\left[S_F^0(q)\right]^{-1} v(q)\,\epsilon_\nu^\star(l)\,\epsilon_\mu(k)\,\frac{d^4k_1}{(2\pi)^4} \\
\mathcal{A}_2^a &= (ie)(ig)\bar{u}(p)\left[S_F^0(p)\right]^{-1} S_F(p,k_1-k_3)\gamma^\lambda\, t^c\, S_F^0(k_1)\gamma^\nu\, S_F(l-k_1,q)\left[S_F^0(q)\right]^{-1} v(q) \\
&\quad \left[G_\lambda^{\ \delta}\right]^{ca}(k_3,k)\left[G_\delta^{0\mu}(k)\right]^{-1}\epsilon_\nu^\star(l)\,\epsilon_\mu(k)\,\frac{d^4k_1}{(2\pi)^4}\frac{d^4k_3}{(2\pi)^4} \\
\mathcal{A}_3^a &= (ie)(ig)\bar{u}(p)\left[S_F^0(p)\right]^{-1} S_F(p,k_1)\,\gamma^\nu S_F(l-k_1,q+k)\,\gamma^\mu\, t^a\, v(q)\,\epsilon_\nu^\star(l)\,\epsilon_\mu(k)\,\frac{d^4k_1}{(2\pi)^4} \\
\mathcal{A}_4^a &= (ie)(ig)\bar{u}(p)\left[S_F^0(p)\right]^{-1} S_F(p,k_1)\,\gamma^\nu\, S_F^0(l-k_1)\gamma^\lambda\, t^c\, S_F(l-k_1-k_3,q)\left[S_F^0(q)\right]^{-1} \\
&\quad \left[G_\lambda^{\ \delta}\right]^{ca}(k_3,k)\left[G_\delta^{0\mu}(k)\right]^{-1}\epsilon_\nu^\star(l)\,\epsilon_\mu(k)\,\frac{d^4k_1}{(2\pi)^4}\frac{d^4k_3}{(2\pi)^4}
\end{aligned} \tag{2}$$

The quark or gluon propagators in the background field (shock wave) are defined as

$$\begin{aligned}
S_F(p,q) &\equiv S_F^0(2\pi)^4\delta^4(p-q) + S_F^0\tau_F(p,q)S_F^0(q) \\
G^{ab}_{\mu\nu}(p,q) &\equiv G^{0,ab}_{\mu\nu}(p)\,\delta^{ab}\,(2\pi)^4\delta^4(p-q) + G^0_{\ \mu\lambda}(p)\tau_g^{ab}(p,q)G^{0,\lambda}{}_{\nu}(q)
\end{aligned} \tag{3}$$

where $S_F^0, G^{0,ab}_{\mu\nu}$ are the free fermion and gluon propagators,

$$S_F^0(p) = \frac{i(\not{p}+m)}{p^2-m^2+i\epsilon} \qquad G^{0,ab}_{\mu\nu}(k) = \frac{id_{\mu\nu}(k)}{k^2+i\epsilon}\,\delta^{ab} \tag{4}$$

with the projector

$$d_{\mu\nu}(k) \equiv \left[-g_{\mu\nu} + \frac{k_\mu n_\nu + k_\nu n_\mu}{n\cdot k}\right] \tag{5}$$

and τ_F, τ_g contain the all-order scattering from the target color field, defined via

$$\begin{aligned}
\tau_F(p,q) &\equiv (2\pi)\delta(p^- - q^-)\gamma^- \int d^2x_t e^{i(p_t-q_t)\cdot x_t}\left\{\theta(p^-)\left[V(x_t)-1\right] - \theta(-p^-)\left[V^\dagger(x_t)-1\right]\right\} \\
\tau_g(p,q) &\equiv -(2\pi)\delta(p^- - q^-)2p^- \int d^2x_t e^{i(p_t-q_t)\cdot x_t}\left\{\theta(p^-)\left[U(x_t)-1\right] - \theta(-p^-)\left[U^\dagger(x_t)-1\right]\right\}
\end{aligned}$$

where V, U are Wilson lines in the fundamental and adjoint representation of $SU(N_c)$,

$$V(x_t) \equiv \hat{P}exp\,\{ig\int dx^- A_a^+(x^-, x_t)\,t_a\}$$
$$U(x_t) \equiv \hat{P}exp\,\{ig\int dx^- A_a^+(x^-, x_t)\,T_a\}$$

and t_a and T_a are the corresponding generators. $\epsilon_\nu(l)$, $\epsilon_\mu(k)$ are the polarization vectors of the incoming virtual photon and the outgoing gluon respectively. We will set the transverse momentum of the virtual photon equal to zero so that $l^+ = -Q^2/2l^-$ without any loss of generality.

Noticing that the couplings, the spinors and the polarization vectors will be common to all 4 diagrams, we define the reduced amplitude A via

$$\mathcal{A} \equiv -eg\,\bar{u}(p)\,[A]^{\mu\nu}\,v(q)\,\epsilon_\nu^\star(l)\,\epsilon_\mu(k) \tag{6}$$

with

$$\begin{aligned}
A_1^{\mu\nu} &= \gamma^\mu t^a S_F^0(p+k)\,\tau_F(p+k,k_1)\,S_F^0(k_1)\gamma^\nu S_F^0(l-k_1)\,\tau_F(l-k_1,q)\,\frac{d^4k_1}{(2\pi)^4}\\
A_2^{\mu\nu} &= \tau_F(p,k_1-k_3)\,S_F^0(k_1-k_3)\gamma^\lambda t^a S_F^0(k_1)\,\gamma^\nu S_F^0(l-k_1)\,\tau_F(l-k_1,q)\,G_\lambda^{0\mu}(k_3)\tau_g^{ac}(k_3,k)\\
&\quad \frac{d^4k_1}{(2\pi)^4}\frac{d^4k_3}{(2\pi)^4}\\
A_3^{\mu\nu} &= \tau_F(p,k_1)\,S_F^0(k_1)\,\gamma^\nu S_F^0(l-k_1)\,\tau_F(l-k_1,q+k)S_F^0(q+k)\,\gamma^\mu t^a\,\frac{d^4k_1}{(2\pi)^4}\\
A_4^{\mu\nu} &= \tau_F(p,k_1)\,S_F^0(k_1)\,\gamma^\nu S_F^0(l-k_1)\,\gamma^\lambda t^c S_F^0(l-k_1-k_3)\tau_F(l-k_1-k_3,q)\,G_\lambda^{0\,\mu}(k_3)\tau_g^{ca}(k_3,k)\\
&\quad \frac{d^4k_1}{(2\pi)^4}\frac{d^4k_3}{(2\pi)^4}
\end{aligned} \tag{7}$$

To proceed further we perform the momentum integrations in the amplitude; to be specific let's consider $A_1^{\mu\nu}$ given in Eq. (7). The integration over d^4k_1 can be performed in two different ways: either by introducing the standard Feynman parameters, or by taking advantage of the delta function in k_1^- and performing the integration over k_1^+ using contour integration on the complex plane. Following the second choice we get

$$\begin{aligned}
A_1^{\mu\nu} &= \frac{i}{2\,l^-}\frac{1}{(p+k)^2}\,\delta(l^- - p^- - q^- - k^-)\int d^2x_t\,d^2y_t e^{-i(p_t+k_t)\cdot x_t}\,e^{-iq_t\cdot y_t}\,V(x_t)\,V^\dagger(y_t)\,\gamma^\mu t^a\,i(\not{p}+\not{k})\\
&\quad \gamma^-\,i\not{k}_1\,\gamma^\nu\,i(\not{l}-\not{k}_1)\,\gamma^-\,K_0\,[L\,(x_t-y_t)]
\end{aligned} \tag{8}$$

where K_0 is the modified Bessel function, $L^2 = \frac{q^-(p^-+k^-)}{l^-l^-}Q^2$ and $k_1^- = p^- + k^- = l^- - q^-$, $k_1^+ = \frac{k_{1t}^2 - i\epsilon}{2(p^-+k^-)}$ and all k_{1t} terms appearing above are to be understood as $-i\partial_{x_t-y_t}$ acting on the Bessel function K_0. We also note the presence of $\frac{1}{(p+k)^2}$ term above which will appear as a collinear divergence when one integrates over the parent quark or gluon in the final state in order to compute the NLO corrections to di-hadron production cross section. This will be identified as one of the contributions to the scale evolution of quark or gluon fragmentation function. Contribution of the $A_3^{\mu\nu}$ proceeds in exactly the same way. On the other hand one can not perform all the integrations in $A_2^{\mu\nu}$ and $A_4^{\mu\nu}$ where one is left with a one-dimensional integral which must be evaluated numerically.

The next step is to square the amplitude and to evaluate the Dirac traces that appear. This is done using FORM and the result is long. Furthermore, the result will involve trace of products of multiple

Wilson lines [15–17] whose evolution equation is known. As a last step one needs to specify the kinematics of the three produced hadrons and convolute the three final state partons with the corresponding parton-hadron fragmentation functions. To avoid issues dealing with di-hadron fragmentation functions we assume the produced hadrons are not too close in azimuthal angle so that using independent fragmentation functions may be justified. Alternatively one can use our results for three-jet (rather than hadron) production which will sidestep the issue of independent hadronization.

3 Summary

Three-parton (hadron) azimuthal angular correlations in DIS are a sensitive probe of gluon saturation dynamics in high energy QCD. They can also be used to calculate the real corrections to NLO corrections to di-hadron production cross section. Work in this direction is in progress and will be reported on elsewhere [14].

Acknowledgement

This is based on work done in collaboration with A. Ayala, M. Hentschinski and M. E. Tejeda-Yeomans. J.J-M. acknowledges support by the DOE Office of Nuclear Physics through Grant No. DE-FG02-09ER41620 and from The City University of New York through the PSC-CUNY Research Award Program, grant 67732-0045.

References

[1] Z. Zhang [H1 and ZEUS Collaborations], arXiv:1511.05402 [hep-ex].

[2] L. N. Lipatov, Sov. J. Nucl. Phys. **23**, 338 (1976) [Yad. Fiz. **23**, 642 (1976)]; E. A. Kuraev, L. N. Lipatov and V. S. Fadin, Sov. Phys. JETP **45**, 199 (1977) [Zh. Eksp. Teor. Fiz. **72**, 377 (1977)]; I. I. Balitsky and L. N. Lipatov, Sov. J. Nucl. Phys. **28**, 822 (1978) [Yad. Fiz. **28**, 1597 (1978)].

[3] L. V. Gribov, E. M. Levin and M. G. Ryskin, Phys. Rept. **100**, 1 (1983).

[4] A. H. Mueller and J. w. Qiu, Nucl. Phys. B **268**, 427 (1986).

[5] E. Iancu and R. Venugopalan, In *Hwa, R.C. (ed.) et al.: Quark gluon plasma* 249-3363; H. Weigert, Prog. Part. Nucl. Phys. **55**, 461 (2005); J. Jalilian-Marian, Y. V. Kovchegov, Prog. Part. Nucl. Phys. **56**, 104-231 (2006); F. Gelis, E. Iancu, J. Jalilian-Marian, R. Venugopalan, Ann. Rev. Nucl. Part. Sci. **60**, 463-489 (2010). [arXiv:1002.0333 [hep-ph]]; J. Jalilian-Marian, Y. V. Kovchegov, Prog. Part. Nucl. Phys. **56**, 104-231 (2006); J. L. Albacete and C. Marquet, Prog. Part. Nucl. Phys. **76**, 1 (2014).

[6] A. Ayala, J. Jalilian-Marian, L. D. McLerran and R. Venugopalan, Phys. Rev. D **52**, 2935 (1995), Phys. Rev. D **53**, 458 (1996); J. Jalilian-Marian, A. Kovner, L. D. McLerran and H. Weigert, Phys. Rev. D **55**, 5414 (1997); J. Jalilian-Marian, A. Kovner, A. Leonidov and H. Weigert, Nucl. Phys. B **504**, 415 (1997), Phys. Rev. D **59**, 014014 (1999), Phys. Rev. D **59**, 014015 (1999), Phys. Rev. D **59**, 034007 (1999), A. Kovner, J. G. Milhano and H. Weigert, Phys. Rev. D **62**, 114005 (2000); A. Kovner and J. G. Milhano, Phys. Rev. D **61**, 014012 (2000); E. Iancu, A. Leonidov and L. D. McLerran, Nucl. Phys. A **692**, 583 (2001), Phys. Lett. B **510**, 133 (2001); E. Ferreiro, E. Iancu, A. Leonidov and L. McLerran, Nucl. Phys. A **703**, 489 (2002); H. Weigert, Nucl. Phys. **A703**, 823-860 (2002); J. -P. Blaizot, E. Iancu, H. Weigert, Nucl. Phys. **A713**, 441-469 (2003).

[7] I. Balitsky, Nucl. Phys. B **463**, 99 (1996); Y. V. Kovchegov, Phys. Rev. D **60**, 034008 (1999), Phys. Rev. D **61**, 074018 (2000).

[8] A. Dumitru and J. Jalilian-Marian, Phys. Rev. Lett. **89**, 022301 (2002); J. Jalilian-Marian, Y. Nara and R. Venugopalan, Phys. Lett. B **577**, 54 (2003); J. Jalilian-Marian, J. Phys. G **30**, S751 (2004), Nucl. Phys. A **748**, 664 (2005), Phys. Rev. C **70**, 027902 (2004), Nucl. Phys. A **739**, 319 (2004); A. Dumitru, A. Hayashigaki and J. Jalilian-Marian, Nucl. Phys. A **765**, 464 (2006), Nucl. Phys. A **770**, 57 (2006); F. Gelis and J. Jalilian-Marian, Phys. Rev. D **76**, 074015 (2007); D. Boer, A. Utermann and E. Wessels, Phys. Rev. D **77**, 054014 (2008); E. R. Cazaroto, V. P. Gonçalves, F. S. Navarra, Nucl. Phys. A **872**, 196 (2011); J. Jalilian-Marian and A. H. Rezaeian, Phys. Rev. D **85**, 014017 (2012), Phys. Rev. D **86**, 034016 (2012); A. H. Rezaeian, Phys. Rev. D **86**, 094016 (2012); J. L. Albacete, A. Dumitru, H. Fujii and Y. Nara, Nucl. Phys. A **897**, 1 (2013); T. Altinoluk and A. Kovner, Phys. Rev. D **83**, 105004 (2011); G. A. Chirilli, B. -W. Xiao and F. Yuan, Phys. Rev. Lett. **108**, 122301 (2012), Phys. Rev. D **86**, 054005 (2012); Z. -B. Kang, I. Vitev and H. Xing, arXiv:1403.5221 [hep-ph].

[9] D. Kharzeev, Y. V. Kovchegov and K. Tuchin, Phys. Rev. D **68**, 094013 (2003); J. L. Albacete, N. Armesto, A. Kovner, C. A. Salgado and U. A. Wiedemann, Phys. Rev. Lett. **92**, 082001 (2004).

[10] C. Marquet, Nucl. Phys. A **796**, 41 (2007).

[11] J. Jalilian-Marian and Y. V. Kovchegov, Phys. Rev. D **70**, 114017 (2004) [Erratum-ibid. D **71**, 079901 (2005)]; A. Kovner and M. Lublinsky, JHEP **0611**, 083 (2006), Phys. Rev. D **84**, 094011 (2011), Phys. Rev. D **83**, 034017 (2011); J. Jalilian-Marian, Eur. Phys. J. C **61**, 789 (2009), Nucl. Phys. A **770**, 210 (2006); J. L. Albacete and C. Marquet, Phys. Rev. Lett. **105**, 162301 (2010); F. Dominguez, C. Marquet, B. -W. Xiao and F. Yuan, Phys. Rev. D **83**, 105005 (2011); A. Stasto, B. -W. Xiao and F. Yuan, Phys. Lett. B **716**, 430 (2012); F. Dominguez, C. Marquet, A. M. Stasto and B. -W. Xiao, Phys. Rev. D **87**, no. 3, 034007 (2013); T. Lappi and H. Mantysaari, Nucl. Phys. A **908**, 51 (2013); E. Iancu and D. N. Triantafyllopoulos, JHEP **1311**, 067 (2013).

[12] H. Xing, Z. B. Kang, I. Vitev and E. Wang, Phys. Rev. D **86**, 094010 (2012); Z. B. Kang, I. Vitev and H. Xing, Phys. Rev. D **85**, 054024 (2012).

[13] I. Balitsky and G. A. Chirilli, Phys. Rev. D **88**, 111501 (2013); A. Kovner, M. Lublinsky and Y. Mulian, arXiv:1405.0418 [hep-ph].

[14] A. Ayala, M. Hentschinski, J. Jalilian-Marian and M. E. Tejeda-Yeomans, work in progress.

[15] Y. V. Kovchegov, J. Kuokkanen, K. Rummukainen and H. Weigert, Nucl. Phys. A **823**, 47 (2009); C. Marquet and H. Weigert, Nucl. Phys. A **843**, 68 (2010); A. Dumitru and J. Jalilian-Marian, Phys. Rev. D **82**, 074023 (2010), Phys. Rev. D **81**, 094015 (2010); A. Dumitru, J. Jalilian-Marian, T. Lappi, B. Schenke and R. Venugopalan, Phys. Lett. B **706**, 219 (2011); E. Iancu and D. N. Triantafyllopoulos, JHEP **1111**, 105 (2011).

[16] F. Dominguez, A. H. Mueller, S. Munier and B. -W. Xiao, Phys. Lett. B **705**, 106 (2011); J. Jalilian-Marian, Phys. Rev. D **85**, 014037 (2012); A. Ayala, E. R. Cazaroto, L. A. Hernández, J. Jalilian-Marian and M. E. Tejeda-Yeomans, Phys. Rev. D **90**, no. 7, 074037 (2014).

[17] S. Jeon and R. Venugopalan, Phys. Rev. D **71**, 125003 (2005); A. Dumitru, J. Jalilian-Marian, E. Petreska, Phys. Rev. **D84**, 014018 (2011); A. Dumitru and E. Petreska, Nucl. Phys. A **879**, 59 (2012); J. Jalilian-Marian, S. Jeon and R. Venugopalan, Phys. Rev. D **63**, 036004 (2001).

EPJ Web of Conferences **112**, 02008 (2016)
DOI: 10.1051/epjconf/201611202008

Numerical precision of the solution to the running-coupling Balitsky-Kovchegov equation

Marek Matas[1,a], Jan Cepila[1], and Jesus Guillermo Contreras Nuno[1]

[1] *Faculty of Nuclear Sciences and Physical Engineering, Czech Technical University in Prague, Prague, Czech Republic*

Abstract. We use the running coupling Balitsky-Kovchegov (rcBK) equation to study the rapidity dependence of saturation in inclusive HERA data and we discuss the behaviour of its numerical solution. The rcBK equation has been solved using Runge-Kutta methods. The influence of the parameters implicit in the numerical evolution has been studied. They include, among others, the order of the Runge-Kutta evolution, the size of the different grids and the step in the numerical evolution. Some suggestions on the minimum value of these parameters are put forward.

1 Introduction

One of the open questions in particle physics is particle production in the high-energy limit of QCD. Evolution equations such as BFKL [1–5] are used to describe the gluon density of the hadron in high-energy collisions. The BFKL evolution equation predicts the emergence of new partons as the energy of the collision increases and in this approach, the gluon density is not bound by unitarity restrictions. The number of partons may be reduced by saturation effects caused by recombination processes inside the hadron. An extension of the BFKL equation was found by Balitsky [6] and Kovchegov [7] (BK) resulting in an equation that generates dynamical balance between radiation and recombination processes driven by a saturation scale. The integro-differential BK evolution equation [8–10] considered in this work, and denoted by rcBK in the following, includes a running-coupling kernel that takes into account two-loop processes and assumes an impact-parameter independent solution [11]. It was derived from the JIMWLK evolution equations in the limit of large number of colors by Kovchegov [7, 12].

Using numerical methods one can solve the rcBK equation [8–10, 13] and calculate a prediction for the proton structure of DIS that accounts for recombination processes within the hadron. Numerical methods used to compute the rcBK solution need to be studied and an optimal setup regarding precision and speed of computation has to be tested in order to estimate the uncertainty coming from the non-analytic solution of rcBK equation.

2 The rcBK evolution equation

The rcBK evolution equation with running coupling kernel reads [14].

[a] e-mail: matas.marek1@gmail.com

$$\frac{\partial N(r,Y)}{\partial Y} = \int d\vec{r_1} K^{run}(\vec{r},\vec{r_1},\vec{r_2})(N(\vec{r_1},Y) + N(\vec{r_2},Y) - N(\vec{r},Y) - N(\vec{r_1},Y)N(\vec{r_2},Y)), \tag{1}$$

where $K^{run}(\vec{r},\vec{r_1},\vec{r_2})$ can be expressed as [15]

$$K^{run}(\vec{r},\vec{r_1},\vec{r_2}) = \frac{N_c\alpha_s(r^2)}{2\pi^2}\left(\frac{r^2}{r_1^2 r_2^2} + \frac{1}{r_1^2}\left(\frac{\alpha_s(r_1^2)}{\alpha_s(r_2^2)} - 1\right) + \frac{1}{r_2^2}\left(\frac{\alpha_s(r_2^2)}{\alpha_s(r_1^2)} - 1\right)\right), \tag{2}$$

$\vec{r_2} = \vec{r} - \vec{r_1}$ and $Y = ln\frac{x_0}{x}$. If we disregard the last term in Eq.1, we obtain a linear equation that can be shown to be equivalent to the BFKL evolution equation [16]. The running coupling depends on the number of considered quark flavors, in this work we use $n_f = 3$

$$\alpha_{s,n_f}(r^2) = \frac{4\pi}{\beta_{0,n_f} ln\left(\frac{4C^2}{r^2\Lambda_{n_f}^2}\right)}, \tag{3}$$

where

$$\beta_{0,n_f} = 11 - \frac{2}{3}n_f. \tag{4}$$

The constant C^2 is the uncertainty coming from the Fourier transformation and is usually fitted to data [10], $\Lambda_{n_f}^2$ is called the scale parameter and its value depends on the value of n_f (we set it to $\Lambda_{n_f} = 0.241$ GeV). Since all dipole sizes are accounted for in the rcBK evolution equation, coupling needs to be reduced after a certain value is reached, so that the maximal value of the coupling constant would not exceed a set limit [10] (in our case $\alpha_s = 0.7$).

We used the MV initial condition [10] with parameters $C = 2.52, \Lambda_{QCD} = 0.241$ GeV, $\gamma = 1.135$, $Q_{s0}^2 = 0.165\,\text{GeV}^2$, $N_c = 3$ [10],

$$N^{MV}(r, x = x_0) = 1 - \exp\left(-\frac{(r^2Q_{s0}^2)^\gamma}{4}\ln\left(\frac{1}{r\Lambda_{QCD}} + e\right)\right), \tag{5}$$

where e is the elementary charge.

3 Solving the rcBK evolution equation

The rcBK equation does not have an analytic solution, so it has to be solved numerically [8–10]. A usual way of solving this equation involves dividing the integral over r_1 into a logarithmic grid with a constant step, then one uses a Runge-Kutta method for solving the differential equation, the Simpson method for numerical integration and linear interpolation for acquiring values of $N(r)$ for an arbitrary point.

The parameters of the numerical computation have been varied and their influence to the resulting scattering amplitude was studied. The proportional change $D(r, Y)$ was computed as

$$D(r,Y) = \frac{|\, N_{orig}(r,Y) - N_{new}(r,Y)\,|}{N_{orig}(r,Y)}. \tag{6}$$

The initial values for the computation were 25 steps over one order of magnitude of the dipole size r over the interval $[10^{-7}, 10^2]$, 10 steps over the interval of $[0;\pi]$ in θ, which is the angle between r and r_1, and a Runge-Kutta method of fourth order with step of 0.01 in rapidity.

In Fig. 1 we can see the integrand for the structure function F_2 (see for example Eq. 9 in [13]) and its dependence on r for $Y = 2$ and $Y = 10$. We can see that the interval that contributes the most to the total value of F_2 is at about $r \sim (0.1, 20)\,\mathrm{GeV}^{-1}$. The highest precision needs to be obtained namely for these values of r.

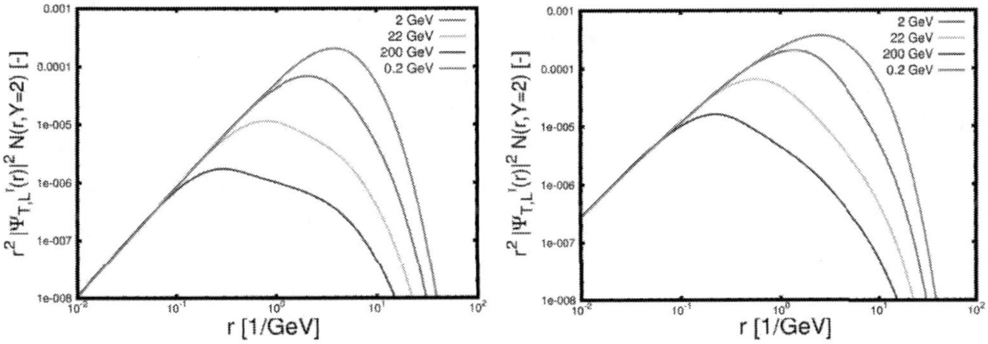

Figure 1. Integrand for the structure function as a function of r for $Y = 2$ (left) and $Y = 10$ (right). Each line corresponds to a different value of the photon virtuality Q^2.

A significant decrease in the running time can be also achieved by modifying the Runge-Kutta method. Since the rcBK evolution equation does not explicitly depend on rapidity, we can split the integration into three terms and then obtain all Runge-Kutta coefficients as their linear combination without the need to recompute the integration itself. The three terms stand as follows [13]

$$\begin{aligned} I_0 &= \int d\vec{r_1} K^{run}(\vec{r_1}, \vec{r_2}, \vec{r}) \\ I_1 &= \int d\vec{r_1} K^{run}(\vec{r_1}, \vec{r_2}, \vec{r})(N(Y, \vec{r_1}) + N(Y, \vec{r_2})) \\ I_2 &= \int d\vec{r_1} K^{run}(\vec{r_1}, \vec{r_2}, \vec{r})(N(Y, \vec{r_1})N(Y, \vec{r_2})) \end{aligned} \tag{7}$$

and the Runge-Kutta coefficients can be expressed as

$$k_1 = I_1 - I_2 - I_0 \cdot N(r, Y) \tag{8}$$

$$k_2 = k_1 + \frac{1}{2}hk_1I_0 - \frac{1}{2}hk_1I_1 - \frac{1}{4}h^2k_1^2I_0 \tag{9}$$

$$k_3 = k_1 + \frac{1}{2}hk_2I_0 - \frac{1}{2}hk_2I_1 - \frac{1}{4}h^2k_2^2I_0 \tag{10}$$

$$k_4 = k_1 + \frac{1}{2}hk_3I_0 - \frac{1}{2}hk_3I_1 - \frac{1}{4}h^2k_3^2I_0 \tag{11}$$

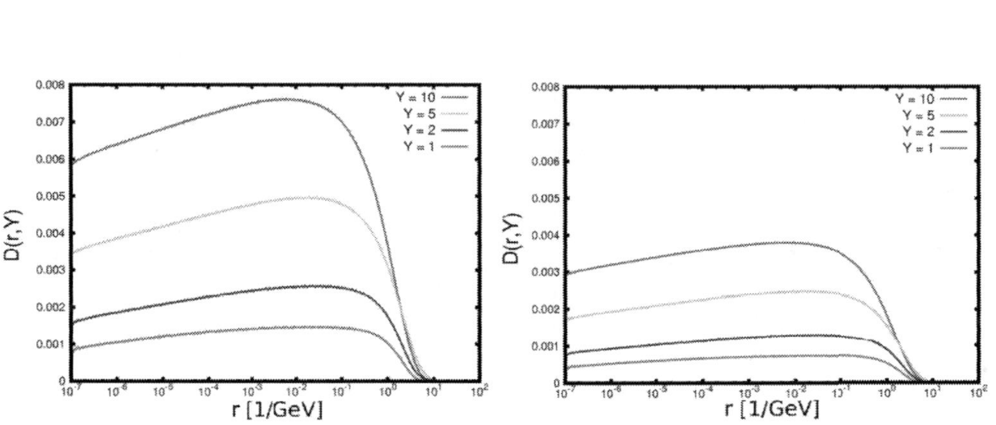

Figure 2. A variation of the Runge-Kutta method. Left - Runge-Kutta method of first order compared to the second, right - second order compared to the fourth.

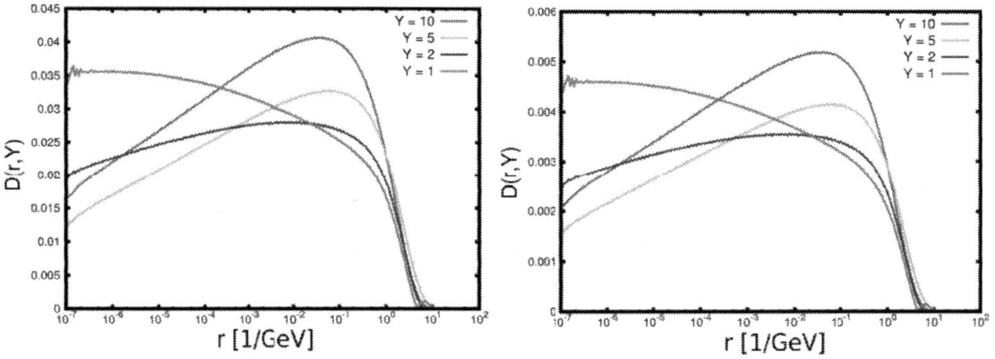

Figure 3. A variation of the rapidity step. Left - step of 0.05 compared to 0.01, right - step of 0.01 compared to 0.005.

4 Results

There is less than 0.7% difference in the rcBK solution when comparing first order Runge-Kutta method to the second order up to $Y = 10$ and less than 0.4% for comparing second order to fourth. Therefore, there is less than 1% uncertainty coming from the choice of the Runge-Kutta method and the uncertainty rises with rapidity. Considering the step in rapidity for the Runge-Kutta method, it can be seen that the difference between the step of 0.5 to the step of 0.1 is up to 4% and the biggest difference is in the region where the integrand dominates. Therefore it can be presumed that the uncertainty will also rise with rapidity. The difference between the step of 0.1 to the step of 0.05 is less than approximately 0.5%.

When the variation of the number of steps for the magnitude of r_1 is done from 10 steps to 25 steps per order of magnitude of r_1, the difference quickly rises with rapidity up to 2.5% at $Y = 10$. Using 50 steps instead of 25 steps does not yield any considerable difference for all rapidities. The results at low values of r have small significance since their contribution to the final structure function is negligible as shown in Fig. 1.

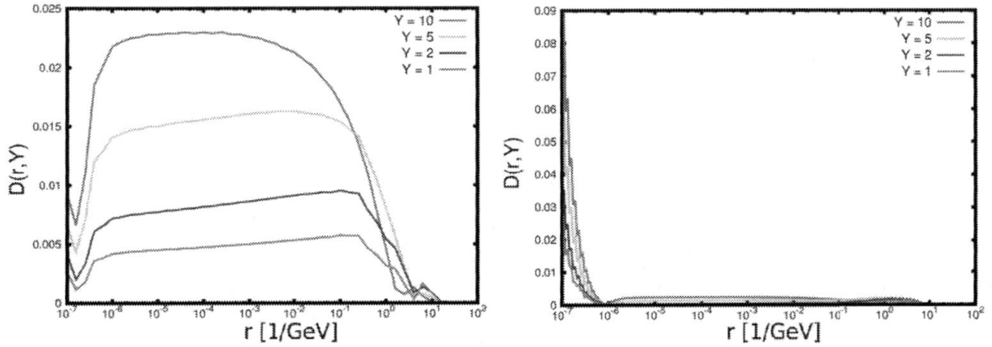

Figure 4. A variation of the step in integration over r, left - the step of 10 over one order of magnitude of r_1 compared to 25, right - the step of 25 compared to the step of 50.

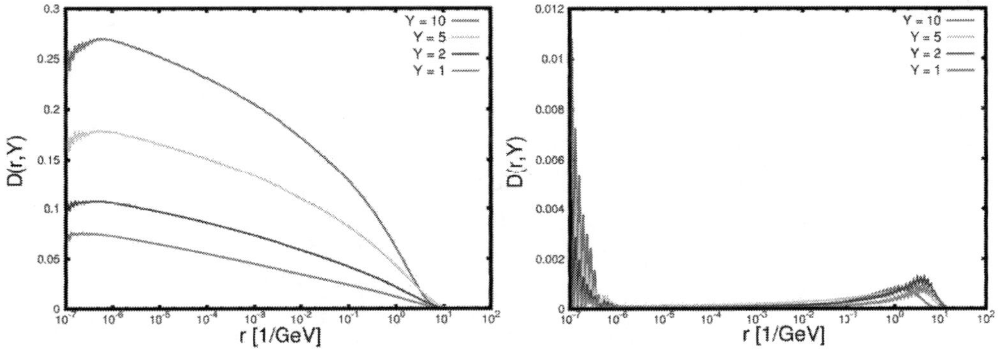

Figure 5. A variation of the step in integration over θ. Left - the step of 5 over the interval of $[0, \pi]$ compared to 10, right - the step of 10 compared to the step of 20.

Also, the variation of the number of steps for the angle θ between $\vec{r_1}$ and $\vec{r}$ from 5 steps to 10 steps over the interval $[0, \pi]$ has been performed resulting in huge difference up to 25% which quickly rises with rapidity. Adding more steps does not yield to any notable difference.

Running time for a fixed scale was about 100 s on a regular personal computer and the mean square variation was below 1.5% of the experimentally measured values of F_2.

5 Conclusion

There are several models that predict the effect of saturation of partons in high-energy collisions. Balitsky-Kovchegov evolution equation does that by modifying the BFKL equation with recombination processes that occur in hadrons at high energies.

The Balitsky-Kovchegov evolution equation was numerically solved using the Runge-Kutta method of fourth order, Simpson rule and linear interpolation. The parameters used for the numerical computation were tested and a suggestion for their setup has been put forward to obtain a good ratio of precision and computing speed since the program for solving the rcBK evolution equation can be very demanding on the running time.

The optimal parameters to compute the rcBK equation prove to be:

- 25 steps per one order of magnitude of r,
- 10 steps in θ over the interval $[0; \pi]$,
- Runge-Kutta method of fourth order,
- A step of 0.01 in rapidity for the Runge-Kutta method.

Acknowledgment

This work was partially supported by the grant of the Grant Agency of Czech Republic n.13-20841S and by grant LK11209 of MŠMT ČR.

References

[1] V.S. Fadin, E.A. Kuraev, L.N. Lipatov, Phys. Lett. **B60**, 50 (1975)
[2] L.N. Lipatov, Sov. J. Nucl. Phys. **23**, 338 (1976), [Yad. Fiz.23,642(1976)]
[3] E.A. Kuraev, L.N. Lipatov, V.S. Fadin, Sov. Phys. JETP **44**, 443 (1976), [Zh. Eksp. Teor. Fiz.71,840(1976)]
[4] E. Kuraev, L. Lipatov, V.S. Fadin, Sov.Phys.JETP **45**, 199 (1977)
[5] I. Balitsky, L. Lipatov, Sov.J.Nucl.Phys. **28**, 822 (1978)
[6] I. Balitsky, Nucl. Phys. **B463**, 99 (1996), hep-ph/9509348
[7] Y.V. Kovchegov, Phys. Rev. **D60**, 034008 (1999), hep-ph/9901281
[8] J.L. Albacete, Y.V. Kovchegov, Phys. Rev. **D75**, 125021 (2007), 0704.0612
[9] J.L. Albacete, N. Armesto, J.G. Milhano, C.A. Salgado, Phys. Rev. **D80**, 034031 (2009), 0902.1112
[10] J.L. Albacete, N. Armesto, J.G. Milhano, P. Quiroga-Arias, C.A. Salgado, Eur. Phys. J. **C71**, 1705 (2011), 1012.4408
[11] M. Kuhlen, Springer Tracts Mod. Phys. **150**, 1 (1999)
[12] Y.V. Kovchegov, Phys. Rev. **D61**, 074018 (2000), hep-ph/9905214
[13] J. Cepila, J.G. Contreras (2015), 1501.06687
[14] I. Balitsky, Phys. Rev. **D75**, 014001 (2007), hep-ph/0609105
[15] Y.V. Kovchegov, H. Weigert, Nucl. Phys. **A784**, 188 (2007), hep-ph/0609090
[16] J.L. Albacete, C. Marquet, Prog. Part. Nucl. Phys. **76**, 1 (2014), 1401.4866
[17] F.D. Aaron et al. (ZEUS, H1), JHEP **01**, 109 (2010), 0911.0884

EPJ Web of Conferences **112**, 02009 (2016)
DOI: 10.1051/epjconf/201611202009

Central Exclusive Production in pp collisions at LHCb

Ronan McNulty[1,a] (on behalf of the LHCb collaboration)

[1] *School of Physics, University College Dublin, Dublin 4, Ireland.*

Abstract.
Central Exclusive Production (CEP) is a unique process at hadron machines in which particles are produced via colourless propagators, and several measurements at the LHC are directly comparable with past and future electron-ion colliders. LHCb have measured the cross-sections for the CEP of vector mesons, $J/\psi, \psi(2S), \Upsilon(1S), \Upsilon(2S)$ and $\Upsilon(3S)$, which are photo-produced. In the double Pomeron exchange process, preliminary measurements have been made of $\chi_{c0}, \chi_{c1}, \chi_{c2}$ meson production while the first observations of the CEP of pairs of charmonia, $J/\psi J/\psi$ and $J/\psi\psi(2S)$, have been made and limits obtained on the pair production of other charmonia.

1 Introduction

Central exclusive production (CEP) at the LHC is characterised by an isolated system of particles surrounded by two rapidity gaps that extend down to the intact colliding protons; see [1] and the papers referenced in [2] for a recent survey of the field. The lack of additional activity signals the presence of colourless propagators: two photons; two Pomerons; or a photon and a Pomeron. LHC results on photon-photon and photon-Pomeron fusion are directly comparable with past and future electron-ion colliders. For example, in J/ψ CEP at LHC central rapidities, parton fractional momenta, x, of about 5×10^{-4} are probed, similar to that examined at HERA. However, at the forward rapidities accessible to the LHCb experiment, this x reach extends to 5×10^{-6}. Looking to the near future, measurements (for example of ρ production at 13 TeV) will probe x values an order of magnitude smaller, where phenomena such as saturation may become observable.

The LHCb detector [3] is suited to studying CEP as it is fully instrumented with tracking, calorimetry and particle identification in the pseudorapidity, η, range between 2 and 5. In addition, charged activity in the backward region, in the approximate range $-3.5 < \eta < -1.5$, can be vetoed due to the presence of a silicon strip detector that surrounds the interaction point. LHCb is designed to trigger on particles produced at low transverse momentum. CEP events are selected by triggering on muons with transverse momentum, p_T, above 400 MeV/c or electromagnetic or hadronic energy above 1000 MeV, in coincidence with a total event charged multiplicity of less than 10 as recorded by a scintillating pad detector. A further advantage of LHCb for CEP is the low number of proton-proton interactions (typically 1.5) per beam crossing. Consequently about 20% of the total luminosity has a single interaction and is suitable for searching for the CEP signature of low multiplicity and large rapidity gaps.

[a]e-mail: ronan.mcnulty@ucd.ie

The LHCb measurements of CEP to date have concentrated on final states with muons. In this report, I concentrate on measurements of the photoproduction of single charmonium and bottomium as well as the production of double charmonia, which is principally produced by double Pomeron exchange. The single charmonium measurements use almost 1 fb^{-1} of data taken at $\sqrt{s}$ = 7 TeV, while the other measurements add an additional 2 fb^{-1} at $\sqrt{s}$ = 8 TeV. Preliminary results, based on 37 pb^{-1} of data, are also available on χ_c production and dimuons produced by the QED process [4].

2 Photoproduction of J/ψ and $\psi(2S)$ mesons

Candidates for J/ψ mesons produced through CEP are selected [5] by requiring two identified muons inside the LHCb acceptance and no photons or additional tracks in either forward or backward directions. The p_T^2 of the dimuon is required to be below 0.8 GeV^2/c^2, and its invariant mass to be within 65 MeV/c^2 of the known J/ψ or $\psi(2S)$ masses. The invariant mass of all candidates (with the mass requirement removed) is shown in Fig.1(a). The non-resonant background, due to the QED production of dimuons via photon propagators, is modelled with an exponential function and is estimated to account for (0.8 ± 0.1)% of the J/ψ and (16 ± 3)% of the $\psi(2S)$ sample. Feed-down backgrounds inside the J/ψ mass window, amounting to (10.1 ± 0.9)%, are due to χ_c or $\psi(2S)$ mesons decaying to J/ψ and photons that are undetected due to being very soft or going outside the detector acceptance. Inelastic J/ψ or $\psi(2S)$ production, in which the proton dissociates but does not produce activity inside the LHCb acceptance, is assessed by fitting the $t \approx p_T^2$ distribution and assuming that $d\sigma/dt$ can be modelled by two exponentials for signal and background, as assumed in Regge theory and observed at HERA [6, 7]. The fitted parameters for the exponentials are consistent with those found at HERA, having corrected for kinematic differences. In total, (59.2 ± 1.2)% of the J/ψ sample and (52 ± 7)% of the $\psi(2S)$ sample is estimated to be exclusively produced.

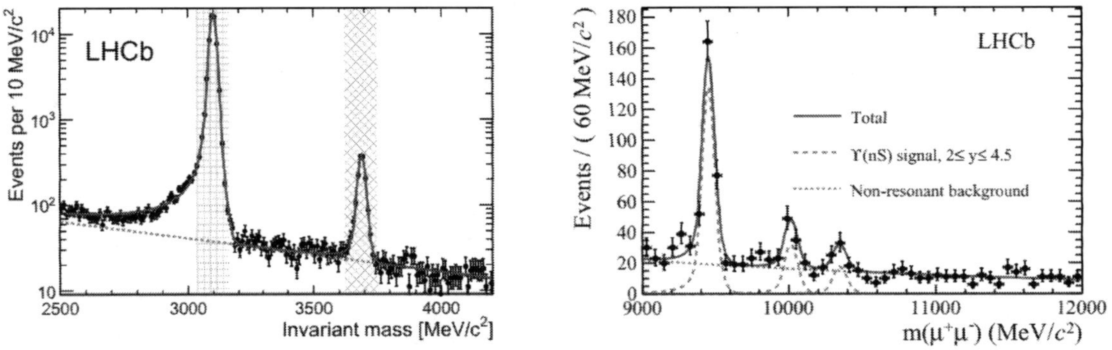

Figure 1. Invariant mass of dimuons in (a, left) the charmonium analysis [5] with the J/ψ and $\psi(2S)$ windows indicated and (b, right) the bottomonium analysis [15].

The cross-section for CEP charmonia is determined from the estimated number of CEP events dividing by the luminosity and correcting for the detector efficiency and acceptance, the former being determined with tag-and-probe techniques in data while the latter is found from simulated events. The total cross-sections are given in Table 1 while differential cross-sections as a function of rapidity are shown in Fig. 2(a,c) compared to LO and approximate NLO predictions from [8]. Results on the total cross-section have been compared to other predictions [9–13] and all agree with the data. A photoproduction cross-section can be derived from these results once rapidity gap corrections and photon flux factors are included. A two-fold ambiguity is present due to not knowing which of the protons the photon was radiated from. This can be resolved in a model-dependent way by assuming the H1 derived power-law [1] for one of the solutions. The (model-dependent) cross-section derived is shown in Fig. 2(b,d), compared to results from HERA, fixed target collisions, and proton-lead collisions at ALICE [14], in which the aforementioned ambiguity can be resolved.

Table 1. Total cross-section results for charmonia and bottomonia states.

Quantity measured	Kinematic region	Measurement (pb)
$\sigma(pp \to pJ/\psi p) \cdot BR(J/\psi \to \mu\mu)$	$2 < \eta_\mu, y_{J/\psi} < 4.5$	$291 \pm 7 \pm 19$ [5]
$\sigma(pp \to p\psi(2S)p) \cdot BR(\psi(2S) \to \mu\mu)$	$2 < \eta_\mu, y_{\psi(2S)} < 4.5$	$6.5 \pm 0.9 \pm 0.4$ [5]
$\sigma(pp \to p\Upsilon(1S)p)$	$2 < \eta_\mu, y_{\Upsilon(1S)} < 4.5$	$9.0 \pm 2.1 \pm 1.7$ [15]
$\sigma(pp \to p\Upsilon(2S)p)$	$2 < \eta_\mu, y_{\Upsilon(2S)} < 4.5$	$1.3 \pm 0.8 \pm 0.3$ [15]
$\sigma(pp \to p\Upsilon(3S)p)$	$2 < \eta_\mu, y_{\Upsilon(3S)} < 4.5$	< 3.4 at 95% c.l. [15]
$\sigma(J/\psi J/\psi)$	$2 < y_{J/\psi J/\psi} < 4.5$	$58 \pm 10 \pm 6$ [20]
$\sigma(J/\psi\psi(2S))$	$2 < y_{J/\psi\psi(2S)} < 4.5$	$63^{+27}_{-18} \pm 10$ [20]
$\sigma(\psi(2S)\psi(2S)$	$2 < y_{\psi(2S)\psi(2S)} < 4.5$	< 237 at 90% c.l. [20]
$\sigma(\chi_{c0}\chi_{c0})$	$2 < y_{\chi_{c0}\chi_{c0}} < 4.5$	< 69000 at 90% c.l. [20]
$\sigma(\chi_{c1}\chi_{c1})$	$2 < y_{\chi_{c1}\chi_{c1}} < 4.5$	< 45 at 90% c.l. [20]
$\sigma(\chi_{c2}\chi_{c2})$	$2 < y_{\chi_{c2}\chi_{c2}} < 4.5$	< 141 at 90% c.l. [20]

3 Photoproduction of Υ mesons

Candidates for Υ mesons produced through CEP are selected [15] by requiring two muons in LHCb and no other charged tracks. The muon pair must have $p_T^2 < 2$ GeV^2/c^2 and an invariant mass between 9 and 20 GeV/c^2, which allows the shape of the QED non-resonant components to be determined at the same time as the fit to the $\Upsilon(1S)$, $\Upsilon(2S)$ and $\Upsilon(3S)$. The distribution of candidates is shown in Fig.1(b).

Feed-down backgrounds coming from the various χ_b states are estimated to contribute 39 ± 7% of the total signal yield. As in the charmonium analysis, inelastic Υ production is assessed by fitting the $t \approx p_T^2$ distribution, but with the signal shape given by the SUPERCHIC generator[13]. After subtraction of the χ_b component, (54 ± 11)% is assessed to be exclusively produced; thus of the total Υ yield, one third is CEP.

After correcting for the detection efficiency, found using simulated events, and taking account of the luminosity, the cross-sections are determined and reported in Table 1. The low statistics and sizeable background impact on the significance of the 2S and 3S states. However, there are sufficient statistics to divide the 1S state into three bins of rapidity, which are plotted in Fig.2(e). Compared to the charmonium analysis, the difference between the LO and NLO predictions is large and the data show a clear preference for the latter. A photoproduction cross-section can be derived from these results. Here, the smaller of the two ambiguous solutions (that contributes between 5% and 20%) is ignored. The result is shown in Fig.2(f) showing good consistency with HERA results [16, 17].

4 Production of pairs of charmonia

The pair production of charmonia has been observed inclusively in a previous LHCb analysis [18], where tetraquark contributions or double parton scattering (DPS) may play an important role. The

[1]The power-law for J/ψ is taken from [6] while for $\psi(2S)$, I use $R(W) = \sigma(\psi(2S))/\sigma(J/\psi) = 0.166$ from [7].

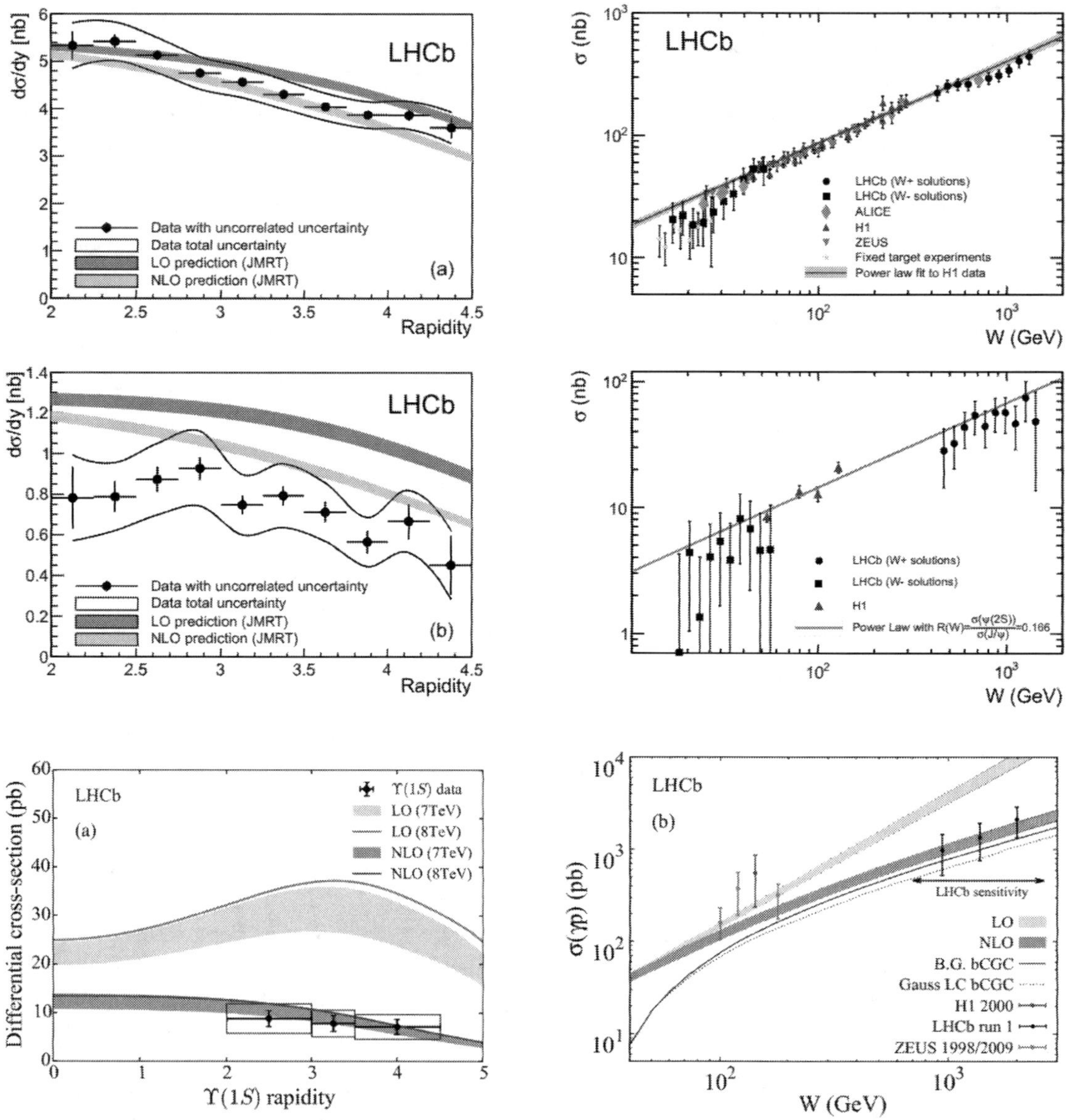

Figure 2. Differential cross-sections for exclusively produced (a, top left) J/ψ, (c, middle left) $\psi(2S)$, and (e, bottom left) Υ, compared to LO and NLO predictions [8]. Compilation of photoproduction cross-sections from various experiments for (b, top right) J/ψ, (d, middle right) $\psi(2S)$, (f, bottom right) Υ. The LHCb results are model-dependent as explained in the text.

contribution of DPS in CEP is minimal and the process proceeds dominantly through double Pomeron exchange. The cross-section for CEP of $J/\psi J/\psi$ inside the LHCb acceptance is predicted to lie in the range 2-20 pb, depending on the model used for the soft survival factor and the gluon PDF that enters with the fourth power [19].

The selection of the CEP of charmonia pairs [20] requires exactly four tracks, at least three of which are identified as muons. The invariant masses of two unlike-sign pairs of muons is required to be consistent with that of the J/ψ or $\psi(2S)$ mesons. The pairwise combinations are shown in Fig.3(a); 37 candidates are consistent with $J/\psi J/\psi$ production and 5 are consistent with $J/\psi\psi(2S)$. The invariant mass of the four tracks is shown in Fig.3(b) and agrees qualitatively with the spectrum observed inclusively [18]. The signal is only seen when there are precisely 4 tracks in the event; it is not present when additional tracks are present in the detector. After accounting for efficiency and luminosity, the cross-section for pairs of charmonia produced in the absence of other charged or neutral activity in LHCb is determined and given in Table 1.

Because of the low statistics, it is difficult to assess how often proton dissociation occurs. A fit to the t distribution suggests $(42 \pm 13)\%$ of the sample is CEP implying an exclusive cross-section for $J/\psi J/\psi$ inside the LHCb acceptance of (24 ± 9) pb, in broad agreement with the predictions.

The analysis is also sensitive to the production of pairs of χ_c mesons, which would be visible through their decays to a J/ψ meson and a photon. Searches are made for additional photons in events with two J/ψ candidates. A single event is consistent with $\chi_{c0}\chi_{c0}$ production but is also consistent with partially reconstructed $J/\psi\psi(2S)$ production. Limits are set on pairs of tensor meson charmonia and are given in Table 1.

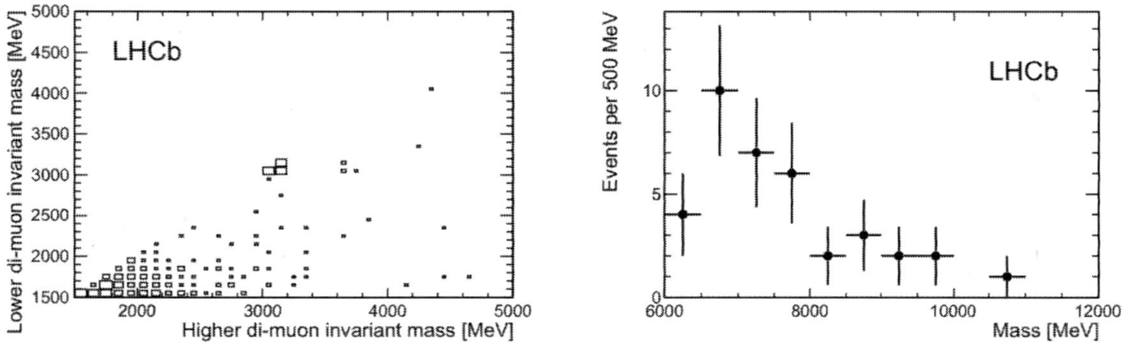

Figure 3. Invariant mass of (a, left) dimuon pairs (b, right) the $J/\psi J/\psi$ system [20].

5 Future propects

Experimentally, LHCb is sensitive to about 5.5 units in pseudorapidity allowing modest rapidity gap sizes to be identified. The identification of CEP events would be significantly aided if the veto region could be increased. To this end, for Run 2 forward shower counters, consisting of five planes of scintillators, have been installed in the LHC tunnel at distances of up to 115 m from the interaction point. These will provide a veto over at least an extra 6 units in pseudorapidity.

The photoproduction predictions have a strong dependence on the gluon PDF at low x, although large scale uncertainties are also present. Recent theoretical work [21] has shown an improvement in understanding these effects and thus CEP of vector mesons, particularly Υ, may provide an important method for constraining the gluon PDF in the poorly understood low-x region [22].

Reductions in the hadronic trigger thresholds for Run 2 will also allow LHCb to extend these studies to investigate the photoproduction of light vector mesons, ρ, ω and ϕ, accessing x values below 10^{-6}. It will also facilitate studies of the DPE production of pairs of pions and kaons, a region of great interest for light meson spectroscopy [23].

References

[1] M.G. Albrow, T.D. Coughlin, J.R. Forshaw, *Prog.Part.Nucl.Phys.* **65** 149 (2010).
[2] M.G. Albrow, *Int.J.Mod. Phys.* **A29** 1402006 (2014).
[3] LHCb collaboration, *JINST* **3** S08005 (2008).
[4] LHCb collaboration, CERN-LHCb-CONF-2011-022 (2011).
[5] LHCb collaboration, *J.Phys.* **G41** 055002 (2014).
[6] H1 collaboration, *Eur.Phys.J.* **C73** 2466 (2013).
[7] H1 collaboration, *Phys.Lett.* **B541** 251 (2002).
[8] S.P. Jones et al., *JHEP* **1311** 085 (2013); *J.Phys.* **G41** 055009 (2014).
[9] V.P. Goncalves, M.V.T. Machado, *Phys. Rev.* **C84** 011902 (2011).
[10] L. Motyka, G. Watt, *Phys. Rev.* **D78** 014023 (2008).
[11] W. Schafer, A. Szczurek, *Phys.Rev.* **D76** 094014 (2007).
[12] S.R. Klein, J. Nystrand, *Phys.Rev.Lett.* **92** 142003 (2004).
[13] L.A. Harland-Lang et al., *Eur.Phys.J.* **C65** 433 (2010).
[14] ALICE collaboration, *Phys.Rev.Lett.* **113** 232504 (2014).
[15] LHCb collaboration, *JHEP* **1509** 084 (2015).
[16] H1 collaboration, *Phys.Lett.* **B483** (2000) 23.
[17] ZEUS collaboration, *Phys.Lett.* **B680** 4 (2009).
[18] LHCb collaboration, *Phys.Lett.* **B707** 52 (2012).
[19] L.A. Harland-Lang et al., *J.Phys.* **G42** 055001 (2015).
[20] LHCb collaboration, *J.Phys.* **G41** 115002 (2014).
[21] S.P. Jones et al., arXiv:1507.06942 (2015).
[22] J. Rojo et al., *J.Phys.* **G42** 103103 (2015).
[23] CDF collaboration, *Phys.Rev.* **D91** 091101 (2015).

EPJ Web of Conferences **112**, 02010 (2016)
DOI: 10.1051/epjconf/201611202010

Leading neutron production at HERA in the color dipole approach

F. Carvalho[1,3,a], V.P. Gonçalves[2,b], D. Spiering[3,c], and F.S. Navarra[3,d]

[1] *Departamento de Ciências Exatas e da Terra, Universidade Federal de São Paulo, Campus Diadema, Rua Prof. Artur Riedel, 275, Jd. Eldorado, 09972-270, Diadema, SP, Brazil.*

[2] *High and Medium Energy Group, Instituto de Física e Matemática, Universidade Federal de Pelotas Caixa Postal 354, 96010-900, Pelotas, RS, Brazil.*

[3] *Instituto de Física, Universidade de São Paulo, C.P. 66318, 05315-970 São Paulo, SP, Brazil.*

Abstract.
In this work we study leading neutron production in $e + p \to e + n + X$ collisions at high energies and calculate the Feynman x_L distribution of these neutrons. The differential cross section is written in terms of the pion flux and of the photon-pion total cross section. We describe this process using the color dipole formalism and, assuming the validity of the additive quark model, we relate the dipole-pion with the well determined dipole-proton cross section. In this formalism we can estimate the impact of the QCD dynamics at high energies as well as the contribution of gluon saturation effects to leading neutron production. With the parameters constrained by other phenomenological information, we are able to reproduce the basic features of the recently released H1 leading neutron spectra.

1 Introduction

Leading neutron production in ep collisions has been studied for many years [1–8]. Monte Carlo studies, using standard deep inelastic scattering (DIS) generators have convincingly shown [9, 10] that in order to fit the data we need to include pion exchange and even the so-called sub-leading contributions, such as reggeon exchange, baryon resonance excitation (and decay) and independent fragmentation [4–8]. As it can be seen in Fig. 1, this process is composed by a pion (or reggeon) emission and by the subsequent photon-pion interaction at high energies. Since the incoming photon interacts with the pion emitted by the proton and then rescatters, interacting also with the emerging neutron we must include absorptive corrections [5–8]. Moreover the corresponding pion-nucleon

[a] e-mail: babi.usp@gmail.com
[b] e-mail: victorpbg@gmail.com
[c] e-mail: diego.spiering@gmail.com
[d] e-mail: navarra@if.usp.br

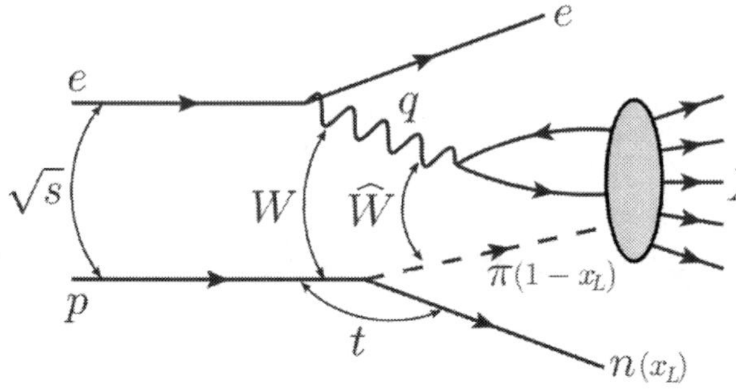

Figure 1. Leading neutron production in the color dipole approach.

vertex must be supplemented with a form factor, which represents the extended nature of hadrons and at the same time regularizes divergent integrals.

An interesting question is the existence or non-existence of the Feynman scaling, which says that the large x_L spectra of secondaries are energy independent. In this kinematical region non-linear effects (due to the high parton density) are expected to change the QCD dynamics. The most appropriate framework to treat QCD at high energies is the Color Glass Condensate (CGC) formalism [11], which predicts gluon saturation at small-x,

In [12] we have extended the color dipole formalism to the study of leading neutron production in $e+p \to e+n+X$ collisions at high energies. We have calculated the Feynman x_F distribution of leading neutrons. We have considered different models for the dipole scattering amplitude and different forms of the pion flux. We have shown that the recently released H1 leading neutron spectra could be well described using the color dipole formalism and that these spectra could help us to observe more clearly gluon saturation effects in future ep colliders. In that work we could also investigate the relation between Feynman scaling (or its violation) and the high energy QCD dynamics. In the near future, Feynman scaling will be investigated experimentally at the LHC by the LHCf Collaboration, as discussed in [12]. In a subsequent paper [13] with the same formalism we calculated the leading neutron spectrum in the exclusive reaction $ep \to \rho \pi n$. In [12] we observed that the inclusive data seemed to prefer one particular form of the pion flux. Later, in the analysis performed in [13], the exclusive measurements seemed to prefer another pion flux. In this note, we revisit the fits made in [12], combining them with the results of [13], and try to make a new fit of the inclusive spectra.

2 Formalism

Th leading neutron production differential cross section reads:

$$\frac{d^2\sigma(W,Q^2,x_L,t)}{dx_L dt} = f_{\pi/p}(x_L,t)\sigma_{\gamma^*\pi}(\hat{W}^2,Q^2) \tag{1}$$

where Q^2 is the virtuality of the exchanged photon and $\hat{W}$ is the center-of-mass energy of the virtual photon-pion system. It can be written as $\hat{W}^2 = (1-x_L)\,W^2$, where W is the center-of-mass energy of the virtual photon-proton system. As it can be seen in Fig. 1, x_L is the proton momentum fraction carried by the neutron and t is the square of the four-momentum of the exchanged pion. The flux of virtual pions emitted by the proton is represented by $f_{\pi/p}$ and $\sigma_{\gamma^*\pi}(\hat{W}^2,Q^2)$ is the cross section of the interaction between the virtual-photon and the virtual-pion at center-of-mass energy $\hat{W}$.

The pion flux $f_{\pi/p}(x_L,t)$ (also called sometimes pion splitting function) is the virtual pion momentum distribution in a physical nucleon (the bare nucleon plus the "pion cloud"). It has been calculated in different ways by several authors [14]. In all the calculations of the pion flux a form factor was introduced to represent the non-pointlike nature of hadrons and hadronic vertices, which contain a cut-off parameter determined by fitting data. The most frequently used parametrizations of the pion flux [1–8] have the following general form:

$$f_{\pi/p}(x_L,t) = \frac{1}{4\pi}\frac{2g^2_{p\pi p}}{4\pi}\frac{-t}{(t-m^2_\pi)^2}(1-x_L)^{1-2\alpha(t)}[F(x_L,t)]^2 \tag{2}$$

where $g^2_{p\pi p}/(4\pi) = 14.4$ is the $\pi^0 pp$ coupling constant, m_π is the pion mass and $\alpha(t)$ will be defined below. The form factor $F(x_L,t)$ accounts for the finite size of the nucleon and pion. We will consider the following parametrizations of the form factor:

$$F_1(x_L,t) = \exp\left[b(t-m^2_\pi)\right] \quad , \quad \alpha(t) = \alpha(t)_\pi \tag{3}$$

from Ref. [1], where $b = 0.3$ GeV^{-2} and

$$F_2(x_L,t) = 1 \quad , \quad \alpha(t) = \alpha(t)_\pi \tag{4}$$

from Ref. [2], where $\alpha_\pi(t) \simeq t$ (with t in GeV^2) is the Regge trajectory of the pion. In what follows, we shall use the two pion fluxes above denoting them by f_1 (calculated with (3)) and f_2 (calculated with (4)). We will include the absorptive corrections introducing a constant factor K which multiplies the uncorrected cross section of leading neutron production.

In order to obtain the photon-pion cross section we will use the color dipole formalism, as usually done in high energy deep inelastic scattering off a nucleon target. In this formalism, the cross section is factorized in terms of the photon wave functions Ψ, which describes the photon splitting in a $q\bar{q}$ pair, and the dipole-pion cross section $\sigma_{d\pi}$. It reads:

$$\sigma_{\gamma^*\pi}(\hat{x},Q^2) = \int_0^1 dz \int d^2\boldsymbol{r} \sum_{L,T} \left|\Psi_{T,L}(z,\boldsymbol{r},Q^2)\right|^2 \sigma_{d\pi}(\hat{x},\boldsymbol{r}) \tag{5}$$

where $\hat{x} = [Q^2 + m_f^2]/[(1-x_L)W^2 + Q^2]$ is the scaled Bjorken variable and the variable $\boldsymbol{r}$ defines the relative transverse separation of the pair (dipole). The photon wave function is easy to compute in QED and its expression can be found in textbooks [15]. As usual z stands for the longitudinal photon momentum fraction carried by the quark and $1-z$ is the longitudinal photon momentum fraction of the antiquark.

The main input in the calculations of $\sigma_{\gamma^*\pi}$ is the dipole-pion cross section. In what follows, for simplicity, we will assume the validity of the additive quark model, which allows us to relate $\sigma_{d\pi}$ with the dipole-proton cross section, usually probed in the typical inclusive and exclusive processes at HERA. Basically, we will assume that

$$\sigma_{d\pi}(x,\boldsymbol{r}) = R_q \cdot \sigma_{dp}(x,\boldsymbol{r}) \tag{6}$$

where $R_q = 2/3$ is the ratio between the number of valence quarks in the hadrons. This assumption is supported by the study of the pion structure function in the low x regime presented in [16]. It also gives a good description of the previous ZEUS leading neutron spectra, as shown in [5, 6]. On the other hand, the direct application of (1) to HERA photoproduction data [17] leads to the result $\sigma^{\text{tot}}_{\gamma\pi}/\sigma^{\text{tot}}_{\gamma p} = 0.32 \pm 0.03$, which is factor 2 lower than the ratio given above. The effective value of R_q

was more recently discussed in [18] on more theoretical grounds with the conclusion that this number could reach the value of $R_q = 0.5$. We shall use relation (6) to estimate $\sigma_{d\pi}$ letting R_q vary from 1/3 to 2/3.

In the eikonal approximation the dipole-proton cross section σ_{dp} is given by:

$$\sigma_{dp}(x, \boldsymbol{r}) = 2 \int d^2\boldsymbol{b}\, \mathcal{N}^p(x, \boldsymbol{r}, \boldsymbol{b}) \,, \tag{7}$$

where $\mathcal{N}^p(x, \boldsymbol{r}, \boldsymbol{b})$ is the imaginary part of the forward amplitude for the scattering between a dipole with transverse size given by the vector $\boldsymbol{r} = \boldsymbol{x} - \boldsymbol{y}$, where $\boldsymbol{x}$ and $\boldsymbol{y}$ are the transverse vectors for the quark and antiquark, respectively, and impact parameter $\boldsymbol{b} = (\boldsymbol{x} + \boldsymbol{y})/2$. Assuming the translational invariance approximation, which implies $\mathcal{N}^p(x, \boldsymbol{r}, \boldsymbol{b}) = \mathcal{N}^p(x, \boldsymbol{r}) S(\boldsymbol{b})$ and $\sigma_{dp}(x, \boldsymbol{r}) = \sigma_0 \cdot \mathcal{N}^p(x, \boldsymbol{r})$, with the normalization of the dipole cross section, σ_0, being fitted to data. In what follows we will use as input the phenomenological model proposed in Ref. [19]. In this parametrization the forward dipole-proton scattering amplitude is given by:

$$\mathcal{N}^p(x, \boldsymbol{r}) = \begin{cases} \mathcal{N}_0 \left(\frac{r Q_s}{2}\right)^{2\left(\gamma_s + \frac{\ln(2/r Q_s)}{\kappa \lambda Y}\right)}, & \text{for } rQ_s(x) \le 2\,, \\ 1 - e^{-a \ln^2 (b\, r Q_s)}\,, & \text{for } rQ_s(x) > 2\,, \end{cases} \tag{8}$$

where a and b are determined by continuity conditions at $rQ_s(x) = 2$, $\gamma_s = 0.7376$, $\kappa = 9.9$, $\mathcal{N}_0 = 0.7$ and Q_s is the saturation scale given by $Q_s^2(x) = Q_0^2 (x_0/x)^\lambda$ with $x_0 = 1.632 \times 10^{-5}$, $\lambda = 0.2197$, $Q_0^2 = 1.0\,\text{GeV}^2$. We will also use the GBW model [20], which captures the basic features of saturation physics and allows for simple analytical estimates. The GBW dipole-proton scattering amplitude is given by:

$$\mathcal{N}^p(x, \boldsymbol{r}) = 1 - \exp\left[-\frac{Q_s^2 r^2}{4}\right] \tag{9}$$

with $Q_0^2 = 1.0\,\text{GeV}^2$, $x_0 = 3 \times 10^{-4}$ and $\lambda = 0.288$. Finally, we will also use the dipole-proton cross section estimated with the help of the DGLAP analysis of the gluon distribution, which is given by [15]:

$$\sigma_{dip}(x, \boldsymbol{r}) = \frac{\pi^2}{3} r^2 \alpha_s x g(x, 10/r^2) \tag{10}$$

where $xg(x, Q^2)$ is the target gluon distribution, for which we use the CTEQ6 parametrization [21]. The above expression represents the linear regime of QCD and is a baseline for comparison with the non linear predictions.

3 Results and discussion

Using the two parametrizations of the pion flux, f_1 and f_2, introduced in the previous section to compute the leading neutron spectra, which are shown in In Fig. 2. The normalization of the spectrum was taken from an experimental analysis, as explained in [12]. The f_2 model has a long and comparatively flat tail in the low x_L region, which makes possible a fit of the data in this region even without including a contribution from other processes. The f_1 model falls too quickly at low x_L. However, the sub-leading contributions mentioned in [12] are more important in the low x_L region of the spectrum and their inclusion combined with f_1 might result in a good fit of the data as well.

In what follows we shall include the effect of absorptive corrections multiplying our spectra by a constant (independent of x_L) factor K which can assume values from 0.5 to 1. In Fig. 3 we show our results for the leading neutron spectra. The bands account for the uncertainty in the choice of

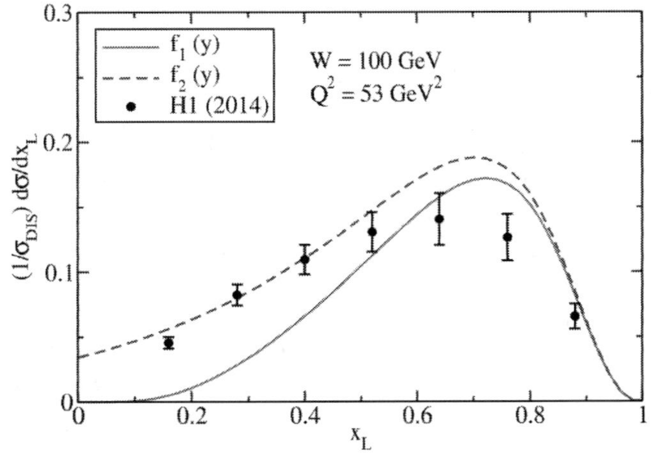

Figure 2. Leading neutron spectra compared with experimental data at $W = 100$ GeV considering the IIMS scattering amplitude and different choices of the pion flux f_1 and f_2.

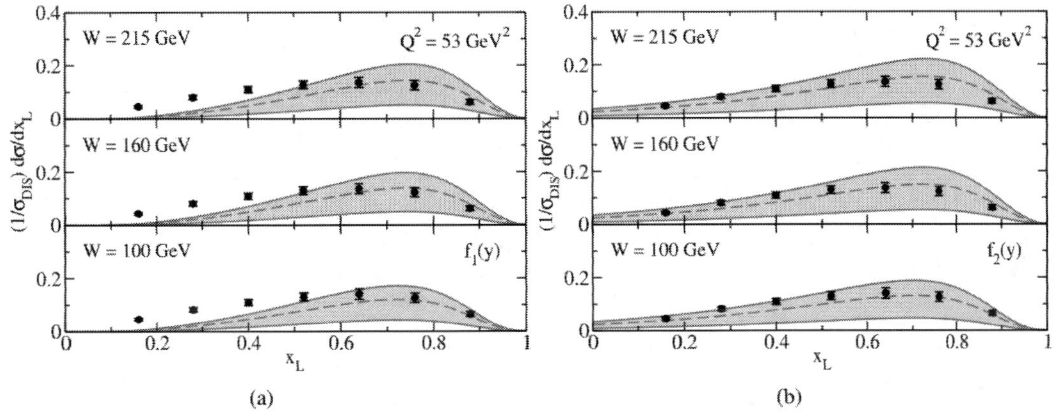

Figure 3. Leading neutron x_L spectrum for three photon-proton energies compared with recent H1 data. a) Calculated with f_1. b) Calculated with f_2. The bands are explained in the text.

the factors R_q and K. The upper limit of each band corresponds to the choice $R_q = 2/3$ and $K = 1$ whereas the lower limit corresponds to the choice $R_q = 1/3$ and $K = 0.5$. The central dashed lines represent the choice where $R_q.K = 0.5$, which could, for example, be realized with $R_q = 0.5$ and $K = 1$. Our results follow the trend dictated by data. For now, it is not possible to be very precise. With a combined analysis of other data on exclusive production with tagged leading neutrons we hope to reduce the bands.

We now address the Feynman scaling in leading neutron spectra and the contribution of non linear effects to this process. The theoretical expectation can be obtained using the GBW model for the scattering amplitude, Eq. (9). In the linear limit, when the dipole radius is very small (or equivalently Q^2 is very large) or the saturation scale is very small (and hence the energy is not very high), we can

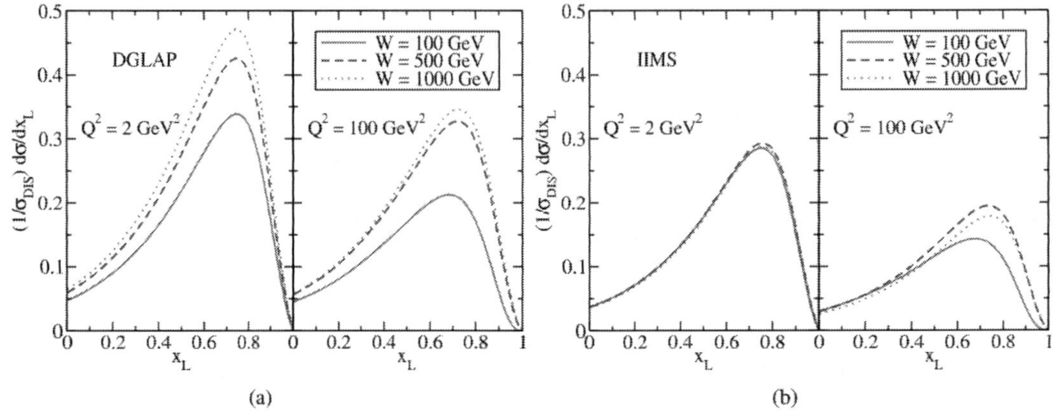

Figure 4. Leading neutron spectra for different energies considering (a) the linear DGLAP model and (b) the nonlinear IIMS model for the dipole-pion cross section.

expand the exponent and obtain

$$\sigma_{d\pi}(r,\hat{x}) \propto \sigma_0\, \mathcal{N}(r,\hat{x}) \simeq \sigma_0\, \frac{Q_s^2(\hat{x})r^2}{4} \simeq \sigma_0\, Q_0^2\, x_0{}^{\lambda} \left[\frac{(1-x_L)W^2+Q^2}{Q^2+m_f^2}\right]^{\lambda}. \tag{11}$$

Consequently, in this regime we see that the leading neutron x_L spectrum will depend on W. In a complementary way, in the nonlinear limit, when the dipole radius is very large (or equivalently Q^2 is very small) or the saturation scale is very large (and hence the energy is very high), we obtain

$$\sigma_{d\pi}(r,\hat{x}) \propto \sigma_0\, \mathcal{N}(r,\hat{x}) \simeq \sigma_0 \tag{12}$$

which is energy independent. The information contained in (11) and (12) indicates the route through which the asymptotic limit is reached and the role played by non linear effects. These expectations can be compared with those obtained using the IIMS and DGLAP models for the dipole-pion cross section. In Fig. 4(a) we show the spectra obtained in a purely linear approach. As expected we see a noticeable energy dependence. In contrast, the non linear predictions presented in Fig. 4(b) show a remarkable suppression of the energy dependence at low values of Q^2, consistent with the expectations. These results indicate that the Feynman scaling (and how it is violated) can be directly related to the QCD dynamics at small-x. As it was shown in [13], it is possible to constrain the unknown numbers and assumptions with the help of more experimental data on exclusive processes with tagged leading neutrons. Our most interesting conclusion is that leading neutron spectra can be used to probe the low x content of the pion target and hence it is a new observable where we can look for gluon saturation effects.

References

[1] B. Kopeliovich, B. Povh and I. Potashnikova, Z. Phys. C **73**, 125 (1996).
[2] M. Bishari, Phys. Lett. B **38**, 510 (1972).

[3] H. Holtmann et al., Phys. Lett. B **338**, 363 (1994); S. Kumano, Phys. Rev. D **43**, 59 (1991); N.N. Nikolaev, W. Schaefer, A. Szczurek and J. Speth, Phys. Rev. D **60**, 014004 (1999).

[4] L.L. Frankfurt, L. Mankiewicz and M.I. Strikman, Z. Phys. A **334**, 343 (1989); T.T. Chou and C.N. Yang, Phys. Rev. D **50**, 590 (1994); K. Golec-Biernat, J. Kwiecinski and A. Szczurek, Phys. Rev. D **56**, 3955 (1997); M. Przybycien, A. Szczurek and G. Ingelman, Z. Phys. C **74**, 509 (1997); W. Melnitchouk, J. Speth and A.W. Thomas, Phys. Rev. D **59**, 014033 (1998); A. Szczurek, N.N. Nikolaev and J. Speth, Phys. Lett. B **428**, 383 (1998).

[5] A.B. Kaidalov,V.A. Khoze, A.D. Martin and M.G. Ryskin, Eur. Phys. J. C **47**, 385 (2006).

[6] V.A. Khoze, A.D. Martin and M.G. Ryskin, Eur. Phys. J. C **48**, 797 (2006).

[7] N.N. Nikolaev, J. Speth and B.G. Zakharov, hep-ph/9708290.

[8] U. D′Alesio and H.J. Pirner, Eur. Phys. J. A **7**, 109 (2000).

[9] J. Olsson [H1 Collaboration], PoS DIS **2014**, 156 (2014); V. Andreev *et al.* [H1 Collaboration], Eur. Phys. J. C **74**, 2915 (2014); F. D. Aaron *et al.* [H1 Collaboration], Eur. Phys. J. C **68**, 381 (2010).

[10] See, for example, ZEUS Collaboration, M. Derrick et al., Phys. Lett. B **384**, 388 (1996).

[11] J. L. Albacete and C. Marquet, Prog. Part. Nucl. Phys. **76**, 1 (2014) and referecnes therein.

[12] F. Carvalho, V. P. Gonçalves, D. Spiering and F. S. Navarra, Phys. Lett. B **752**, 76 (2016).

[13] V. P. Gonçalves, F. S. Navarra and D. Spiering, arXiv:1512.06594 [hep-ph].

[14] J.D. Sullivan, Phys. Rev. D **5**, 1732 (1972); A.W. Thomas, Phys. Lett. B **126**, 97 (1983); J. Speth and A. W. Thomas, Adv. Nucl. Phys. **24**, 83 (1998); S. Kumano, Phys. Rep. **303**, 183 (1998); F. S. Navarra, M. Nielsen, C. A. A. Nunes and M. Teixeira, Phys. Rev. D **54**, 842 (1996); S. Paiva, M. Nielsen, F. S. Navarra, F. O. Durães and L. L. Barz, Mod. Phys. Lett. A **13**, 2715 (1998); W. Melnitchouk and A.W. Thomas, Phys. Lett. B **414**, 134 (1997); F. Carvalho, F. O. Duraes, F. S. Navarra, M. Nielsen and F. M. Steffens, Eur. Phys. J. C **18**, 127 (2000).

[15] V. Barone and E. Predazzi, *High-Energy Particle Diffraction*, Springer-Verlag, Berlin Heidelberg, (2002).

[16] N.N. Nikolaev, J. Speth and V.R. Zoller, Phys. Lett. B **473**, 157 (2000).

[17] S. Chekanov *et al.* [ZEUS Collaboration], Nucl. Phys. B **637**, 3 (2002).

[18] B. Z. Kopeliovich, I. K. Potashnikova, B. Povh and I. Schmidt, Phys. Rev. D **85**, 114025 (2012).

[19] G. Soyez, Phys. Lett. B **655**, 32 (2007).

[20] K. J. Golec-Biernat, M. Wusthoff, Phys. Rev. D **59**, 014017 (1998).

[21] H. L. Lai, J. Huston, Z. Li, P. Nadolsky, J. Pumplin, D. Stump and C.-P. Yuan, Phys. Rev. D **82**, 054021 (2010).

EPJ Web of Conferences **112**, 02011 (2016)
DOI: 10.1051/epjconf/201611202011

Interaction region design and auxiliary detector systems for an EIC

R. Petti[1,a]

[1] *Physics Department, Brookhaven National Laboratory, Upton, New York 11793-5000, USA*

Abstract. There are a number of exciting physics opportunities at a future electron-ion collider facility. One possible design for such a facility is eRHIC, where the current RHIC facility located at Brookhaven National Lab would be transformed into an electron-ion collider. It is imperative for a seamless integration of auxiliary detector systems into the interaction region design to have a machine that meets the needs for the planned physics analyses, as well as take into account the space constraints due to the tunnel geometry and the necessary beam line elements. In this talk, we describe the current ideas for integrating a luminosity detector, electron polarimeter, roman pots, and a low Q^2-tagger into the interaction region for eRHIC.

1 Introduction

An extremely exciting physics program is being developed for a future electron-ion collider facility, which will definitively answer current open questions in the field (such as the proton spin puzzle and providing concrete evidence for gluon saturation) [1]. This ongoing effort involves both machine development of the collider facility and infrastructure, as well as the development and design of the detector apparatus on which the physics measurements will be performed. In addition to the main detector that is required to track and identify charged and neutral particles, auxiliary detector systems are also necessary to monitor beam condition as well as provide other supporting functionality to the experiment.

In this proceedings, we focus on the possible design and integration into the machine lattice of a luminosity monitoring system, an electron beam polarimeter, as well as forward and backward detectors designed to measure protons and electrons that scatter at small angle (less than 1°), which will miss the main detector acceptance. It is critical for the implementation of these subsystems to be studied in coincidence with the machine lattice development, as various constraints due to space, beam optics and backgrounds have to be considered and optimized together to get a functioning system.

In the following sections, we start by describing the software tools currently being used to perform the simulation studies. Then we will discuss the aforementioned subsystems, both their function and form. Finally we close with a summary.

[a] e-mail: rpetti@bnl.gov

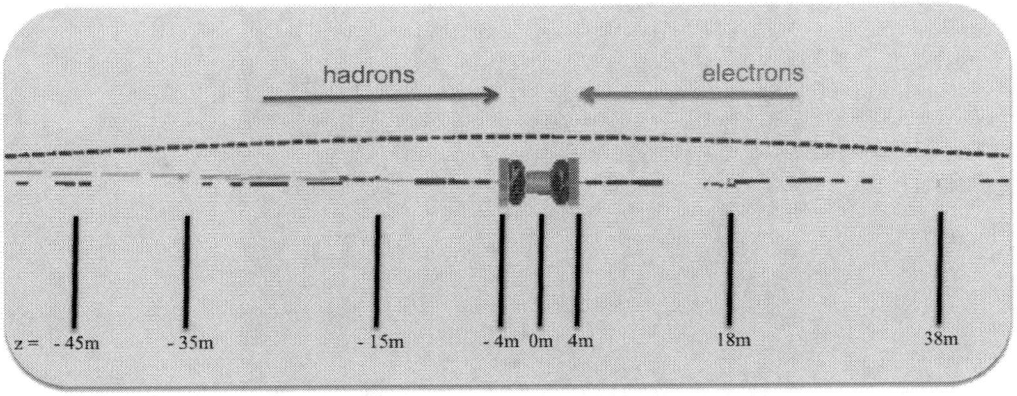

Figure 1. A picture of the interaction region from within the simulation framework.

2 Framework for detector simulations, EicRoot

A simulation framework has been developed at Brookhaven National Laboratory to aide in the design and optimization of detector components. The framework is called EicRoot [3]. EicRoot is an adaptation of the FairRoot software framework. The simulation package uses ROOT [4] for the detector and machine element geometry descriptions, GEANT3 or GEANT4 [5, 6] for the particle transport and interaction with materials, and also includes an event display module, code to digitize the registered GEANT signal and perform rudimentary reconstruction via cluster formation and tracking. The detector geometry is produced with scripts and the output is encoded in ROOT files so that components can easily be swapped in and out of the simulation setup, allowing for a quick turnaround in the simulation studies.

Additionally, the machine magnet geometry and fields can also be imported into the simulation, allowing for a full simulation of the IR with the detector and machine lattice integrated together. An example of the IR in the context of the simulation is shown in figure 1. The main detector is shown in the center. The red boxes along the beam lines represent the electron beam line magnets and the blue boxes along the beam lines represent the hadron beam line magnets. Locations of note are indicted by the distance markers. Various auxiliary detector systems have been integrated and will be discussed in the next sections.

3 The luminosity monitoring system

Essential to any measurement at an EIC is precise knowledge of the delivered luminosity of the beams, which can be monitored during running by measuring the elastic process $e + p \rightarrow e + p + \gamma$. This process can act as a useful diagnostic of the luminosity because it is a well-known calculable pure QED process [7]. Additionally, this process has a large cross-section and so statistics can be collected in a short time. Another desirable feature about this process is that the photons produced in the scattering are emitted in a very narrow cone in the direction of the electron beam, making measurement with a compact device possible.

The luminosity (L) can be measured if one measures the rate of photons hitting the detector (N_γ) while also knowing the true cross-section of the process (σ), and the acceptance correction factor for any photons that may not have been measured (A) via the relation $L = N_\gamma / A\sigma$.

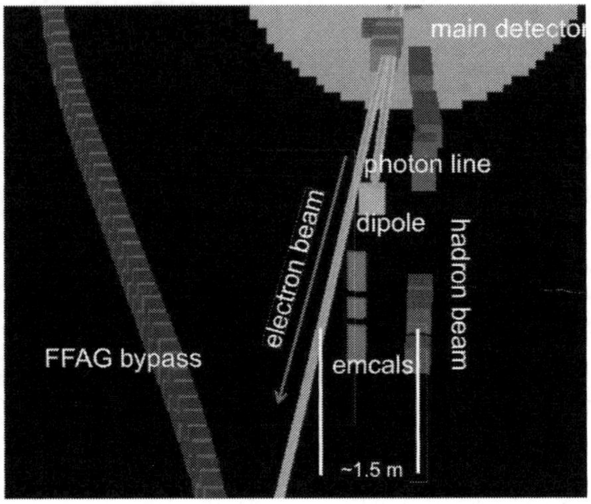

Figure 2. A display of the luminosity monitor in the simulation framework.

A dedicated luminosity monitoring system has been implemented and studied in simulation. It has been found that the current IR design of the machine leaves sufficient space for the setup and acceptance of the photons from the scattering process. A display of the setup in the simulation is shown in figure 2. Major components and landmarks are labeled in the figure.

The luminosity monitor is planned to be installed after the dipole bending magnet along the outgoing electron beam line where the beam is separated away from the bremsstrahlung photon from the process. The installation of two lumi systems is being considered. One system will feature an electromagnetic calorimeter in the plane of the beam (called the lzdc for the lumi-zero degree calorimeter from here on). This calorimeter will measure bremsstrahlung photons from the Bethe-Heitler process directly. The second system will feature a pair spectrometer, with a converter end-cap at the end of the photon transport line converting a fraction of the bremsstrahlung photons to e^+e^- pairs. These di-electrons will then pass through a dipole magnet (illustrated by the yellow box labeled "dipole" in the figure, located about 35m from the IP), in which the field is set to bend the electrons in the vertical direction and are measured by a set of calorimeters situated above and below the lzdc. The calorimeter systems are shown by the blue boxes labeled "emcals" in the figure and are currently placed about 45m from the IP. This setup has been utilized successfully in the ZEUS experiment [8].

This dual setup has the distinct advantage of allowing consistency checks within the luminosity measurement while also providing a well defined way to evaluate the systematic uncertainties. The contributions to the systematic uncertainties in each measurement will be substantially different. The main contributors to the systematic uncertainty utilizing either setup are the backgrounds in the detectors, the pileup of events from the high luminosity and the knowledge of the acceptance of the detector in the face of varying beam conditions. The backgrounds will be substantially different in the two setups, given the the lzdc resides in the plane of the primary synchrotron radiation fan of the bending electron beam, while the pair spectrometer calorimeters do not. Additionally, pile-up effects due to high rates will also be very different, since the converter in the pair spectrometer will reduce the overall rate in those calorimeters. The acceptance correction will also be very different in the two setups.

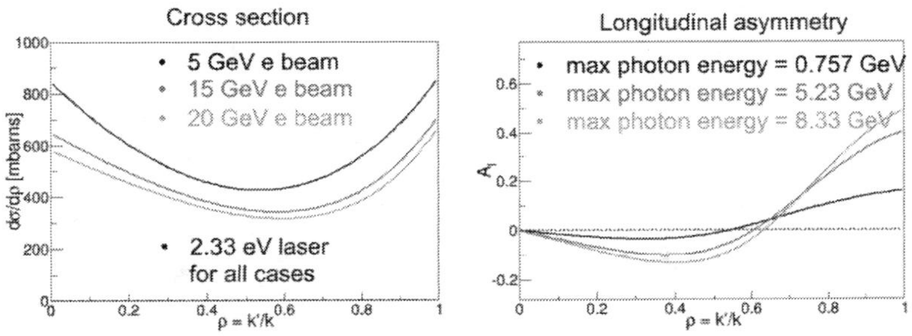

Figure 3. Left: The differential cross section for Compton scattering of electrons off of circularly polarized photons with different electron energies. Right: The longitudinal asymmetry, A_l for different electron energies. Both quantities are plotted as a function of ρ, or the scattered photon energy relative to the maximum allowed photon energy from kinematics (Compton edge).

4 Electron polarimetry

Precise knowledge of the polarization fraction of the electron beam is essential to the physics program at an EIC. The same level of precision for the luminosity measurement is needed for the polarization of the electron beam, better than 1%. This is especially important since the polarization of the beam will enter the calculation for the luminosity (as the Compton cross-section is dependent on polarization). The electrons are to be longitudinally polarized at the IP. It is necessary not only to determine the longitudinal polarization fraction of the beam, but also to determine the transverse polarization component to ensure that the spin is fully rotated to longitudinal. The Compton backscattering process will be used to measure the polarization of the electron beam.

The key to measuring the polarization of the beam is to measure an asymmetry between collisions with two different helicity states. The Compton events are produced by impinging a laser beam onto the electron beam at some location, and measuring the scattered electron or photon in the process. The polarization state of the laser will be flipped from +1 or -1, giving rise to two electron/photon helicity states in which the asymmetry can be measured. The Compton cross section, as well as the asymmetry expected, can be calculated exactly in QED. In the case of the longitudinal polarization component, the important parameters are related by the following expression: $A_{exp} = P_e P_\gamma A_L$. In the equation, A_{exp} is the experimentally measured asymmetry, P_e is the polarization of the electron beam, P_γ is the polarization of the laser beam, and A_L is the theoretical asymmetry calculated from QED.

The polarimeter will be placed on the incoming electron beam far from the IP. The system needs to be placed before the IP, but after the spin rotators. A natural location for the polarimeter is found where the orbit shift occurs to bring the electron beam into the IP. The magnets used for this purpose can also be used to separate scattered photons and electrons from the main electron beam. There is space both for detectors for the scattered photon, as well as for the scattered electron. We are currently investigating both options.

A possible photon detector would be a simple segmented calorimeter. Longitudinally polarized electron beams colliding with circularly polarized laser results in an energy asymmetry, while a transversely polarized electron beam results in an energy dependent position asymmetry due to the spin component of the electron breaking azimuthal symmetry. A well calibrated, segmented calorimeter would be able to make both measurements. Simulations on this topic are ongoing.

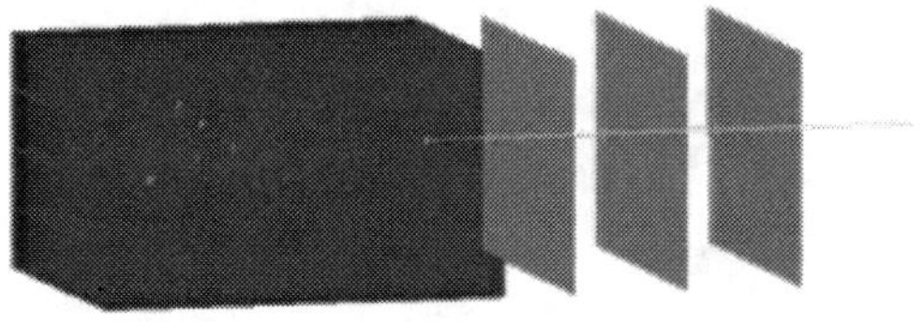

Figure 4. The current design of the low Q^2-tagger.

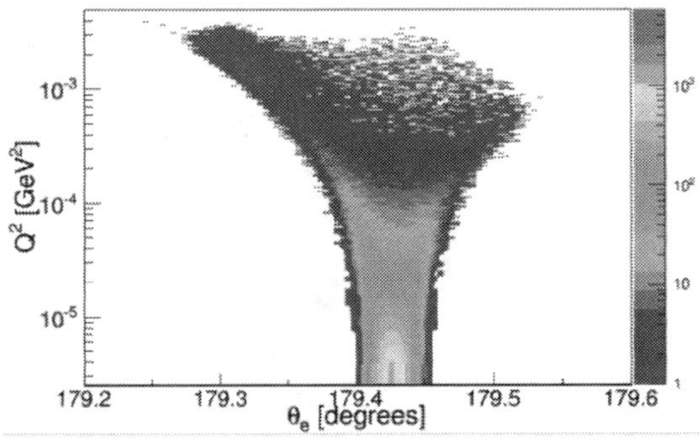

Figure 5. The range of acceptance in Q^2 and scattering angle, θ, of electrons from PYTHIA events.

5 Low Q^2-tagger

It is planned to install a dedicated low Q^2-tagger to measure electrons that scatter at small angle from low Q^2 events. These electrons will miss the main detector, and so installing an auxiliary device is essential for low Q^2 physics. The current concept of the low Q^2-tagger is shown in figure 4. It consists of three tracking layers followed by an electromagnetic calorimeter. Initial simulation studies show that this type of device can help to reconstruct the scattering angle of the electron and hence measure the Q^2 of the event. The detector is placed near the outgoing electron beam roughly 15m from the IP after the first set of bending dipole magnets where the scattered electrons will separate from the main beam. Figure 5 shows the Q^2 acceptance to electrons from PYTHIA [9] events. The coverage goes to a low Q^2 of about 10^{-5}.

6 Forward proton tagger

Another essential detector for the planned physics program at an EIC is a forward proton tagger. Like the electrons that scatter at small angle, there will be protons that scatter at small angle and will escape the main detector. Exclusive measurements need to capture these protons and so forward tracking systems are needed. In order to maximize acceptance of these protons, it is necessary to place the tracking sensors as close the the proton beam as possible. One possible implementation of the tagger currently being considered is roman pots. With this type of setup, the tracking sensors would be installed inside bellows with a movable arm so that the sensors can be retracted during fill time and moved into position during running. Multiple stations located at various distances from the IP would be needed in order to gain as much acceptance as possible (since protons that scatter and lose less energy will take a greater distance to be separated from the main beam). Figure 6 shows the acceptance of scattered protons from deeply virtual compton scattering (DVCS) events generated from the MILOU generator [10] for 20x250GeV e+p collisions. The top row shows the hit distribution in the x-y plane on the surface of the sensor. The top left is for low |t| events, while the top right is for low |t| events (where |t| is the momentum transfer at the proton vertex and goes as p_T^2). The bottom plot shows the acceptance (number hitting sensor/number generated) as a function of |t|. The acceptance is good at moderate |t|, but suffers at high |t|. It has been found that protons with high |t| are lost in the yoke of the first superconducting magnet. Thus we are also investigating placing some type of forward sensor very close to the main detector.

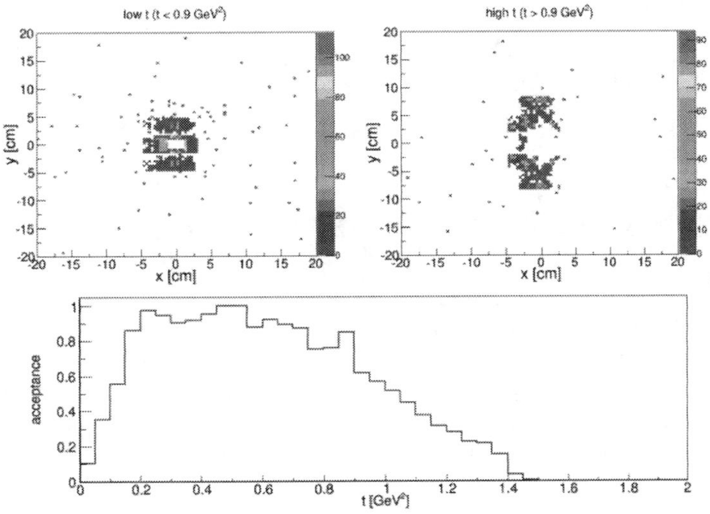

Figure 6. The acceptance of a roman pot station in |t| installed at 18m from the IP.

7 Summary

The status of the interaction region design and integration of auxiliary detector subsystems at eRHIC has been discussed in this proceedings. The detectors focus on far forward and backward detectors and consist of a luminosity monitoring system, electron polarimeter system, low Q^2-tagger for electron detection and a forward proton tagger. The basic location integrated into the IR has been determined and overall the current IR design is very compatible with the needs arising from the physics requirements on the program. Simulation studies in this regards are ongoing.

References

[1] E. Aschenauer et al., *Electron Ion Collider: The Next QCD Frontier - Understanding the glue that binds us all* (2014), `http://arxiv.org/pdf/1212.1701v3.pdf`

[2] A. Accardi et al., *eRHIC Design Study - An Electron-Ion Collider at BNL* (2014), `http://arxiv.org/ftp/arxiv/papers/1409/1409.1633.pdf`

[3] `https://wiki.bnl.gov/eic/index.php/Eicroot`

[4] `https://root.cern.ch/`

[5] R. Brun, F. Bruyant, M. Maire, A.C. McPherson, P. Zanarini (1987)

[6] `http://www.geant4.org/geant4/`

[7] H. Bethe, W. Heitler, Proc. Phys. Soc. Long. **146**, 83 (1934)

[8] L. Adamczyk et al. (ZEUS), Nucl. Instrum. Meth. **A744**, 80 (2014), `1306.1391`

[9] `https://wiki.bnl.gov/eic/index.php/PYTHIA`

[10] E. Perez, L. Schoeffel, L. Favart (2004), `hep-ph/0411389`

EPJ Web of Conferences **112**, 02012 (2016)
DOI: 10.1051/epjconf/201611202012

Di-photon versus di-hadron correlations in p+A collisions

Alex Kovner[1] and Amir H. Rezaeian[2,3,a]

[1]*Dept. of Physics, University of Connecticut, High, Storrs, CT 06269, USA*
[2]*Departamento de Física, Universidad Técnica Federico Santa María, Avda. España 1680, Casilla 110-V, Valparaiso, Chile*
[3]*Centro Científico Tecnológico de Valparaíso (CCTVal), Universidad Técnica Federico Santa María, Casilla 110-V, Valparaíso, Chile*

Abstract. We study the di-photon correlations in quark-nucleus and proton-nucleus collisions at RHIC and the LHC. We show that the single fragmentation di-photon produces the away side correlations peak, and the double fragmentation component of prompt di-photon is responsible for the near-side peak, and the long-range in rapidity near-side azimuthal collimation, the so-called "ridge" structure. We show that while di-photon ridge exists at the LHC in quark-nucleus collisions, the effect disappears in proton-nucleus collisions at the LHC. We show that the di-photon correlation measurments in p+A collisions can help to discriminate among models and understand the true origin of the observed di-hadron ridge in p+A collisions.

1 Introduction

It is believed that prompt photons, do not participate in the final-state interactions and can be considered as a golden channel to probe intrinsic correlations of partons in the hadronic and nuclear wavefunctions. Moreover, in contrast to gluons (and hadrons) production, prompt photons do not scatter on the target gluon field, but rather decohere from the projectile wavefunction due to scattering of the quarks, see Fig. 1. In this sense, the underlying mechanism of di-photon and di-hadron productions are quite different and di-photon can provide vital information to address whether it is the initial or final state effects that play dominant role in formation of the ridge collimation in p+p(A) collisions.

The observation of the so-called ridge phenomenon, namely the long-range in rapidity near-side ($\Delta\phi \approx 0$) di-hadron correlations in high-multiplicity events selection in both proton-proton (p+p) and proton (duetron)+nucleus collisions at the LHC and RHIC [1], triggered an on-going hot debate about the underlying dynamics of high-multiplicity events. It is an open question whether the ridge phenomenon in p+p(A) collisions mainly come from initial-state Color-Glass-Condensate (CGC) [2–6] or final-state hydrodynamic [7] approaches. Another interesting question is that if the ridge is a universal phenomenon, for all different two-particle productions such as two photons production in p+p(A) collisions.

[a]e-mail: Amir.Rezaeian@usm.cl

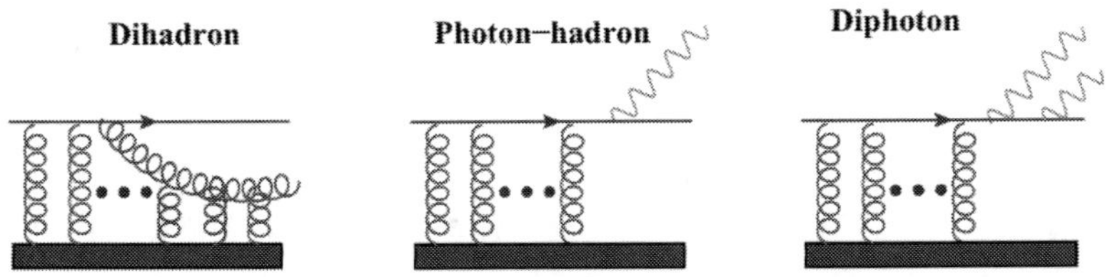

Figure 1. Typical CGC diagrams contributing to the inclusive gluon-quark (left), photon-quark (center) and prompt di-photon-quark (right) production at leading-order in proton-nucleus collisions.

2 Main formulation: the factorization

The inclusive prompt di-photon production $h + A \to \gamma_1 + \gamma_2 + X$ in high-energy dilute-dense scatterings was recently calculated in the CGC approach [8, 9] where in the leading order approximation, at forward rapidity, a valence quark of the projectile hadron emits two photons with momenta k_1 and k_2 via Bremsstrahlung and the produced di-photon+jet is then put on shell by interacting coherently over the whole longitudinal extent of the target, see Fig. 1. The inclusive di-photon differential cross-section at leading-log approximation can be written in the following factorization form [8, 9],

$$\frac{d\sigma^{pA\to\gamma(k_1)\gamma(k_2)X}}{d^2\mathbf{b}d^2\mathbf{k}_{1T}d\eta_1 d^2\mathbf{k}_{2T}d\eta_2} = \alpha_{em}^2 \int_{x_q^{min}}^{1} dx_q f(x_q,\mu_I^2) \int d^2\mathbf{l}_T \mathcal{H}\left(\mathbf{k}_{1T},\mathbf{k}_{2T},\mathbf{l}_T,\zeta_1,\zeta_2\right) N_F(l_T,x_g), \quad (1)$$

where the function $N_F(l_T, x_g)$ is the correlator of two light-like fundamental Wilson lines in the background of the color fields of the target, and it encodes the small-x dynamics of the target. The dipole amplitude $N_F(l_T, x_g)$ depends on the total transfer transverse momentum to target l_T and the fraction of light-cone momentum of target taken away by the exchange gluons x_g. At large N_c the dipole amplitude is obtained via the Balitsky-Kovchegov evolution equation [10]. The matrix function $\mathcal{H}$ depends on the transverse momenta of produced photons $\mathbf{k}_{1T}$, $\mathbf{k}_{2T}$ and the light-cone fractions of momentum of projectile taken away by two photons ζ_1, ζ_2. The matrix functional $\mathcal{H}$ is fairly lengthy and can be found in Ref. [9]. In Eq. (1), $f_q(x_q, \mu_I^2)$ denotes the parton distributions function (PDF) in the projectile hadron which depends on factorization scale μ and the parameter x_q defined as the ratio of the incoming quark to the projectile nucleon energy. The summation over quark and anti-quark flavor is implicit in Eq. (1). The matrix functional $\mathcal{H}$ includes the single and double fragmentation photon contributions, corresponding to the kinematics where only one emitted photon or both emitted photons are almost collinear with the outgoing quark. Note that both single and double fragmentation di-photon contributions, as well direct di-photon part are sensitive to the saturation dynamics via a convolution with the dipole amplitude $N_F(l_T, x_g)$, see Eq. (1). The lower limit of integral x_q^{min} in the factorization Eq. (1) and light-cone variables x_g, ζ_1, ζ_2 are obtained via the energy-momentum conservation relations [8, 9].

The azimuthal correlation of the produced di-photon is defined via

$$C_2(\Delta\phi, k_{1T}, k_{2T}, \eta_1, \eta_2) = \frac{d\sigma^{pA\to\gamma(k_1)\gamma(k_2)X}}{d^2\mathbf{k}_{1T}d\eta_{\gamma_1}d^2\mathbf{k}_{2T}d\eta_{\gamma_2}}[\Delta\phi] / \int_0^{2\pi} d\Delta\phi \frac{d\sigma^{pA\to\gamma(k_1)\gamma(k_2)X}}{d^2\mathbf{k}_{1T}d\eta_{\gamma_1}d^2\mathbf{k}_{2T}d\eta_{\gamma_2}} - C_{ZYAM}, \quad (2)$$

where $\Delta\phi$ denotes the azimuthal angle between the two produced photons in the plane transverse to the collision axis.

In our hybrid approach for di-photon production [8, 9], the projectile proton is assumed to be in dilute regime (as one may expect to be so at forward rapidities) and is consequently treated in standard

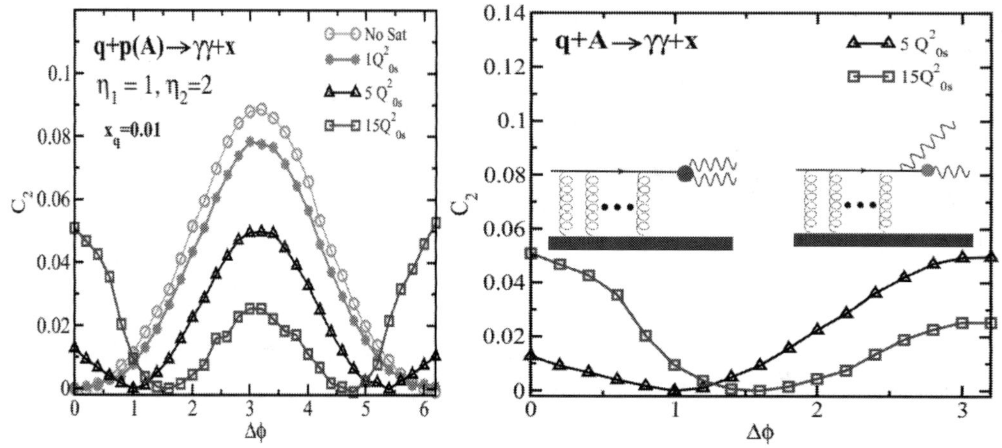

Figure 2. Prompt di-photon correlations C_2 as a function of angle between two photons $\Delta\phi$ in q+Pb collisions at the LHC ($\sqrt{s}$ = 5.02 TeV) with $\eta_1 = 1, \eta_2 = 2$ and $x_q = 0.01$ at different initial saturation scales (corresponding to different centralities). A schematic contribution of different typical CGC diagrams to the double inclusive prompt photon production at different angle is also shown. The plots are taken from [9].

parton model approach, while the target is treated as a CGC object at forward rapidities. Therefore, we have here only one saturation scale. This is one of main differences between our approach and the approach of Ref. [5] where the existence of two saturation scales in the problem (for the projectile proton and the target) in p+Pb collisions at forward rapidities, is essential in order to find a good fit to the di-hadron ridge data.

3 Results and Conclusion

In Fig. 2, we show the initial saturation scale dependence of C_2 at a fixed $x_q = 0.01$ in q+p(A) collisions at 5.02 TeV with rapidities of the produced photon $\eta_1 = 1$ and $\eta_2 = 2$. The effect of size and density of the target is simulated by taking different initial saturation scale for the running-coupling BK (rcBK) evolution equation. Namely we use the different solutions of the rcBK evolution equation with different initial saturation scales for the target $Q^2_{0sA} = NQ^2_{0s}$ and $N = 1, 5, 15$. We recall that an initial saturation scale Q_{0s} describes a proton target in minimum-bias electron-proton or proton-proton collisions [12], while an initial saturation scale of about $5 \div 7Q_{0s}$ describes a heavy nuclear target in minimum-bias proton-nucleus collisions [11]. We also compare with the non-saturation model (labeled by "No Sat" in the plot). It is seen that increasing the saturation scale in the system increases the near-side correlations leading to near-side collimation, while it decreases away-side correlation leading to the away-side decorrelation. Note that if momentum transfer from the target is small enough, the photon collinear to the outgoing quark, and the photon emerging from the initial photon-quark state have opposite transverse momenta, leading to back-to-back correlation at $\Delta\phi = \pi$ due to the single fragmentation di-photon contribution. Now increasing the exchanges to the target by increasing the saturation scale, unbalances the back-to-back correlations. The double fragmentation and direct di-photon contributions on the other hand, are restricted to kinematics where the transverse momentum of the outgoing quark jet is relatively large. By increasing the saturation scale, the main

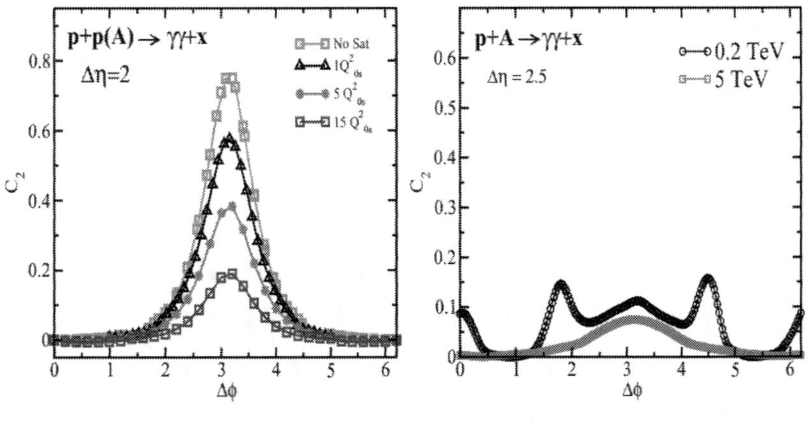

Figure 3. Left: The initial-saturation-scale dependence of prompt di-photon correlations C_2 in p+p(A) collisions at the LHC 5.02 TeV. Right: Comparison of C_2 at RHIC and the LHC for fixed transverse momenta of di-photon at $k_{1T} = k_{1T} = 1$ GeV with $\eta_1 = 1$ and $\eta_2 = 3.5$ in minimum-bias p+A collisions. The plots are taken from [9].

contributions of integrand is shifted to higher transverse momentum, leading to the enhancement of double-fragmentation contributions at the near-side $\Delta\phi = 0$.

In order to compute the correlations in proton-nucleus collisions, we should convolute the partonic cross-section with the PDFs of quarks and anti-quarks, and perform the integral over x_q, see Eq. (1). Note that the near-side and away-side correlations are enhanced by decreasing x_q light-cone variable. For a large x_q, all correlations both at near-side and away-side disappear, albeit this effect is more pronounced for the near-side correlations [9]. Therefore, a priori, it is not obvious how the integration over x_q will affect the di-photon correlations in p+A collisions. In Fig. 3, we show the initial-saturation-scale dependence of prompt di-photon correlations defined via C_2 in p+p(A) collisions at 5.02 TeV and $\eta_1 = 1$, $\eta_2 = 3$ with fixed transverse momenta of photons $k_{1T} = k_{1T} = 2$ GeV. It is seen at the LHC energy 5.02 TeV while the away-side correlation decreases by increasing the saturation scale, the near-side correlations defined via Eq. (2) almost disappears at all different centralities (corresponding to different saturation scales). In Fig. 3, we also compare the correlation C_2 at RHIC and the LHC at fixed transverse momenta of di-photon $k_{1T} = k_{1T} = 1$ GeV with rapidities $\eta_1 = 1$ and $\eta_2 = 3.5$ in minimum-bias proton-nucleus collisions. It is interesting that at RHIC in stark contrast to the LHC, a ridge-like structure appears at low transverse momentum of the produced photons $0.2 < k_T[\text{GeV}] < 3$. The near-side collimation at RHIC survives up to rapidity interval $\Delta\eta \approx 3$. It is also seen in Fig. 3 that the prompt di-photon correlations at forward rapidities in the away side region exhibit single or double peak structure depending on the kinematics. The double-peak structure is universal feature of Electromagnetic probes [9] and similar effect was also reported for photon-hadron [13] and dilepton-hadron [14] correlations in high-energy p+Pb collisions. This feature more clearly shows up at RHIC energy [9, 13, 14].

To conclude: we showed that in high-energy collisions, prompt di-photon production exhibits long-range correlations in quark-nucleus collisions at RHIC and the LHC albeit the physics of these correlations is significantly different than that of di-hadron correlations discussed previously [2–6] in the CGC framework. In particular, the correlations decrease for moderate rapidity separation between the two photons about $\Delta\eta \approx 1 \div 3$. We found that di-photon correlations defined via Eq. (2) exhibits ridge-like structure at low transverse momenta in p+A collisions at RHIC while this effect disappears in p+A collisions at the LHC 5.02 TeV. We showed that away-side di-photon de-correlations by increasing the saturation scale is somehow similar to the di-hadron de-correlations observed at the RHIC [15], see also Ref. [16]. However, di-photon correlations can exhibit double-peak structure at away-side which is an unique feature of Electromagnetic probes [9, 13, 14].

Acknowledgements The work of A.H.R. is supported in part by Fondecyt grant 1150135.

References

[1] V. Khachatryan *et al.* (CMS Collaboration), JHEP **09**, 091 (2010); S. Chatrchyan *et al.* (CMS Collaboration), Phys. Lett. **B718**, 795 (2013); B. Abelev *et al.* (ALICE Collaboration), Phys. Lett. **B719**, 29 (2013); G. Aad *et al.* (ATLAS Collaboration), Phys. Rev. Lett. **110**,182302 (2013); B. B. Abelev *et al.* (ALICE Collaboration), Phys. Lett. **B726**, 164 (2013); A. Adare *et al.* (PHENIX Collaboration), Phys. Rev. Lett. **111**, 212301 (2013), 1303.1794; L. Adamczyk *et al.* (STAR Collaboration) (2014), accepted by Phys. Lett.B, 1412.8437; 1502.07652.

[2] A. Dumitru, K. Dusling, F. Gelis, J. Jalilian-Marian, T. Lappi and Venugopalan, Phys. Lett. **B697**, 21 (2011).

[3] A. Kovner and M. Lublinsky, Phys. Rev. **D83**, 034017 (2011).

[4] E. Levin and A. H. Rezaeian, Phys. Rev. **D84**, 034031 (2011).

[5] K. Dusling and R. Venugopalan, Phys. Rev. **D87**, 094034 (2013); **D87**, 054014 (2013).

[6] A. Kovner and M. Lublinsky, Int. J. Mod. Phys. **E22**, 1330001 (2013) and references therein.

[7] P. Bozek, Eur. Phys. J. **C71**, 1530 (2011); Phys. Rev. **C88**, 014903 (2013); P. Bozek and W. Broniowski, Phys. Lett. **B718**, 1557 (2013); P. Bozek, W. Broniowski and G. Torrieri, Phys. Rev. Lett. **111**, 172303 (2013).

[8] A. Kovner and A. H. Rezaeian, Phys. Rev. **D90**, 014031 (2014).

[9] A. Kovner and A. H. Rezaeian, Phys. Rev. **D92**, 074045 (2015), arXiv:1508.02412.

[10] I. Balitsky, Nucl. Phys. **B463**, 99 (1996); Y. V. Kovchegov, Phys. Rev. **D60**, 034008 (1999); Phys. Rev. **D61**, 074018 (2000).

[11] A. H. Rezaeian, Phys. Lett. **B718**, 1058 (2013): J. Jalilian-Marian and A. H. Rezaeian, Phys. Rev. **D85**, 014017 (2012); E. Levin and A. H. Rezaeian, Phys. Rev. **D82**, 014022 (2010); Phys. Rev. **D82**, 054003 (2010); Phys. Rev. **D83**, 114001 (2011); A. H. Rezaeian, Phys. Rev. **D85**, 014028 (2012); A. H. Rezaeian, M. Siddikov, M. Van de Klundert and R. Venugopalan, Phys. Rev. **D87**, 034002 (2013); A. H. Rezaeian and I. Schmidt, Phys. Rev. **D88**, 074016 (2013); N. Armesto and A. H. Rezaeian, Phys. Rev. **D90**, 054003 (2014); A. H. Rezaeian, Phys. Lett. **B727**, 218 (2013).

[12] J. L. Albacete, N. Armesto, J. G. Milhano, P. Quiroga Arias and C. A. Salgado, Eur. Phys. J. **C71**, 1705 (2011).

[13] A. H. Rezaeian, Phys. Rev. **D86**, 094016 (2012); J. Jalilian-Marian and A. H. Rezaeian, Phys. Rev. **D86**, 034016 (2012).

[14] A. Stasto, B-W. Xiao and D. Zaslavsky, Phys. Rev. D86, 014009 (2012).

[15] A. Adare et al. [PHENIX Collaboration], Phys. Rev. Lett. **107**, 172301 (2011); E. Braidot, for the STAR Collaboration, Nucl. Phys. **A854**, 168 (2011); E. Braidot, Ph.D. thesis, arXiv:1102.0931.

[16] J. L. Albacete and C. Marquet, Phys. Rev. Lett. **105**, 162301 (2010).

EPJ Web of Conferences **112**, 02013 (2016)
DOI: 10.1051/epjconf/201611202013

Linear vs non-linear QCD evolution in the neutrino-nucleon cross section

Javier L. Albacete[1,a], José I. Illana[1,b] and Alba Soto-Ontoso[1,c]

[1] *CAFPE and Departamento de Física Teórica y del Cosmos, Universidad de Granada, E-18071 Campus de Fuentenueva, Granada, Spain.*

Abstract. Evidence for an extraterrestrial flux of ultra-high-energy neutrinos, in the order of PeV, has opened a new era in Neutrino Astronomy. An essential ingredient for the determination of neutrino fluxes from the number of observed events is the precise knowledge of the neutrino-nucleon cross section. In this work, based on [1], we present a quantitative study of $\sigma_{\nu N}$ in the neutrino energy range $10^4 < E_\nu < 10^{14}$ GeV within two transversal QCD approaches: NLO DGLAP evolution using different sets of PDFs and BK small-x evolution with running coupling and kinematical corrections. Further, we translate this theoretical uncertainty into upper bounds for the ultra-high-energy neutrino flux for different experiments.

1 Introduction

In 2013, the IceCube Collaboration [2] announced the detection of 28 neutrino events of astrophysical origin from TeV to PeV energies. To probe this region of UHE a new generation of experiments is already operating or being deployed. Thus, the Auger collaboration has recently reported on the limits of neutrino fluxes from 10^8 to 10^{10} GeV in [3]. Also, the LUNASKA project [4] will perform related measurements at even higher energies ($E_\nu \geq 10^{12}$ GeV) using the Square Kilometre Array [5]. Among the possible astrophysical scenarios, cosmogenic neutrinos could be produced in the scattering of protons off cosmic microwave background photons that would result in a non-negligible flux of neutrinos impinging on Earth up to energies of $E_\nu \sim 10^{12}$ GeV. Neutrinos with higher energies may arise in top-down models of UHE cosmic rays.

The inelastic interaction of neutrinos with nucleons is described in terms of charged current (CC) and neutral current (NC) interactions, which proceed through $W^\pm$ and Z^0 exchanges, respectively. The expression of the differential cross section in the fixed-target frame is:

$$\frac{\mathrm{d}^2\sigma_{\nu N}^{\mathrm{CC,NC}}}{\mathrm{d}x\mathrm{d}y} = \frac{G_F^2 s}{\pi}\left(\frac{M_i^2}{M_i^2+Q^2}\right)^2 \times\left[xy^2F_1^{\mathrm{CC,NC}}(x,Q^2)+(1-y)F_2^{\mathrm{CC,NC}}(x,Q^2)+xy\left(1-\frac{y}{2}\right)F_3^{\mathrm{CC,NC}}(x,Q^2)\right] \quad (1)$$

[a] e-mail: albacete@ugr.es
[b] e-mail: jillana@ugr.es
[c] e-mail: aontoso@ugr.es

where G_F is the Fermi coupling constant, E_ν is the neutrino energy, $s = 2M_N E_\nu$ with M_N the nucleon mass and M_i denotes the mass of the charged or neutral gauge boson exchanged. The information about the partonic structure of the nucleon is encoded in the structure functions $F_i(x, Q^2)$. The kinematics of this process is described in terms of the virtuality of the gauge boson Q^2, the fraction of the nucleon's momentum carried by the struck quark Bjorken-x and the inelasticity $y = Q^2/(xs)$. The integral of the differential cross section Eq. 1 is dominated by values of $Q^2 \sim M_i^2 = 10^4$ GeV^2. Hence the typical x value probed is $x \sim M_i^2/2M_N E_\nu$. For $E_\nu > 10^8$ GeV this translates into $x < 10^{-5}$, that is, we are going to explore the region of small-x and Q^2 values of $\sim M_{Z,W}^2$. This kinematic region hasn't been probed so far in ground based accelerators such as HERA or the LHC. Thus, the study of $\sigma_{\nu N}$ provides a test for the different QCD frameworks that describe the scale dependence of parton distribution functions. In particular, we have computed $\sigma_{\nu N}$ within two orthogonal QCD approaches: NLO DGLAP and running coupling BK evolution. Clearly, the reliability of either approach in this intermediate kinematic region cannot be determined on a priori theoretical arguments. This is so because, at a parametric level, one expects that both large logarithmic corrections $\sim \ln Q^2$ and $\ln 1/x$ resumed in either case to be relevant in that kinematic regime. In our view the use of one or another QCD evolution scheme should be considered as a systematic uncertainty associated to the theoretical estimate of the νN cross section.

2 Improved parton model and DGLAP evolution

In the QCD improved parton model the structure functions $F_i(x, Q^2)$ are linear combinations of parton distribution functions. We have chosen two sets of PDFs, `NNPDF3.0` [6] and `MSTW08` [7]. Both sets correspond to DGLAP evolution at NLO accuracy.

In our previous discussion about the kinematical region explored in the neutrino-nucleon interaction we have indicated that high E_ν is equivalent to small values of x. The most recent update of the experimental DIS data available [6] explores x down to $\sim 10^{-5} - 10^{-6}$. The PDFs are essentially unconstrained below this barrier and these uncertainties are fully propagated to the neutrino-nucleon cross section. This uncertainty, however, is not universal for all the PDFs collaborations even though they use an analogous procedure: an initial parametrization for the PDF that is fitted to data followed by DGLAP evolution. Not only the error bands but also the central value changes from one set to another, as it is displayed in Fig. 1 [1]. In the left plot of Fig. 1 it is shown the neutrino-nucleon cross section calculated with the `MSTW08` PDF set. It can be seen how the error bands increase while rising the neutrino energy due to the increasing contribution of small-x values reaching a 30% uncertainty for the highest energy. The right plot shows the ratio between the cross section obtained with the `MSTW08` and `NNPDF3.0` PDF sets, both for charged and neutral currents. The prediction with the `NNPDF3.0` set is higher for $E_\nu \gtrsim 10^8$ GeV becoming incompatible with the `MSTW08` set within the error band at the highest neutrino energy. This illustrates the decreasing predictive power of the approaches based on DGLAP when extended in the x-direction to values smaller than those included in their fitting data sets.

3 Dipole model and BK evolution

In the dipole model of DIS at low-x, the neutrino-nucleon cross section affords the following physics interpretation: long before reaching the target, the exchanged electroweak boson fluctuates into a colourless quark-antiquark dipole. Subsequently, the $q\bar{q}$ dipole scatters off the hadronic target via multiple gluon exchanges. The DIS structure functions for neutrino-nucleon scattering in the dipole

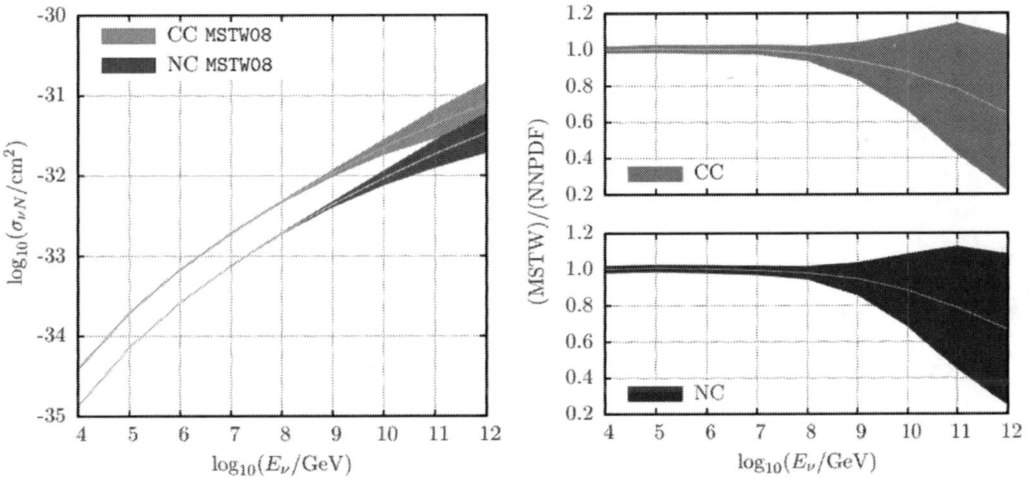

Figure 1. Left: the neutrino-nucleon cross section at NLO with MSTW08 PDFs. Right: comparison between the results for $\sigma_{\nu N}$ obtained with MSTW08 and NNPDF3.0.

model are given by

$$F_{\mathrm{T,L}}^{\mathrm{CC,NC}}(x,Q^2) = \frac{Q^2}{4\pi}\int d^2\mathbf{b}d^2\mathbf{r}\int_0^1 dz\left|\psi_{\mathrm{T,L}}^{W^\pm,Z^0}(z,Q^2,\mathbf{r})\right|^2 \mathcal{N}_{\mathrm{dip}}(x,\mathbf{r},\mathbf{b}). \tag{2}$$

All the information about the strong interactions in Eq. (2) is encoded in $\mathcal{N}_{\mathrm{dip}}(x,\mathbf{r},\mathbf{b})$, the dipole-hadron scattering matrix. Although of ultimate non-perturbative origin, the evolution of the dipole scattering amplitude towards smaller values of x can be studied perturbatively via the BK equation. It reads

$$\frac{\partial\mathcal{N}(r,x)}{\partial\ln(x_0/x)} = \int d^2\mathbf{r}_1\mathcal{K}(r,r_1,r_2)\left[\mathcal{N}(r_1,x)+\mathcal{N}(r_2,x)-\mathcal{N}(r,x)-\mathcal{N}(r_1,x)\mathcal{N}(r_2,x)\right] \tag{3}$$

where x_0 is the initial value of Bjorken-x and $\mathcal{K}$ is the evolution kernel, which plays an analogous role to the DGLAP splitting functions. The evolution kernel is currently known to NLO accuracy in $\alpha_s \ln 1/x$ [8] and also at running coupling accuracy [9–11]. However, NLO BK evolution turns unstable for a large class of initial conditions [12]. Therefore, we have solved the BK equation either at running coupling accuracy and also adding double logarithmic corrections recently calculated in [13, 14] to the evolution kernel. We have used the parametrizations of the dipole scattering amplitude obtained in [15, 16] and [17] after fitting all available data on the reduced cross sections measured in ep collisions at HERA with $x < 10^{-2}$. The kernel was evaluated at running coupling accuracy (RUN) and including large double transverse logarithmic corrections (DLA). For comparison, we shall also calculate the neutrino-nucleon cross section using the GBW [18], IIM [19] and Soyez [20] phenomenological models for the quark-antiquark dipole cross section (PHENO).

We present in Fig. 2 our results for the νN total cross section as a function of the neutrino energy calculated within the dipole model. We observe that the DLA parametrizations yield a systematically smaller cross section than the RUN ones. The differences between the two are of a factor $\lesssim 2$ for $E_\nu = 10^{14}$ GeV. This is an expected result, since the main role of DLA corrections is to further reduce the phase-space for small-x gluon emission, thus resulting in a slower growth of the dipole amplitude with decreasing values of x. Finally, the results obtained from the PHENO parametrizations yield an

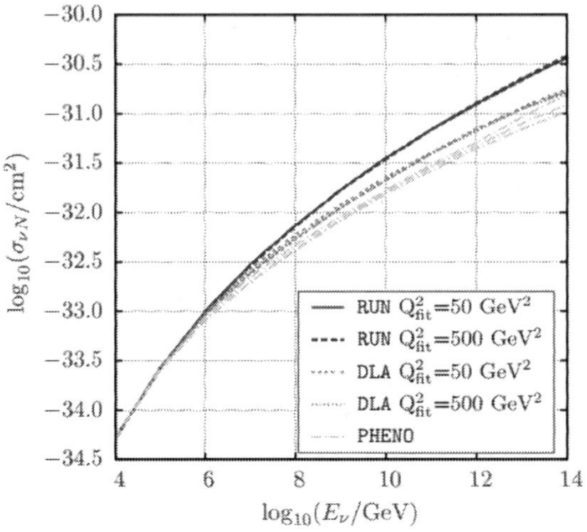

Figure 2. The neutrino-nucleon cross section calculated within the dipole model using different parametrizations for the dipole cross section, $\sigma_{q\bar{q}}$: RUN, DLA and PHENO.

even smaller value of the cross section at high energies. Once the evolution kernel for BK is fixed, the value of $\sigma_{\nu N}$ is almost insensitive to the initial conditions for the BK equations. This is seen by comparing the predictions for either DLA or RUN parametrizations fitted to different subsets of HERA data, up to $Q^2_{\rm fit} = 50$ and 500 GeV^2 respectively. We conclude that the main systematic uncertainty in the calculation of $\sigma_{\nu N}$ within the dipole model stems from the precise dynamical input embodied in the evolution kernel, and not from the initial conditions for the evolution.

4 DGLAP vs BK

The main result of our study is shown in Fig. 3. In this plot we compare the values for $\sigma_{\nu N}$ obtained under the DGLAP and BK QCD evolution schemes. The band for the DGLAP based calculation originates from the uncertainties of the NNPDF3.0 PDF set used here. In turn, the uncertainty band associated to the BK-based results is related to the choice of running coupling or running coupling plus DLA corrections evolution schemes for the high-energy extrapolation. We see that the DGLAP and BK approaches yield similar results for $\sigma_{\nu N}$ up to energies $E_\nu \sim 10^8$ GeV. For higher neutrino energies, the BK-based approaches lead to a systematically smaller value of the cross section than the DGLAP based calculation. That difference between DGLAP and BK approaches become as high as a factor ~ 4.5 at the highest neutrino energy $E_\nu = 10^{14}$ GeV.

5 Astrophysical implications

In absence of a unified theoretical framework for the description of QCD scattering in the full kinematic (x, Q^2)-plane, the differences in the theoretical prediction of the total neutrino-nucleon cross section induce an uncertainty in the determination of the upper bounds of astrophysical neutrino fluxes. The event rate at a neutrino telescope is given by

$$N_{\rm evt} = \sum_{i=1,2,3} \sum_{\nu_i,\bar{\nu}_i} \int dE_\nu \frac{d\phi_\nu}{dE_\nu} \omega_{\nu_i}(E_\nu) \int_0^1 dy \mathcal{E}(yE_\nu) \frac{d\sigma_{\nu_i N}}{dy}. \tag{4}$$

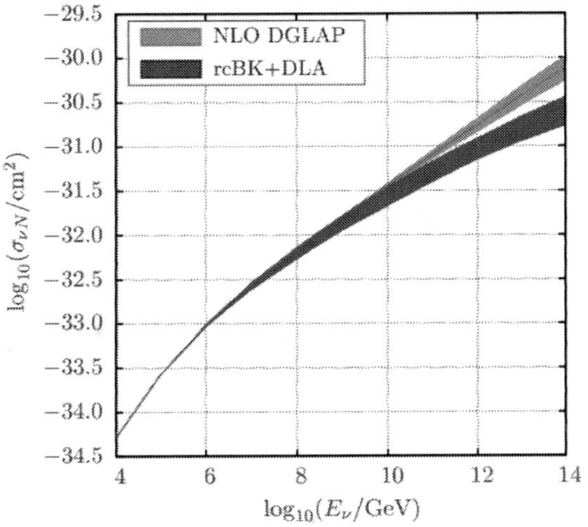

Figure 3. Neutrino-nucleon cross section calculated with NLO DGLAP with `NNPDF3.0` and with running coupling BK.

Table 1. Theoretical uncertainty in $\sigma_{\nu N}$ for different experiments covering the whole range of energies considered in this work.

Experiment	Energy range [GeV]	$\sigma_{\nu N}^{\mathrm{DGLAP}}/\sigma_{\nu N}^{\mathrm{BK}}$
Auger [3]	$10^8 - 10^{10}$	$\sim 1.3 - 1.7$
ANITA-2 [21]	$10^9 - 10^{14}$	$\sim 1.5 - 4.5$
LUNASKA [22]	$\geq 10^{12}$	$\gtrsim 2.5$

It is proportional to the flux (ϕ_ν), the exposure ($\mathcal{E}$) and the cross section ($\sigma_{\nu N}$). Importantly, the exposures can also depend on the neutrino-nucleon cross section. The current upper bounds on the UHE neutrino flux have been obtained adopting the DGLAp ν-nucleon cross sections. In table 1 we show how these bounds could be enlarged for different experiments due to the theoretical uncertainty coming from $\sigma_{\nu N}$. Neutrino radio-detection experiments based on the Askaryan effect, such as ANITA-2 and LUNASKA, explore the extremely high energy range. They are the best playground to test our predictions that start to differ significantly from those of DGLAP for $E_\nu \gtrsim 10^8$ GeV. We have shown that for the energies explored by the LUNASKA experiments, $E_\nu \sim 10^{14}$ GeV, the current upper bounds could be enlarged by a factor of up to ~ 4.5.

6 Conclusions

In summary, we have presented a quantitative study of the neutrino-nucleon total cross section in two different QCD approaches: NLO DGLAP and running coupling BK evolution. We have used state-of-the-art parametrizations of the parton distribution functions, in the DGLAP case, and of the dipole-nucleon scattering amplitude in the BK case. These parametrizations have been successfully tested against the available experimental data from HERA and the LHC. The cross sections obtained in the BK framework are systematically smaller than those calculated within the DGLAP framework, the difference between these two approaches bein as large as a factor 4.5 at the hightest neutrino energy studied in this work $E_\nu = 10^{14}$ GeV. These systematic differences in the theoretical calculation of the total neutrino-nucleon cross section affect directly the experimental determination of the neutrino fluxes of astrophysical origin. Our current understanding of the hadronic structure would be improved by an electron-ion collider thus reducing the theoretical uncertainty of $\sigma_{\nu N}$.

Acknowledgements

This work has been supported by the Spanish MINECO (FPA2013-47836 and Consolider-Ingenio MultiDark CSD2009-00064) and by Junta de Andalucía (FQM101, 3048, 6552). The research of J.L.A. is also funded by a Ramón y Cajal contract of MINECO and by the *QCDense* Career Integration Grant of the FP7 Framework Program for Research of the European Commission, reference number CIG/631558.

References

[1] J.L. Albacete, J.I. Illana, A. Soto-Ontoso, Phys. Rev. **D92**, 014027 (2015), 1505.06583
[2] M. Aartsen et al. (IceCube), Phys.Rev.Lett. **111**, 021103 (2013), 1304.5356
[3] A. Aab et al. (Pierre Auger), Phys.Rev.D (2015), 1504.05397
[4] C. James, R. Ekers, J. Alvarez-Muniz, J. Bray, R. McFadden et al., Phys.Rev. **D81**, 042003 (2010), 0911.3009
[5] C.L. Carilli, S. Rawlings, New Astron.Rev. **48**, 979 (2004), astro-ph/0409274
[6] R.D. Ball et al. (NNPDF), JHEP **1504**, 040 (2015), 1410.8849
[7] A. Martin, W. Stirling, R. Thorne, G. Watt, Eur.Phys.J. **C63**, 189 (2009), 0901.0002
[8] I. Balitsky, G.A. Chirilli, Phys. Rev. **D77**, 014019 (2008), 0710.4330
[9] I.I. Balitsky, Phys. Rev. D **75**, 014001 (2007), hep-ph/0609105
[10] Y. Kovchegov, H. Weigert, Nucl. Phys. **A 784**, 188 (2007), hep-ph/0609090
[11] J.L. Albacete, Y.V. Kovchegov, Phys. Rev. **D75**, 125021 (2007), arXiv:0704.0612 [hep-ph]
[12] T. Lappi, H. Mäntysaari, Phys.Rev. **D91**, 074016 (2015), 1502.02400
[13] G. Beuf, Phys.Rev. **D89**, 074039 (2014), 1401.0313
[14] E. Iancu, J. Madrigal, A. Mueller, G. Soyez, D. Triantafyllopoulos, Phys.Lett. **B744**, 293 (2015), 1502.05642
[15] J.L. Albacete, N. Armesto, J.G. Milhano, C.A. Salgado, Phys. Rev. **D80**, 034031 (2009), 0902.1112
[16] J.L. Albacete, N. Armesto, J.G. Milhano, P. Quiroga Arias, C.A. Salgado, Eur.Phys.J. **C71**, 1705 (2011), 1012.4408
[17] J.L. Albacete (2015), 1507.07120
[18] K. Golec-Biernat, M. Wüsthoff, Phys. Rev. **D59**, 014017 (1998), hep-ph/9807513
[19] E. Iancu, K. Itakura, S. Munier, Phys. Lett. **B590**, 199 (2004), hep-ph/0310338
[20] G. Soyez, Phys. Lett. **B655**, 32 (2007), 0705.3672
[21] P. Gorham et al. (ANITA), Phys.Rev. **D82**, 022004 (2010), 1003.2961
[22] J. Bray, R. Ekers, P. Roberts, J. Reynolds, C. James et al., Phys.Rev. **D91**, 063002 (2015), 1502.03313

EPJ Web of Conferences **112**, 02014 (2016)
DOI: 10.1051/epjconf/201611202014

Prospects for forward photon measurements at LHC

Marco van Leeuwen[1,2,a] *for the ALICE-FOCAL Collaboration*

[1] *National Institute for Subatomic Physics Nikhef, PO Box 41882, 1009 DB Amsterdam, Netherlands*
[2] *Institute for Subatomic Physics, Utrecht University, PO Box 80000, 3508 TA Utrecht, Netherlands*

Abstract. We present the opportunities to experimentally probe the gluon density at small x in nuclei to explore non-linear gluon evolution, saturation and the physics of the Color Glass Condensate by measuring photon production at forward rapidity in proton-nucleus collisions at the LHC. Performance studies for π^0 and direct photon measurements based on simulations of a Forward Calorimeter (FoCal), which is under consideration as an upgrade for the ALICE detector, are presented. Other aspects of the FoCal physics program for pp, p+Pb and Pb+Pb collisions are briefly discussed as well.

1 Introduction

The large gluon density of the proton at small fractional momentum x is expected to give rise to saturation which can be described by a non-linear evolution of the gluon Parton Distribution Functions (PDFs). This saturation is expected to set in at larger x (or larger Q^2 at the same x) in nuclei than in protons, leading to a suppression of the particle yield in p-A and d-A collisions [1–3]. In addition, multi-gluon effects are expected to become important, leading to a reduced recoil yield in di-hadron correlation measurements, since the recoil momentum is carried by multiple partons, in contrast to the more conventional $2 \to 2$ scatterings in leading order collinear QCD [4, 5].

Both the single-particle suppression and the suppression of recoil yields have been observed at forward rapidities at RHIC [6–9]. However, the observed suppression is at low p_T, which raises some doubts about the validity of the separation between long and short range physics and in particular about the expected scaling with the number of binary nucleon-nucleon collisions N_{coll}. Moreover, the forward measurements at RHIC are close to the kinematic limit, where other effects, such as energy loss and transverse momentum broadening may also affect the yields [10, 11].

Forward measurements at the LHC provide the opportunity to probe smaller x than at RHIC while also being further from the kinematic limit, thus probing deeper into the expected saturation regime. Photon measurements are of particular interest since they are directly sensitive to the gluon density at low x via the quark-gluon Compton process. Here we present the physics case and performance studies for a forward electromagnetic calorimeter (FoCal) which is being considered as a possible upgrade for the ALICE experiment at LHC.

[a] e-mail: marco.van.leeuwen@nikhef.nl

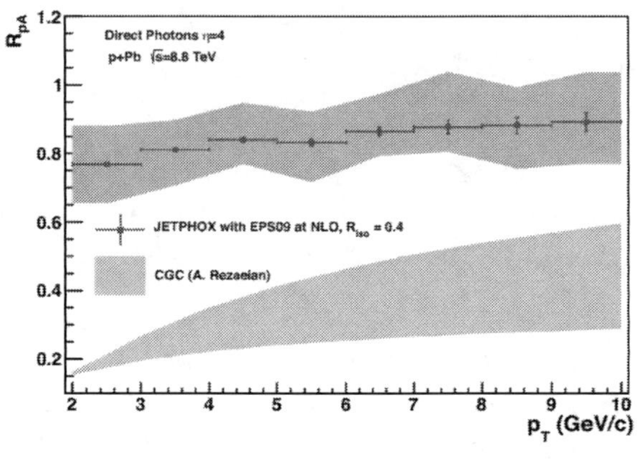

Figure 1. Expected nuclear modification factor R_{pPb} for photons at $y = 4$ at LHC. The blue area shows the result of linear evolution with the EPS09 nuclear PDFs [16] while the yellow area shows the result of a CGC calculation [17].

2 Forward photons at the LHC

Forward particle production at rapidity 3 – 5 is being explored by LHCb and ALICE at the LHC [12, 13]. Using the back-of-the-envelope estimate that $x_2 \approx \frac{p_T}{\sqrt{s}} \exp(-y)$, it is found that at $p_T = 4$ GeV$/c$ and $y = 4$, one reaches $x \approx 5 \cdot 10^{-6}$ at $\sqrt{s} = 14$ TeV. Photons are an ideal probe because they are directly sensitive to the gluon density via the (dominant) quark-gluon Compton process (qg $\rightarrow$ qγ), and the photon is directly produced in the hard process, so that the lowest x are probed at a given p_T. A more in-depth study of the x-ranges probed by direct photon and π^0 production at LHC, including Next-to-Leading Order (NLO) QCD effects, is given in [14].

Figure 1 shows the nuclear modification factor R_{pPb} for direct photons, as expected from a NLO collinear QCD calculation using JETPHOX [15], which includes nuclear effects via the EPS09 nuclear Parton Density Functions (nPDFs) [16] and a calculation using the rcBK framework for non-linear evolution [17]. The effect of gluon shadowing at small x in the nPDF-based calculation is a mild suppression $R_{\mathrm{pPb}} \approx 0.8$, while the rcBK calculation shows a strong suppression $R_{\mathrm{pPb}} < 0.5$ over the entire p_T range. It is expected that at even higher p_T, corresponding to larger scales Q, saturation effects are reduced and R_{pPb} in the rcBK framework approaches the nPDF results [18].

In addition to single inclusive measurements, two-particle correlation measurements can be used to probe the gluon density in the nucleus. A suppression of recoil yields associated with moderate to high-p_T π^0 and photons is expected due to multi-gluon recoil effects [5, 17, 19].

3 The FOCAL detector

The ALICE collaboration is considering to build a forward calorimeter FoCal, to measure the forward production of neutral pions and photons. This detector can be placed at about 7m from the interaction point, where it will cover about two units of pseudo-rapidity $3.3 < \eta < 5.3$. The detector would consist of a highly segmented electromagnetic calorimeter section, FoCal-E, which is capable of separating photons with a two-photon resolution of a few mm to identify decay photons, and a hadronic section FoCal-H, to improve photon-hadron separation and the measurement of isolation energy in a cone

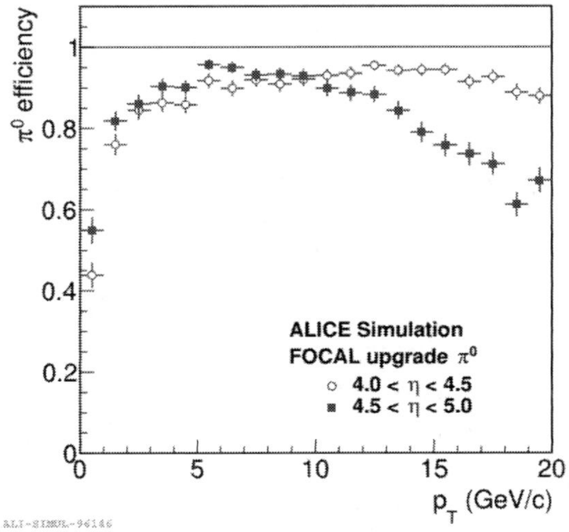

Figure 2. Efficiency for π^0 reconstruction in simulated events with a single π^0 including the simulated FoCal-E detector response and reconstruction algorithms in two different pseudo-rapidity ranges.

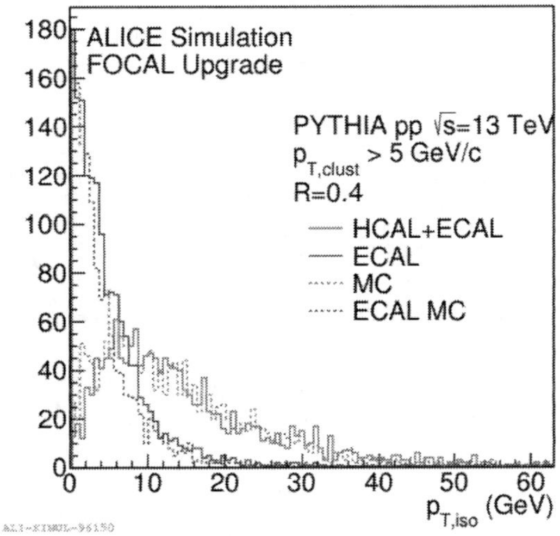

Figure 3. Distribution of the total transverse momentum $p_{\mathrm{T,iso}}$ in a cone with $R = 0.4$ in $\eta - \varphi$ around clusters with $p_{\mathrm{T}} > 5$ GeV/c from simulated p+p events. The blue curves show results using only the electromagnetic section of the FoCal, and the red curves show results with both FoCal-E and FoCal-H. The dashed curves show the particle-level results, while the solid lines show the results from full detector simulation and reconstruction.

around the photons, which is used to suppress contributions from decay photons and fragmentation photons.

The FoCal-E will consist of layers of Tungsten with interspersed layers of Silicon sensor material as sensitive elements. The reference design consists of 20 layers of 3.5 mm W, about 1 interaction length per layer, with 18 Low Granularity Layers (LGL) instrumented with pads of 1×1 cm^2 with analog readout, while 2 High Granularity Layers (HGL) with pixel sensors are used to improve the shower separation. The HGL are layer 5 and layer 10, which are around the shower maximum. In simulations, the granularity of these layers is 50 μm, with digital readout, which are further summed into 1×1 mm^2 macropixels.

A prototype detector with 20 layers of full pixel read-out using Monolithic Active Pixel Sensors (MAPS) has been constructed [20, 21] and tested using beams at DESY, and CERN (PS and SPS). The results of these beams tests are very encouraging [22]. Several options for the pad technology under consideration are also being tested [23].

4 FOCAL performance

To illustrate the performance of the FoCal calorimeter, we carried out detector simulations with the FoCal in a realistic environment of the full ALICE detector using the Geant3 detector simulation package [24]. Figure 2 shows the efficiency for π^0 reconstruction as a function of transverse momentum for two forward rapidity ranges: $4 < \eta < 4.5$ and $4.5 < \eta < 5$. The efficiency here is defined as the fraction of simulated π^0 that are fully reconstructed, by calculating the invariant mass of pairs of clusters and obtaining a value that is close to the expected mass (a window $0.07 < m_{\gamma\gamma} < 0.18\mathrm{GeV}/c^2$ was used). The results were obtained with simulated events that contain a single π^0 per event for technical convenience. It was verified that the results for PYTHIA pp events with a centre-of-mass energy $\sqrt{s} = 13$ TeV are very similar. The reconstruction efficiency is extremely good (> 0.9) over a broad range of momenta, from a few GeV/c up to $p_\mathrm{T} = 20$ GeV/c for the lower rapidity range and up to $p_\mathrm{T} = 10$ GeV/c for the higher rapidity range. The high efficiency is achieved by a reconstruction algorithm that uses the high-resolution information from the HGL to split energy from the LGL into contributions from two photons. This works well for photon separations of a few mm and larger. The high efficiency will make it possible to perform precise measurements of the π^0 p_T-spectra in both pp and p+Pb collisions.

The high π^0 reconstruction efficiency is also crucial for the measurement of direct photons, since this makes it possible to tag decay photons from π^0 decay, which are the main background for direct photons.

The second observable that is commonly used to identify direct photons is the isolation energy or momentum. Figure 3 shows the total transverse momentum in a cone of $R = 0.4$ in $\eta - \varphi$ around a photon candidate cluster with $p_\mathrm{T} > 5$ GeV/c from simulated PYTHIA pp events with $\sqrt{s} = 13$ TeV with full detector response simulation. The sample of photon candidate clusters consists mostly of decay photons. The figure shows results using only the electromagnetic calorimeter (blue lines) as well as the combined FoCal-E and FoCal-H result (red lines). It can be seen in the figure that adding the hadronic calorimeter response increases the in-cone energy for decay photons significantly, which improves the effectiveness of an isolation condition to reject decay photons. We have chosen to use a cut $p_\mathrm{T,iso} < 3$ GeV/c for further performance studies.

Figure 4 shows the signal fraction, i.e. the fraction of direct photon clusters in fully simulated pp events as a function of p_T. The solid blue circle markers show the signal fraction without background rejection, which is a few per cent at $p_\mathrm{T} = 5$ GeV/c and increases to about 5% at 15 GeV/c. This ratio is basically given by the production cross sections of direct photons, π^0 and the decay kinematics. Using the direct rejection of decay photons with the invariant mass method or the isolation selection individually, increases the direct photon fraction by a factor 5-8; except at the lowest $p_\mathrm{T} = 4$ GeV/c, where the isolation cut is not very effective. The combined effect of applying both the direct decay photon rejection and the isolation cut is an improvement of the signal fraction by a factor 10, which brings the signal fraction above 0.1 over the entire range, and close to 0.5 at higher p_T. This makes it possible to measure the direct photon cross section with a relative uncertainty of about 5%.

The projected uncertainty for a direct photon measurement at forward rapidity with FOCAL is shown in Fig. 5. The dominant uncertainty in the direct photon measurement at low momentum is due the background subtraction, for which the uncertainty is given by the uncertainty on the π^0 spectrum. The uncertainties shown in the figure are obtained with a projected 5% uncertainty on the background subtraction. At high $p_\mathrm{T} \gtrsim 10$ GeV/c, the total systematic uncertainty is dominated by the uncertainty in the efficiency and energy scale calibration for direct photons, which is (also) assumed to be 5%, while at lower p_T the background strongly increases, leading to larger uncertainties. The dependence of the final uncertainty in that region on the γ/π^0 ratio is illustrated by showing the uncertainty estimate based on the γ/π^0 from PYTHIA (light band, larger uncertainties) and JETPHOX (dark band, smaller uncertainties).

5 FoCal program in p+p and p+Pb: other observables

Besides the direct photon measurement for which the performance has been worked out in detail in the preceding section, there are a number of other measurements which are part of the FoCal

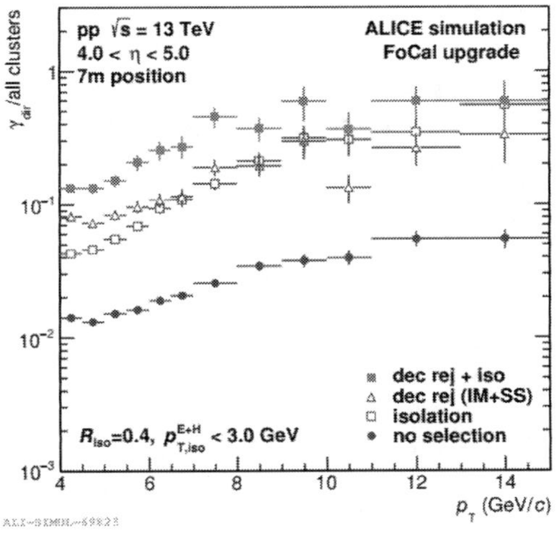

Figure 4. Projected γ_{dir} signal fraction in reconstructed clusters in simulated pp events with $\sqrt{s} = 13$ TeV. The solid blue points indicate the ratio without any selection cuts and the red solid points give the ratio after using both the direct rejection of decay photons based on the reconstructed invariant mass and a cut on the isolation energy $p_{\text{T,iso}} < 3\text{GeV}/c$. The open points indicated results with only the direct rejection or only the isolation cut.

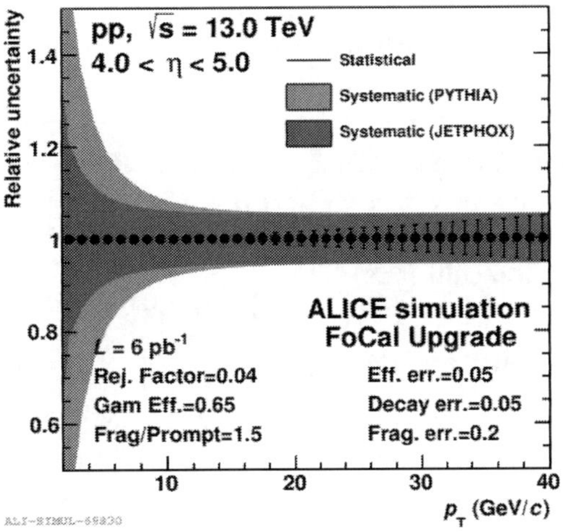

Figure 5. Projected relative uncertainty for a direct photon measurement with FoCal in pp collisions with $\sqrt{s} = 13$ TeV, at forward rapidity $4 < \eta < 5$. Statistical and systematic uncertainties are indicated separately. The lighter band at $p_{\text{T}} \lesssim 10$ GeV/c indicate the projected uncertainty using the $\gamma_{\text{dir}}/\pi^0$ ratio from PYTHIA instead of JETPHOX (dark band).

program in pp and p+Pb collisions. The excellent π^0 efficiency will allow to measure both inclusive spectra and di-hadron correlations of neutral pions in the forward direction. The inclusive spectra are also sensitive to the low-x gluon density in the target, but probe larger x than photons at the same momentum. Di-hadron measurements probe the di-jet structure. Earlier measurements at RHIC have shown a broadening and depletion of the recoil yield [8, 9], which is attributed to CGC effects, where the recoil consists of multiple softer gluons instead of a single hard gluon, as expected from collinear QCD calculations [4, 5]. Exploring this phenomenon at LHC is also a key part of the FoCal physics program.

Another observable that has been suggested to explore CGC dynamics are photon-hadron correlations [17]. These will also be explored with FoCal, but the precision may be limited due to the large background of decay photons.

Finally, we can also explore longer range correlations, i.e. the near and away-side *ridge* with correlations between charged particles at mid-rapidity and neutral pions (and photons) at forward rapidity. Such correlation measurements explore dual parton scattering in the CGC-evolution, which have a non-trivial angular distribution due to interference effects [25, 26].

6 Heavy ion program for FoCal

In heavy-ion collisions, the density of produced particles in the forward direction is much larger than in pp and p+Pb collisions. In this environment, FoCal will still be able to reconstruct neutral pions, to measure the nuclear modification factor R_{AA} at forward rapidity and study the longitudinal evolution of the medium density in such collisions as well as the effect of the larger quark jet fraction that is expected at forward rapidity [27]. These studies are complementary to measurements of J/ψ and Υ, which are found to be suppressed at forward rapidities. An independent measurement of the medium density at forward rapidity is crucial to disentangle the various mechanisms for quarkonium suppression that are being discussed in the literature, such as energy loss [28], co-movers [29] and melting due to screening effects [30, 31]. Di-hadron correlation measurements with two π^0s will also be explored, which will provide further insight into the time evolution of the medium density. In addition, we are also exploring the potential for direct photon measurements at forward rapidity in heavy ion collisions.

References

[1] D. Kharzeev, E. Levin, L. McLerran, Phys. Lett. **B561**, 93 (2003), hep-ph/0210332

[2] J.L. Albacete, N. Armesto, A. Kovner, C.A. Salgado, U.A. Wiedemann, Phys. Rev. Lett. **92**, 082001 (2004), hep-ph/0307179

[3] D. Kharzeev, Y.V. Kovchegov, K. Tuchin, Phys. Lett. **B599**, 23 (2004), hep-ph/0405045

[4] D. Kharzeev, E. Levin, L. McLerran, Nucl. Phys. **A748**, 627 (2005), hep-ph/0403271

[5] J.L. Albacete, C. Marquet, Phys. Rev. Lett. **105**, 162301 (2010), [arXiv]1005.4065

[6] I. Arsene et al. (BRAHMS), Phys. Rev. Lett. **93**, 242303 (2004), nucl-ex/0403005

[7] J. Adams et al. (STAR), Phys. Rev. Lett. **97**, 152302 (2006), nucl-ex/0602011

[8] E. Braidot (STAR), *Suppression of Forward Pion Correlations in d+Au Interactions at STAR*, in *Proceedings, 45th Rencontres de Moriond on QCD and High Energy Interactions* (2010), [arXiv]1005.2378

[9] A. Adare et al. (PHENIX), Phys. Rev. Lett. **107**, 172301 (2011), [arXiv]1105.5112

[10] I. Vitev, Phys. Lett. **B562**, 36 (2003), nucl-th/0302002

[11] Z.B. Kang, I. Vitev, H. Xing, Phys. Rev. **D85**, 054024 (2012), [arXiv]1112.6021

[12] R. Aaij et al. (LHCb), Eur. Phys. J. **C71**, 1645 (2011), [arXiv]1103.0423

[13] B.B. Abelev et al. (ALICE), JHEP **02**, 073 (2014), [arXiv]1308.6726

[14] I. Helenius, K.J. Eskola, H. Paukkunen, JHEP **09**, 138 (2014), [arXiv]1406.1689

[15] P. Aurenche, M. Fontannaz, J.P. Guillet, E. Pilon, M. Werlen, Phys. Rev. **D73**, 094007 (2006), hep-ph/0602133

[16] K.J. Eskola, H. Paukkunen, C.A. Salgado, JHEP **04**, 065 (2009), [arXiv]0902.4154

[17] A.H. Rezaeian, Phys. Rev. **D86**, 094016 (2012), [arXiv]1209.0478

[18] A.M. Staśto, B.W. Xiao, F. Yuan, D. Zaslavsky, Phys. Rev. **D90**, 014047 (2014), [arXiv]1405.6311

[19] A. van Hameren, P. Kotko, K. Kutak, C. Marquet, S. Sapeta, Phys. Rev. **D89**, 094014 (2014), [arXiv]1402.5065

[20] G. Nooren, PoS **RD11**, 026 (2011)

[21] D. Fehlker, J. Alme, A. Van Den Brink, A.P. De Haas, G.J. Nooren, M. Reicher, D. Röhrich, M. Rossewij, K. Ullaland, S. Yang, JINST **8**, P03015 (2013)

[22] G. Nooren, E. Rocco, PoS **RD13**, 017 (2014)

[23] S. Muhuri, S. Mukhopadhyay, V.B. Chandratre, M. Sukhwani, S. Jena, S.A. Khan, T.K. Nayak, J. Saini, R.N. Singaraju, Nucl. Instrum. Meth. **A764**, 24 (2014), [arXiv]1407.5724

[24] R. Brun, F. Bruyant, M. Maire, A.C. McPherson, P. Zanarini (1987)

[25] K. Dusling, R. Venugopalan, Phys. Rev. **D87**, 094034 (2013), [arXiv]1302.7018

[26] A. Kovner, A.H. Rezaeian, Phys. Rev. **D92**, 074045 (2015), [arXiv]1508.02412
[27] T. Renk (2014), [arXiv]1406.6784
[28] F. Arleo, S. Peigné, JHEP **10**, 73 (2014), [arXiv]1407.5054
[29] E.G. Ferreiro, Phys. Lett. **B731**, 57 (2014), [arXiv]1210.3209
[30] Y.p. Liu, Z. Qu, N. Xu, P.f. Zhuang, Phys. Lett. **B678**, 72 (2009), [arXiv]0901.2757
[31] X. Zhao, R. Rapp, Nucl. Phys. **A859**, 114 (2011), [arXiv]1102.2194

EPJ Web of Conferences **112**, 02015 (2016)
DOI: 10.1051/epjconf/201611202015

High-energy resummation effects in the production of Mueller-Navelet dijets at the LHC

B. Ducloué[1,2,a], L. Szymanowski[3,b], and S. Wallon[4,5,c]

[1] *Department of Physics, University of Jyväskylä, P.O. Box 35, 40014 University of Jyväskylä, Finland*
[2] *Helsinki Institute of Physics, P.O. Box 64, 00014 University of Helsinki, Finland*
[3] *National Centre for Nuclear Research (NCBJ), Warsaw, Poland*
[4] *Laboratoire de Physique Théorique, UMR 8627, CNRS, Univ. Paris Sud, Université Paris-Saclay, 91405 Orsay, France*
[5] *UPMC Univ. Paris 06, Faculté de Physique,75252 Paris, France*

Abstract. We study the production of two forward jets with a large interval of rapidity at hadron colliders, which was proposed by Mueller and Navelet as a possible test of the high energy dynamics of QCD, within a complete next-to-leading logarithm framework. We show that using the Brodsky-Lepage-Mackenzie procedure to fix the renormalization scale leads to a very good description of the recent CMS data at the LHC for the azimuthal correlations of the jets. We show that the inclusion of next-to-leading order corrections to the jet vertex significantly reduces the importance of energy-momentum non-conservation which is inherent to the BFKL approach, for an asymmetric jet configuration. Finally, we argue that the double parton scattering contribution is negligible in the kinematics of actual CMS measurements.

1 Introduction

The dijet production with large rapidity separation, as proposed by Mueller and Navelet [1], is one of the most promising observables in order to reveal the high energy dynamics of QCD, described by the Balitsky-Fadin-Kuraev-Lipatov (BFKL) approach [2]. We here report on our study of this process in a next-to-leading logarithmic (NLL) BFKL resummation, which includes the NLL corrections both to the Green's function [3, 4] and to the jet vertex [5, 6]. It has demonstrated that the NLL corrections to the jet vertex have a very large effect, leading to a lower cross section and a much larger azimuthal correlation [7]. These findings are however very dependent on the choice of the scales, especially the renormalization scale μ_R and the factorization scale μ_F, a fact which remains true when using realistic kinematical cuts for LHC experiments [8]. This dependency can be reduced after using the Brodsky, Lepage and Mackenzie (BLM) scheme [9], leading to a very satisfactory description [10] of the most recent LHC data extracted by the CMS collaboration for the azimuthal correlations of these jets [11]. Our results are not affected, neither by energy-momentum conservation issues nor by potential contribution of multiparton interaction (MPI).

[a] e-mail: bertrand.b.ducloue@jyu.fi
[b] e-mail: Lech.Szymanowski@ncbj.gov.pl
[c] e-mail: Samuel.Wallon@th.u-psud.fr

2 BFKL approach

The production of two jets of transverse momenta $\mathbf{k}_{J,1}$, $\mathbf{k}_{J,2}$ and rapidities $y_{J,1}$, $y_{J,2}$ is described by the differential cross-section

$$\frac{\mathrm{d}\sigma}{\mathrm{d}|\mathbf{k}_{J,1}|\,\mathrm{d}|\mathbf{k}_{J,2}|\,\mathrm{d}y_{J,1}\,\mathrm{d}y_{J,2}} = \sum_{\mathrm{a,b}} \int_0^1 \mathrm{d}x_1 \int_0^1 \mathrm{d}x_2\, f_{\mathrm{a}}(x_1) f_{\mathrm{b}}(x_2) \frac{\mathrm{d}\hat{\sigma}_{\mathrm{ab}}}{\mathrm{d}|\mathbf{k}_{J,1}|\,\mathrm{d}|\mathbf{k}_{J,2}|\,\mathrm{d}y_{J,1}\,\mathrm{d}y_{J,2}}, \tag{1}$$

where $f_{\mathrm{a,b}}$ are the usual collinear partonic distributions (PDF). In the BFKL framework, the partonic cross-section reads

$$\frac{\mathrm{d}\hat{\sigma}_{\mathrm{ab}}}{\mathrm{d}|\mathbf{k}_{J,1}|\,\mathrm{d}|\mathbf{k}_{J,2}|\,\mathrm{d}y_{J,1}\,\mathrm{d}y_{J,2}} = \int \mathrm{d}\phi_{J,1}\,\mathrm{d}\phi_{J,2} \int \mathrm{d}^2\mathbf{k}_1\,\mathrm{d}^2\mathbf{k}_2\, V_{\mathrm{a}}(-\mathbf{k}_1, x_1)\, G(\mathbf{k}_1, \mathbf{k}_2, \hat{s})\, V_{\mathrm{b}}(\mathbf{k}_2, x_2), \tag{2}$$

where $V_{\mathrm{a,b}}$ and G are respectively the jet vertices and the BFKL Green's function. Besides the cross section, the azimuthal correlation of the two jets is another relevant observable sensitive to resummation effects [12, 13]. Defining the relative azimuthal angle φ such that $\varphi = 0$ corresponds to the back-to-back configuration, the moments of the distribution $\langle\cos(n\varphi)\rangle$ are of special interest. Even at NLL accuracy, these observables depend strongly on the choice of the scales, and in particular the renormalization scale μ_R. An optimization procedure to fix the renormalization scale allows to reduce this dependency. We use the BLM procedure [9], here adapted to the context of BFKL [14], which is a way of absorbing the non conformal terms of the perturbative series in a redefinition of the coupling constant, to improve the convergence of the perturbative series.

3 Results: symmetric configuration and CMS data

We compare our results with the CMS data on the azimuthal correlations of Mueller-Navelet jets at the LHC at a center of mass energy $\sqrt{s} = 7$ TeV [11]. The two jets have transverse momenta larger than 35 GeV and rapidities $0 < y_1 < 4.7$ and $-4.7 < y_2 < 0$. The improvement due to the BLM

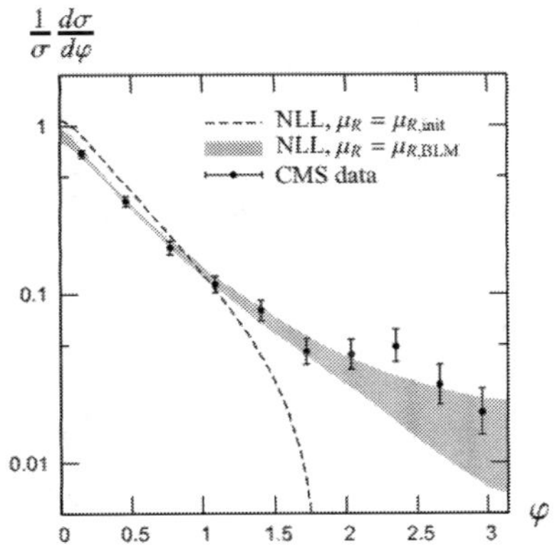

Figure 1. Symmetric configuration. Azimuthal distribution at NLL accuracy compared with CMS data.

procedure is most clearly seen through the azimuthal distribution of the jets

$$\frac{1}{\sigma}\frac{d\sigma}{d\varphi} = \frac{1}{2\pi}\left\{1 + 2\sum_{n=1}^{\infty} \cos(n\varphi)\,\langle\cos(n\varphi)\rangle\right\}, \tag{3}$$

as displayed in fig. 1. On the plot we show the CMS data (black dots with error bars), the NLL BFKL result using the "natural" scale choice $\mu_R = \sqrt{|\mathbf{k}_{J,1}| \cdot |\mathbf{k}_{J,2}|}$ (dashed black line) and the NLL BFKL results using the BLM scale setting (gray error band). We refer to Refs.[15–20] for similar studies.

It is well known that fixed order calculations are unstable when the lower cut on the transverse momenta of both jets is the same, which is the situation encountered by the above CMS measurement [11]. This is due to the fact that in the very peculiar situation where the two jets are almost back-to-back, resummation effects à la Sudakov should be considered, to stabilize the calculation. In the BFKL approach, although this back-to-back limit is stable, the azimuthal distribution can be significantly affected by such resummation effects. These have been obtained recently in the LL approximation [21]. However, this is not available neither at NLL BFKL nor in the fixed order approach. To evade this instability, it would thus be very useful to measure dijet production in a slightly asymmetric configuration, for which the ratio $\langle \cos 2\varphi \rangle / \langle \cos \varphi \rangle$ is expected [10] to sizeably differ between NLL BFKL and next-to-leading fixed order (NLO) [22] approaches.

4 Energy-momentum conservation

Energy-momentum conservation is not satisfied in the BFKL approach, being formally a sub-leading effect. It was proposed [23] to evaluate the importance of this effect by comparing the results of an

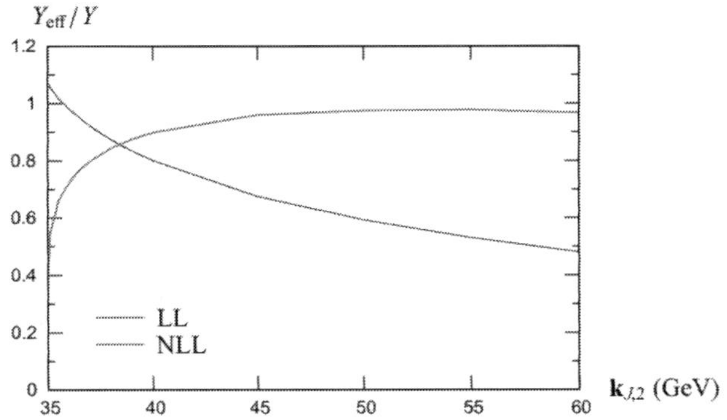

Figure 2. Variation of $Y_{\rm eff}/Y$ as defined in eq. (4) as a function of $\mathbf{k}_{J,2}$ at fixed $\mathbf{k}_{J,1} = 35$ GeV for $Y = 8$ and $\sqrt{s} = 7$ TeV at leading logarithmic (blue) and next-to-leading logarithmic (brown) accuracy.

exact $O(\alpha_s^3)$ calculation with the BFKL result, expanded in powers of α_s and truncated to order α_s^3. Having in mind that adding corrections beyond the LL approximation should reduce the violation of energy-momentum conservation, we here also include NLL corrections to the jet vertices [24]. Consider the effective rapidity $Y_{\rm eff}$ [23]

$$Y_{\rm eff} \equiv Y \frac{C_m^{2\to3}}{C_m^{\rm BFKL,\mathcal{O}(\alpha_s^3)}}, \tag{4}$$

where $C_m^{2\to3}$ is the exact $O(\alpha_s^3)$ result obtained by studying the reaction $gg \to ggg$, while $C_m^{\rm BFKL,\mathcal{O}(\alpha_s^3)}$ is the BFKL result expanded in powers of α_s and truncated to order $O(\alpha_s^3)$. The value of $Y_{\rm eff}$ indicates how valid the BFKL approximation is: a value close to Y means that this approximation is valid, whereas a value significantly different from Y means that it is a too strong assumption in the kinematics under study. On fig. 2 we show the values obtained for $Y_{\rm eff}/Y$ as a function of $\mathbf{k}_{J,2}$ for fixed $\mathbf{k}_{J,1} = 35$

GeV at a center of mass energy $\sqrt{s} = 7$ TeV and for a rapidity separation $Y = 8$, in the LL and NLL approximation. As found in ref. [23], the LL calculation strongly overestimates the cross section for transverse momenta of the jets not too close, while at NLL accuracy, the situation is much improved for significantly different jet transverse momenta, where instabilities are not expected, see above.

5 Double parton scattering contribution to MN jets

At high energies and low transverse momenta where BFKL effects are expected to be enhanced, parton densities can become large enough that contributions where several partons from the same incoming hadron take part in the interaction could become important. We restrict ourselves to the case of double parton scattering where there are at most two subscatterings and where both these scatterings are hard, as illustrated in fig. 3. For simplicity, the order of magnitude of this contribution is evaluated at LL,

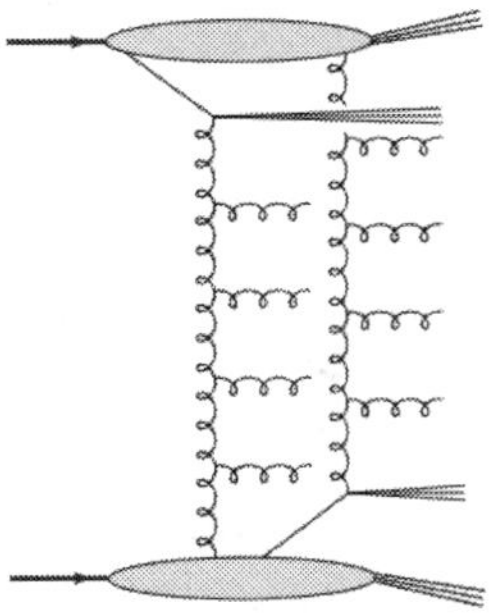

Figure 3. The DPS contribution.

which we compare with our prediction involving single parton scattering in the BFKL LL and NLL approaches. We use a simple factorized ansatz to compute the DPS contribution according to

$$\sigma_{\rm DPS} = \frac{\sigma_{\rm fwd}\sigma_{\rm bwd}}{\sigma_{\rm eff}}\,, \tag{5}$$

where $\sigma_{\rm fwd(bwd)}$ is the inclusive cross section for one jet in the forward (backward) direction and $\sigma_{\rm eff}$ is a phenomenological quantity related to the density of the proton in the transverse plane. We vary $\sigma_{\rm eff}$ between 10 and 20 mb, to be consistent with the measurements at the Tevatron [25–27] and at the LHC [28, 29]. The inclusive cross section for one jet in the forward or backward direction is built as the convolution of the leading order jet vertex with unintegrated gluon distributions (UGD), the global normalization being fitted with CMS [30] data (see ref. [31] for more details). We focus on four choices of kinematical cuts:

(a) $\sqrt{s} = 7$ TeV, $|\mathbf{k}_{J,1}| = |\mathbf{k}_{J,2}| = 35$ GeV,
(b) $\sqrt{s} = 14$ TeV, $|\mathbf{k}_{J,1}| = |\mathbf{k}_{J,2}| = 35$ GeV,
(c) $\sqrt{s} = 14$ TeV, $|\mathbf{k}_{J,1}| = |\mathbf{k}_{J,2}| = 20$ GeV,
(d) $\sqrt{s} = 14$ TeV, $|\mathbf{k}_{J,1}| = |\mathbf{k}_{J,2}| = 10$ GeV.

The first choice is similar to the cuts used by the CMS analysis of azimuthal correlations of Mueller-Navelet jets at the LHC [11]. The other three choices correspond to the higher center of mass energy that the LHC is expected to reach soon. The last two choices correspond to lower transverse momenta at which measurements could become possible in the future, and are particularly relevant since MPI are expected to become more and more important at lower transverse momenta. The rapidities of the jets are restricted according to $0 < y_{J,1} < 4.7$ and $-4.7 < y_{J,2} < 0$. We use the MSTW 2008 parametrization [32] for collinear parton densities. To estimate the uncertainty associated with the

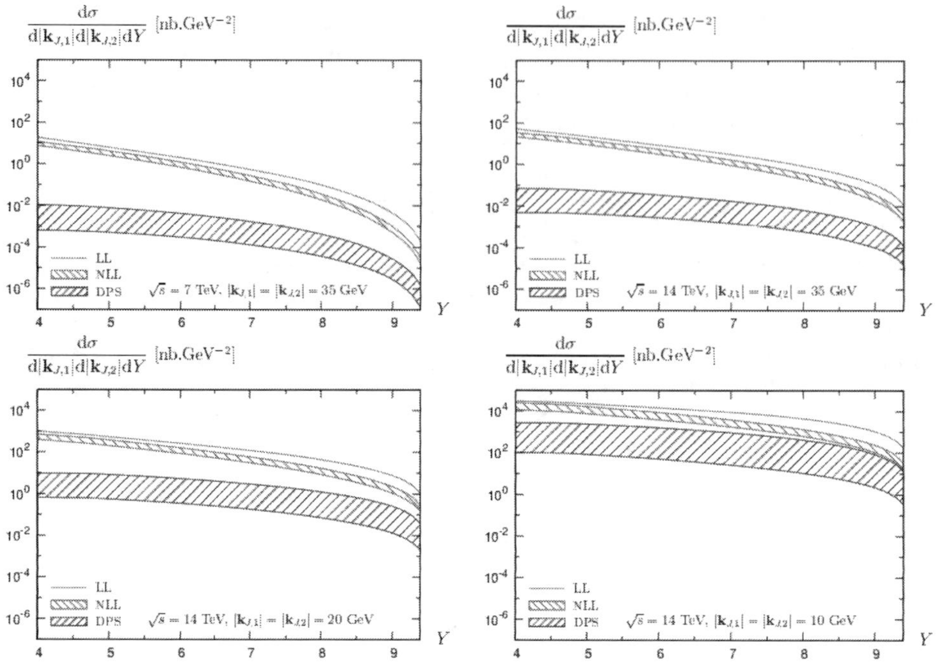

Figure 4. Comparison of the differential cross section obtained at LL (green) and NLL (red) accuracy in the BFKL approach and the DPS cross section (blue) for the four kinematical cuts described in the text.

choice of the UGD parametrization needed to compute the DPS cross section, we use four different parametrizations [33–36]. As shown in fig. 4, the DPS cross section is always smaller than the SPS one in the LHC kinematics we considered here. The same conclusion can be reached for the impact of double parton scattering on the angular correlation between the jets [31]. It is only for the set of parameters giving the largest DPS contribution, *i.e.* at low transverse momenta and large rapidity separations, that the effect of DPS can become larger than the uncertainty on the NLL BFKL calculation.

6 Conclusions

The azimuthal correlations of Mueller-Navelet jets recently extracted by the CMS collaboration can be well described by a full NLL BFKL calculation supplemented by the use of the BLM renormalization scale fixing procedure. We have shown that two effects which are claimed to have a potential significant impact in this picture do not affect our results. First, the effect of the absence of strict energy-momentum conservation in a BFKL calculation is expected to be tiny at NLL accuracy for significantly different values of transverse momenta of the tagged jets. Second, the order of magnitude of DPS contributions is negligible for the kinematics which is under consideration at the LHC.

Acknowledgments

B. Ducloué acknowledges support from the Academy of Finland, Project No. 273464. This work was done using computing resources from CSC – IT Center for Science in Espoo, Finland. L. Szymanowski was partially supported by a French Government Scholarship and by a grant of National

Science Center, Poland, No. 2015/17/B/ST2/01838. This work is partially supported by the French Grant ANR PARTONS No. ANR-12-MONU- 0008-01.

References

[1] A.H. Mueller, H. Navelet, Nucl. Phys. **B282**, 727 (1987)
[2] V.S. Fadin, E. Kuraev, L. Lipatov, Phys. Lett. **B60**, 50 (1975); Sov. Phys. JETP **44**, 443 (1976); E. Kuraev, L. Lipatov, V.S. Fadin, Sov. Phys. JETP **45**, 199 (1977); I. Balitsky, L. Lipatov, Sov. J. Nucl. Phys. **28**, 822 (1978)
[3] V.S. Fadin, L.N. Lipatov, Phys. Lett. **B429**, 127 (1998), hep-ph/9802290
[4] M. Ciafaloni, G. Camici, Phys. Lett. **B430**, 349 (1998), hep-ph/9803389
[5] J. Bartels, D. Colferai, G.P. Vacca, Eur. Phys. J. **C24**, 83 (2002), hep-ph/0112283
[6] J. Bartels, D. Colferai, G.P. Vacca, Eur. Phys. J. **C29**, 235 (2003), hep-ph/0206290
[7] D. Colferai, F. Schwennsen, L. Szymanowski, S. Wallon, JHEP **1012**, 026 (2010), 1002.1365
[8] B. Ducloué, L. Szymanowski, S. Wallon, JHEP **1305**, 096 (2013), 1302.7012
[9] S.J. Brodsky, G.P. Lepage, P.B. Mackenzie, Phys. Rev. **D28**, 228 (1983)
[10] B. Ducloué, L. Szymanowski, S. Wallon, Phys. Rev. Lett. **112**, 082003 (2014), 1309.3229
[11] V. Khachatryan et al. (CMS) (2016), 1601.06713
[12] V. Del Duca, C.R. Schmidt, Phys. Rev. **D49**, 4510 (1994), hep-ph/9311290
[13] W.J. Stirling, Nucl. Phys. **B423**, 56 (1994), hep-ph/9401266
[14] S. J. Brodsky, F. Hautmann, D. E. Soper, *Phys. Rev. Lett.* **78**, 803 (1997). [Erratum: Phys. Rev. Lett.79,3544(1997)]; *Phys. Rev.* **D56**, 6957 (1997); S. J. Brodsky, V. S. Fadin, V. T. Kim, L. N. Lipatov, G. B. Pivovarov, *JETP Lett.* **70**, 155 (1999); *JETP Lett.* **76**, 249 (2002).
[15] F. Caporale, D.Yu. Ivanov, B. Murdaca, A. Papa, Nucl. Phys. **B877**, 73 (2013), 1211.7225
[16] F. Caporale, B. Murdaca, A. Sabio Vera, C. Salas, Nucl. Phys. **B875**, 134 (2013), 1305.4620
[17] F. Caporale, D.Yu. Ivanov, B. Murdaca, A. Papa, Eur. Phys. J. **C74**, 3084 (2014), 1407.8431
[18] F. Caporale, D.Yu. Ivanov, B. Murdaca, A. Papa, Phys. Rev. **D91**, 114009 (2015), 1504.06471
[19] F.G. Celiberto, D.Yu. Ivanov, B. Murdaca, A. Papa, Eur. Phys. J. **C75**, 292 (2015), 1504.08233
[20] F. G. Celiberto, D.Yu. Ivanov, B. Murdaca and A. Papa, 1601.07847
[21] A.H. Mueller, L. Szymanowski, S. Wallon, B.W. Xiao, F. Yuan (2015), 1512.07127
[22] P. Aurenche, R. Basu, M. Fontannaz, Eur. Phys. J. **C57**, 681 (2008), 0807.2133
[23] V. Del Duca, C.R. Schmidt, Phys. Rev. **D51**, 2150 (1995), hep-ph/9407359
[24] B. Ducloué, L. Szymanowski, S. Wallon, Phys. Lett. **B738**, 311 (2014), 1407.6593
[25] CDF Collaboration (CDF), Phys. Rev. **D47**, 4857 (1993); **D56**, 3811 (1997)
[26] D0 Collaboration (D0), Phys. Rev. **D81**, 052012 (2010), 0912.5104
[27] D0 Collabration (D0), Phys. Rev. **D89**, 072006 (2014), 1402.1550
[28] ATLAS Collaboration (ATLAS), New J. Phys. **15**, 033038 (2013), 1301.6872
[29] CMS Collaboration (CMS), JHEP **1403**, 032 (2014), 1312.5729
[30] CMS Collaboration (CMS), JHEP **1206**, 036 (2012), 1202.0704
[31] B. Ducloué, L. Szymanowski, S. Wallon, Phys. Rev. **D92**, 076002 (2015), 1507.04735
[32] A. Martin, W. Stirling, R. Thorne, G. Watt, Eur. Phys. J. **C63**, 189 (2009), 0901.0002
[33] M.A. Kimber, A.D. Martin, M.G. Ryskin, Phys. Rev. **D63**, 114027 (2001), hep-ph/0101348
[34] M. Hansson, H. Jung, proceedings of DIS 03, hep-ph/0309009
[35] K. Kutak, S. Sapeta, Phys. Rev. **D86**, 094043 (2012), 1205.5035
[36] F. Hautmann, H. Jung, Nucl. Phys. **B883**, 1 (2014), 1312.7875

EPJ Web of Conferences **112**, 02016 (2016)
DOI: 10.1051/epjconf/201611202016

JIMWLK and beyond: From concepts to observables

Heribert Weigert[1], Tuomas Lappi[2,4], Andrecia Ramnath[2], and Kari Rummukainen[3,4]

[1]*Department of Physics, University of Cape Town, Private Bag X3, Rondebosch 7701, South Africa*
[2]*Department of Physics, University of Jyväskylä, P.O. Box 35, 40014 University of Jyväskylä, Finland*
[3]*Department of Physics, University of Helsinki, 00014 University of Helsinki, Finland*
[4]*Helsinki Institute of Physics, P.O. Box 64, 00014 University of Helsinki, Finland*

Abstract. The Color Glass Condensate and its associated evolution equation, the JIMWLK equation have applications to many observables far beyond totally inclusive observables. The phenomenology is so rich that little has been done to explore beyond scaling behavior of correlators. We show first examples that exemplify the considerations necessary to access additional information both experimentally and theoretically and demonstrate that the Wilson line correlators appearing throughout make it imperative to consistently take into account that one is dealing with correlators of group elements and demonstrate how this imposes physical and phenomenological constraints. Similar considerations apply also to jet observables at finite N_c.

1 Introduction

Over the last decade, the notion of the Color Glass Condensate and its associated evolution equation, the JIMWLK equation, has become a mature tool to quantitatively describe QCD cross sections and the initial state entering heavy ion collisions at high energy. The key ingredient to its derivation is the observation that in a suitably asymmetric high energy collision, where a Fock space expansion of a dilute target in terms of its parton constituents is meaningful the interaction of these constituents with the target gluon field is fully eikonal. Light like Wilson lines along the projectile constituent trajectories then fully summarize the interaction; their correlators fully determine the cross sections. Of course, Wilson lines appear in many other QCD observables in which a separation of hard and soft degrees of freedom is possible. A well established example are non-global jet observables [1] where evolution inside predefined jet cones maps directly onto the structure of the JIMWLK equation (e.g. [2]) via a stereographic projection [3], a fact that has been exploited to infer the NLO contributions of JIMWLK from their jet counterparts [4]. The physics discussions of CGC applications at the leading order all center on scaling phenomena, the behavior of the saturation scale $Q_s(Y)$ and how it impacts the total cross section for deep inelastic scattering and similarly inclusive phenomena such as total multiplicities in heavy ion collisions. In this contribution we attempt to broaden that picture somewhat and give a glimpse at a richer set of accessible phenomena that will show radically different behavior. To do so one needs a systematic way of including degrees of freedom beyond the pomeron. Naturally, the first glimpse at qualitatively new behavior comes from a consistent inclusion of odderon degrees of freedom. Since the gist of this is already present at leading order (LO) we will keep the discussion on this level– it *does* generalize to full next to leading order (NLO).

2 Observables, hierarchies, and parametrizations

The simplest example for Wilson line correlators in cross sections is the expression for the DIS total cross section at small x to leading order. Its diagrammatic expression reads

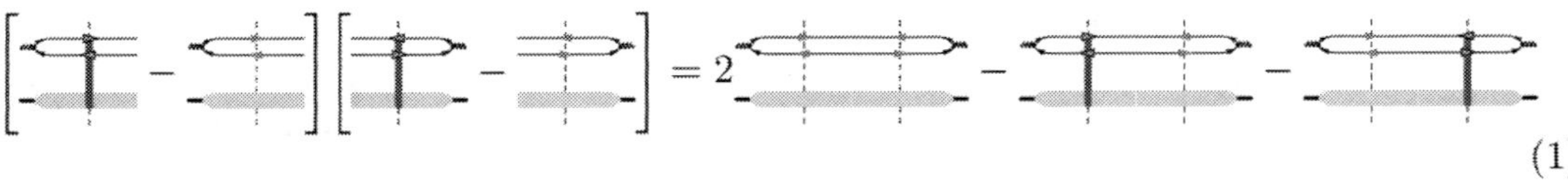

(1)

The Wilson line correlators appear in the last two terms. Since this is an expression for the total cross section they explicitly combine into the real part of this correlator

$$\langle \mathrm{tr}(U_{\boldsymbol{x}}U^\dagger_{\boldsymbol{y}})\rangle(Y) + \langle \mathrm{tr}(U^\dagger_{\boldsymbol{x}}U_{\boldsymbol{y}})\rangle(Y) = 2\langle \mathrm{Re}\,\mathrm{tr}(U_{\boldsymbol{x}}U^\dagger_{\boldsymbol{y}})\rangle(Y) \tag{2}$$

Average displayed is explicitly $Y = \ln(1/x)$ dependent and to be understood as a QCD average in the presence of the target wave function.

The Wilson line correlator itself *does* have both a real and an imaginary part although one needs a more exclusive observable to probe it. A suitable example is the Single Transverse Spin Asymmetry (STSA), the Sivers' function, for which Kovchegov and Sievert [5] have suggested a new mechanism at small x that is triggered by this imaginary part. Diagrammatically

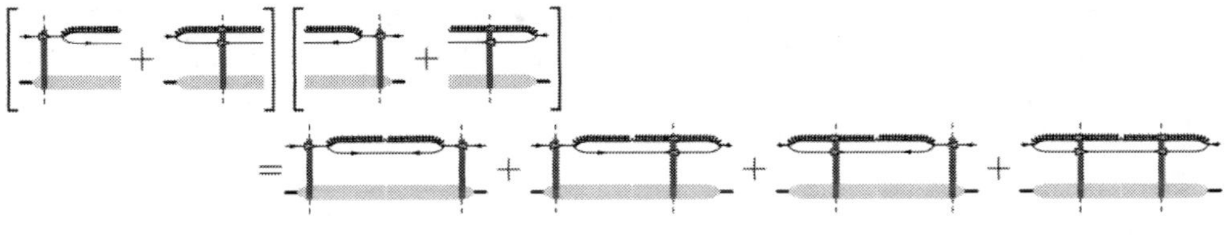

(3)

where the momentum and spin of the quark in the final state is tagged and the color in the initial state is summed over. Tagging the quark momentum leads to different coordinates on the corresponding Wilson lines in amplitude and complex conjugate amplitude. The gluon momentum is integrated over so that the gluon Wilson line in the last term cancels between the two sides of the cut. As a result the color correlators in Eq. (3) –in term by term correspondence to the diagrams– are

$$\mathcal{I}^{(q)} = \left\langle \frac{\mathrm{tr}(U_{\boldsymbol{z}}U^\dagger_{\boldsymbol{y}})}{N_c} - \frac{1}{d_A}\tilde{U}^{ab}_{\boldsymbol{x}}\mathrm{tr}(t^a U_{\boldsymbol{z}} t^b U^\dagger_{\boldsymbol{w}}) - \frac{1}{d_A}\tilde{U}^{ab}_{\boldsymbol{x}}\mathrm{tr}(t^a U_{\boldsymbol{u}} t^b U^\dagger_{\boldsymbol{y}}) + \frac{\mathrm{tr}(U_{\boldsymbol{u}}U^\dagger_{\boldsymbol{w}})}{N_c} \right\rangle \tag{4}$$

The STSA is driven by the contributions that are anti-symmetric under exchange of the quark and anti-quark coordinates $\boldsymbol{z} \leftrightarrow \boldsymbol{y}$ and thus the imaginary part of, for example the first term.

Focusing on the the dipole correlator, we parametrize this complex number in terms of two real degrees of freedom

$$S_{\boldsymbol{xy}}(Y) := \langle \mathrm{tr}(U_{\boldsymbol{x}}U^\dagger_{\boldsymbol{y}})\rangle(Y)/N_c = 1 - P_{\boldsymbol{xy}}(Y) + iO_{\boldsymbol{xy}}(Y) = e^{-C_f(\mathcal{P}_{\boldsymbol{xy}}+i\mathcal{O}_{\boldsymbol{xy}})(Y)} \tag{5}$$

either directly through real and imaginary parts ($1 - P_{\boldsymbol{xy}}$ and $O_{\boldsymbol{xy}}$ respectively) or exponentially via a logarithmic modulus and phase ($\mathcal{P}_{\boldsymbol{xy}}$ and $\mathcal{O}_{\boldsymbol{xy}}$ respectively). Noting that complex conjugation simply swaps the coordinates on S, $S^*_{\boldsymbol{xy}} = S_{\boldsymbol{yx}}$, implies that P and $\mathcal{P}$ are symmetric, while O and $\mathcal{O}$ are anti-symmetric under the exchange of $\boldsymbol{x}$ and $\boldsymbol{y}$. This symmetry property links them to the pomeron and odderon respectively.

The three point function of the form $\tilde{U}^{ab}_{\boldsymbol{z}}\mathrm{tr}(t^a U_{\boldsymbol{x}} t^b U^\dagger_{\boldsymbol{y}})$ in Eq. (4) –it also appears in the evolution equation of the $q\bar{q}$ dipole operator– is slightly more complicated but intimately related to the $q\bar{q}$ and gg dipole operators in different coincidence limits. Schematically

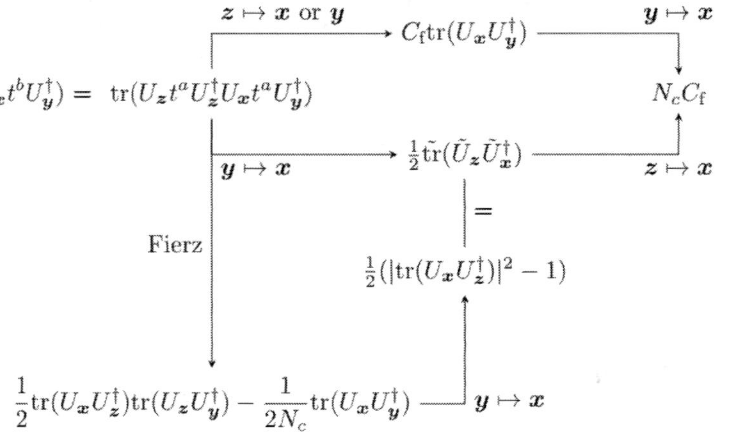

(6)

It is well known that the JIMWLK equation, already at LO, is equivalent to infinite coupled hierarchies of evolution equations for Wilson line correlators. It is worth emphasizing that it governs the Y dependence of both real and imaginary parts of all of these correlators although many applications are blind to the imaginary parts: It turns out that they may be set to zero consistently within the initial conditions of the evolution equation if not needed phenomenologically.

A general parametrization that manifestly respects the relationships in (6) is a generalization of the Gaussian truncation used already in [6–9] where we truncate at the level of three point functions:

$$\begin{aligned}\langle \ldots \rangle(Y) = \Big\langle P_Y \exp\Big\{ \int_{Y_0}^{Y} dy \Big[-\frac{1}{2} \int_{\boldsymbol{uv}} G_{\boldsymbol{uv}}(y) i\nabla^a_{\boldsymbol{u}} i\nabla^a_{\boldsymbol{v}} \\ - \frac{1}{3!} \int_{\boldsymbol{uvw}} G_{\boldsymbol{uvw}}(y)\; d^{abc} i\nabla^a_{\boldsymbol{u}} i\nabla^b_{\boldsymbol{v}} i\nabla^c_{\boldsymbol{w}} + \text{etc.} \Big] \Big\} \ldots \Big\rangle(Y_0)\,. \end{aligned} \tag{7}$$

This can be generalized systematically at any n-point level, the number of singlets of n objects in the adjoint representation determines the number of n-point functions appearing in the Y-ordered exponential. For convenience we choose $G_{\boldsymbol{uu}} = 0$ and $G_{\boldsymbol{uuu}} = 0$ and do not display the three point function associated with f^{abc} as it does not enter the correlators discussed here.

Eq. (7) induces explicit expressions for all correlators in Eqns. (2) and (4), namely

$$\frac{1}{N_c}\langle \mathrm{tr}(U_{\boldsymbol{x}}U^\dagger_{\boldsymbol{y}})\rangle = e^{-C_f(\mathcal{P}+i\mathcal{O})_{\boldsymbol{xy}}} \tag{8}$$

$$\frac{1}{2N_cC_f}\langle U^{ab}_{\boldsymbol{z}}\mathrm{tr}(t^aU_{\boldsymbol{x}}t^bU^\dagger_{\boldsymbol{y}})\rangle = e^{-\{[\frac{N_c}{2}((\mathcal{P}+i\mathcal{O})_{\boldsymbol{xz}}+(\mathcal{P}+i\mathcal{O})_{\boldsymbol{zy}})-(\mathcal{P}+i\mathcal{O})_{\boldsymbol{xy}})-C_{\mathrm{f}}(\mathcal{P}+i\mathcal{O})_{\boldsymbol{xy}}\}(Y)} \tag{9}$$

where

$$\mathcal{P}_{\boldsymbol{xy}} := \int^{Y} dy \left(G_{\boldsymbol{xy}} - \frac{1}{2}(G_{\boldsymbol{xx}} + G_{\boldsymbol{yy}}) \right) = \mathcal{P}_{\boldsymbol{yx}} \qquad i\mathcal{O}_{\boldsymbol{xy}} := \frac{C_d}{4} \int^{Y} dy \left(G_{\boldsymbol{yxx}} - G_{\boldsymbol{yyx}} \right) = -i\mathcal{O}_{\boldsymbol{yx}} \tag{10}$$

Strikingly, the observables considered here do *not* allow access to $G_{\boldsymbol{uvw}}$ with all three coordinates independent! The effective two point functions however *do* show the required symmetry properties; one obtains a consistent *gauge invariant* truncation of the associated Balitsky hierarchy with the evolution equation

$$\frac{d}{dY}\mathcal{G}_{\boldsymbol{xy}}(Y) = \frac{\alpha_s}{\pi^2} \int d^2z K_{\boldsymbol{xzy}} \left(1 - e^{-\frac{N_c}{2}[\mathcal{G}_{\boldsymbol{xz}} + \mathcal{G}_{\boldsymbol{zy}} - \mathcal{G}_{\boldsymbol{xy}}](Y)} \right) \qquad \mathcal{G}_{\boldsymbol{xy}}(Y) = (\mathcal{P} + i\mathcal{O})_{\boldsymbol{xy}}(Y) \tag{11}$$

which obviously couples real and imaginary parts. Eq. (11) generalizes the the large N_c results of [10, 11]. Curiously, Eq. (11) differs from the BK equation for the total cross section in the Gaussian truncation only through a non-vanishing imaginary part $i\mathcal{O} \neq 0$ (this is one of the differences between our treatment and the discussion in [10, 11]). Indeed $i\mathcal{O} = 0$ and $\mathcal{P} \in [0, 1]$ is a consistent solution to this equation which leads to a successful phenomenology for HERA data at small x [9]: If the initial condition for $\mathcal{G}$ is real, the equation never generates an imaginary part. This behavior is numerically stable as can be seen from analyzing the coupled equation for the real and imaginary parts explicitly:

$$\frac{d}{dY}\mathcal{P}_{\boldsymbol{xy}}(Y) = \frac{\alpha_s}{\pi^2} \int d^2z K_{\boldsymbol{xzy}} \left(1 - e^{-\frac{N_c}{2}[\mathcal{P}_{\boldsymbol{xz}} + \mathcal{P}_{\boldsymbol{zy}} - \mathcal{P}_{\boldsymbol{xy}}](Y)} \cos\left(\frac{N_c}{2} \left[\mathcal{O}_{\boldsymbol{xz}} + \mathcal{O}_{\boldsymbol{zy}} - \mathcal{O}_{\boldsymbol{xy}} \right](Y) \right) \right) \tag{12}$$

$$\frac{d}{dY}\mathcal{O}_{\boldsymbol{xy}}(Y) = \frac{\alpha_s}{\pi^2} \int d^2z K_{\boldsymbol{xzy}} \left(e^{-\frac{N_c}{2}[\mathcal{P}_{\boldsymbol{xz}} + \mathcal{P}_{\boldsymbol{zy}} - \mathcal{P}_{\boldsymbol{xy}}](Y)} \sin\left(\frac{N_c}{2} \left[\mathcal{O}_{\boldsymbol{xz}} + \mathcal{O}_{\boldsymbol{zy}} - \mathcal{O}_{\boldsymbol{xy}} \right](Y) \right) \right) \tag{13}$$

The odderon contribution will naturally shrink no matter which initial condition is used for it.

The choice of a such an initial condition for the pair of $\mathcal{P}$ and $\mathcal{O}$, one that is physically meaningful, needs some thought. At short distance, light cone perturbation theory leads to $\mathcal{P}_{\boldsymbol{xy}} \propto |\boldsymbol{x} - \boldsymbol{y}|^2$, while $\mathcal{O}_{\boldsymbol{xy}} \propto |\boldsymbol{x} - \boldsymbol{y}|^3$ [10], but the symmetry properties imposed by complex conjugation and evidenced in Eq. (10) require the presence of an additional transverse vector $\hat{\boldsymbol{s}}$ – to be able to see a scalar odderon function in an experiment one needs to break rotational symmetry in the transverse plane such as is done by measuring a spin asymmetry in STSA. We thus expect a short distance behavior of the form

$$\mathcal{P}_{\boldsymbol{xy}} \propto \boldsymbol{r}^2 Q_0^2 \,, \qquad \mathcal{O}_{\boldsymbol{xy}} \propto |\boldsymbol{r}|^3 \kappa Q_0^2 \hat{\boldsymbol{r}} \cdot \hat{\boldsymbol{s}} \qquad \text{where} \qquad \boldsymbol{r} = \boldsymbol{x} - \boldsymbol{y} \,. \tag{14}$$

Q_0 is of the order of $Q_s(Y_0)$, the well know saturation scale of the initial condition, but κ, the scale ratio relating the scales of the two contributions, is not yet constrained.

A qualitative and even semi-quantitative constraint comes from an elementary group theoretical condition on the dipole correlator $\operatorname{tr}(U_{\boldsymbol{x}} U_{\boldsymbol{y}}^\dagger)$. As an element of SU($N_c$) any Wilson line configuration $U_{\boldsymbol{x}} U_{\boldsymbol{y}}^\dagger$ has eigenvalues on the unit circle $e^{i\phi_i}$ which are subject to the determinant condition $\sum_{i=1}^{N_c} \phi_i = 2\pi n; n \in \mathbb{Z}$, so that the trace of any individual configuration can

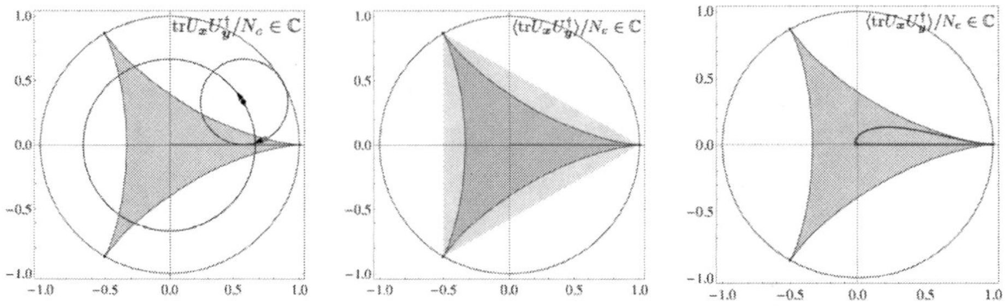

Figure 1. Left: The trace of a given group element of SU(3) such as a given dipole configuration $\mathrm{tr}(U_{\boldsymbol{x}}U_{\boldsymbol{y}}^\dagger)/N_c$ falls into the hypocycloid shown in grey. Its boundary is traced out by a point on the smaller circle shown in red rolling counter-clockwise inside the outer circle. Middle: The expectation value or average of such configurations $\langle\mathrm{tr}(U_{\boldsymbol{x}}U_{\boldsymbol{y}}^\dagger)\rangle(Y)/N_c$ is slightly less constrained: It may fall outside the hypocycloid but must remain inside a polygon connecting the cusps. Right: Perturbative initial conditions for real and imaginary parts of $\langle\mathrm{tr}(U_{\boldsymbol{x}}U_{\boldsymbol{y}}^\dagger)\rangle(Y)/N_c$ for ensembles with or without a net odderon (blue and red lines respectively). The blue line represents the extreme case assuming there is no non-perturbative mechanism to pull the average outside the hypocycloid.

be parametrized as

$$\mathrm{tr}(U_{\boldsymbol{x}}U_{\boldsymbol{y}}^\dagger)/N_c = \frac{1}{N_c}\left(\sum_{i=1}^{N_c-1} e^{i\phi_i} + e^{-i\sum\limits_{i=1}^{N_c-1}\phi_i}\right) \tag{15}$$

with $N_c - 1$ distinct phases ϕ_i, all $\boldsymbol{x}$ and $\boldsymbol{y}$ dependent. (15) falls into a hypocycloid in the complex plane with N_c cusps, each representing one of the center elements of SU(N_c) as shown for $N_c = 3$ in Fig 1. (This is a textbook result, see [12] for a recent discussion.) The expectation value or average of such configurations, $\langle\mathrm{tr}(U_{\boldsymbol{x}}U_{\boldsymbol{y}}^\dagger)\rangle(Y)/N_c$, is slightly less constrained. It may fall outside the hypocycloid but must remain inside a polygon connecting the cusps. Note, however, that the cusps corresponding to the nontrivial center elements (at $e^{\pm 2\pi i/3}$ for $N_c = 3$) also correspond to non-perturbative configurations – only the vicinity of the unit element corresponds to a short distance correlator. Hence, one needs a non-perturbative mechanism to push the average outside the hypocycloid. In either case, this puts severe constraints on an initial condition that extrapolates the short distance limit through exponentiation in the form (Q_0 is of the order of $Q_s(Y_0)$, the real part is nothing but the Golec-Biernat Wüsthoff model)

$$\langle\mathrm{tr}(U_{\boldsymbol{x}}U_{\boldsymbol{y}}^\dagger)\rangle(Y_0)/N_c = e^{-r^2Q_0^2 + i\kappa r^3 Q_0^3 \hat{\boldsymbol{r}}\cdot\hat{\boldsymbol{s}}} \tag{16}$$

For the first case $\kappa \le 1/3$ while for the second we get an approximate bound of $\kappa \le .98$. The first case is shown in Fig. 1, right panel. Real and imaginary parts of this expression are shown in Fig. 2 for both extreme cases. Real and imaginary parts show modulation only near $Q_s(Y_0)$ and thus provide perturbatively consistent starting points for evolution in a calculation where $Q_s(Y_0)$ is assumed to be in the perturbative domain. Note that the size of the odderon peak (the maximum of imaginary parts shown in red in Fig. 2) is severely limited by the bounds on κ. This is very different from the BFKL odderon in which there is only a phenomenological reason to assume a small odderon contribution in the initial condition.

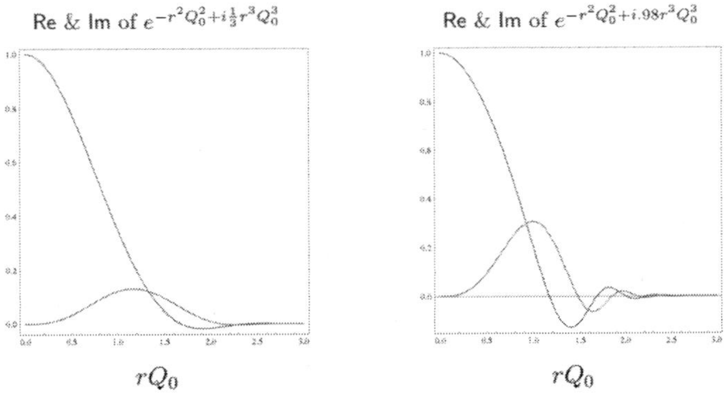

Figure 2. Perturbatively motivated initial conditions for both real and imaginary parts (blue and red respectively) of $\langle \mathrm{tr}(U_{\boldsymbol{x}}U_{\boldsymbol{y}}^\dagger)\rangle(Y_0)/N_c = e^{-r^2Q_0^2+i\kappa r^3Q_0^3\hat{\boldsymbol{r}}\cdot\hat{\boldsymbol{s}}}$ at $\hat{\boldsymbol{r}}\cdot\hat{\boldsymbol{s}}=1$ assuming no non-perturbative mechanism to drive the correlator outside the hypocycloid (left) and relaxing this condition (right). In both cases real and imaginary parts show modulation only near $Q_s(Y_0)$.

In the non-linear case *both* the pomeron and odderon are subject to group constraints and severely limited in magnitude.

3 Qualitative simulation results

The easiest way to study the qualitative energy behavior of the coupled pomeron and odderon configurations is a full Langevin simulation of the JIMWLK equation as first done in [13] and developed further for example in [14, 15]. This simulation does not even require a modification of the evolution code: The only ingredient requiring modification is the initial condition. One must start from an ensemble in which, on average, the imaginary part of $\langle \mathrm{tr}(U_{\boldsymbol{x}}U_{\boldsymbol{y}}^\dagger)\rangle(Y_0)/N_c$ is nonzero. As discussed above this requires that rotational invariance is broken. We have implemented this by distorting a symmetric initial condition to show a bias along one of the coordinate axes of the transverse plane.

Fig. 3 shows the result for a pure pomeron ensemble in the top row and an ensemble with an initial odderon admixture in the bottom row, clearly demonstrating that any odderon admixture (the imaginary part) is erased quickly and efficiently with only minimal feedback into the behavior of the pomeron (the real part). The behavior of the two is strikingly different: While the pomeron shows clear scaling behavior in both cases, the odderon (if present in the initial condition) does not scale – it is erased in place, the maximum falls to zero exponentially without moving much as Y increases.

4 Conclusions

The fact that JIMWLK evolution deals with observables entirely expressed in terms of Wilson line correlators –a feature shared with a large set of jet evolution observables– gives a strong handle on both structure and possible initial conditions of such correlators. Coincidence limits reveal simpler correlators with fewer Wilson lines appearing in any given n-point function, the web of relationships for the $q\bar{q}g$ correlator shown in Eq. (6) is but a simple example for

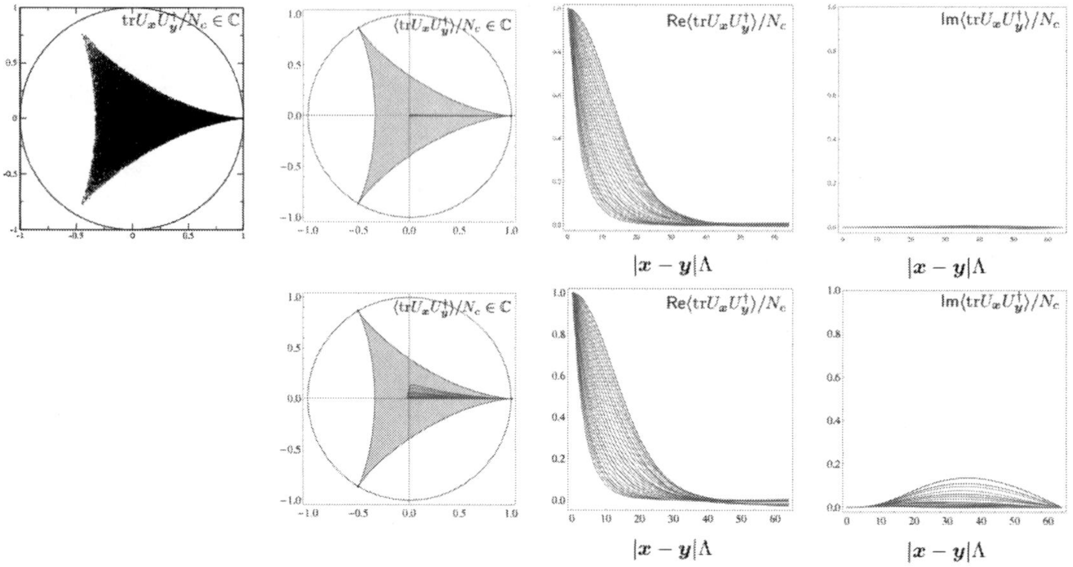

Figure 3. Leading order Langevin simulations with curves for different Y values. The top row shows evolution with zero odderon in the initial condition, the bottom row shows an ensemble with an odderon admixture in the initial condition. The first colum shows a scatterplot of individual traces from the simulation to its right. The non-perturbative cusps are not populated. The second column shows the average $\langle \mathrm{tr}U_{\boldsymbol{x}}U_{\boldsymbol{y}}^{\dagger}\rangle/N_c$ in the complex plane, $|\boldsymbol{x}-\boldsymbol{y}|$ grows as one follows any individual curve from 1 to 0 in $\mathbb{C}$. The third column shows the real parts (evolution moves these from "right to left"), the pomeron, and the fourth column provides the imaginary parts, the odderon (evolution moves these downwards). Both pomeron and odderon contributions are plotted as a function of $|\boldsymbol{r}| = |\boldsymbol{x}-\boldsymbol{y}|$ in arbitrary units – units can only be determined through a fit to data. Real and imaginary parts show different qualitative behavior: while the real parts exhibit scaling behavior the imaginary parts do not move, instead they are erased in place.

the general case. To consistently truncate the hierarchies of evolution equations on the n-point level, one needs a parametrization that respects the coincidence limits the Wilson line correlators satisfy by construction. The examples shown give a first glimpse at a classification of correlators of Wilson lines, be it at small x or for jet observables, that is quite analogous to the discussion of Generalized Parton Distributions (GPDs) and what types of measurements allow access to them. First qualitative simulations using a LO Langevin code demonstrate clearly that different degrees of freedom can behave qualitatively different: the pomeron contributions to the cross sections approach a scaling behavior with a sliding scale dependent on the evolution variable, while the odderon contributions simply are erased in place. This qualitatively different behavior of the odderon contribution under JIMWLK evolution is a genuinely new result. What we find most intriguing, however, is the fact that one can proceed to include new degrees of freedom into the formalism in a consistent manner and that the results are striking. The interplay of group theoretical properties with features imprinted by unresummed perturbation theory have surprisingly strong implications, here the conclusion that the size of the odderon contribution is limited from the outset by consistency of the formalism, not only by phenomenology. We are confident that such features hold also for higher order correlators.

Acknowledgements: Heribert Weigert was supported by NRF under CPRR grant nr 90509 and is grateful for support through labex P2IO during his stay at CPhT Ecole Polytechnique. Tuomas Lappi is supported by Academy of Finland projects 267321 and 273464.

References

[1] A. Banfi, G. Marchesini, G. Smye, JHEP **08**, 006 (2002), hep-ph/0206076
[2] H. Weigert, Nucl. Phys. **B685**, 321 (2004), hep-ph/0312050
[3] Y. Hatta, JHEP **11**, 057 (2008), 0810.0889
[4] S. Caron-Huot (2015), 1501.03754
[5] Y.V. Kovchegov, M.D. Sievert, Phys. Rev. **D89**, 054035 (2014), 1310.5028
[6] H. Weigert, J. Kuokkanen, K. Rummukainen, AIP Conf. Proc. **1105**, 394 (2009)
[7] Y.V. Kovchegov, J. Kuokkanen, K. Rummukainen, H. Weigert, Nucl. Phys. **A823**, 47 (2009), 0812.3238
[8] C. Marquet, H. Weigert, Nucl. Phys. **A843**, 68 (2010), 1003.0813
[9] J. Kuokkanen, K. Rummukainen, H. Weigert, Nucl. Phys. **A875**, 29 (2012), 1108.1867
[10] Y.V. Kovchegov, L. Szymanowski, S. Wallon, Phys. Lett. **B586**, 267 (2004), hep-ph/0309281
[11] Y. Hatta, E. Iancu, K. Itakura, L. McLerran, Nucl. Phys. **A760**, 172 (2005), hep-ph/0501171
[12] N. Kaiser (2009), math-ph/0609082
[13] K. Rummukainen, H. Weigert, Nucl. Phys. **A739**, 183 (2004), hep-ph/0309306
[14] T. Lappi, H. Mäntysaari, Eur. Phys. J. **C73**, 2307 (2013), 1212.4825
[15] A. Dumitru, T. Lappi, Y. Nara, Phys. Lett. **B734**, 7 (2014), 1401.4124

EPJ Web of Conferences **112**, 03001 (2016)
DOI: 10.1051/epjconf/201611203001

Tagged spectator deep-inelastic scattering off the deuteron as a tool to study neutron structure

Wim Cosyn[1,a] and Misak Sargsian[2,b]

[1] *Department of Physics and Astronomy, Proeftuinstraat 86, Ghent University, 9000 Ghent, Belgium*
[2] *Department of Physics, Florida International University, Miami FL 33199, U.S.A.*

Abstract. We give an overview of a model to describe deep-inelastic scattering (DIS) off the deuteron with a spectator proton, based on the virtual nucleon approximation (VNA). The model accounts for the final-state interactions (FSI) of the DIS debris with the spectator proton. Values of the rescattering cross section are obtained by fits to high-momentum spectator data. By using the so-called "pole extrapolation" method, free neutron structure functions can be obtained by extrapolating low-momentum spectator proton data to the on-shell neutron pole. We apply this method to the BONuS data set and find a surprising Bjorken x dependence, indicating a possible rise of the neutron to proton structure function ratio at high x.

1 Introduction

Tagged proton spectator DIS off the deuteron ($e+d \to e'+X+p_s$) can be used in several ways to study the role of QCD dynamics at nucleonic length scales. At low spectator momenta, one has access to the structure functions of an almost on-shell neutron. Together with proton structure functions, flavor separation can be performed and the d and u quark densities can be determined. At higher proton spectator momenta, the deuteron is in a high-density configuration and medium modifications of the structure functions can be quantified. In kinematics where FSI mechanisms are large, one can use the reaction to constrain models for the X–proton rescattering. Recently, two experiments at Jefferson Lab have measured the reaction in different kinematics: Deeps at higher spectator momenta of a few hundred MeV [1] and, BONuS at lower ones around 100 MeV [2]. In the near future, the BONuS12 experiment will run [3] and the reaction is considered as a method to extract neutron structure at a future electron-ion collider [4]. In order to provide meaningful interpretations of the measured data, theoretical models that quantify the importance of the FSI in the reaction are needed [5–10]. The major difficulty in doing this is that one lacks detailed information about the composition and space-time evolution of the hadronic system produced in the deep-inelastic scattering and how this changes as a function of Bjorken x and Q^2. Here, we present work in a model that has been used to account for FSI mechanisms at moderate and high Bjorken x in $e + d \to e' + X + p_s$ [11–13] as well as inclusive deuteron DIS [14, 15].

[a] e-mail: wim.cosyn@ugent.be
[b] e-mail: sargsian@fiu.edu

2 Formalism

The model introduced in Ref. [11] accounted for FSI effects in DIS from the deuteron based on general properties of high-energy diffractive scattering. The underlying assumption was that due to the restricted phase space (finite values of W and Q^2), the minimal Fock state component of the wave function can be used to describe DIS from the bound nucleon. In this case the scattered state consists of three outgoing valence quarks. The model uses the virtual nucleon approximation (VNA) [6, 7, 16] and only considers the proton-neutron component of the deuteron wave function. The effects of FSI are encoded in an effective rescattering amplitude between the produced hadronic mass X and the spectator proton. Given the GeV range of the energies in the experiments, the rescatterings will occur over small angles and be highly diffractive. This allows us to model the rescattering amplitude using the generalized eikonal approximation (GEA) [17]. In the GEA, the eikonal rescattering amplitude $f_{X'N,XN}$ is parametrized using three parameters (total rescattering cross section σ_{XN}, slope factor B, and real part ϵ), that have a dependence on the Q^2 and Bjorken x (or equivalently invariant mass W of hadronic products X):

$$f_{X'N,XN} = \sigma(Q^2, x)\left[i + \epsilon(Q^2, x)\right] e^{\frac{B(Q^2,x)}{2}t}. \tag{1}$$

The cross sections including FSI are obtained in a factorized approach, whereby the DIS interaction of the virtual photon with the off-shell nucleon (encoded in the structure functions F_{iN}) is taken out of the integration over the momentum of the on-shell intermediate spectator nucleon. Finally, the four structure functions of the tagged spectator DIS process can be written in the following form

$$F_i^{SI} = (\alpha(i)F_{1N} + \beta(i)F_{2N})\, S^D(p_s)(2\pi)^3 2E_s, \qquad \text{with } i = \text{L,T,LT,TT}. \tag{2}$$

Here, $\alpha(i),\beta(i)$ are kinematical factors (see Ref. [11] for detailed expressions), and $S^D(p_s)$ is the distorted deuteron momentum distribution that contains an impulse approximation (IA) contribution and an FSI contribution including the rescattering amplitude of Eq. (1) [11].

Even for very low spectator momenta, the struck neutron is still off-shell in the deuteron. One can still perform an analytical continuation, however, of the amplitude into the unphysical kinematical region and extrapolate it to the on-shell point $p_i^2 = (p_D - p_s)^2 = m_n^2$. The IA part of the amplitude has a pole at this value of p_i^2, while the FSI part does not due to the additional integration over the loop momentum. This is the so-called "no-loop theorem" [7]. This means carrying out the pole extrapolation automatically gets rid of nuclear effects and allows one to extract the on-shell neutron structure function. This procedure is similar to the ones first proposed by Chew and Lowin hadron-proton scattering [18] . Due to the small binding energy of the deuteron, the actual extrapolation length to the on-shell pole is very small, and there are also no issues regarding a zero crossing for p_i^2. The pole extrapolation procedure for the extraction of F_{2n} consists of multiplying the measured deuteron structure function, $F_{2D}^{SI,\text{EXP}}$ by the factor $I(\alpha_s, \mathbf{p}_{s\perp}, t)$ [7], which cancels the singularity of the IA amplitude and is normalized such that

$$F_{2n}^{\text{extr}}(Q^2, x, p_i^2) = I(\alpha_s, \mathbf{p}_{s\perp}, p_i^2) \cdot F_{2D}^{SI,\text{EXP}}(Q^2, x, \alpha_s, \mathbf{p}_{s\perp}) \tag{3}$$

approaches the free $F_{2n}(Q^2, x)$ in the $p_i^2 \to m_n^2$ limit with FSI effects being diminished.

3 Results

Figure 1 shows a comparison between one data set from the high spectator momentum deeps experiment [1] with our model calculations. The most striking feature of the data is the large excess at forward spectator angles. This is markedly different from the quasi-elastic breakup case where the

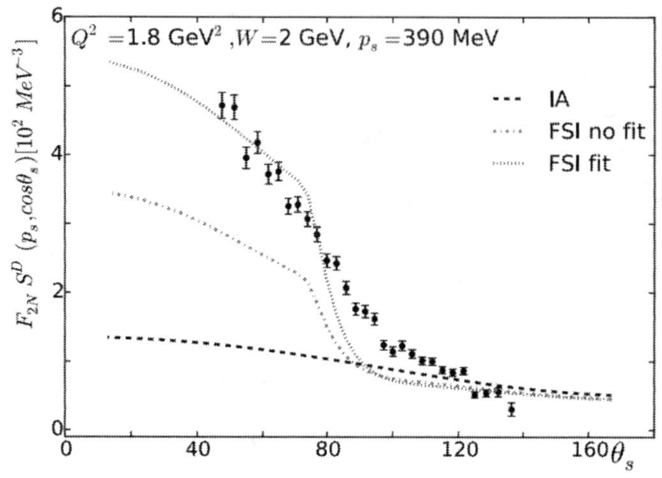

Figure 1. Comparison between deeps data [1], the plane-wave IA calculation (dashed black curve), an unfitted FSI calculation using standard NN scattering values of σ = 40 mb, B = 8 GeV^{-2}, $\epsilon = -0.5$ (red dash-dotted), and a FSI calculation where σ = 55.8 was mb was fitted (blue dotted).

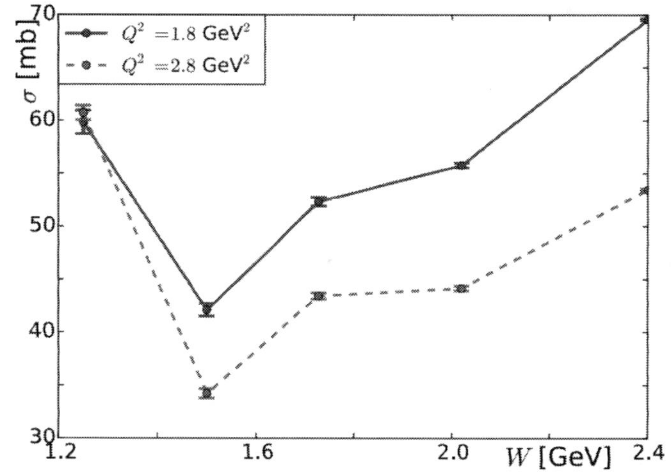

Figure 2. Values of the fitted σ rescattering parameter in Eq. (1) for Q^2 = 1.8 GeV^2 (dashed green curve) and Q^2 = 2.8 GeV^2 (full blue curve), obtained by fitting the high momentum spectator deeps data.

cross section peaks in the perpendicular region [19]. It is clear the IA calculation has an almost flat dependence on the spectator angle and can in no way reproduce the full range of the data. An FSI calculation with values of the rescattering parameters inspired by the known NN ones is able to reproduce the angular dependence better but still underestimates the data. Intuitively, this is no surprise as for the value of W = 2 GeV in Fig. 1, we are well above the nucleon mass. Therefore, we have fitted a value of σ for each $\{W, Q^2\}$ setting of the data. The calculation using this fitted parameter is also shown for the kinematics of Fig. 1. For a comparison with the full deeps data set, we refer to Ref. [11].

In Fig. 2, we show the W and Q^2 dependence of the fitted values of σ from our analysis of the deeps data. The parameter σ has a value of about 60 mb at the Δ-resonance, and then clearly increases with increasing invariant mass W of the hadronic products. More interesting is the dependence on Q^2, where there is a significant decrease in the σ parameter with increasing Q^2. This could be a sign of a color transparency signal in this reaction, where the hadronic products are produced in a colorless point-like configuration at higher momentum transfers, which experiences reduced strong interactions with the remaining hadrons. More measurements at higher Q^2 values are needed, however, to further investigate this behavior.

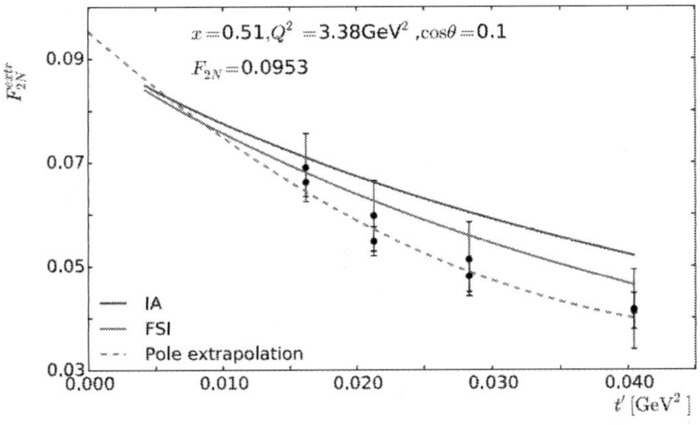

Figure 3. Example of the pole extrapolation method using the renormalized BONuS data (black circles) with the quadratic pole extrapolation curve (red dashed curve) as a function of $t'^2 = p_i^2 - m_n^2$. The IA (full blue curve) and FSI (full green curve) calculations are shown for comparison.

In Ref. [12], the pole extrapolation method was applied to the high momentum spectator deeps data. Given the large extrapolation length, no robust results were obtained for F_{2n}. With the more recent low spectator momentum BONuS data, which measured down to proton momenta of 70 MeV, the extrapolation length is very short and pole extrapolation became applicable. One problem with the BONuS data is that the measurements for different spectator momenta happened at different and not well known efficiencies. In the original analysis, the data was therefore normalized to an IA model [20] in the backward spectator region. In Ref. [13], we renormalized the BONuS data using our FSI model at high W, Q^2 (where the neutron structure function is quite well known), using the rescattering parameter values extracted in our previous fit to the the deeps data. Taking these renormalized data, we implemented the pole extrapolation method for the BONuS data. Figure 3 shows an example for one kinematical setting. The two data points at each t' value correspond to data taken at two initial beam energy values and one can clearly observe they are consistent with each other. Plotting our model calculations with the data, it is also clear that through implementing Eq. (3), FSI effects indeed become smaller as one moves closer to the on-shell pole. As the extrapolation curve shows in Fig. 3, the extrapolation distance to the on-shell point is reasonable.

For each kinematical setting of the BONuS data, pole extrapolation was carried out as in Fig. 3, and a weighted average was taken over all measured spectator angle bins. The final result is shown in Fig. 4 as the ratio of F_{2n}/F_{2p} where the parametrization of Ref. [21] was used for the F_{2p} values. We repeat that the values obtained in this manner are free of Fermi motion and nuclear medium modification effects. The systematic errors were estimated by taking all uncertainties in the data and our normalization procedure (deuteron wave function, structure function parametrization, renormalization fit) into account in a Monte Carlo simulation [13]. The most striking feature in Fig. 4 is the rise of the ratio at large Bjorken x. The results at lower Bjorken $x < 0.5$ are in agreement with existing estimates. It is worth noting that the upward trend of the ratio at higher x is not due to our renormalization of the data. Indeed, the values of the ratio are higher without the renormalization applied. As the W ($\approx$ 1.18 GeV) values corresponding to the highest x values in Fig. 4 are in the sub-DIS regime, one cannot directly relate the rise of the ratio to the underlying properties of the u and d quark pdf's. Such a connection can only be made using duality arguments. It is worth noting, however, that the duality analysis of the BONuS data [22], concluded that the Δ-resonance contributes to the duality, within 20-30% accuracy. One possible explanation of the rise could be the presence of isosinglet qq short-range correlations (SRCs) in the nucleon at $x \to 1$. Such a correlation will result in the same momentum sharing effect, which is observed recently in asymmetric nuclei in the NN SRC region [23].

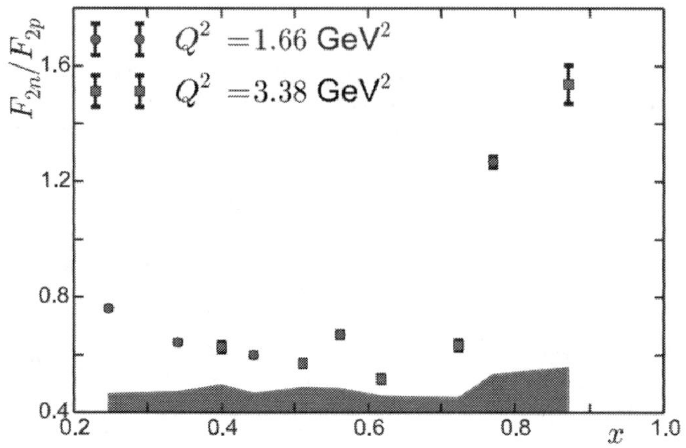

Figure 4. F_{2n} to F_{2p} ratio obtained using the pole extrapolation applied to the renormalized BONuS data. The grey band denotes estimated systematic errors in the procedure.

4 Conclusion

We have presented a selection of results obtained in a model used to describe the tagged spectator DIS process off the deuteron including the effect of FSI. We obtained a good description of data at higher spectator momenta and used these data to fit the Q^2- and W-dependence of the rescattering parameter σ. For data at lower spectator momenta, we applied the pole extrapolation method, which allows to extract the free neutron structure function in a way that eliminates nuclear effects. We observe an intriguing rise of the F_{2n}/F_{2p} ratio at high Bjorken x.

Acknowledgements

The computational resources (Stevin Supercomputer Infrastructure) and services used in this work were provided by Ghent University, the Hercules Foundation and the Flemish Government – department EWI. This work is supported by the Research Foundation Flanders as well as by the U.S. Department of Energy Grant under Contract DE-FG02-01ER41172.

References

[1] A.V. Klimenko et al. (CLAS), Phys. Rev. **C73**, 035212 (2006), nucl-ex/0510032

[2] N. Baillie, S. Tkachenko, J. Zhang, P. Bosted, S. Bültmann, M.E. Christy, H. Fenker, K.A. Griffioen, C.E. Keppel, S.E. Kuhn et al. (CLAS Collaboration), Phys.Rev.Lett. **108**, 199902 (2012), 1110.2770

[3] S. Bueltmann (2006), jLAB-PR12-06-113

[4] W. Cosyn, V. Guzey, D.W. Higinbotham, C. Hyde, S. Kuhn, P. Nadel-Turonski, K. Park, M. Sargsian, M. Strikman, C. Weiss, J. Phys. Conf. Ser. **543**, 012007 (2014), 1409.5768

[5] S. Simula, Phys. Lett. **B387**, 245 (1996), nucl-th/9605024

[6] W. Melnitchouk, M. Sargsian, M.I. Strikman, Z. Phys. **A359**, 99 (1997), nucl-th/9609048

[7] M. Sargsian, M. Strikman, Phys. Lett. **B639**, 223 (2006), hep-ph/0511054

[8] C. Ciofi degli Atti, L.P. Kaptari, S. Scopetta, Eur. Phys. J. **A5**, 191 (1999), hep-ph/9904486

[9] C. Ciofi degli Atti, B.Z. Kopeliovich, Eur. Phys. J. **A17**, 133 (2003), nucl-th/0207001

[10] V. Palli, C. Ciofi degli Atti, L.P. Kaptari, C.B. Mezzetti, M. Alvioli, Phys. Rev. **C80**, 054610 (2009), 0911.1377

[11] W. Cosyn, M. Sargsian, Phys. Rev. **C84**, 014601 (2011), 1012.0293
[12] W. Cosyn, M. Sargsian, AIP Conf.Proc. **1369**, 121 (2011), 1101.1258
[13] W. Cosyn, M.M. Sargsian (2015), 1506.01067
[14] W. Cosyn, W. Melnitchouk, M. Sargsian, Phys.Rev. **C89**, 014612 (2014), 1311.3550
[15] W. Cosyn, M. Sargsian, J. Phys. Conf. Ser. **543**, 012006 (2014), 1407.1653
[16] M.M. Sargsian, S. Simula, M.I. Strikman, Phys. Rev. **C66**, 024001 (2002), nucl-th/0105052
[17] L.L. Frankfurt, M.M. Sargsian, M.I. Strikman, Phys. Rev. **C56**, 1124 (1997), nucl-th/9603018
[18] G.F. Chew, F.E. Low, Phys. Rev. **113**, 1640 (1959)
[19] M.M. Sargsian, Phys. Rev. **C82**, 014612 (2010), 0910.2016
[20] S. Tkachenko et al. (CLAS), Phys. Rev. **C89**, 045206 (2014), [Addendum: Phys. Rev.C90,059901(2014)], 1402.2477
[21] M. Christy, P.E. Bosted, Phys.Rev. **C81**, 055213 (2010), 0712.3731
[22] I. Niculescu et al., Phys. Rev. **C91**, 055206 (2015), 1501.02203
[23] M.M. Sargsian, Phys. Rev. **C89**, 034305 (2014), 1210.3280

EPJ Web of Conferences **112**, 03002 (2016)
DOI: 10.1051/epjconf/201611203002

Nuclear Parton Distributions with the LHeC

Max Klein[1,2,a]

[1] *University of Liverpool, Department of Physics, L697ZE Liverpool, Oxford Street, United Kingdom*
[2] *CERN, 1211 Geneva 23, Switzerland*

Abstract. Nuclear parton distributions are far from being known today because of an infant experimental base. Based on design studies of the LHeC and using new simulations, of the inclusive neutral and charged current cross section measurements and of the strange, charm and beauty densities in nuclei, it is demonstrated how that energy frontier electron-ion collider would unfold the complete set of nuclear PDFs in a hugely extended kinematic range of deep inelastic scattering, extending in Bjorken x down to values near to 10^{-6} in the perturbative domain. Together with a very precise and complete set of proton PDFs, the LHeC nPDFs will thoroughly change the theoretical understanding of parton dynamics and structure inside hadrons.

1 Introduction

Nuclear parton distribution functions (nPDF) are considered through the factorisation theorem [1] of pQCD as universal descriptions of ion-ion (pA or AA) and lepton-hadron (lA) interactions [2]. They are currently poorly determined while the physics of AA scattering with RHIC and the LHC has made striking progress. The theory of nuclear structure, of parton dynamics inside nuclei and of the evolution of the Quark-Gluon Plasma state require a much deeper understanding of nPDFs than has been possible to achieve, also including available LHC pPb data [3], which to much surprise even did exhibit collective phenomena.

PDFs are determined in deep inelastic lepton-hadron scattering, with sometimes useful constraints derived from hadron-hadron interactions. The situation of nPDFs is that unsatisfactory because HERA was not operated with ion beams despite serious proposals [4] to make that happen. Therefore, one is in a situation which resembles the status ante HERA in the determination of proton PDFs: the gluon is essentially unknown, heavy quark distributions are undetermined, up and down light quarks only marginally distinguished and the range in Bjorken x and 4-momentum transfer squared, Q^2, is very limited. These experimental deficiencies are particularly disturbing because the nucleus, unlike the proton, is a much more complex environment for the dynamics of partons to unfold. The nuclear PDFs, $f_A^i(x, Q^2)$, comprise not only the genuine probabilities for the parton densities but account as well for nuclear binding effects known, for decreasing x, as Fermi motion, EMC effect, anti-shadowing and shadowing. It therefore is not only desirable but necessary to determine the full set of PDFs, $i = xg,\ u_v,\ d_v,\ \bar{u},\ \bar{d},\ s,\ c$ and b, with nuclei as well as with protons in order to then determine experimentally, with pQCD assisting, the ratio $R_A^i = f_p^i / f_A^i$ parton by parton in a maximum

[a] e-mail: max.klein@cern.ch

range of x and Q^2. That is the core of the eA measurement programme of the LHeC. Special attention is also given to electron-deuteron scattering because the deuteron is a very special nucleus providing the probably cleanest information on the structure of the neutron.

This brief summary, presented at the POETIC workshop at Paris in September 2015, relies on the layout of the eA programme in the LHeC Conceptual Design Report (CDR) [5], see also the LHeC web page [6] for subsequent presentations. The present contribution also illustrates a new attempt to fit nPDFs based on newly simulated LHeC cross section data as well as novel simulations of the determination of the strange and heavy quark PDFs in nuclei. The LHeC is designed to collide electrons of 60 GeV beam energy off protons of 7 TeV energy (at luminosities up to 1 ab^{-1}) and off lead (A=207, Z=82) ions of $2.7 = 7 \cdot Z/A$ TeV energy (at an estimated integrated luminosity corresponding to 1 fb^{-1}). This opens the prospect for an unprecedented new base and range of unfolding nuclear PDFs and determining the various nuclear binding effects individually for each parton. At this collider one will also be able to control and exclude higher twist effects which, appearing $\propto 1/Q^2$, result from final state rescattering effects and are likely to be amplified $\propto A^{1/3}$ disturbing the determination of nPDFs from low energy experiments even more.

This contribution is organised as follows: Section 2 summarises the status of the current nPDF determinations and presents a summary of the LHeC data simulation. Section 3 briefly summarises initial results of a study of the determination of PDFs in electron-deuteron scattering. Section 4 presents the nPDF simulation using LHeC data performed within an adapted EPS09 pQCD framework. Section 5 discusses the gluon distribution and the possible search for saturation of the rise of the gluon density towards low x. Section 6 includes the determination of the strange, charm and beauty distributions in nuclei from a future eA operation of the LHeC. A brief summary is presented in Section 7.

2 nPDFs Today and their Derivation from the LHeC

Figure 1 presents a summary of presently available sets of nuclear parton distributions, from the HKN (2007), EPS (2009), DSSZ (2012) and nCTEQ (2015) groups with their references and basic characteristics. The data used in all sets come from various fixed target DIS experiments and include dilepton data from Drell-Yan hadron-hadron scattering. Most of the sets use also pion production DAu data for constraining the gluon distribution to some extent. There are three main characteristics of these sets: i) the nPDFs are obtained by using a baseline set of proton PDFs to which a parameterised nuclear correction is fitted. That assumes factorisation of the nuclear and genuine PDF properties, and it introduces an undesirable dependence of the nPDFs on the proton PDF base; ii) owing to the restricted data input, the sets provide only a minimum number of nuclear PDFs employing symmetry assumptions, usually just one valence quark, a combined sea quark and the gluon distribution. The recent nCTEQ analysis removed the $u_v = d_v$ assumption inherent in all previous analyses and, not surprisingly, large variations appeared with a softer up and a harder down quark distribution; iii) as in the so-called global analyses of proton PDFs, the data fitted are not really consistent among each other. Consequently, the analyses introduce some χ^2-tolerance criteria and mimic what a one standard deviation may amount to. All three major deficiencies will be overcome through the LHeC.

The LHeC provides measurements of eA scattering cross sections in the deep inelastic scattering region $Q^2 > 1\,\text{GeV}^2$ reaching to Q^2 up to about $5 \cdot 10^5\,\text{GeV}^2$ and corresponding x values between a few times 10^{-6} and near to $x = 1$. This enables the determination of a complete set of nPDFs in electron-lead scattering at the LHC from the inclusive neutral and charged current cross sections with a clean separation of up and down valence and sea quark distributions. The very high Q^2 which reaches much beyond the W mass squared makes the CC measurements extremely valuable for the separation of different flavours when taken together with the NC, from photon and Z boson exchange.

Table 1. Summary of assumed uncertainties for future inclusive cross section measurements at the LHeC.

Source of uncertainty	Error on the source or cross section
scattered electron energy scale	0.1 %
scattered electron polar angle	0.1 mrad
hadronic energy scale	0.5 %
calorimeter noise ($y < 0.01$)	1-3 %
radiative corrections	1-2 %
photoproduction background	1 %
global efficiency error	0.7 %

As will be illustrated below, charm tagging in CC determines the anti-strange quark distribution in a wide kinematic range to typically 10 – 20 % precision, while charm and beauty tagging in NC provide high precision determinations of xc and xb from nuclei. Using coherent data from just this, one experiment the uncertainties of these nPDFs will follow from a straightforward $\Delta\chi^2 = 1$ criterion. Since, with much larger luminosity, a comparative PDF set will be measured in ep scattering at the LHeC, the ratios $R^i_A = f^i_p/f^i_A$ can be expected to be measured to few or better than percent accuracy as functions of x and Q^2, separately for all i. It is furthermore expected that the nPDFs are not universal in the sense that they depend on the type of boson (γ, W^+, W^-, Z) with which they are probed [7]. Effects like that may only be experimentally established at DIS energies much exceeding HERA as it requires precision measurements of PDFs with disentangled electromagnetic and weak contributions. It is obvious that the LHeC eA programme is expected to change the theoretical description of nPDFs in QCD thoroughly, in diversity, accuracy and range.

The QCD analyses illustrated subsequently of pseudo LHeC cross section data employ sets of simulated NC and CC measurements under assumptions which are summarised in Table 1. The cross section simulation was done numerically employing derivative formulae from [8] and found to compare well to a detailed Monte Carlo simulation when tested for the conditions of the H1 experiment. The assumptions made are all reasonable when comparing with the H1 achievements, which shall probably be exceeded owing to new detector techniques and higher statistics. The control of radiative corrections in eA scattering is a special challenge as these grow $\propto Z^2$. The LHeC detector thus needs to be equipped with reliable photon detectors and the exploitation of the energy-momentum conservation, via the $E - p_z$ cut, should further reduce the effect of photon radiation to a few per cent level. It is also to be noted that the semi-inclusive measurements of the s, c and b quark distributions carry additional uncertainties for tagging, acceptance and background influences.

3 Deuteron PDFs from the LHeC

Electron-deuteron scattering at the LHeC is of very special interest because it enables to access the neutron structure and study the binding effects in the simplest bound proton-neutron configuration. For the derivation of PDFs, eD is the obvious closest partner to ep scattering in measuring an independent up-down quark combination. This permits to disentangle u and d in the valence and the sea region, assisting and cross checking the effect of adding CC and NC data in ep. Figure 2 illustrates the very high precision one would obtain in eD scattering alone, assuming a luminosity of 0.3 fb^{-1} and systematic errors as listed above. It is visible that the gluon and sea distributions are determined to a few per cent precision and with larger relative uncertainty in the valence quark region $x > 0.3$ where both are very small. The valence quarks are measured very well in the medium to large x range. At smaller $x < 0.01$ their relative uncertainty increases while the distributions fall. Details of a similar fit to simulated *ep* data have been released through [9] and the resulting proton PDFs with their expected full experimental uncertainty are available in the LHAPDF data base.

This remarkable sensitivity is obtained in an *eD* only study. The combination of *ep* and *eD* data resolves any $u - d$ ambiguity, within the uncertainty of nuclear corrections. These, however, can be controlled experimentally by two novel means: i) the high resolution tagging of protons in *en* scattering and ii) through Gribov's relation of shadowing and diffraction, see [4] for a discussion. It shall also be stressed that in *eD* a singlet combination of PDFs is measured which stabilises the QCD evolution of PDFs and thus supports the ambitious extraction of α_s to per mille accuracy [5]. There is no deuteron beam yet foreseen for the LHC, which, however, is a straightforward project of reasonable effort.

4 Study of Nuclear PDFs from LHeC eA Pseudodata

A first analysis has been pursued within the EPS09 framework [10]. It uses the EPS09 data (see Figure 1) and adds for comparison the LHeC neutral current data, keeping the three basic distributions for both valence, the light sea quarks and the gluon. The analysis uses CTEQ6.6 as baseline and the SACOT heavy quark treatment. The default result is shown as dashed band in Figure 3 (top). Accounting for the fact that there are no *lA* DIS data available for $x < 0.01$ (indicated by the vertical line in R_G^A), while the LHeC would cover $x > 10^{-5}$ the assumption on shadowing to approach a constant for $x \to 0$ was replaced by a functional form, see [11], with three free parameters. This provides a more

item	HKN07	EPS09	DSSZ	nCTEQ	LHeC
Reference	Phys. Rev. C76 (2007) 065207	JHEP 0904 (2009) 065	Phys.Rev. D85 (2012) 074028	arXiv: 1509.00792	Workshops + this talk PRD(2030+)
Order pQCD	LO & NLO	LO & NLO	NLO	NLO	NNLO
NC e+A / e+d DIS	√	√	√	√	NC
Drell-Yan ll in p+A / p+d	√	√	√	√	--
RHIC pions in d+Au / p+p		√	√	√	--
Neutrino-nucleus DIS			√		CC
$\sqrt{Q^2}$ cut in DIS	1 GeV	1.3 GeV	1 GeV	2 GeV	free
# of data points	1241	929	1579	740	many
Free parameters	12	15	25	17	O(20)
Error sets available		√	√	√	(y)
Error tolerance $\Delta\chi^2$	13.7	50	30	35	1
Baseline	MRST98	CTEQ6.1	MSTW08	CTEQ6M	None – or ep+eD+eA
Heavy quark treatment	ZM_VFNS	ZM_VFNS	GM_VFNS	GM_VFNS	s,c,b data

Figure 1. Sets of available nuclear parton distributions and, right column, characteristics of the future LHeC determination of nPDFs.

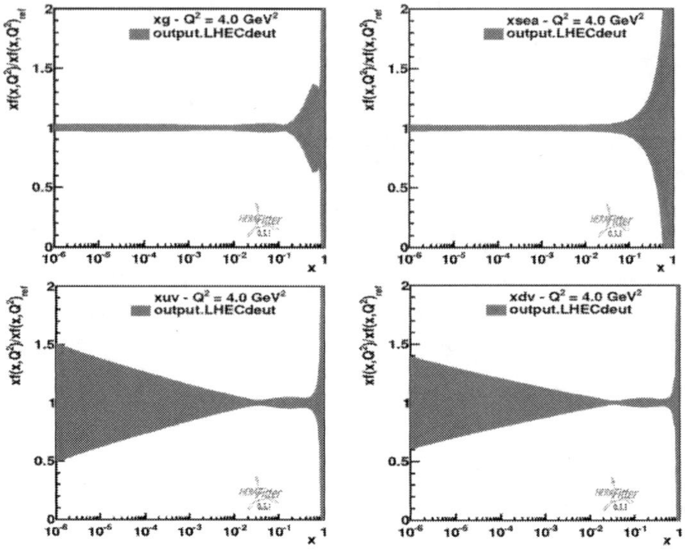

Figure 2. Gluon (left top), sea (right top) and valence quark (down) distribution uncertainties, comprising statistical and experimental uncertainties, from a simulation of electron-deuteron scattering, see text.

appropriate illustration on the inferior control of all R^A's at small x, given by the outer error bands presented in Figure 3. The addition of the LHeC NC data, obtained in e^-A scattering as characterised in Section 2, leads to a striking improvement on all ratios, especially for the sea and gluon distributions. This illustration is only a first step in a detailed study of the potential for determining nPDFs with the LHeC. It had not included the CC data and neither the direct determinations of the heavy quark distributions. Moreover, as illustrated for the case of deuterons above, with such a complete data set, as will be provided with the LHeC, the nPDF determination will not need to employ a proton PDF base anymore but rather and directly determine the nuclear PDFs f_i^A. Then a ratio R will follow from the independent fits of the complete sets f_i^A and of f_i^p. Unlike now these flavour dependent ratios will be fixed with data down to below $x = 10^{-5}$. In this region novel effects are predicted to exist, such as non-linear parton interactions including a possible saturation [12] of the density of gluons at small x.

5 On the Saturation of the Gluon Density at Low x

The prediction of non-linear parton-parton dynamics has long been made in the BFKL formalism. At small x, terms in the expansion containing logarithms of x are calculated to be large and deviations from the linear evolution in Q^2, as described by the DGLAP equations, were expected to be observed but surprisingly not established at HERA. There was, however, a rise of the sea quark and gluon distributions discovered towards low x. HERA had a cms energy squared $s = Q^2_{max}$ available of $10^5\,\text{GeV}^2$.

For the determination of the gluon density xg at small x in DIS one needs to have a lever arm available between, about, 2 and $10\,\text{GeV}^2$ because it is the logarithmic derivative of the structure function $F_2(x, Q^2)$ which determines xg. HERA thus could probe xg to $x > 10^{-4}$, actually twice larger only for one knows from a Taylor expansion of the splitting function that a measurement of F_2 at

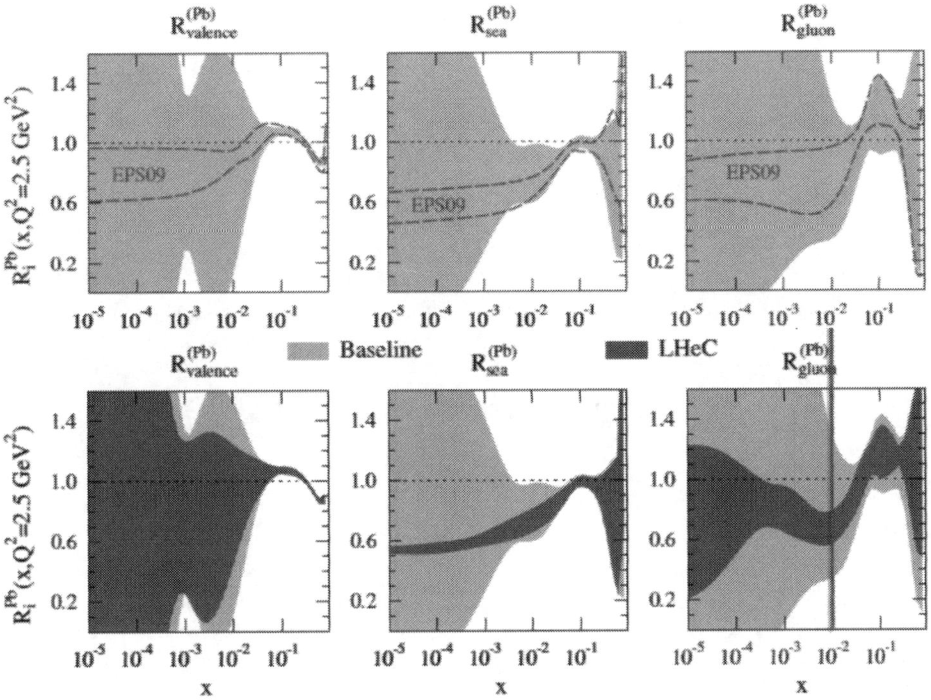

Figure 3. Ratios R_i^{Pb} of i = valence, sea and gluon distributions (from left to right). Top: inner dashed band: estimated uncertainty of R in the default EPS09 framework; outer area: uncertainty estimate with relaxed assumptions (see text). Bottom: outer area as above. inner dark (red) area: uncertainty band with added pseudo LHeC data (NC only), from [11]. The vertical line in R_G^{Pb} at $x_c = 0.01$ indicates that there are no nuclear DIS data available at smaller $x < x_c$ which makes the low x uncertainties drawn in the top panel no more than educated guesses with more (dashed) or less (outer bands) strict assumptions on how much nuclear and proton PDFs may be allowed to differ.

certain value of x provides a constraint on xg at $2x$ [13]. It therefore is required to go to higher beam energies than at HERA if one wanted to establish the effect of gluon saturation [1] in DIS. The most promising procedure has been studied in the CDR of the LHeC: One needs high precision ep data on F_2 but also on F_L in order to fix the gluon distribution and preventing a simple parameterisation effect to screen a saturation phenomenon. Then one needs to measure xg also in eA scattering, as has been illustrated above. From the three independent xg determinations, using F_2^p, F_L^p and F_2^A, one can indeed hope to be able to extract the genuine gluon density $xg(x, Q^2)$, any non-linear corrections and shadowing effects accurately. It should be noted that these measurements and analyses are experimentally and theoretically most demanding and cannot be replaced by interesting control measurements as of diffraction or vector meson production even though these are principally sensitive to $(xg)^2$ while resting on theoretically inferior ground.

[1] At the low energy EIC one often argues that gluon saturation may be seen in eA scattering for the gluon density was expected to be amplified $\propto A^{1/3}$. There are two observations which weaken that argument: i) the gluon density at initial scales Q^2 and low x is observed to be small, of valence type, and an amplification of a very low density hardly can lead to its saturation; ii) if indeed there was a peculiar, saturation-like effect observed in xg at medium x, different from xg as derived in ep at HERA, then one would not know reliably whether that behaviour was resulting from the nuclear environment, a shadowing effect, or a genuine non-linear gg interaction.

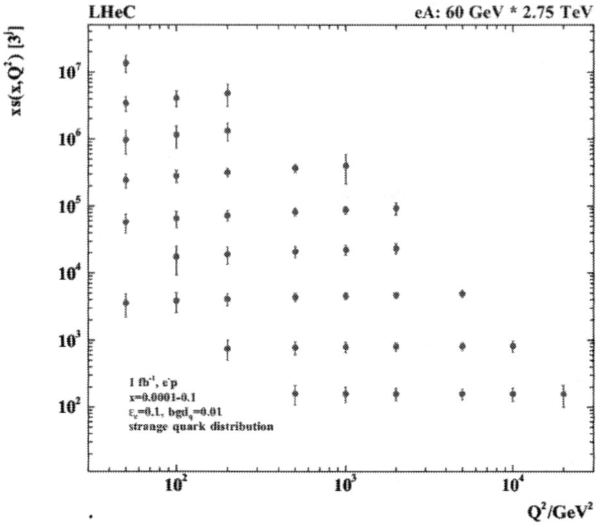

Figure 4. Simulation of the measurement of the (anti)-strange quark distribution $x\bar{s}(x, Q^2)$ in charged current eA scattering through the t-channel reaction $W^- \bar{s} \rightarrow c$. The data are plotted with full systematic and statistical errors added in quadrature.

6 Strange and Heavy Quark PDFs from the LHeC in eA Scattering

The LHeC is the ideal environment to determine the strange, charm and beauty (also the top) PDFs. The principal technique is charm tagging (in CC for xs, in NC for xc) and beauty tagging (in NC for xb). The beam spot of the LHeC has the transverse extension of about $(7\,\mu\text{m})^2$. Modern Silicon detectors have a resolution of a few microns to be compared with typical decay lengths of charm and beauty particles of hundreds of μm. The experimental challenges then are the beam pipe radius, coping at the LHeC with strong synchrotron radiation effects, and the forward tagging acceptance, similar to the HL-LHC challenges.

A study was made of the possible measurements of the anti-strange density (Figure 4) using impact parameter tagging in eA CC scattering, and of the charm and beauty structure functions in NC (Figures 5, 6). Following experience on heavy flavour tagging at HERA and ATLAS, assumptions were made on the charm and beauty tagging efficiencies, to be 10 % and 60 %, respectively. The light quark background in the charm analysis is assumed to be controllable to per cent level, while the charm background in the beauty tagging sample is assumed to be 10 %. The tagging efficiencies and background contaminations affect the statistical error. Moreover, an additional systematic error is assumed in the simulated NC (CC) measurements of 3 (5) %. This results in very promising measurements of the heavier quark distributions: to about 10 − 20 % total uncertainty on the strange and 3 − 5 % on the charm and beauty measurements, for typically x between 10^{-4} and 0.1 and Q^2 extending from below threshold m_Q^2 up to a few times $10^4\,\text{GeV}^2$. The knowledge of the heavy quark densities is of prime relevance for understanding nuclear structure and the development of QCD as has often been emphasised.

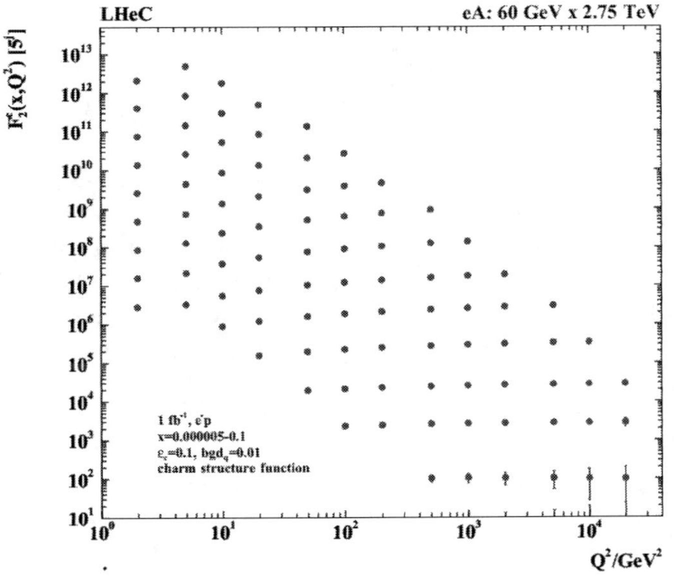

Figure 5. Simulation of the measurement of the charm quark distribution expressed as $F_2^c = e_c^2 x(c + \bar{c})$ in neutral current eA scattering. The data are plotted with full systematic and statistical errors added in quadrature.

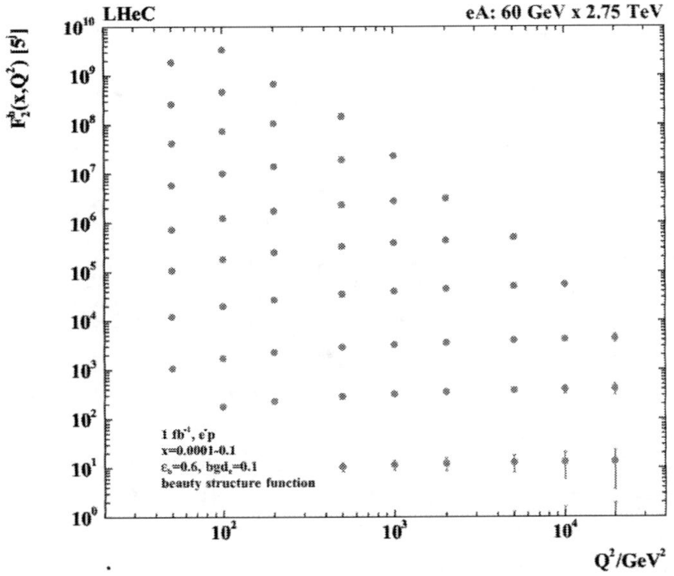

Figure 6. Simulation of the measurement of the bottom quark distribution expressed as $F_2^b = e_b^2 x(b + \bar{b})$ in neutral current eA scattering. The data are plotted with full systematic and statistical errors added in quadrature.

7 Summary

The field of nuclear PDFs is in its infancy while the LHC probes heavy ion physics in *PbPb* and *pPb* highest energy collisions, surprisingly revealing similar collective phenomena. PDFs are determined in deep inelastic scattering. The LHeC will surpass the HERA information in various crucial areas: it will extend the kinematic range in *ep* scattering by a factor of 20, the luminosity by a factor of 10^{2-3} and provide data on electron-lead, and possibly electron deuteron, collisions which extend the current kinematic range in lepton-nucleus scattering by $3-4$ orders of magnitude. This programme will change the whole of nuclear substructure, parton dynamics and the QGP physics dramatically. It has been demonstrated here briefly how large the potential for a novel, complete and accurate determination of full PDF sets in *eD* and *eA* scattering is. The salient features of the LHeC, in contrast to lower energy EIC configurations, are i) the very high Q^2 (and huge range) which makes weak neutral and charged currents a prime tool for separating different flavours in *ep*, *en* and *eA* scattering leading to an unprecedented level of determining nuclear binding effects over 5 orders of magnitude in Bjorken x; and ii) the very low x region covered in the perturbative domain which will discover the long awaited gluon saturation phenomenon, or further defer it to beyond sooner accessible ranges, to possibly the even larger energies considered now with the FCC-eh configuration. It is for these reasons and the here not covered huge variety of important nuclear physics phenomena accessible at energy frontier *eA* scattering, partially presented at this workshop, that the LHeC has been put on the long range map for European nuclear physics [14] a few years ago.

Acknowledgement The present summary describes work jointly performed with Nestor Armesto, Peter Kostka, Hannu Paukkunen and Voica Radescu most gratefully acknowledged. Thanks are due to a large LHeC Collaboration of physicists and engineers engaged in the study of future deep inelastic scattering at the energy frontier. A special *merci* to the organisers of POETIC à Paris.

References

[1] J.C. Collins, D.E. Soper and G.F. Sterman, Adv. Ser. Direct. High Energy Phys. **5** (1989) 1.

[2] For a recent review see: H. Paukkunen, Nucl. Phys. A **926** (2014) 24.

[3] N. Armesto, H. Paukkunen, J. M. Penin, C. A. Salgado and P. Zurita, arXiv:1512.01528.

[4] T. Alexopoulos *et al.*, eD Scattering with H1, A Letter of Intent, DESY 03-194;
H. Abramowicz *et al.*, A New Experiment for HERA, MPP-2003-62;
M. Klein and T. Greenshaw, *J. Phys. G Nucl. Part. Phys.* 28:2505, 2002.

[5] J. Abelleira Fernandez et al., LHeC Study Group,
A Large Hadron Electron Collider at CERN, *J. Phys. G Nucl. Part. Phys.*, 39(7):075001, 2012.

[6] LHeC Webpage: lhec.web.cern.ch.

[7] S.J. Brodsky, LHeC Physics Highlights, LHeC Workshop, Chavannes-de-Bogis, 2015, see [6].
S.J. Brodsky, I. Schmidt and J.J. Yang, Phys. Rev. D **70** (2004) 116003 [hep-ph/0409279].

[8] J. Blümlein and M. Klein, PHE-90-19, unpublished.

[9] M. Klein and V. Radescu, Partons from the LHeC, LHeC-Note-2013-002 PHY, Contribution to Snowmass 2013, unpublished.

[10] K. J. Eskola, H. Paukkunen and C. A. Salgado, JHEP **0904** (2009) 065.

[11] N. Armesto, presented at LHeC Workshop, Chavannes-de-Bogis, 2015, see [6].

[12] L. V. Gribov, E. M. Levin and M. G. Ryskin, Phys. Rept. **100** (1983) 1.

[13] K. Prytz, Phys. Lett. B **311** (1993) 286.

[14] NuPECC: Long Range Plan 2010: Perpectives of Nuclear Physics in Europe, www.nupecc.org

EPJ Web of Conferences **112**, 03003 (2016)
DOI: 10.1051/epjconf/201611203003

Progress on nuclear modifications of structure functions

S. Kumano[1,2]

[1] *KEK Theory Center, Institute of Particle and Nuclear Studies, KEK, 1-1, Ooho, Tsukuba, Ibaraki, 305-0801, Japan*
[2] *J-PARC Branch, KEK Theory Center, Institute of Particle and Nuclear Studies, KEK, and Theory Group, Particle and Nuclear Physics Division, J-PARC Center, 203-1, Shirakata, Tokai, Ibaraki, 319-1106, Japan*

Abstract. We report progress on nuclear structure functions, especially on their nuclear modifications and a new tensor structure function for the deuteron. To understand nuclear structure functions is an important step toward describing nuclei and QCD matters from low to high densities and from low to high energies in terms of fundamental quark and gluon degrees of freedom beyond conventional hadron and nuclear physics. It is also practically important for understanding new phenomena in high-energy heavy-ion collisions at RHIC and LHC. Furthermore, since systematic errors of current neutrino-oscillation experiments are dominated by uncertainties of neutrino-nucleus interactions, such studies are valuable for finding new physics beyond current framework. Next, a new tensor-polarized structure function b_1 is discussed for the deuteron. There was a measurement by HERMES; however, its data are inconsistent with the conventional convolution estimate based on the standard deuteron model with D-state admixture. This fact suggests that a new hadronic phenomenon should exist in the tensor-polarized deuteron at high energies, and it will be experimentally investigated at JLab from the end of 2010's.

1 Introduction

Nuclear structure functions are different from corresponding ones of the nucleon. Such nuclear modifications have been measured from relatively small x (~ 0.005) to large x (~ 0.7) mainly in charged-lepton deep inelastic scattering (DIS). Physics mechanisms for the nuclear effects are different in each x region [1]. At small x, a virtual photon becomes $q\bar{q}$ (or vector mesons), which interacts with a nucleus by strong interactions. It leads to nuclear shadowing phenomena due to multiple interactions in a nucleus. At medium x, there are negative contributions from nuclear binding and possible internal nucleon modifications. The modifications are positive at large x due to nucleon's Fermi motion in a nucleus. Although the major nuclear-modification mechanisms are known, we cannot calculate the structure functions precisely, typically less than 10% accuracy, by the theoretical models. For actual application to heavy-ion physics and neutrino reactions, it is practically not appropriate to reply on such models. For example, neutrino-nucleus cross sections need to be calculated within 5% accuracy in future leptonic CP violation measurements [2]. Therefore, it is desirable to use global analysis results as a model for the nuclear parton distribution functions (NPDFs) [3, 4], in the same way as the global analysis PDFs are used for calculating cross sections at LHC to find physics beyond the standard model.

The NPDFs are determined by analyzing world data on high-energy nuclear reactions, including charged-lepton DIS, neutrino DIS, Drell-Yan, and hadron productions. Currently, there are two major issues in the NPDF studies. One is that gluon shadowing is not determined from the current measurements [3, 4], and the other is that modifications suggested by the neutrino DIS data could be different from the ones which are inferred from the charged-lepton DIS [5]. On the first point, the issue is due partly to the fact that a lepton-nucleus collider similar to HERA does not exist to observe scaling violation at small x. However, the situation could improve because LHC started producing data, which are sensitive to small-x physics. On the second point, somewhat conflicting results are obtained among different analysis groups. It is desirable to clarify the situation by another serious analysis to discuss the details of analysis conditions and data handling. In this report, we discuss the situation of these studies in Sec. 2.

As the second topic, we explain tensor structure functions of the deuteron in charged-lepton DIS with a polarized deuteron [6]. The deuteron structure is well known at low energies and it is described by a proton-neutron S-wave bound state with small D-wave admixture. Since the deuteron is a spin-1 hadron, it has new polarized structure functions, b_1, b_2, b_3, and b_4 in addition to the four structure functions, F_1, F_2, g_1, and g_2, which exist for the spin-1/2 nucleon in the charged-lepton DIS. The structure functions b_1 and b_2 are leading-twist ones, which are related by the Callan-Gross like relation $2xb_1 = b_2$ in the scaling limit. The b_3 and b_4 are higher-twist ones.

There are only a few b_1 data measured by the HERMES collaboration [7]. Although much accurate measurements are needed, there is already an interesting hint in the data toward a new discovery. The most conventional way to estimate b_1 theoretically is to use a convolution model, where b_1 is calculated by an unpolarized structure function of the nucleon with nucleon-momentum distributions including the D wave. The HERMES data are an order-of-magnitude larger than this standard model predication. It indicates that b_1 cannot be understood by the conventional model, and a new hadronic physics should be introduced for interpretation of the HERMES data. Fortunately, an experimental proposal to JLab was approved for b_1, and its measurement will start around the year of 2019 [8]. It is a good opportunity to find a new exotic hadronic phenomenon in the simplest nucleus, deuteron. The deuteron tensor structure used to play an important role in low-energy nuclear physics. It is now time to understand the tensor structure in terms of quark and gluon degrees of freedom with new hadron physics. We explain this current situation as the second topic in Sec. 3.

2 Nuclear parton distribution functions

Nuclear parton distributions are modified from the ones for the nucleon. The space is limited for this article, so that we do not address ourselves to physics mechanisms. An interested reader may read summary articles in Ref. [1]. We focus our discussions on the status of NPDFs and an application to neutrino physics. The NPDFs are determined by a global analysis of world data on high-energy lepton-nucleus, proton-nucleus, and nucleus-nucleus reactions. The distributions are parametrized, and optimum parameters are determined by a χ^2 analysis. There are a few groups which investigate the NPDFs. There are two types for the parametrization. One is to obtain modifications from typical nucleonic PDFs, and the other is to obtain the NPDFs directly. The latter one is a general way of analysis on the same footing with the nucleonic PDF analysis, whereas it is easier to impose physical constrains in the former one because the determined NPDFs could become unphysical distributions due to the lack of data especially at extreme kinematical conditions. If we were to have sufficient experimental data, both results should agree with each other.

There are two major issues in the determination of NPDFs. First, nuclear modifications of the gluon distribution cannot be fixed due to the lack of data which are sensitive to them. Second, some inconsistencies were pointed out for the modifications between charged-lepton and neutrino scattering processes [5], although some other analysis do not show such inconsistency clearly. In the following, we first explain a connection to neutrino physics and it is partially related to the second issue.

Precise NPDFs are needed for finding any new physics in high-energy nuclear reactions such as properties of quark-gluon matters in heavy-ion collisions at RHIC and LHC. There is another important application to neutrino physics. The statistical errors in neutrino-oscillation measurements become smaller and smaller, but their systematic errors are still dominated by neutrino-nucleus interactions, which is an obstacle for a new discovery. There are three major kinematical regions as shown in Fig. 1: quasi-elastic (QE), resonance (RES), and deep inelastic scattering (DIS). There are not definite boundaries in the sense that W^2 and Q^2 cut values depend on researchers. However, a usual choice is the cut $W^2 \geq 4$ GeV^2 for the DIS, and Q^2 should be large enough, typically $Q^2 \geq 1$ GeV^2. These three regions have been investigated by different types of physicists. In the current neutrino experiments, except for the ultra-high energy Ice-Cube experiment, typical energies are from several hundred MeV to a few dozen GeV. Therefore, all of these kinematical cross sections should be precisely known for reducing the systematical uncertainties in neutrino oscillation measurements. For future leptonic CP violation measurements, the cross sections need to be understood within the 5% level. A project is in progress to provide a code to calculate the cross section in any kinematical range by combining theories of different regions at the J-PARC branch of the KEK theory center [2]. In the following, we discuss only the DIS studies and such a unification effort will be explained elsewhere.

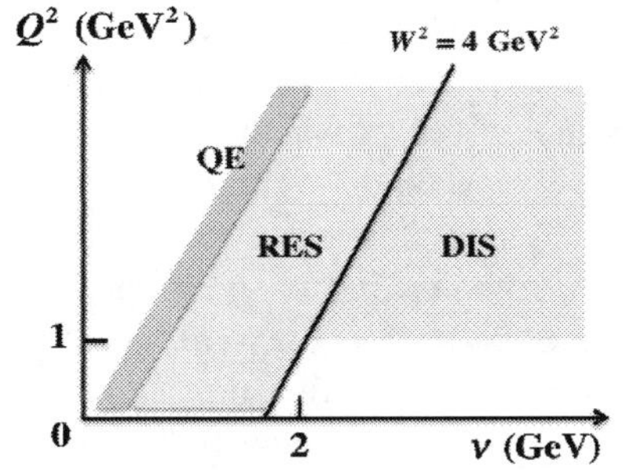

Figure 1. Kinematical regions of neutrino-nucleus scattering. Here, QE, RES, and DIS indicate quasi-elastic, resonance, and deep inelastic scattering.

If there were no nuclear modification, the NPDFs are simple addition of proton and neutron contributions. However, from experimental measurements, nuclear medium effects are typically 10-20% depending on a nuclear size. Such medium effects are parametrized by the function w_i in Ref. [3] as

$$f_i^A(x, Q_0^2) = w_i(x, A, Z) \frac{1}{A} \left[Z f_i^p(x, Q_0^2) + N f_i^n(x, Q_0^2) \right], \tag{1}$$

where p and n indicate the proton and neutron, A, Z, and N are mass number, atomic number, and neutron number, i indicates a type of distribution ($i = u_v,\ d_v,\ \bar{q},\ g$), and Q_0^2 is the initial Q^2 scale. The functions w_i are expressed by a number of parameters, which are then determined by a global analysis. The kinematical range of x is $0 < x < A$ for a nucleus. Therefore, the functional form of Eq. (1) cannot describe the region $x > 1$. However, there is no DIS data in such a large-x region, so it is not a serious problem at this stage. In the HKN analysis [3], the nuclear modification functions are expressed in terms of the parameters, $\alpha, \beta, a_i, b_i, c_i$, and d_i as

$$w_i(x, A, Z) = 1 + \left(1 - \frac{1}{A^\alpha}\right) \frac{a_i + b_i x + c_i x^2 + d_i x^3}{(1 - x)^\beta}, \tag{2}$$

at $Q_0^2 = 1$ GeV^2. The parameters are determined by a global χ^2 analysis with the standard DGLAP evolution equation in comparison with the world data.

As an example, nuclear modifications of ^{40}Ca PDFs and their uncertainties are shown for the next-to-leading (NLO) order HKN parametrization [3] in Fig. 2 at $Q^2 = 1$ GeV^2. There are some differences in the NPDFs from other groups [4] at large x especially for the antiquark and gluon distributions because there are few data which constrain them. In addition, large differences exist in the gluon distributions at small x. The nucleonic gluon distribution is constrained by the scaling violation of F_2 from the HERA ep-collider experiments, whereas there is no such a measurement for nuclei at small x. However, the situation may improve in the near future due to LHC measurements with heavy ions. Using the obtained NPDFs together with DGLAP evolution equations, one can calculate nuclear modifications of structure functions in neutrino scattering. Some inconsistencies were reported between the charged-lepton DIS and neutrino DIS structure functions in Ref. [5]; however, the differences are not clear in other groups [4]. Our analysis is now in progress by including neutrino data for clarifying this issue as an independent trial.

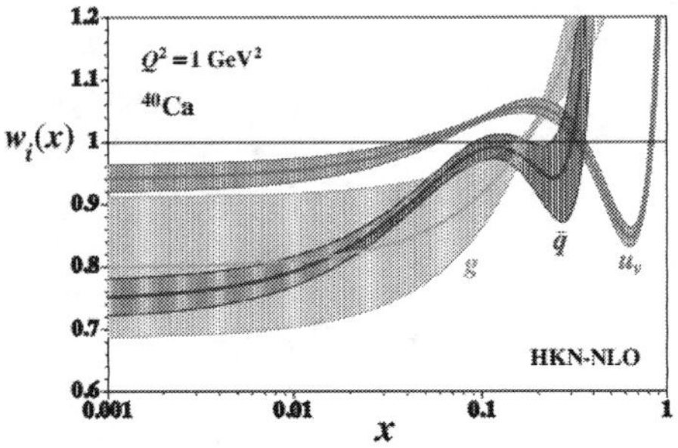

Figure 2. HKN nuclear modifications for ^{40}Ca at $Q^2 = 1$ GeV^2.

3 New spin structure function b_1 for deuteron

There are four new structure functions, b_1, b_2, b_3, and b_4, for the spin-1 hadron in charged-lepton deep inelastic scattering. They are defined in the hadron tensor as [6]:

$$W_{\mu\nu}^{\lambda_f\lambda_i} = -F_1\,\hat{g}_{\mu\nu} + \frac{F_2}{M\nu}\hat{p}_\mu\hat{p}_\nu + \frac{ig_1}{\nu}\epsilon_{\mu\nu\lambda\sigma}\,q^\lambda s^\sigma + \frac{ig_2}{M\nu^2}\epsilon_{\mu\nu\lambda\sigma}\,q^\lambda\,(p\cdot qs^\sigma - s\cdot qp^\sigma)$$
$$- b_1 r_{\mu\nu} + \frac{1}{6}b_2(s_{\mu\nu} + t_{\mu\nu} + u_{\mu\nu}) + \frac{1}{2}b_3(s_{\mu\nu} - u_{\mu\nu}) + \frac{1}{2}b_4(s_{\mu\nu} - t_{\mu\nu}), \quad (3)$$

where p and q are spin-1 hadron and virtual-photon momenta, ν is the energy transfer, and $r_{\mu\nu}$, $s_{\mu\nu}$, $t_{\mu\nu}$, and $u_{\mu\nu}$ are defined by the polarization vector of the spin-one hadron E^μ in Ref. [6]. The spin vector s is expressed by the polarization vector as $(s_{\lambda_f\lambda_i})^\mu = -i\epsilon^{\mu\nu\alpha\beta}E^*_\nu(\lambda_f)E_\alpha(\lambda_i)p_\beta/M$ where the initial and final polarization vectors are denoted as $E^\mu(\lambda_i)$ and $E^\mu(\lambda_f)$ with the spin states λ_i and λ_f, and M is the mass of the spin-1 hadron. The coefficients of b_1, b_2, b_3 and b_4 are defined as symmetric under $\mu \leftrightarrow \nu$ in Eq. (3), and they vanish under the spin average. They are also defined so that b_1 and b_2 are twist-two functions to satisfy the Callan-Gross type relation, $2xb_1 = b_2$. The b_3 and b_4 are higher-twist structure functions. The twist-two structure function b_1 is expressed in terms of the tensor-polarized parton distributions $\delta_T q$ and $\delta_T \bar{q}$ as

$$b_1(x, Q^2) = \frac{1}{2}\sum_i e_i^2\left[\delta_T q_i(x, Q^2) + \delta_T \bar{q}_i(x, Q^2)\right], \quad \delta_T q_i \equiv q_i^0 - \frac{q_i^{+1} + q_i^{-1}}{2}. \quad (4)$$

Here, i indicates the flavor of a quark, e_i is a quark charge, and q_i^λ indicates an unpolarized-quark distribution in the hadron spin state λ. The tensor-polarized distribution is much different from the longitudinally-polarized one $\Delta q = q_\uparrow - q_\downarrow$, where $\uparrow$ and $\downarrow$ indicate the polarization of a quark along the hadron-spin direction, in the sense that it is "unpolarized"-quark distribution in a tensor-polarized spin-1 hadron.

The structure function b_1 can be calculated by the standard convolution picture for the deuteron in terms of parton-momentum distributions convoluted with nucleon momentum distributions. The b_1 is associated with tensor structure of the deuteron, so that D-wave admixture should be properly considered for the convolution estimation. However, it is surprising to find an order-of-magnitude difference between such a standard-model estimate and experimental data obtained by the HERMES collaboration, although the data have large errors. There are other theoretical-model calculations; however, we do not step into such theoretical models in this article, and we discuss probable tensor-polarized distributions from the HERMES data.

Such a parametrization was studied in Ref. [9]. We consider the following constraint in the form of b_1 sum rule [10], which was obtained in a parton model and it is similar to the Gottfried sum:

$$\int dx\, b_1(x) = 0 + \frac{1}{9}\int dx\left[4\,\delta_T\bar{u}(x) + \delta_T\bar{d}(x) + \delta_T\bar{s}(x)\right],$$

$$\int \frac{dx}{x}\left[F_2^p(x) - F_2^n(x)\right] = \frac{1}{3} + \frac{2}{3}\int dx\left[\bar{u}(x) - \bar{d}(x)\right]. \tag{5}$$

As the Gottfried-sum-rule violation created a field of flavor-symmetric light-antiquark distributions ($\bar{u} \neq \bar{d}$) [11], a finite b_1 sum could indicate a finite tensor-polarized antiquark distribution, which cannot be expected from ordinary theoretical ideas. For the time being as the first-step study, we neglect such exotic contributions and consider the condition $\int dx b_1(x) = 0$ for constraining the tensor distributions.

As the first trial, we consider the case that a certain fraction ($\delta_T w$) of unpolarized PDFs is tensor polarized in the deuteron [9]: $\delta_T q_{iv}^D(x) = \delta_T w(x)\, q_{iv}^D(x)$, $\delta_T \bar{q}_i^D(x) = \alpha_{\bar{q}}\,\delta_T w(x)\,\bar{q}_i^D(x)$ for the tensor-polarized parton distributions. Here, we assume a common function $\delta_T w(x)$ for both quarks and antiquarks except for a different overall constant $\alpha_{\bar{q}}$. Now, the problem becomes how to determine $\delta_T w(x)$. We assume that the valence-quark distributions should satisfy the sum $\int dx(b_1)_{\text{valence}} = 0$. Then, the function $\delta_T w(x)$ should have a node at least, so we may take the parametrization $\delta_T w(x) = ax^b(1-x)^c(x_0 - x)$. Here, a, b, and c are the parameters determined by the analysis, and the node position x_0 is expressed by the other parameters as $x_0 = \int dx\, x^{b+1}(1-x)^c(u_v+d_v)/\int dx\, x^b(1-x)^c(u_v+d_v)$ due to the sum rule. Such a node also exists in b_1 calculated by the convolution model with the D-state admixture.

We tried two scenarios depending on whether or not the antiquark tensor polarization exists:

- Set 1: Tensor-polarized antiquark distributions are terminated ($\alpha_{\bar{q}} = 0$).
- Set 2: Finite tensor-polarized antiquark distributions are allowed ($\alpha_{\bar{q}}$ is a parameter).

The parameters are determined by a χ^2 analysis in each case. Determined b_1 and tensor-polarized PDFs are shown in Figs. 3 and 4. There are two curves in Fig. 3 for the sets 1 and 2. It is obvious from this figure, within the restriction of the used parametrization, the HERMES data cannot be explained without the antiquark tensor polarization. Namely, χ^2 is much larger for the set-1 analysis. The obtained tensor distributions are shown in Fig. 4. The dashed curve indicates the valence-quark distribution in the set 1. The solid and dotted curves indicate the valence-quark and antiquark distributions in the set 2. The valence-quark distribution is negative at large x (> 0.3) and then it turns into

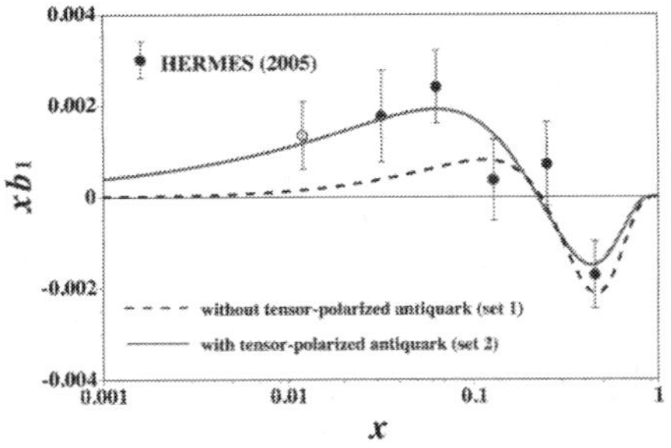

Figure 3. Obtained b_1 and HERMES data [9].

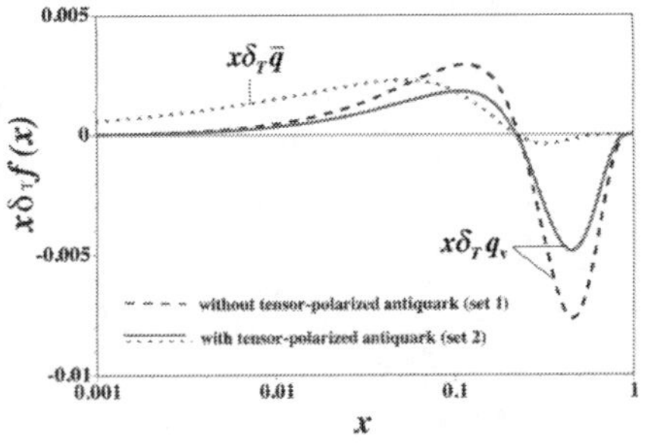

Figure 4. Determined tensor-polarized distributions [9].

positive at $x \simeq 0.2$. The antiquark distributions are especially large at small x (< 0.1). Since there are only a few data points at this stage, there are large uncertainties for the distributions.

In future, we need theoretical and experimental efforts on b_1 physics. On the theory side, we need to understand the obtained tensor-polarized distributions in Fig. 4 by appropriate theoretical models. The standard deuteron model with the D-state admixture cannot explain the distributions because a typical convolution-model estimate is one-order-of-magnitude smaller than the values in Fig. 4. We need to investigate more exotic hadronic mechanisms. At the same time, experimental efforts are needed to measure b_1 much accurately because there are only a few data points as shown in Fig. 3 with the large errors. However, it is fortunate that the JLab b_1 experiment was approved, and the measurement will start in a few years [8]. On the other hand, there are possibilities for investigating the antiquark tensor polarization by polarized Drell-Yan processes, such as $p + \vec{d} \to \mu^+\mu^- + X$ and $\pi + \vec{d} \to \mu^+\mu^- + X$, at hadron facilities, such as Fermilab-MI, RHIC, CERN-COMPASS, J-PARC, GSI-FAIR, and U70 [12]. For example, a tensor spin symmetry in $p\vec{d}$ Drell-Yan is given by

$$A_{UQ_0}(\text{large } x_F) \simeq \frac{\sum_i e_i^2 \, \bar{q}_i(x_1) \, \delta_T q_i(x_2)}{\sum_i e_i^2 \, \bar{q}_i(x_1) \, q_i(x_2)}, \tag{6}$$

so that $\delta_T q_i$ can be measured directly. It is similar to the case that $\bar{u}/\bar{d} \neq 1$ was clarified by the Drell-Yan experiment although it was suggested by the Gottfried-sum-rule violation [11].

Acknowledgements

This work was supported by Ministry of Education, Culture, Sports, Science and Technology (MEXT) KAKENHI Grant No. 25105010.

References

[1] D. F. Geesaman, K. Saito, and A. W. Thomas, Ann. Rev. Nucl. Part. Sci. **45**, 337 (1995); L. Frankfurt, V. Gusey, and M. Strikman, Phys. Rept. **512**, 255 (2012).

[2] See http://nuint.kek.jp/ for activities of neutrino-nucleon interaction collaboration at J-PARC branch of KEK theory center; Y. Hayato, M. Hirai, W. Horiuchi, H. Kamano, S. Kumano, T. Murata, S. Nakamura, K. Saito, M. Sakuda, and T. Sato, to be submitted for publication.

[3] M. Hirai, S. Kumano, and T. -H. Nagai, Phys. Rev. C **76**, 065207 (2007).

[4] K. J. Eskola, H. Paukkunen, and C. A. Salgado, JHEP **04**, 065 (2009); D. de Florian, R. Sassot, P. Zurita, and M. Stratmann, Phys. Rev. D **85**, 074028 (2012); K. Kovarik *et al.*, arXiv:1509.00792.

[5] I. Schienbein *et al.*, Phys. Rev. D **77**, 054013 (2008); D **80**, 094004 (2009); K. Kovarik *et al.*, Phys. Rev. Lett. **106**, 122301 (2011).

[6] S. Kumano, J. Phys.: Conf. Series **543**, 012001 (2014).

[7] A. Airapetian *et al.* (HERMES Collaboration), Phys. Rev. Lett. **95**, 242001 (2005);

[8] Proposal to Jefferson Lab PAC-38, J.-P. Chen *et al.* (2011); K. Slifer, talk at the Tensor Polarized Solid Target Workshop, March 10-12, 2014, JLab, Newport News, USA.

[9] S. Kumano, Phys. Rev. D **82**, 017501 (2010); T.-Y. Kimura and S. Kumano, Phys. Rev. D **78**, 117505 (2008).

[10] F. E. Close and S. Kumano, Phys. Rev. D **42**, 2377 (1990).

[11] S. Kumano, Phys. Rept. **303**, 183 (1998); G. T. Garvey and J.-C. Peng, Prog. Part. Nucl. Phys. **47**, 203 (2001); J.-C. Peng and J.-W. Qiu, Prog. Part. Nucl. Phys. **76**, 43 (2014).

[12] X. Jiang and D. Keller, personal communications; S. Hino and S. Kumano, Phys. Rev. D **59**, 094026 (1999); **60**, 054018 (1999); S. Kumano and M. Miyama, Phys. Lett. B **479**, 149 (2000).

EPJ Web of Conferences **112**, 03004 (2016)
DOI: 10.1051/epjconf/201611203004

Electroweak and BSM Physics at the EIC

K.S. Kumar[1,a], A. Deshpande[1], J. Huang[2], S. Riordan[1], and Y.X. Zhao[1]

[1]*Department of Physics and Astronomy, Stony Brook University, NY 11794-3800, USA*
[2]*Physics Department, Brookhaven National Laboratory, Upton, NY 11973-5000, USA*

Abstract. We discuss the QCD and electroweak physics that becomes accessible by the analysis of semi-leptonic neutral weak amplitudes in polarized electron-light ion collisions at an EIC. Specifically, we discuss the reach for precise weak mixing angle measurements at much higher Q^2 than fixed target measurements, new neutral current spin-independent and -dependent interference structure functions, and searches for $e - \tau$ charged lepton flavor violation.

1 Introduction

The EIC offers a unique new combination of experimental probes given the high center-of-mass energy, high luminosity and the ability to polarize the electron and hadron beams. It is therefore natural to ask whether the envisioned machine parameters could enable new discoveries in the broad subfield of Fundamental Symmetries, which explores the origin and evolution of visible matter in the early universe. One aveune is studies of lepton-hadron electroweak interactions at progressively higher sensitivity. By comparing the measured interaction amplitudes with theoretical predictions within the framework of the Standard Model (SM) of strong, weak and electromagnetic interactions, insights are gained into the symmetries and interactions of matter in the universe at its earliest moments of existence, indirectly accessing energy scales similar to, and sometimes beyond the reach of, the highest energy accelerators currently operational.

Electroweak interaction studies at the EIC can also be used to probe novel aspects of nucleon structure via measurements of spin observables constructed from weak interaction amplitudes mediated by the W and Z bosons. Indeed, some parity-violating observables become accessible that have never before been measured.

In this article, we explore specific observables involving neutral current interactions that relate to the two broad physics themes discussed above.

2 Weak Neutral Currents at low Q^2

A comprehensive strategy to indirectly probe for new high energy dynamics via sensitive tests of electroweak interactions at the intensity frontier must also include precision measurements of flavor-diagonal weak neutral current interactions mediated by the Z^0 boson. For electron-hadron interactions at $Q^2 \ll M_Z^2$, weak neutral current amplitudes are accessed via parity violating asymmetries

[a]e-mail: krishna.kumar@stonybrook.edu

A_{PV}, since pseudoscalar observables sensitive to weak-electromagnetic interference terms can be constructed from the product of vector and axial-vector electron and quark electroweak currents. The parity-violating part of the electron-hadron interaction at $Q^2 \ll M_Z^2$ can be given in terms of phenomenological couplings C_{ij}

$$\mathcal{L}^{PV} = \frac{G_F}{\sqrt{2}}[\bar{e}\gamma^\mu\gamma_5 e(C_{1u}\bar{u}\gamma_\mu u + C_{1d}\bar{d}\gamma_\mu d) + \bar{e}\gamma^\mu e(C_{2u}\bar{u}\gamma_\mu\gamma_5 u + C_{2d}\bar{d}\gamma_\mu\gamma_5 d)]$$

with additional terms as required for the heavy quarks. Here C_{1j} (C_{2j}) gives the vector (axial-vector) coupling to the jth quark.

Within the SM context, each coupling constant is precisely predicted since they are all functions of the weak mixing angle $\sin^2\theta_W$. Over the past two decades, the C_{1i}'s have been measured with steadily improving precision in table-top atomic parity violation experiments and in fixed target parity-violating electron scattering experiments, both in elastic scattering and deep inelastic scattering (DIS), most recently at Jefferson Laboratory (JLab) [1, 2]. Comparing these measurements to SM predictions has produced strong constraints on new high energy dynamics, such as limits on TeV-scale heavy Z' bosons and certain classes of interactions in supersymmetric theories, in a manner complementary to direct searches at colliders [3–6].

A unique feature of DIS A_{PV} measurements is the sensitivity to the C_{2i} coupling constants which involve the amplitudes with axial-vector quark currents. While the C_{2i}'s are kinematically accessible at large scattering angle measurements in fixed target elastic electron scattering, axial-hadronic radiative correction uncertainties cloud the interpretation of the measurements in terms of fundamental electroweak physics. Parity-violating DIS using ^{2}H is the only practical way to measure one combination accurately, namely $2C_{2u} - C_{2d}$. The first ever non-zero measurement of this combination was made recently at JLab [1], and a new experiment called SOLID has been proposed to improve on this measurement by nearly an order of magnitude [7]. Since there is no other way to make accurate measurements of axial-quark couplings, it is worth investigating whether one can make additional measurements at the EIC.

One example of the importance of achieving sensitive constraints on the C_{2i} couplings is a heavy Z' boson (predicted in many SM extensions), which could introduce an additional amplitude and induce a deviation in the measured C_{2i} couplings [8, 9]. A remarkable feature of this amplitude is the fact it is sensitive to the Z' boson even in the case that it might not couple to leptons (so-called lepto-phobic Z'). The limits on the existence of such bosons from other precision weak neutral current measurements as well as from colliders is very weak because all signatures require non-zero lepton-Z' couplings. Note that this amplitude cannot contribute to any tree-level amplitudes nor amplitudes involving the C_{1i} couplings at the quantum loop level. The projected uncertainty from the JLab measurements will be sensitive to a lepto-phobic Z' with a mass $\sim$ 150 GeV, significantly better than the current limit from indirect searches when there is no significant Z-Z' mixing.

The JLab extraction will rely on a simultaneous fit of electroweak couplings, higher-twist effects and violation of charge symmetry to a series of A_{PV} measurements in narrow x and Q^2 bins. It is highly motivated to find ways to improve the sensitivity to the C_{2i} couplings further, given its unique sensitivity for TeV-scale dynamics such as the aforementioned Z' bosons. Apart from statistical reach, the EIC measurements will have the added advantage of being at significantly higher range of Q^2 so that higher-twist effects should be totally negligible.

In the next section, we explore how well we can measure semi-leptonic weak neutral current couplings at the EIC in a more general framework that explores new parity-violating observables comprehensively.

3 Electroweak Physics at the EIC

3.1 Neutral Current Structure Functions

One can write down the most general hadronic tensor lepton-nucleon DIS under the assumption of CP-symmetry including the spin of both the lepton and hadron. For purely electromagnetic scattering at leading twist and without parity-violation, there are four structure functions F_1^γ, F_2^γ, g_1^γ and g_2^γ. With the addition of the neutral weak interaction and parity-violation but with the additional assumption of the Callan-Gross relation, one can now access the interference structure functions $F_1^{\gamma Z}$, $F_3^{\gamma Z}$, $g_1^{\gamma Z}$ and $g_5^{\gamma Z}$ [10–13]. Charged weak interactions can be similarly analyzed and indeed the observables at the EIC from which various W-exchange structure functions can be extracted have been studied [14]. We have begun the analysis required to carry out a similar analysis for weak-electromagnetic interference neutral current structure functions, as discussed below.

3.1.1 Predicted Asymmetries

The DJANGOH generator can be used to simulate DIS lepton-proton (nuclei) scattering with QED and QCD radiative effects treated at at next-to-leading order [15]; this has already been used for the EIC charged current study mentioned above [14]. The R-L asymmetry for inclusive scattering of longitudinal polarized electrons off unpolarized hadrons can be written as (neglecting the pure Z-exchange amplitude):

$$A_{beam} = \frac{G_F Q^2}{2\sqrt{2}\pi\alpha}\Big[g_A^e \frac{F_1^{\gamma Z}}{F_1^\gamma} + g_V^e \frac{Y_-}{2Y_+}\frac{F_3^{\gamma Z}}{F_1^\gamma}\Big]. \tag{1}$$

Here, g_V^e and g_A^e are the vector and axial-vector couplings of the electron to the Z^0 boson. The (+)-(-) asymmetry [1] for inclusive scattering of an unpolarized electron beam on longitudinally polarized hadron can be written as:

$$A_L = \frac{G_F Q^2}{2\sqrt{2}\pi\alpha}\Big[g_V^e \frac{g_5^{\gamma Z}}{F_1^\gamma} + g_A^e \frac{Y_-}{Y_+}\frac{g_1^{\gamma Z}}{F_1^\gamma}\Big], \tag{2}$$

$$g_A^e = -0.5, \qquad g_V^e = -0.5 + 2\sin^2\theta_W, \tag{3}$$

$$Y_- = 2y - y^2, \qquad Y_+ = y^2 - 2y + 2. \tag{4}$$

In the quark-parton model, the structure functions can be written as linear combinations of the unpolarized and polarized parton distribution functions (pdfs). The nice feature of the interference structure functions is that they carry different weights for u-type and d-type quarks as compared to the purely electromagnetic structure functions and therefore it will help in 6-flavor global fits trying to unfold the individual unpolarized and polarized pdfs.

The data in each (x, Q^2) bin can be binned in kinematic variable y and then be fitted to a slope and intercept to cleanly extract $F_1^{\gamma Z}$, $F_3^{\gamma Z}$ (or $g_1^{\gamma Z}$, $g_5^{\gamma Z}$) structure functions. Electron-proton collisions were used to do the projections for the structure functions, while electron-deuteron collisions were used for $\sin^2\theta_W$ projections (since structure function uncertainties cancel for ^{2}H). The projections have taken into account the radiative corrections. Bin migration due to radiative effects was also taken into account with an unfolding technique. The cuts were $Q^2 > 1$ GeV^2, $W_h > 2$ GeV, $y > 0.1$ for structure function studies, with $Q^2 < 6400$ GeV^2 and $x > 0.2$ in addition for $\sin^2\theta_W$ projections. All

[1] The electron direction is designated as the spin (+) direction.

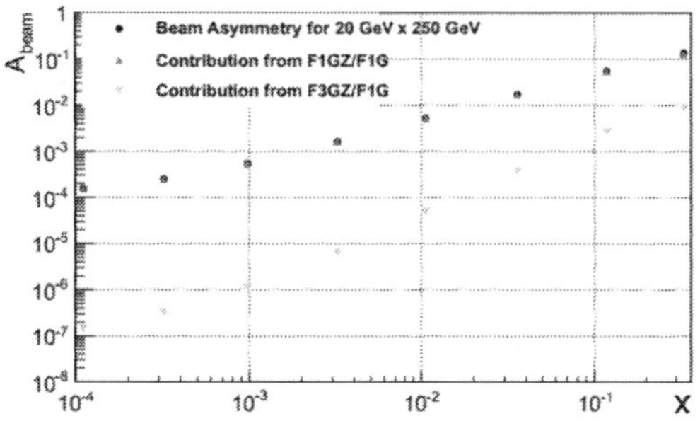

Figure 1. The right-left asymmetry for 20 GeV polarized electrons on 250 GeV unpolarized protons.

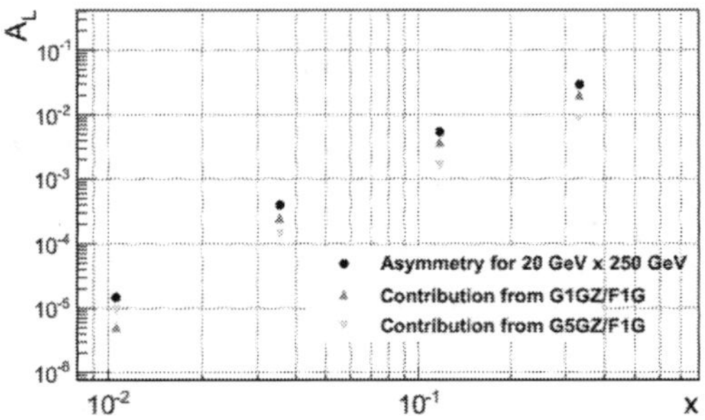

Figure 2. The right-left asymmetry for 20 GeV unpolarized electrons on 250 GeV polarized protons.

the projections were assuming integrated luminosity as 100 fb^{-1} (per nucleon for e-D collisions) and beam (target) polarization as 80 %. We studied four different energy configurations for the electron and hadron energies: 10 GeV x 100 GeV, 10 GeV x 250 GeV, 20 GeV x 250 GeV and 20 GeV x 325 GeV.

Figure 1 and Fig. 2 show the asymmetries as a function of x for the 20 x 250 case for the case of flipping the electron beam and hadron beam helicities respectively. An interesting feature is the fact that while the electron beam asymmetry is mostly sensitive to $F_1^{\gamma Z}$, the hadron beam asymmetry has significant contribution from both $g_1^{\gamma Z}$ and $g_5^{\gamma Z}$.

3.2 The Weak Mixing Angle

The observable with the best sensitivity to cleanly measure coupling constants without significant theoretical uncertainty is A_{beam} in e–^{2}H collisions, averaging over the hadron polarization. From the extracted interference structure function ratios, clean separation of two linear combination of couplings namely $2C_{1u} - C_{1d}$ and $2C_{2u} - C_{2d}$ becomes feasible as a function of Q^2. For this first study, we show the extracted measurement of the weak mixing angle at the appropriate values of Q for three different beam energy scenarios, as shown in Fig. 3 for an integrated luminosity of 400 fb^{-1} per nucleon.

Such projected errors might turn out to be very important when the EIC turns on depending on the outcome of other future measurements indicated in Fig. 3. For example, it is quite possible that the dark matter in the universe has weak couplings to ordinary matter via dark photons, a subject of great current interest. If such dark photons have a small admixture with the Z^0 boson, it is possible that the theoretical prediction shown in Fig. 3 is modified [16]; depending on the mass of the dark boson, this deviation could begin to occur at the Q range accessed by the EIC measurements.

The next steps are to see if the projected errors can be improved by studying electron-proton collisions using different luminosities. In this case, structure function uncertainties might become important. However, it is possible that these uncertainties are small enough at high x so that the required precision can be obtained with significantly lower luminosity and without the requirement of a deuteron beam.

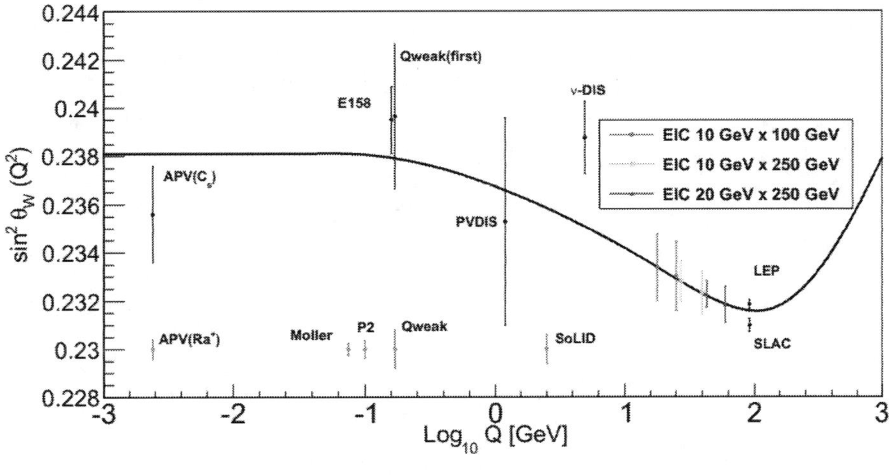

Figure 3. The projected errors on the weak mixing angle are shown at the appropriate average Q values for the case of 400 fb^{-1} per nucleon in e-^{2}H collisions for 3 different beam energy combinations. Also shown are other projected determinations at lower Q anticipated over the next decade.

3.3 Polarized Quark Distributions

From the hadron spin-flip asymmetries, polarized structure functions measurements can be extracted. It can be seen from Figs. 4 and 5 that for an integrated luminosity of 100 fb^{-1}, significant new constraints on the interference structure functions become feasible at $x > 0.1$. The next steps are combine these constraints with the γ and W structure functions and determine whether the new constraints from the inteference structure functions alleviates the need to use semi-inclusive scattering to accomplish a clean 6-flavor separation of the parton distribution functions including polarization.

4 Charged Lepton Flavor Violation

With the discovery of neutrino oscillations, it is natural to ask whether lepton flavor non-conservation can be observed in *charged* lepton interactions. Speculative new theories of the early universe that predict Majorana neutrinos often also predict observable rates of charged lepton flavor violation (CLFV). The most sensitive CLFV searches to date have come from searches for the neutrinoless conversion of stopped muons to electrons in nuclei, searches for the rare decay of a free muon to an electron and photon, and searches for the rare decay of a kaon to an electron and muon. The limits from these processes, though extremely sensitive, all involve the $e \leftrightarrow \mu$ transition. Speculative CLFV theories can predict enhanced rates for $e \leftrightarrow \tau$ transitions.

In lepton-hadron interactions, one could search for the rare cases where an electron converts to a muon or tau lepton, or a muon converts to a tau lepton. The only successful such searches for $e \rightarrow \tau$ transitions have been carried out at the HERA electron-hadron collider experiments ZEUS and H1. The CLFV process could be mediated by a hypothesized new heavy boson known as a leptoquark, which carries both lepton and baryon quantum numbers and appears naturally in many SM extensions such as Grand Unified Theories, supersymmetry, and compositeness and technicolor models (for a concise review, see [17]). The most recent published search by H1 finds no evidence for CLFV $e \rightarrow \tau$

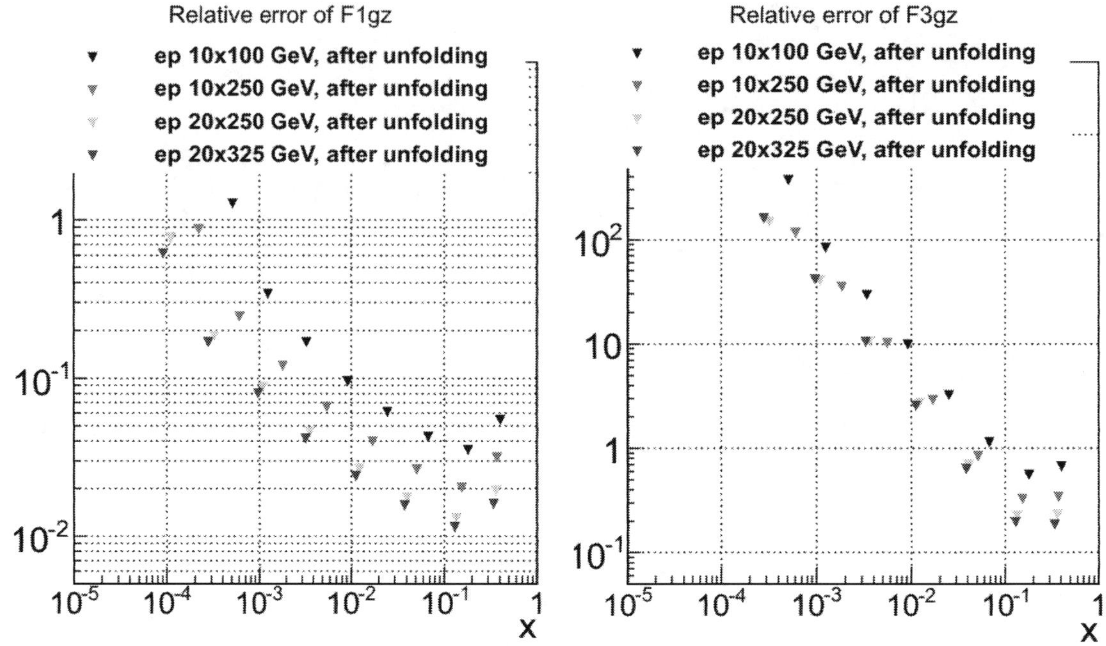

Figure 4. The projected fractional errors on $F_1^{\gamma Z}$ (left) and $F_3^{\gamma Z}$ from A_{beam} measurements using $e-p$ collisions with an integrated luminosity of 100 fb^{-1}.

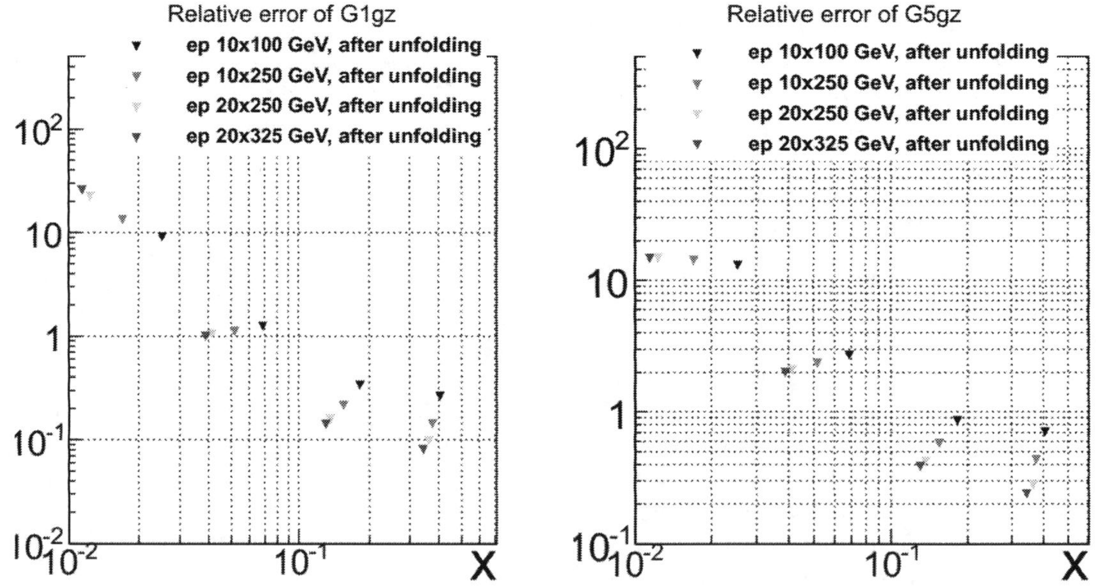

Figure 5. The projected fractional errors on $g_1^{\gamma Z}$ (left) and $g_5^{\gamma Z}$ from A_L measurements using $e-p$ collisions with an integrated luminosity of 100 fb^{-1}.

transitions [18], which can in turn be converted to a limit on the mass and the couplings of leptoquarks in specific SM extensions.

A high energy, high luminosity EIC, with 100 to 1000 times the accumulated luminosity of HERA experiments would allow a large increase in sensitivity. A recent study has shown that an EIC with 90 GeV center-of-mass energy could surpass the current limits with an integrated luminosity of $10fb^{-1}$ [19]. The study also showed that the EIC could compete or surpass the updated leptoquark limits from rare CLFV tau decays for a subset of quark flavor-diagonal couplings. Once the neutral current structure function study is completed, it is planned to carry out full detector simulations and carry out a feasibility study of how to select high energy *tau*-jets in the EIC detector, exploiting both jet reconstruction variables as well as vertex reconstruction to identify displaced vertices.

References

[1] D. Wang et al. (PVDIS), Nature **506**, 67 (2014)
[2] D. Androic et al. (Qweak), Phys. Rev. Lett. **111**, 141803 (2013), 1307.5275
[3] M.J. Ramsey-Musolf, Phys. Rev. **C60**, 015501 (1999), hep-ph/9903264
[4] R.D. Young, R.D. Carlini, A.W. Thomas, J. Roche, Phys. Rev. Lett. **99**, 122003 (2007), 0704.2618
[5] K.S. Kumar, S. Mantry, W.J. Marciano, P.A. Souder, Ann. Rev. Nucl. Part. Sci. **63**, 237 (2013), 1302.6263
[6] J. Erler, C.J. Horowitz, S. Mantry, P.A. Souder, Ann. Rev. Nucl. Part. Sci. **64**, 269 (2014), 1401.6199
[7] J.P. Chen, H. Gao, T.K. Hemmick, Z.E. Meziani, P.A. Souder (SoLID) (2014), 1409.7741
[8] M.R. Buckley, M.J. Ramsey-Musolf, Phys. Lett. **B712**, 261 (2012), 1203.1102
[9] M. González-Alonso, M.J. Ramsey-Musolf, Phys. Rev. **D87**, 055013 (2013), 1211.4581
[10] X.D. Ji, Nucl. Phys. **B402**, 217 (1993)
[11] M. Anselmino, A. Efremov, E. Leader, Phys. Rept. **261**, 1 (1995), [Erratum: Phys. Rept.281,399(1997)], hep-ph/9501369
[12] M. Stratmann, A. Weber, W. Vogelsang, Phys. Rev. **D53**, 138 (1996), hep-ph/9509236
[13] J. Blumlein, N. Kochelev, Nucl. Phys. **B498**, 285 (1997), hep-ph/9612318
[14] E.C. Aschenauer, T. Burton, T. Martini, H. Spiesberger, M. Stratmann, Phys. Rev. **D88**, 114025 (2013), 1309.5327
[15] K. Charchula, G.A. Schuler, H. Spiesberger, Comput. Phys. Commun. **81**, 381 (1994)
[16] H. Davoudiasl, H.S. Lee, W.J. Marciano, Phys. Rev. **D92**, 055005 (2015), 1507.00352
[17] C.H. Albright, M.C. Chen, Phys. Rev. **D77**, 113010 (2008), 0802.4228
[18] F.D. Aaron et al. (H1), Phys. Lett. **B701**, 20 (2011), 1103.4938
[19] M. Gonderinger, M.J. Ramsey-Musolf, JHEP **11**, 045 (2010), [Erratum: JHEP05,047(2012)], 1006.5063

EPJ Web of Conferences **112**, 03005 (2016)
DOI: 10.1051/epjconf/201611203005

Intrinsic bottom and its impact on heavy new physics at the LHC

Florian Lyonnet[1,a]

[1] *Southern Methodist University, Dallas, TX, USA*

Abstract. Heavy quark parton distribution functions (PDFs) play an important role in several Standard Model and New Physics processes. Most analyses rely on the assumption that the charm and bottom PDFs are generated perturbatively by gluon splitting and do not involve any non-perturbative degrees of freedom. On the other hand, non- perturbative, intrinsic heavy quark parton distributions have been predicted in the literature. We demonstrate that to a very good approximation the scale-evolution of the intrinsic heavy quark content of the nucleon is governed by non-singlet evolution equations. This allows to analyze the intrinsic heavy quark distributions without having to resort to a full-fledged global analysis of parton distribution functions. We exploit this freedom to model intrinsic bottom distributions which are so far missing in the literature. We estimate the impact of the non-perturbative contribution to the charm and bottom-quark PDFs and on several important parton-parton luminosities at the LHC.

1 Introduction

Several Standard Model (SM) and New Physics (NP) processes at the CERN Large Hadron Collider (LHC) crucially depend on heavy quark parton distribution functions (PDFs); see for instance [1] for key processes involving the bottom quark PDF.

In the standard approach, the heavy quark distributions are generated *radiatively* via the DGLAP evolution equations, starting with a perturbatively calculable boundary condition. However, a purely perturbative, *extrinsic*, treatment for the heavy quarks might *not* be adequate; several models indeed postulate a non-perturbative, *intrinsic*, heavy quark component e.g. light-cone [2, 3] and meson cloud models [4–6].

Along the years, various global fits have been performed to estimate the amount of intrinsic charm (IC) allowed in the nucleon, see [7] for a recent review. Among them, the two most recent analyses [8, 9] set significantly different limits on the allowed IC contribution highlighting the utility of the techniques discussed in this paper as we can freely adjust the amount of IC/IB contributions without having to regenerate a complete global analysis for each case.

In this contribution, we summarize a technique introduced in [10] which can provide IB PDFs for any generic non-IB PDF set. This approach relies on the fact that the intrinsic bottom PDF evolves (to an excellent precision) according to a standalone non-singlet evolution equation. It is then easy to produce a matched set of IB and non-IB PDFs. Note that obtaining information on the IB content of the nucleon from a global fit is doomed because the data entering global analyses of proton PDFs do not constrain the IB PDFs.

[a] e-mail: flyonnet@smu.edu

Sec. 2, demonstrates that to a good approximation the scale-evolution of the intrinsic PDF is governed by a non-singlet evolution equation; it also provide a set of matched IC/IB PDFs. The IB PDFs are then used in Sec. 3 to obtain predictions for parton–parton luminosities relevant at the LHC. Finally, our results and conclusions are summarized in Sec. 4.

2 Parton distribution functions for intrinsic heavy quarks

2.1 Definition and Evolution

In the context of a global analysis of PDFs the different parton flavors are specified via a boundary condition at the input scale μ_0 which is typically of the order $O(1\text{ GeV})$. The PDFs at higher scales $\mu > \mu_0$ are then obtained by solving the DGLAP evolution equations with these boundary conditions. Thus, a non-perturbative (intrinsic) heavy quark distribution Q_1 can be defined at the input scale μ_0 as the difference of the full boundary condition for the heavy quark PDF Q and the perturbatively calculable (extrinsic) boundary condition Q_0.

The Dokshitzer-Gribov-Lipatov-Altarelli-Parisi (DGLAP) evolution equations read $\dot{f}_i = P_{ij} \otimes f_j$, where $f_i = \begin{pmatrix} g \\ q \\ b \end{pmatrix}$

$$
\begin{aligned}
\dot{g} &= P_{gg} \otimes g + P_{gq} \otimes q + P_{gQ} \otimes Q_0 + \cancel{P_{gQ} \otimes Q_1}\,, && (1)\\
\dot{q} &= P_{qg} \otimes g + P_{qq} \otimes q + P_{qQ} \otimes Q_0 + \cancel{P_{qQ} \otimes Q_1}\,, && (2)\\
\dot{Q}_0 + \dot{Q}_1 &= P_{Qg} \otimes g + P_{Qq} \otimes q + P_{QQ} \otimes Q_0 + P_{QQ} \otimes Q_1\,, && (3)
\end{aligned}
$$

where vector of light quarks is denoted 'q' and the heavy quark distribution by 'Q' (where $Q = c$ or $Q = b$). Note that we substituted $Q = Q_0 + Q_1$ where Q_0 denotes the usual radiatively generated extrinsic heavy quark component and Q_1 is the non-perturbative intrinsic heavy quark distribution.[1]

If one neglects the tiny contribution coming from the crossed out terms the system of evolution equations can be separated into two independent parts.

For the system of gluon, light quarks and extrinsic heavy quark (g, q, Q_0) one recovers the same evolution equations as in the standard approach without an intrinsic heavy quark component while the intrinsic heavy quark distribution, Q_1, follows a standalone non-singlet evolution equation, $\dot{Q}_1 = P_{QQ} \otimes Q_1$.

It is then apparent that it is possible to entirely decouple the analysis of the intrinsic heavy quark distribution from the rest of the system by allowing for a small violation of the sum rule. Therefore, PDFs for the gluon, the light quarks and the extrinsic heavy quark can be taken from a global analysis in the standard approach where they already saturate the momentum sum rule. On top of these PDFs the intrinsic heavy quark PDF can be determined in a standalone analysis using the non-singlet evolution equation.

The violation of the momentum sum rule induced by the term $\int_0^1 \mathrm{d}x\, x \left(Q_1 + \bar{Q}_1\right)$ is, however, very small for bottom quarks[2]. Numerical checks of the validity of our approximations are performed below.

[1]Strictly speaking, the decomposition of Q into Q_0 and Q_1 is defined at the input scale where the calculable boundary condition for Q_0 is known. Consequently, $Q_1 := Q - Q_0$ is known as well. Only due to the approximations in Eqs. (1) and (2) it is possible to entirely decouple Q_0 from Q_1 so that the decomposition becomes meaningful at any scale.

[2]It is also acceptable in case of charm provided that the allowed normalization of IC is not too big.

2.2 Boundary condition for intrinsic heavy quarks

The x-dependence of the intrinsic charm (IC) parton distribution function is predicted by the BHPS model [2]to be:

$$c_1(x) = \bar{c}_1(x) \propto x^2[6x(1+x)\ln x + (1-x)(1+10x+x^2)] \,. \tag{4}$$

We expect the x-shape of the intrinsic bottom distribution $b_1(x)$ to be very similar to the one of the intrinsic charm distribution. Furthermore, the normalization of IB is expected to be parametrically suppressed with respect to IC by a factor $m_c^2/m_b^2 \simeq 0.1$. To fix the freedom related to the scale of the boundary condition we use in the following the *Same Scales* boundary condition, which remains valid at any scale Q: $b_1(x, m_c) = \frac{m_c^2}{m_b^2} c_1(x, m_c)$. Let us also note that in this approach it would be no problem to work with asymmetric boundary conditions, $\bar{c}_1(x) \neq c_1(x)$ and $\bar{b}_1(x) \neq b_1(x)$, as predicted for example by meson cloud models [5].

2.3 Intrinsic heavy quark PDFs from non-singlet evolution

The initial x-dependence at the scale of the charm mass is defined via Eq. (4) and the normalization is fixed to match the one predicted by the CTEQ6.6c0 fit. The IB PDF was generated using the *Same Scales* boundary conditions (see above) together with the same x-dependent input of Eq. (4). Both PDFs were then evolved according to the non-singlet evolution equation and the corresponding grids were produced. Note that because in our approximation, the evolution of the intrinsic charm and bottom PDFs is completely decoupled, the normalization of our PDFs can be easily changed by means of simple rescaling.

2.4 Numerical validation

We use the CTEQ6.6c series of intrinsic charm fits to test the ideas presented above, and in particular CTEQ6.6c0 and CTEQ6.6c1 which employ the BHPS model with 1% and 3.5% IC probability, respectively.[3] In the following we compare CTEQ6.6c0 and CTEQ6.6c1 sets where IC has been obtained from global analysis to our approximate IC PDFs supplemented with the central CTEQ6.6 fit, which has a radiatively generated charm distribution.

In Fig. 1(a), one can see that at low Q^2 the difference between the sum $c_0 + c_1$ and the CTEQ6.6c0 charm distribution is tiny, and smaller than 5% at the higher Q^2.

The inclusion of the intrinsic charm distribution will alter the other parton distributions, most notably the gluon PDF. In order to gauge this effect, we compare in Fig. 1(b) the gluon distribution from the standard CTEQ6.6 fit with the one from the CTEQ6.6c0 analysis. For small x ($x < 0.1$) the gluon PDF is not affected by the presence of a BHPS-like intrinsic charm component which is concentrated at large x. We note that at large-x, where most of the difference lies, the gluon distribution is already quite small and the uncertainty of the gluon PDF is sizable (of order of 40 – 50% for the CTEQ6.6 set).

We conclude that for most applications, adding a standalone intrinsic charm distribution to an existing standard global analysis of PDFs is internally consistent and leads only to a small error. Moreover, for the case of intrinsic bottom which is additionally suppressed, the accuracy of the approximation will be even better. For completeness we also provided similar validation using the parton–parton luminosities in [10] to which we refer the reader for further details.

[3]This corresponds to the values of 0.01 and 0.035 of the first moment of the charm PDF, $\int dx\, c(x)$, calculated at the input scale $Q_0 = m_c = 1.3$ GeV.

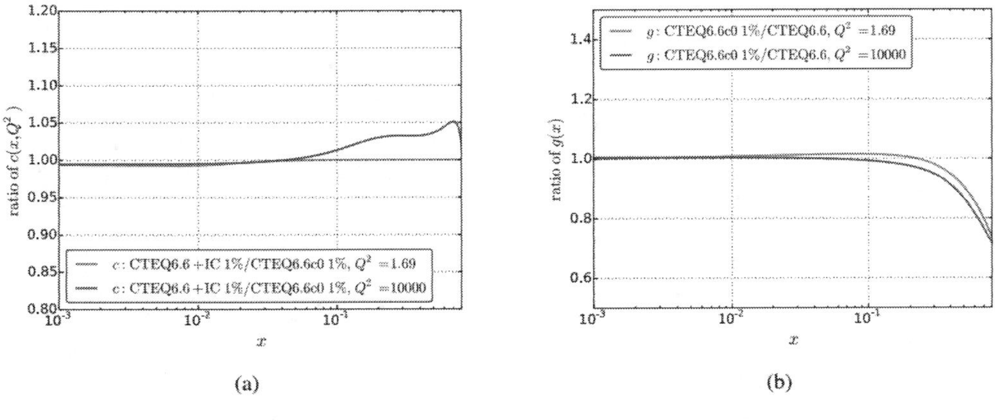

Figure 1. (a) Ratio of the CTEQ6.6 + IC 1% and CTEQ6.6c0 (1%) charm distributions. (b) Ratio of the CTEQ6.6c0 and the CTEQ6.6 gluon distributions. The results are shown as function of x for two scales, Q^2 = 1.69 and Q^2 = 10000 GeV^2.

3 LHC observables: possible effects of IC/IB

We study the effects of both IC and IB on parton–parton luminosities, $\frac{d\mathcal{L}_{ij}}{d\tau}$, at LHC with $\sqrt{S} = 14$ TeV. We define the luminosity as

$$\frac{d\mathcal{L}_{ij}}{d\tau}(\tau,\mu) = \frac{1}{1+\delta_{ij}}\frac{1}{\sqrt{S}}\int_\tau^1 \frac{dx}{x}\left[f_i(x,\mu)f_j(\tau/x,\mu) + f_j(x,\mu)f_i(\tau/x,\mu)\right], \tag{5}$$

where $\tau = x_1 x_2$ and f_i are parton distribution functions. This allows us to assess the relevance of a non-perturbative heavy quark component for the production of new heavy particles coupling to the SM fermions.

To explore how the presence of IC and IB would affect physics observables with a non-negligible heavy quark initiated subprocesses, in Fig. 2 we show the ratios of luminosities for charm and bottom with and without an intrinsic contribution for 1% and 3.5% normalizations. Furthermore, since there are no experimental constraints on the IB normalization, in Fig. 2 (right), we also include an extreme scenario where we remove the usual m_c^2/m_b^2 factor; thus, the first moment of the IB is 1% at the initial scale m_c. As expected, in the case of IB the effect is smaller. However, for the $b\bar{b}$ luminosity a 3.5% normalization of IB leads to a curve which lies clearly above the error band of the purely perturbative result. In the extreme scenario (which is not likely but by no means excluded) the IB component has a big effect on the $b\bar{b}$.

Note that similar results for the cg and bg luminosities have been obtained but are not shown here, see [10].

4 Conclusions

This contribution presented a new method to generate a matched IC/IB distributions for any PDF set without the need for a complete global re-analysis. This renders easy to carry out a consistent analysis including intrinsic heavy quark effects. In addition, because the evolution equation for the intrinsic heavy quarks decouples, one can freely adjust the normalization of the IC/IB PDFs.

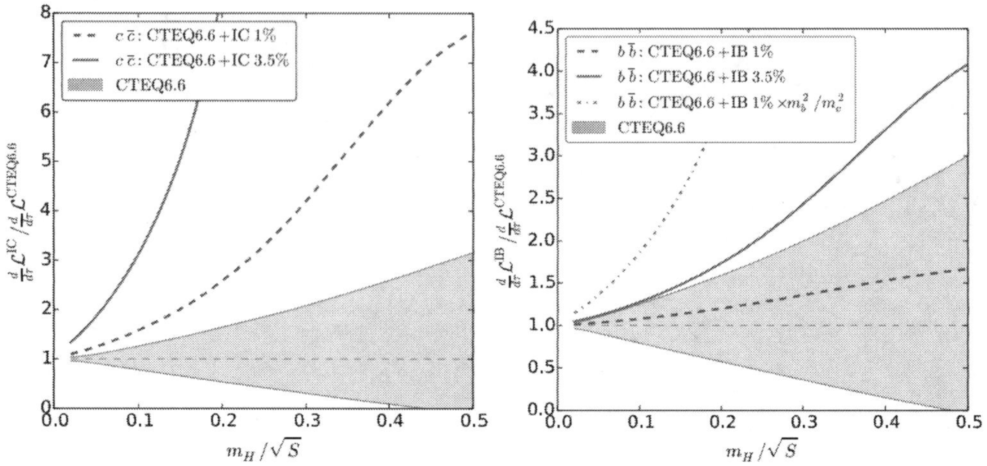

Figure 2. Ratio of $c\bar{c}$ luminosities (left) and $b\bar{b}$ luminosities (right) at the LHC14 for charm(bottom)-quark PDF sets with and without an intrinsic component as a function of $\sqrt{\tau} = m_H/\sqrt{S}$. The ratio for the $c\bar{c}$ ($b\bar{b}$) luminosity (solid, green line) in the left (right) figure reaches values of 50 (17) at $\sqrt{\tau} = 0.5$. In addition to the curves with 1% normalization (red, dashed lines) we include the results for the 3.5% normalization (green, solid lines) which was found to be still compatible with the current data [11].

We showed that our approximation holds to a very good precision for the IB. For the IC, the error is larger (because the IC increases), yet our method is still useful. Indeed, for an IC normalization of 1-2%, the error is less than the PDF uncertainties at the large-x where the IC is relevant. If the normalization is larger, although the error may be the same order as the PDF uncertainties, the IC effects also grow and can be separately distinguished from the case without IC. In any case, the IC/IB represents a non-perturbative systematic effect which should be accounted for.

The method presented here greatly simplifies our ability to estimate the impact of the intrinsic heavy quark effects on the new physics searches. It can also be very useful in searching for and constraining the intrinsic charm and bottom components of the nucleon by itself. In particular in the future facilities such as an Electron Ion Collider (EIC), the Large Hadron-Electron collider (LHeC), or AFTER@LHC.

The PDF sets for intrinsic charm and intrinsic bottom discussed in this analysis (1% IC, 3.5% IC, 1% IB, 3.5% IB) are available from the authors upon request.

References

[1] F. Maltoni, G. Ridolfi, M. Ubiali, JHEP **1207**, 022 (2012), 1203.6393

[2] S.J. Brodsky, P. Hoyer, C. Peterson, N. Sakai, Phys. Lett. **B93**, 451 (1980)

[3] S.J. Brodsky, C. Peterson, N. Sakai, Phys.Rev. **D23**, 2745 (1981)

[4] F.S. Navarra, M. Nielsen, C.A.A. Nunes, M. Teixeira, Phys. Rev. **D54**, 842 (1996), hep-ph/9504388

[5] S. Paiva, M. Nielsen, F.S. Navarra, F.O. Duraes, L.L. Barz, Mod. Phys. Lett. **A13**, 2715 (1998), hep-ph/9610310

[6] W. Melnitchouk, A.W. Thomas, Phys. Lett. **B414**, 134 (1997), hep-ph/9707387

[7] S. Brodsky, A. Kusina, F. Lyonnet, I. Schienbein, H. Spiesberger et al. (2015), 1504.06287

[8] J. Gao, M. Guzzi, J. Huston, H.L. Lai, Z. Li et al., Phys.Rev. **D89**, 033009 (2014), 1302.6246
[9] P. Jimenez-Delgado, T. Hobbs, J. Londergan, W. Melnitchouk, Phys.Rev.Lett. **114**, 082002 (2015), 1408.1708
[10] F. Lyonnet, A. Kusina, T. Ježo, K. Kovařík, F. Olness et al. (2015), 1504.05156
[11] P.M. Nadolsky et al., Phys. Rev. **D78**, 013004 (2008), 0802.0007

EPJ Web of Conferences **112**, 03006 (2016)
DOI: 10.1051/epjconf/201611203006

Frontiers of QCD with Precision nPDFs *

Aleksander Kusina[1,a], Florian Lyonnet[2,b], Fredrick I. Olness[2,c,d], and Ingo Schienbein[1,e]

[1] *LPSC, Université Joseph Fourier/CNRS-IN2P3/INPG, UMR5821, Grenoble, F-38026, France*
[2] *Southern Methodist University, Dallas, TX 75275, USA*

Abstract. Searches for new physics will increasingly depend on identifying deviations from precision Standard Model (SM) predictions. Quantum Chromodynamics (QCD) will necessarily play a central role in this endeavor as it provides the framework for the parton model. However, as we move to higher orders and into extreme kinematic regions, we begin to see the full complexities of the QCD theory. Recent theoretical developments improve our ability to analyze both proton and nuclear PDFs across the full kinematic range. These developments are incorporated into the new nCTEQ15 PDFs, and we review these developments with respect to future measurements, and identify areas where additional effort is required.

1 The Character and Consequences of QCD[1]

"QCD is our most perfect physical theory." [2] Its scope is wide ranging spanning nuclear physics, particle physics, cosmology, and astrophysics. It also contains a wealth of phenomena including: "radiative corrections, running couplings, confinement, spontaneous (chiral) symmetry breaking, anomalies, and instantons"[1].

QCD is both a force whose properties we want to study, and also a tool we can use to link experimentally measured hadronic cross sections to theoretically calculable cross sections in the context of the parton model. The basic statement of the parton model is: $\sigma_{A\to X} = f_{a/A} \otimes \widehat{\sigma}_{a\to X}$ where we factorize the complex non-perturbative QCD interactions into the Parton Distribution Functions (PDFs) $f_{a/A}$ and the theoretical partonic cross section $\widehat{\sigma}_{a\to X}$. Here, $\sigma_{A\to X}$ is the physically measurable cross section, and A can represent either a proton or a nucleon.

As we push our calculations to higher precision and into extreme kinematic regions (such as low-Q^2 or hi/low-x) we may encounter a variety of phenomena such as those illustrated schematically in Fig.1. While many of these phenomena can be neglected at lower orders and lower precision, if

---

*Based on work in collaboration with D. B. Clark, E. Godat, T. Ježo, C. Keppel, K. Kovařík, J.G. Morfín, P. Nadolsky, J.F. Owens, & J.Y. Yu. This work is supported in part by the U.S. Department of Energy under grant DE-FG02-04ER41299, and by Projet international de cooperation scientifique PICS05854 between France and the U.S.

[a] e-mail: kusina@lpsc.in2p3.fr
[b] e-mail: flyonnet@smu.edu
[c] e-mail: olness@smu.edu
[d] Presenter
[e] e-mail: schien@lpsc.in2p3.fr

[1] In this short presentation we provide only an abbreviated set of references.
[2] *What QCD Tells Us About Nature – and Why We Should Listen*, by Frank Wilczek.[1]

we are to properly analyze the new generation of high-precision high-statistics experiments, we must consider the potential impact of these manifestations of the QCD theory.

The recent performance of the LHC has exceeded expectations and produced an unprecedented number of events to be analyzed. However, to advance our understanding of these processes and the inherent aspects of the QCD theory requires both improved theoretical tools *and* complementary data from other experiments such as an Electron Ion Collider (EIC) and Large Hadron electron Collider (LHeC). In the same manner that the combination of the HERA ep and Tevatron $p\bar{p}$ data provided maximum knowledge of the underlying physics, the combination of the LHC and future lepton-hadron data can further expand our search for new physics and our understanding of QCD.

To carry out this program requires both the above data, and also improved theoretical tools to perform the analysis. We briefly review some of the recent progress, and identify areas where additional effort is required.

1.1 Nuclear PDFs

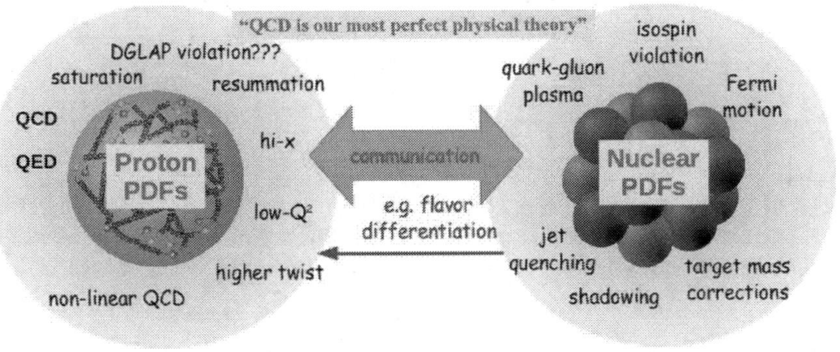

Figure 1. We schematically represent the variety of phenomena which may enter when considering proton and nuclear PDFs at high precision. We emphasize that the nuclear PDFs and corrections are important for flavor differentiation.

Historically, the nuclear PDFs (nPDFs) and the associated nuclear correction factors have played a key role in the flavor differentiation of the proton PDFs. The common processes used to extract the proton PDFs were Deeply Inelastic Scattering (DIS), Drell-Yan (DY) lepton-hadron scattering, and jet production. These processes involve different linear combinations of the partonic flavors, and the neutrino-nucleon DIS processes are especially valuable as we can extract the four quantities $\{F_2^{\nu N}, F_2^{\bar{\nu}N}, F_3^{\nu N}, F_3^{\bar{\nu}N}\}$. Due to the small neutrino cross section, these measurements are usually performed on heavy nuclear targets; hence, a detailed understanding of the nPDFs and their corrections are essential to constraining the flavor structure of both the nuclear and proton PDFs.

To comprehensively analyze both proton and nuclear data in a single unified framework, we have developed the nCTEQ project which can combine both data sets (including uncertainties) into the fit. The nCTEQ15 set of nPDFs[2] is illustrated in Fig.2, and this extends the CTEQ proton PDFs fits to include the nuclear dependence across the full A range from proton up to ^{208}Pb.

1.2 The Strange Quark

To illustrate the challenges of fitting the nPDFs, in Fig.3 we display the effective nuclear correction factors R on iron for neutral-current charged-lepton ($\ell^{\pm}$) DIS and charged-current neutrino ($\nu, \bar{\nu}$) DIS.

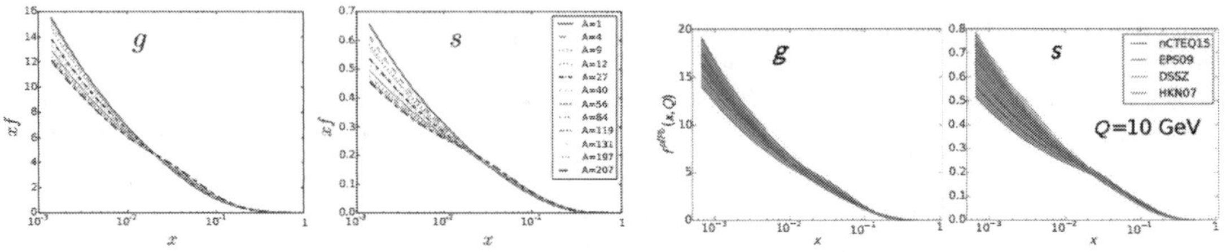

Figure 2. a) The nCTEQ15 PDFs for a range of nuclear A values.[2] b) The nPDF uncertainty bands, together with HKN07,[3] EPS09,[4] and DSSZ.[5]

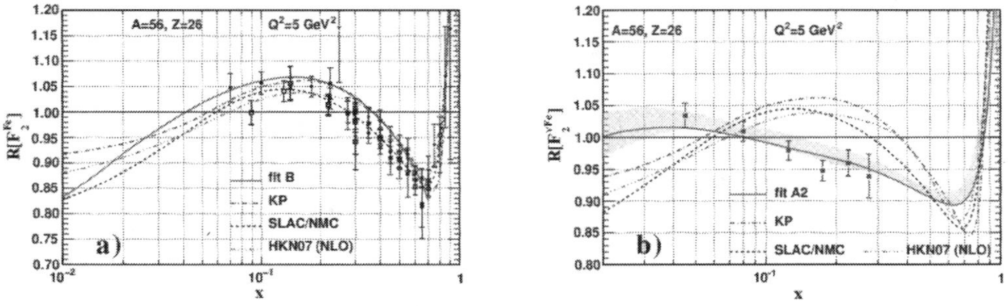

Figure 3. The computed nuclear correction ratio, $R = F_2^{Fe}/F_2^{D}$ as a function of x. Figure-a) shows the fit using the neutral-current lepton ($\ell^{\pm}N$) DIS data (KP, SLAC/NMC, HKN07). Figure-b) shows the fit using the charged-current neutrino (νN) DIS data; the data are from the NuTeV experiment. See Ref. [6]

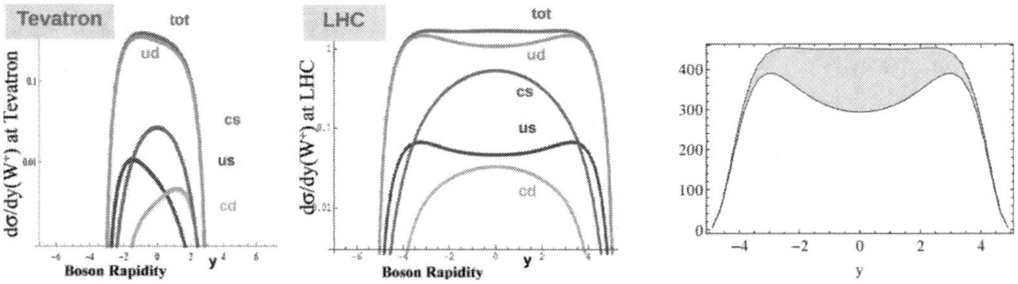

Figure 4. a) b) The leading-order (LO) differential cross section ($d\sigma/dy$) for W^{+}production at the Tevatron (2 TeV) and the LHC (14 TeV) as a function of rapidity. The partonic contributions are also displayed for $\{u\bar{d}, c\bar{s}, u\bar{s}, c\bar{d}\}$. The vertical scales are logarithmic. c) The strange quark contribution (yellow) as a fraction of the total $d^2\sigma/dM/dy$ in pb/GeV for pp to W^{+} production at the LHC for 14 TeV with CTEQ6.6 using the VRAP program at NNLO. C.f. Ref. [7] for details.

The R values differ, and this uncertainty then feeds into the individual flavors such as the strange quark $s(x)$. Prior to the LHC data, our knowledge of $s(x)$ came primarily from νFe DIS, so the uncertainty of the nuclear correction was an important consideration. Consequently, improvements in the nuclear correction factor R will yield improved proton PDFs.

At the higher energy scales of the LHC, the (relatively) heavier parton flavors (such as the strange) play a more prominent role. This is illustrated in Fig.4 where we find the $c\bar{s} \to W^{+}$ contribution is much larger than at the Tevatron (for example) where the $u\bar{d}$ process dominated. We can then turn this issue around and use the LHC data to impose constraints on the strange quark.[8]

In this instance, the combination of fixed-target νN DIS data together with LHC W/Z production data can improve constraints of both the strange PDF and the nuclear correction factors.

1.3 Charm and Bottom Quark

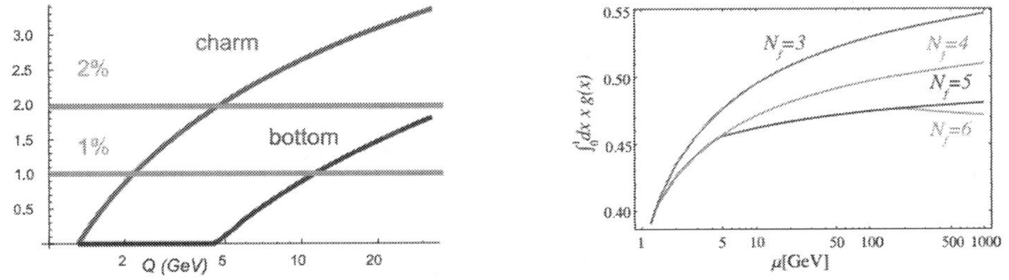

Figure 5. a) Momentum fraction of the charm and bottom PDFs as a function of the scale. b) Momentum fraction of the gluon for $N_F = \{3, 4, 5, 6\}$ as a function of the scale.

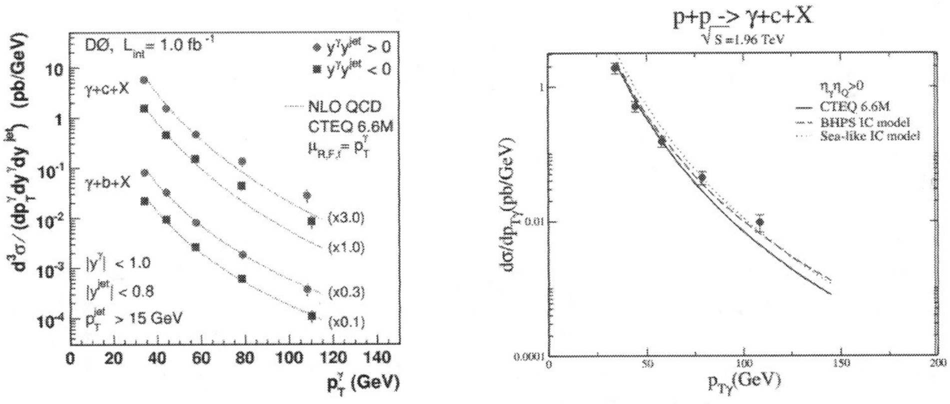

Figure 6. a) Cross section for associated production of γ and c/b.[9] b) Same with an intrinsic charm component.[10]

Considering the heavier charm and bottom quarks, these enter the PDF for energy scales above the quark mass values; as the c and b PDFs are primarily produced by gluon splitting ($g \to c\bar{c}$ and $g \to b\bar{b}$), these affect the momentum fraction carried by the gluon (*c.f.*, Fig.5).

While we generally assume that there is no primordial component of the c and b PDFs for energy scales below the quark mass, it has been postulated that there could be an additional "intrinsic" component which would be added to the usual "extrinsic" component (which comes from gluon splitting). There are a number of measurements where the "intrinsic" component could manifest itself; see Ref. [11] for a recent review.

Fig.6-a shows the data for $\gamma + c \to X$ is above the theory prediction at the largest p_T^γ values; an additional "intrinsic" component (Fig.6-b) can shift the theory in the direction of the data at large p_T^γ.

While this observation is intriguing, additional precision and analysis is required to definitively claim evidence of an "intrinsic" component.

On a related topic, recent theoretical work demonstrated that (to a good approximation) the evolution of the intrinsic heavy quark component is governed by non-singlet evolution equations which can be decoupled from the evolution of the remaining PDFs.[12] The observation allows an (approximate) intrinsic component to easily be included with any calculation without the need for a specially evolved PDF set; this greatly simplifies our ability to search for, and place constraints upon, intrinsic charm and bottom components of the nucleon.

1.4 Higher Order Calculations

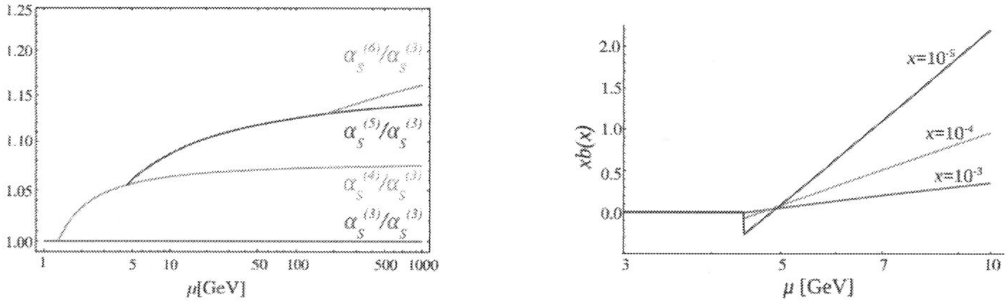

Figure 7. a) Ratios of $\alpha_s(\mu)$ for $N_F = \{3,4,5,6\}$. b) The discontinuity in the b-quark PDF $f_b(x,\mu)$ at NNLO *vs.* μ for $x = \{10^{-3}, 10^{-4}, 10^{-5}\}$.

Recent advances in the automation of higher order calculations have tremendously increased the breadth of processes which are available at NNLO and beyond. As we move to these higher orders, our PDFs and $\alpha_s(\mu)$ functions are discontinuous[3] when we shift the number of flavors from N_F to $N_F + 1$; *c.f.* Fig.7. An elegant property of QCD is that despite the (fixed order) discontinuity of the PDFs and α_s, the discontinuity of any physical quantity will systematically decrease order by order in perturbation theory; that is to say, at order k we have the relation: $\sigma \simeq f_{N_F} \otimes \hat{\sigma}_{N_F} = f_{N_F+1} \otimes \hat{\sigma}_{N_F+1} + O(\alpha_s^{k+1})$. This is certainly a remarkable result.

1.5 Incisive Analysis Tools

Finally, as we advance our analysis methods it is important to develop new tools and metrics that can insightfully and effectively convey the essence of the physics. In Fig.8 we display the correlation cosine $\cos\phi$ *vs.* x for a variety of experiments. The plot effectively conveys which experiments influence the PDFs in different x regions, and how they are correlated. To emphasize this point we compare the u-valence (left) with the flipped d-valence figure (right); by flipping the figure we see that u-valence and d-valence are anti-correlated. Thus, as the fit works to find a minimum χ^2 value, it has to balance these experimental data sets which are pulling the u-valence and d-valence in opposite directions.

[3] $\alpha_s(\mu)$ has discontinuities at order $\alpha_s(\mu)^3$, but these are not apparent in Fig.7-a.

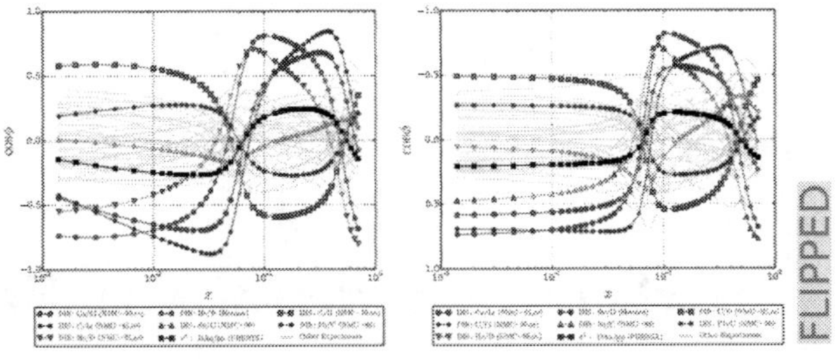

Figure 8. Correlation measures ($\cos\phi$) for u-valence (left) and d-valence (right) for lead at $Q = 10$GeV. Eight selected experiments are highlighted with symbols. To emphasize the anti-correlation between u-valence and d-valence we have flipped the d-valence plot vertically. See Ref.[2] for the original figures and additional details.

1.6 Conclusions

Recent theoretical developments provide improved precision for both proton and nuclear targets across a wide kinematic range. Further improvements will rely on a combination of additional theoretical advances, and a broad range of experimental measurements. The prospect of future data from EIC and LHeC facilities would be a valuable complement to the LHC and advance both our search for new physics and our understanding of QCD.

References

[1] F. Wilczek, Nucl. Phys. A **663**, 3 (2000) [hep-ph/9907340].

[2] K. Kovarik *et al.*, arXiv:1509.00792 [hep-ph].

[3] M. Hirai, S. Kumano and T.-H. Nagai, Phys. Rev. C **76**, 065207 (2007) doi:10.1103/PhysRevC.76.065207 [arXiv:0709.3038 [hep-ph]].

[4] K. J. Eskola, H. Paukkunen and C. A. Salgado, JHEP **0904**, 065 (2009) doi:10.1088/1126-6708/2009/04/065 [arXiv:0902.4154 [hep-ph]].

[5] D. de Florian, R. Sassot, P. Zurita and M. Stratmann, Phys. Rev. D **85**, 074028 (2012) doi:10.1103/PhysRevD.85.074028 [arXiv:1112.6324 [hep-ph]].

[6] K. Kovarik, I. Schienbein, F. I. Olness, J. Y. Yu, C. Keppel, J. G. Morfin, J. F. Owens and T. Stavreva, Phys. Rev. Lett. **106**, 122301 (2011) doi:10.1103/PhysRevLett.106.122301 [arXiv:1012.0286 [hep-ph]].

[7] A. Kusina *et al.*, Phys. Rev. D **85**, 094028 (2012) doi:10.1103/PhysRevD.85.094028 [arXiv:1203.1290 [hep-ph]].

[8] G. Aad *et al.* [ATLAS Collaboration], Phys. Rev. Lett. **109**, 012001 (2012) doi:10.1103/PhysRevLett.109.012001 [arXiv:1203.4051 [hep-ex]].

[9] D. Duggan, arXiv:0906.0136 [hep-ex].

[10] T. P. Stavreva and J. F. Owens, Phys. Rev. D **79**, 054017 (2009) doi:10.1103/PhysRevD.79.054017 [arXiv:0901.3791 [hep-ph]].

[11] S. J. Brodsky, A. Kusina, F. Lyonnet, I. Schienbein, H. Spiesberger and R. Vogt, Adv. High Energy Phys. **2015**, 231547 (2015) doi:10.1155/2015/231547 [arXiv:1504.06287 [hep-ph]].

[12] F. Lyonnet, A. Kusina, T. Ježo, K. Kovarík, F. Olness, I. Schienbein and J. Y. Yu, JHEP **1507**, 141 (2015) doi:10.1007/JHEP07(2015)141 [arXiv:1504.05156 [hep-ph]].

EPJ Web of Conferences **112**, 03007 (2016)
DOI: 10.1051/epjconf/201611203007

New advances in the statistical parton distributions approach *

Jacques Soffer[1,a] and Claude Bourrely[2,b]

[1] *Physics Department, Temple University,*
1925 N, 12th Street, Philadelphia, PA 19122-1801, USA
[2] *Aix Marseille Université, Université de Toulon, CNRS, CPT, UMR 7332*
13288 Marseille, Cedex 09, France

Abstract. The quantum statistical parton distributions approach proposed more than one decade ago is revisited by considering a larger set of recent and accurate Deep Inelastic Scattering experimental results. It enables us to improve the description of the data by means of a new determination of the parton distributions. This global next-to-leading order QCD analysis leads to a good description of several structure functions, involving unpolarized parton distributions and helicity distributions, in terms of a rather small number of free parameters. There are many serious challenging issues. The predictions of this theoretical approach will be tested for single-jet production and charge asymmetry in $W^{\pm}$ production in $\bar{p}p$ and pp collisions up to LHC energies, using recent data and also for forthcoming experimental results.

1 Introduction

Deep Inelastic Scattering (DIS) of leptons and nucleons is indeed our main source of information to study the internal nucleon structure in terms of parton distributions. Several years ago a new set of parton distribution functions (PDFs) was constructed in the framework of a statistical approach of the nucleon [1]. For quarks (antiquarks), the building blocks are the helicity dependent distributions $q^{\pm}(x)$ ($\bar{q}^{\pm}(x)$). This allows to describe simultaneously the unpolarized distributions $q(x) = q^{+}(x) + q^{-}(x)$ and the helicity distributions $\Delta q(x) = q^{+}(x) - q^{-}(x)$ (similarly for antiquarks). At the initial energy scale Q_0^2, these distributions are given by the sum of two terms, a quasi Fermi-Dirac function and a helicity independent diffractive contribution. The flavor asymmetry for the light sea, *i.e.* $\bar{d}(x) > \bar{u}(x)$, observed in the data is built in. This is simply understood in terms of the Pauli exclusion principle, based on the fact that the proton contains two up-quarks and only one down-quark. We predict that $\bar{d}(x)/\bar{u}(x)$ must remain above one for all x values and this is a real challenge for our approach, in particular in the large x region which is under experimental investigation at the moment. The flattening out of the ratio $d(x)/u(x)$ in the high x region, predicted by the statistical approach, is another interesting challenge worth mentioning. The chiral properties of QCD lead to strong relations between $q(x)$ and $\bar{q}(x)$. For example, it is found that the well established result $\Delta u(x) > 0$ implies $\Delta\bar{u}(x) > 0$ and similarly $\Delta d(x) < 0$ leads to $\Delta\bar{d}(x) < 0$. This earlier prediction was confirmed by

* Presented by J. Soffer at POETIC 2015
[a] e-mail: jacques.soffer@gmail.com
[b] e-mail: claude.bourrely@univ-amu.fr

recent polarized DIS data and it was also demonstrated that the magnitude predicted by the statistical approach is compatible with recent BNL-RHIC data on $W^{\pm}$ production [2]. In addition we found the approximate equality of the flavor asymmetries, namely $\bar{d}(x) - \bar{u}(x) \sim \Delta\bar{u}(x) - \Delta\bar{d}(x)$. Concerning the gluon, the unpolarized distribution $G(x, Q_0^2)$ is given in terms of a quasi Bose-Einstein function, with only *one free parameter*. The new analysis of a larger set of recent accurate DIS data leads to the emergence of a large positive gluon helicity distribution, giving a significant contribution to the proton spin, a major point which was emphasized in a recent letter [3].
It is crucial to note that the quantum-statistical approach differs from the usual global parton fitting methodology for the following reasons:
i) It incorporates physical principles to reduce the number of free parameters which have a physical interpretation
ii) It has very specific predictions, so far confirmed by the data
iii) It is an attempt to reach a more physical picture on our knowledge of the nucleon structure, the ultimate goal being to solve the problem of confinement
iv) Treating simultaneously unpolarized distributions and helicity distributions, a unique siuation in the literature, has the advantage to give access to a vast set of experimental data, in particular up to LHC energies

2 Review of the statistical parton distributions

Let us now recall the main features of the statistical approach for building up the PDFs, as opposed to the standard polynomial type parameterizations of the PDF, based on Regge theory at low x and on counting rules at large x. The fermion distributions are given by the sum of two terms, a quasi Fermi-Dirac function and a helicity independent diffractive contribution, at the input energy scale $Q_0^2 = 1\text{GeV}^2$,

$$xq^h(x, Q_0^2) = \frac{A_q X_{0q}^h x^{b_q}}{\exp[(x - X_{0q}^h)/\bar{x}] + 1} + \frac{\tilde{A}_q x^{\tilde{b}_q}}{\exp(x/\bar{x}) + 1}, \tag{1}$$

$$x\bar{q}^h(x, Q_0^2) = \frac{\bar{A}_q (X_{0q}^{-h})^{-1} x^{\bar{b}_q}}{\exp[(x + X_{0q}^{-h})/\bar{x}] + 1} + \frac{\tilde{A}_q x^{\tilde{b}_q}}{\exp(x/\bar{x}) + 1}. \tag{2}$$

We note that the universal diffractive term is absent in the quark helicity distribution Δq and in the quark valence contribution $q - \bar{q}$.
In Eqs. (1,2) the multiplicative factors X_{0q}^h and $(X_{0q}^{-h})^{-1}$ in the numerators of the non-diffractive parts of the q's and $\bar{q}$'s distributions, imply a modification of the quantum statistical form, we were led to propose in order to agree with experimental data. The presence of these multiplicative factors was justified in our earlier attempt to generate the transverse momentum dependence (TMD) [4, 5]. The parameter $\bar{x}$ plays the role of a *universal temperature* and $X_{0q}^{\pm}$ are the two *thermodynamical potentials* of the quark q, with helicity $h = \pm$. They represent the fundamental characteristics of the model. Notice the change of sign of the potentials and helicity for the antiquarks [1].
For a given flavor q the corresponding quark and antiquark distributions involve *eight* free parameters: $X_{0q}^{\pm}$, A_q, $\bar{A}_q$, $\tilde{A}_q$, b_q, $\bar{b}_q$ and $\tilde{b}_q$. It reduces to *seven* since one of them is fixed by the valence sum rule, $\int(q(x) - \bar{q}(x))dx = N_q$, where $N_q = 2, 1, 0$ for u, d, s, respectively.

[1] At variance with statistical mechanics where the distributions are expressed in terms of the energy, here one uses x which is clearly the natural variable entering in all the sum rules of the parton model.

For the light quarks $q = u, d$, the total number of free parameters is reduced to *eight* by taking, as in Ref. [1], $A_u = A_d$, $\bar{A}_u = \bar{A}_d$, $\tilde{A}_u = \tilde{A}_d$, $b_u = b_d$, $\bar{b}_u = \bar{b}_d$ and $\tilde{b}_u = \tilde{b}_d$. For the strange quark and antiquark distributions, the simple choice made in Ref. [1] was improved in Ref. [6], but here they are expressed in terms of *seven* free parameters.
For the gluons we consider the black-body inspired expression

$$xG(x, Q_0^2) = \frac{A_G x^{b_G}}{\exp(x/\bar{x}) - 1}\,, \tag{3}$$

a quasi Bose-Einstein function, with b_G being the only free parameter, since A_G is determined by the momentum sum rule. In our earlier works [1, 7], we were assuming that, at the input energy scale, the helicity gluon distribution vanishes, so

$$x\Delta G(x, Q_0^2) = 0\,. \tag{4}$$

However as a result of the present analysis of a much larger set of very accurate unpolarized and polarized DIS data, we must give up this simplifying assumption. We are now taking

$$x\Delta G(x, Q_0^2) = \frac{\tilde{A}_G x^{\tilde{b}_G}}{(1 + c_G x^{d_G})} \cdot \frac{1}{\exp(x/\bar{x} - 1)}\,. \tag{5}$$

To summarize the new determination of all PDFs involves a total of *twenty one* free parameters: in addition to the temperature $\bar{x}$ and the exponent b_G of the gluon distribution, we have *eight* free parameters for the light quarks (u, d), *seven* free parameters for the strange quarks and *four* free parameters for the gluon helicity distribution. These parameters have been determined from a next-to-leading order (NLO) QCD fit of a large set of accurate DIS data, unpolarized and polarized structure functions [8].

3 A selection of results

Some selected experimental tests for the unpolarized PDFs have been considered from μN and eN DIS, for which several experiments have yielded a large number of data points on the structure functions $F_2^N(x, Q^2)$, N stands for either a proton or a deuterium target. We have used fixed target measurements which probe a rather limited kinematic region in Q^2 and x and also HERA data which cover a very large Q^2 range and probe the very low x region, dominated by a fast rising behavior, consistent with our diffractive term (See Eq. (1)).
For illustration of the quality of our fit and, as an example, we show in Fig. 1, our results for $F_2^p(x, Q^2)$ on different fixed proton targets, together with H1 and ZEUS data. We note that the analysis of the scaling violations leads to a gluon distribution $xG(x, Q^2)$, in fairly good agreement with our simple parameterization (See Eq. (3)).

We now turn to the important issue concerning the asymmetries $A_1^{p,d,n}(x, Q^2)$, measured in polarized DIS. We recall the definition of the asymmetry $A_1(x, Q^2)$, namely

$$A_1(x, Q^2) = \frac{[g_1(x, Q^2) - \gamma^2(x, Q^2) g_2(x, Q^2)]}{F_2(x, Q^2)} \frac{2x[1 + R(x, Q^2)]}{[1 + \gamma^2(x, Q^2)]}\,, \tag{6}$$

where $g_{1,2}(x, Q^2)$ are the polarized structure functions, $\gamma^2(x, Q^2) = 4x^2 M_p^2/Q^2$ and $R(x, Q^2)$ is the ratio between the longitudinal and transverse photoabsorption cross sections. When $x \to 1$ for $Q^2 =$ 4 GeV^2, R is the order of 0.30 or less and $\gamma^2(x, Q^2)$ is close to 1, so if the u quark dominates, we have $A_1 \sim 0.6\Delta u(x)/u(x)$. Therefore it is unlikely to find $A_1 \to 1$, as required by the counting rules prescription, which we don't impose. We display in Fig. 2 the world data on $A_1^p(x, Q^2)$ (*Left*) and $A_1^n(x, Q^2)$ (*Right*), with the results of the statistical approach at $Q^2 = 4\text{GeV}^2$, up to $x = 1$. Indeed we find that $A_1^{p,n} < 1$.

Finally one important outcome of this new analysis of the polarized DIS data in the framework of the statistical approach, is the confirmation of a large positive gluon helicity distribution, which gives a significant contribution to the proton spin [3].

The NLO QCD calculations at $O(\alpha_s^3)$ of the cross section for the production of a single-jet of rapidity y and transverse momentum p_T, in a pp or $\bar{p}p$ collision, were done using a code based on a semi-analytical method within the "small-cone approximation"', improved recently with a jet algorithm for a better definition [2]. In Fig. 3(*Left*) our results are compared with the data from STAR experiment at BNL-RHIC and this prediction agrees very well with the data.
Now we would like to test, in a pure hadronic collision, our new positive gluon helicity distribution, mentioned above. In a recent paper, the STAR experiment at BNL-RHIC has reported the observation, in single-jet inclusive production, of a non-vanishing positive double-helicity asymmetry A_{LL}^{jet} for $5 \leq p_T \leq 30\text{GeV}$, in the near-forward rapidity region [10]. We show in Fig. 3(*Right*) our prediction compared with these high-statistics data points and the agreement is very reasonable.

There are several data sets for the cross section of single-jet production, which allow us to test our predictions, in particular the results from ATLAS and CMS displayed in Fig. 4 at $\sqrt{s} = 7\text{TeV}$.

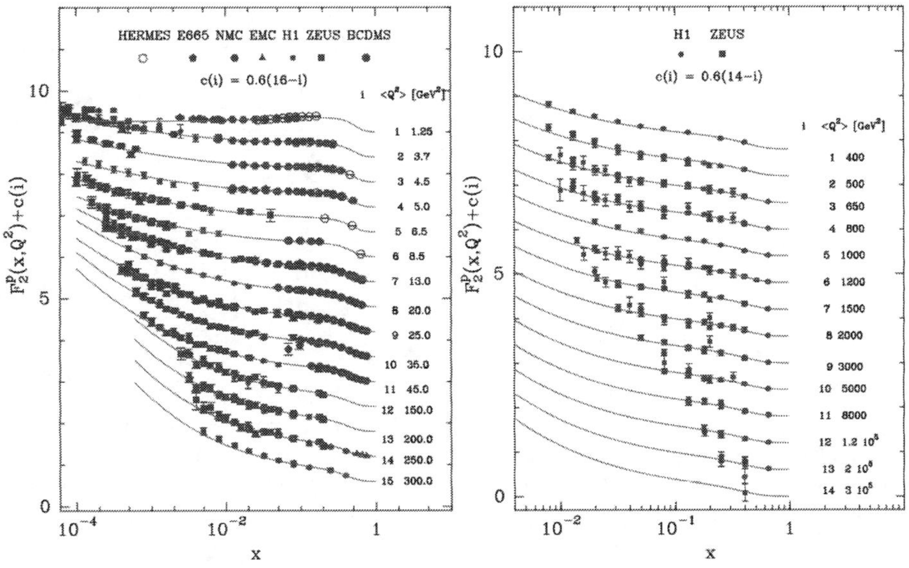

Figure 1. $F_2^p(x, Q^2)$ as a function of x for fixed $\langle Q^2 \rangle$ and data from HERMES, E665, NMC, EMC, H1, ZEUS, BCDMS. *Left*: The function $c(i) = 0.6(16 - i)$, $i = 1$ corresponds to $\langle Q^2 \rangle = 1.25\text{GeV}^2$. *Right*: The function $c(i) = 0.6(14 - i)$, $i = 1$ corresponds to $\langle Q^2 \rangle = 400\text{GeV}^2$. The curves are the results of the statistical approach.

[2]We thank Werner Vogelsang for providing us with the code to make this calculation.

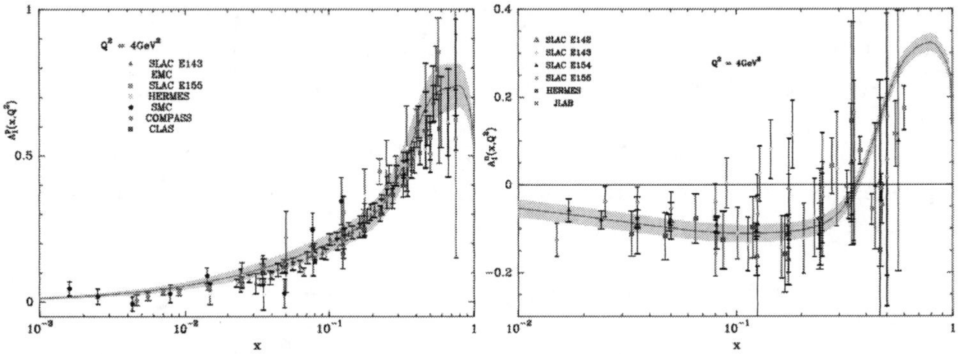

Figure 2. *Left* : Comparison of the world data on $A_1^p(x, Q^2)$ with the result of the statistical approach at Q^2 = 4 GeV^2, including the corresponding error band. *Right*: Comparison of the world data on $A_1^n(x, Q^2)$ with the result of the statistical approach at Q^2 = 4 GeV^2, including the corresponding error band.

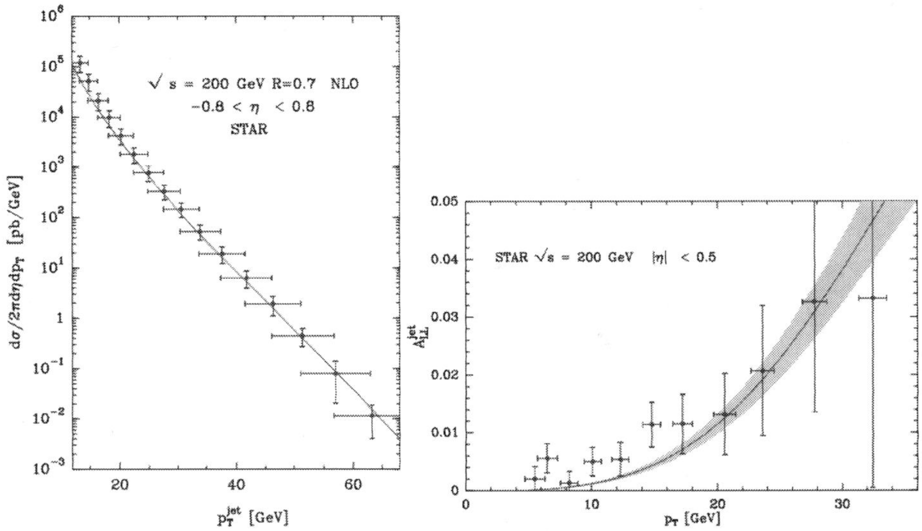

Figure 3. *Left*: Double-differential inclusive single-jet cross section in pp collisions at $\sqrt{s}$ = 200GeV, versus p_T^{jet}, with jet radius parameter R=0.7, for $-0.8 < \eta < 0.8$, from STAR data, obtained with an integrated luminosity of 5.39pb^{-1} [9] and the prediction from the statistical approach. *Right*: Our predicted double-helicity asymmetry A_{LL}^{jet} for single-jet production at BNL-RHIC in the near-forward rapidity region, versus p_T and the data points from STAR [10], with the corresponding error band.

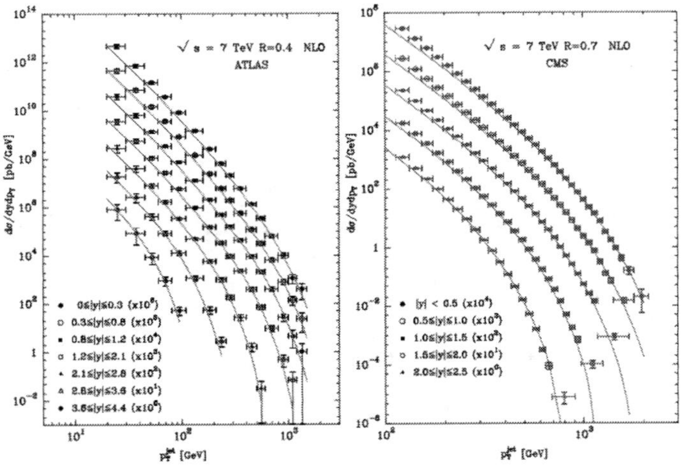

Figure 4. *Left*: Double-differential inclusive single-jet cross section in pp collisions at $\sqrt{s}$ = 7TeV, versus p_T^{jet}, with jet radius parameter R = 0.4, for different rapidity bins from ATLAS [11] and the predictions from the statistical approach
Right: Same from CMS [12], with R = 0.7.

Acknowledgements

J. Soffer is very grateful to the organizers, and in particular to Bernard Pire, for their kind invitation to POETIC 2015.

References

[1] C. Bourrely, F. Buccella and J. Soffer, Eur. Phys. J. C **23**, 487 (2002)
[2] C. Bourrely, F. Buccella and J. Soffer, Phys. Lett. B **726**, 296 (2013)
[3] C. Bourrely and J. Soffer, Phys. Lett. B **740**, 168 (2015)
[4] C. Bourrely, F. Buccella and J. Soffer, Phys. Rev. D **83**, 074008 (2011)
[5] C. Bourrely, F. Buccella and J. Soffer, Int. J. of Mod. Phys. A **28**, 1350026 (2013)
[6] C. Bourrely, F. Buccella and J. Soffer, Phys. Lett. B **648**, 39 (2007)
[7] C. Bourrely, F. Buccella and J. Soffer, Eur. Phys. J. C **41**, 327 (2005)
[8] C. Bourrely and J. Soffer, Nucl. Phys. A **941**, 307 (2015)
[9] B. Abelev *et al.* (STAR Collaboration), Phys. Rev. Lett. **97**, 252001 (2006); see also T. Sakuma, MIT Thesis (2009)
[10] L. Adamczyk *et al.* (STAR Collaboration), Phys. Rev. Lett. **115**, 9 092002 (2015)
[11] G. Aad *et al.* (ATLAS Collaboration), JHEP 1502, 153 (2015); 1509, 141 (2015)
[12] S. Chatrchyan *et al.* (CMS Collaboration), Phys. Rev. D **87**, 112002 (2013); Erratum-ibid, 119902

EPJ Web of Conferences **112**, 03008 (2016)
DOI: 10.1051/epjconf/201611203008

COMPASS Hadron Multiplicity Measurements and Fragmentation Functions

M. Stolarski[1,a,b] on behalf of the COMPASS Collaboration

[1] *LIP, Av. Elias Garcia 14 - 1º 1000-149 Lisbon, Portugal*

Abstract. COMPASS preliminary results on hadron, pion and kaon multiplicities are presented. The hadron and pion data show a good agreement with (N)LO QCD expectations and some of these preliminary data have been already successfully incorporated in the global NLO QCD fits to world data. However, the results for kaon multiplicities, are different from the expectations of the DSS fit. There is also a tension between COMPASS and HERMES results, the only other experiment which measured kaon multiplicities in SIDIS.

1 Introduction

In recent years there is a growing interest in analysis of the Semi-Inclusive Deep Inelastic scattering data (SIDIS). In order to interpret these measurements in the context of perturbative quantum chromodynamics (QCD), a new non-perturbative object has to be extracted from data, namely the fragmentation functions (D_q^i). In leading order (LO) the QCD D_q^i describe the probability density for a quark of flavour q to fragment into a hadron of type i. The cleanest way to access D_q^i is the measurement of hadrons produced in e^+e^- annihilation. However, in such a measurement only the quark anti-quark sum of fragmentation functions, $D_q^i + D_{\bar{q}}^i$, can be accessed. In addition, in e^+e^- measurements full flavour separation is not possible. The extraction of D_q^i from SIDIS data is more complex, as the fragmentation functions are convoluted with the parton distribution functions. On the other hand one can separately extract D_q^i and $D_{\bar{q}}^i$ as well as perform a full flavour separation. In this proceeding the preliminary results of COMPASS multiplicity measurements for $h^{\pm}$, $\pi^{\pm}$ and $K^{\pm}$ and the extraction of π and K fragmentation functions are presented.

2 COMPASS experiment

COMPASS is an experiment located at CERN SPS accelerator. A detailed description of the COMPASS spectrometer can be found elsewhere [1]. For the results presented in this paper a 160 GeV positive muon beam was impinging on ^{6}LiD target. The COMPASS spectrometer was designed to reconstruct scattered muons and charged hadrons in a wide kinematic range. The angular acceptance of the COMPASS spectrometer is about ±180mrad. Three muon filters along the spectrometer provide

[a] e-mail: mstolars@cern.ch

[b] supported by Portuguese Fundação para a Ciência e Tecnologia

very good hadron/muon separation. Hadrons are identified in the Ring Imaging CHerenkov counter (RICH) from 3 GeV/c, 9 GeV/c and 18 GeV/c, for pions, kaons and protons respectively and up to about 50 GeV/c.

3 Data Selection and Analysis

In the analysis only data from DIS region are used: it is required that the negative four momentum transfer Q^2 is larger than 1 $(\mathrm{GeV/c})^2$ and the mass, W, of the hadronic system is larger than 5 GeV/c^2. The Bjorken scaling variable, x, is selected to be between $0.004 < x < 0.4$. The beam energy fraction carried by the virtual photon, y, is kept between $0.1 < y < 0.7$. The lower limit of the latter is related with the precision and stability of the μ' reconstruction, while the higher y cut excludes the region where radiative corrections are large. The total number of selected DIS events is about 13 millions.

For the analysis a hadron candidate must have a measured momentum and must not to be identified as a muon. The fraction of the virtual-photon energy carried by the hadron candidate, z, should be $0.2 < z < 0.85$. The lower limit ensures that the so called current fragmentation region is analysed, while the upper one removes the region where the contribution from non-DIS processes like diffraction is sizable. Stability of the RICH performance limited the hadron momenta to the interval 12-40 GeV/c.

The multiplicity is defined as the number of hadrons per DIS event. In LO the observed multiplicity is related in the following way with the parton distribution function $q(x)$, and fragmentation functions D_q^i

$$\frac{dM^i(x,Q^2,z)}{d(x,Q^2,z)} = \frac{\sum_q e_q^2 q(x,Q^2) D_q^i(z,Q^2)}{\sum_q e_q^2 q(x,Q^2)}. \tag{1}$$

Here e_q is the electric charge of quark flavour q. In this analysis the multiplicities are measured as functions of x, y and z; Q^2 is not used due to a strong correlation between x and Q^2. Measured raw multiplicities are corrected by spectrometer acceptance, the RICH efficiency and particle misidentification probability, the contribution from decay products of diffractive mesons, and finally radiative corrections. The COMPASS acceptance is high, between 40% and 70%. It was evaluated using the LEPTO generator and a COMPASS spectrometer simulation program based on GEANT3. The RICH efficiency and purity were evaluated on real data using using decay products of K^0, ϕ and Λ. The efficiency for π is high, above 90% for $p < 30$ GeV/c, while the misidentification probability is low. For kaon the efficiency is above 90% in the whole momentum range used. Since there are much more produced pions than kaons the misidentification probability that a π is identified as a K is very important. In COMPASS it is below 2%. The contribution from decay products of diffractively produced mesons is estimated using HEPGEN Monte Carlo program [2]. This correction is most important for π, where a sizable contribution from $\rho^0 \to \pi^+\pi^-$ exists. The correction is largest for low Q^2 data and can reach 0.55 at high z. The (x, y) dependent radiative corrections are calculated using the programme TERAD [3].

4 Results

The preliminary results of pion multiplicities are presented in figure 1, together with the LO QCD fit performed by COMPASS and described in section 5. The data are well described by the fit. The closer inspection reveals that in the whole kinematic range the multiplicity of π^+ is higher than that of π^-. Remembering that absolute charges of u, $\bar{u}$ quarks are higher than those of d, $\bar{d}$, the above mentioned fact confirms that $D_{fav} > D_{unf}$, *i.e.* the fragmentation into quarks which are valence quarks of a hadron X, (D^X_{fav}) is higher than into quarks which belong to the sea of a hadron X (D^X_{unf}).

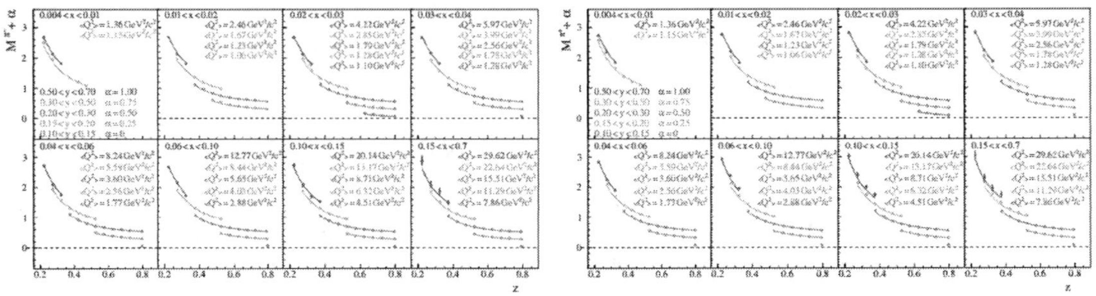

Figure 1. The pion multiplicity as a function of (x, y, z) for π^+ (left) and π^- (right). The results of COMPASS LO QCD fit (cf. section 5) are also shown.

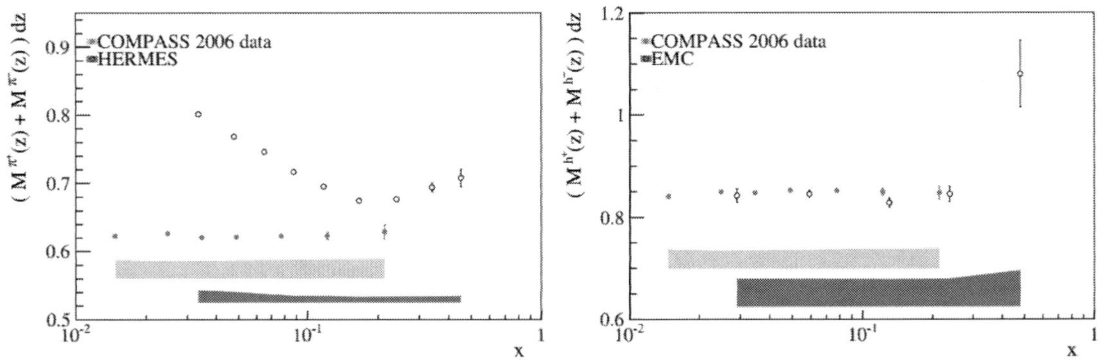

Figure 2. The z integrated pion multiplicity sums from COMPASS and HERMES [6] (left) The z integrated hadron multiplicity sums from COMPASS and EMC [5] (right). The data points are shown with statistical errors only, the systematic is presented as a shaded band below.

A very interesting observation can be made while studying the z-integrated sum of π^+ and π^- multiplicities, $\mathcal{M}^\pi$, as a function of x. In LO QDC the relation between $\mathcal{M}^\pi$ and the fragmentation functions for an isoscalar target is the following:

$$\mathcal{M}^\pi = \int D^\pi_{fav} + \int D^\pi_{unf} - \frac{2S}{5Q + 2S}\left(\int D^\pi_{fav} - \int D^\pi_{unf}\right) \approx \int D^\pi_{fav} + \int D^\pi_{unf} \approx const, \quad (2)$$

where $Q = u + \bar{u} + d + \bar{d}$ and $S = s + \bar{s}$. The $\int D^\pi_{fav}$ and $\int D^\pi_{unf}$ are the z-integrated favoured and unfavoured pion fragmentation functions. In equation (2) a possible x and y dependencies were neglected for simplicity. In fixed target experiments x is correlated with Q^2, but the Q^2 dependence of the z-integrated $D^\pi_{fav} + D^\pi_{unf}$ is rather weak (3% according to Ref. [4] in the COMPASS Q^2 range). Thus, the $\mathcal{M}^\pi$ distribution is expected to be approximately flat. A similar expectation holds for unidentified hadrons. The COMPASS data are presented in figure 2. Indeed the distribution of $\mathcal{M}^\pi$ and $\mathcal{M}^h$ are flat as expected in LO QCD. The same conclusion can be reached for the EMC results of unidentified hadron multiplicities [5], presented on the right panel of figure 2. However, it is interesting to notice that the shape of HERMES data, [6], is very different from the COMPASS (and EMC) results.

The kaon multiplicities are presented in figure 3 as functions of (x, y, z). Again the results are compared to the COMPASS LO QCD fit described in section 5. For kaons the $\mathcal{M}^K$ is even more

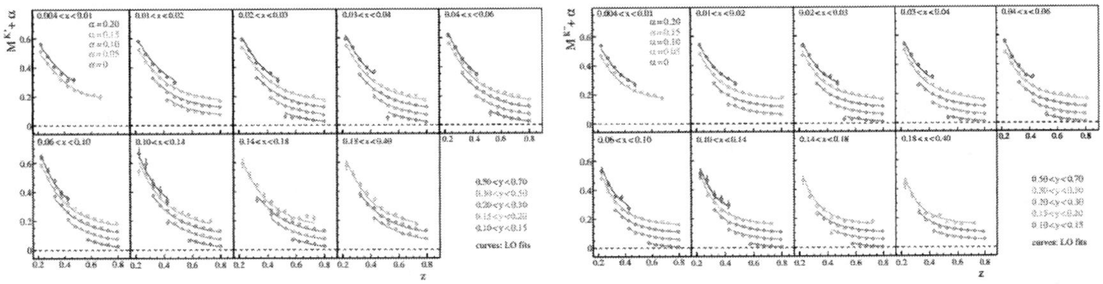

Figure 3. The kaon multiplicity as a function of (x, y, z) for K^+ (left) and K^- (right). The results of COMPASS LO QCD fit (cf. section 5) are also shown.

interesting than for the pion case. As pointed out in [7] in LO QCD the following relation holds

$$5\mathcal{M}^K \approx \int D_Q^K + S/Q \int D_{str}^K. \tag{3}$$

Here, $\int D_Q^K$ is the z-integrated value of $4D_u^K + 4D_{\bar{u}}^K + D_d^K + D_{\bar{d}}^K \approx 4D_{fav} + 6D_{unf}$ and $\int D_{str} = D_{s+\bar{s}}^K$. In the high x region the value of S/Q is small. Therefore, in this region the $\int D_Q^K$ can be extracted. The COMPASS result, presented in figure 4, points towards $\int_{0.2}^{0.85} D_Q^K \approx 0.70$, while the expectation of the DSS fit is only 0.43 . The COMPASS results on $\mathcal{M}^K$ suggest that D_{fav} and/or D_{unf} are larger than those from the DSS fit. On the other hand, for low x values the last therm in Eq. (3) cannot be neglected. Again assuming $S(x)$ as in MSTW08L, [8], and D_{str} from DSS parametrisation, [4], an increase of the $\mathcal{M}^K$ by about 50% should be seen comparing the low and high x regions. Such an increase is not observed in the COMPASS data, which points towards smaller value of D_{str}/D_{fav} ratio than in the DSS fit (or $S(x)$ is smaller than expected from MSTW08L). Such a result strongly influences the extracted polarisation of the strange sea in SIDIS analyses. The lower values of D_{str}/D_{fav} with respect to DSS parametrisation can explain the so called strange quark polarisation puzzle [9, 10]. It should be noted that even the shapes of $\mathcal{M}^K(x)$ of COMPASS and HERMES do not agree, as was already seen in the pion case. In addition in the COMPASS case much more kaons are produced at high x with respect to HERMES. Due to the beam energy difference and to RICH restrictions the HERMES and COMPASS kinematics for charged kaons are not overlapping. However, for the neutral K^0, the RICH momentum cut $12 < p < 40$ GeV/c is not needed. This allows COMPASS to study K^0 production in the lower energy region, closer to the HERMES kinematics. The analysis of K^0 multiplicities is ongoing.

Another interesting observation can be done when inspecting figure 5, where the multiplicity ratios of π^+, π^- and K^+, K^- are shown for COMPASS, HERMES and JLab. Despite the different shape of $\mathcal{M}^\pi$ in COMPASS and HERMES, ratios of π^+/π^- multiplicities agree very well. This is not the case for the ratios of K^+/K^- multiplicities. It should be also noted that from the experimental point of view the ratio of X^+/X^- is known with much better precision than the individual values, as a considerable fraction of experimental systematic uncertainties cancel in the X^+/X^- ratio.

5 COMPASS LO QCD fit to fragmentation functions

COMPASS performed a LO QCD fit of the fragmentation functions using the presented multiplicities. A simple functional form was used, $N_i z^{\alpha_i}(1-z)^{\beta_i}$, for parametrisation of fragmentation functions at

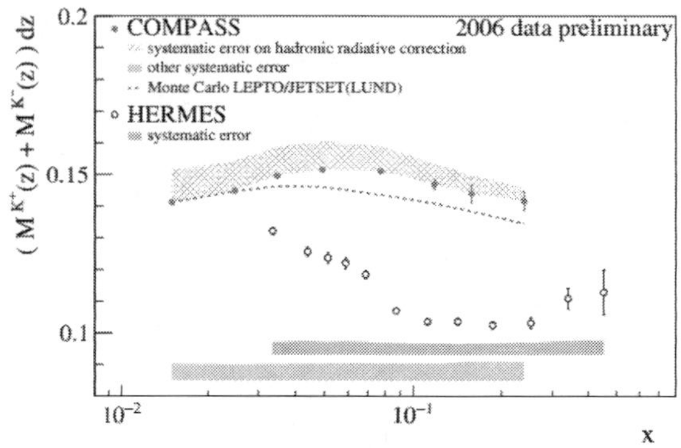

Figure 4. The z integrated kaon multiplicity sums for COMPASS and HERMES [6] experiments. The data points are shown with statistical errors only, the systematic uncertainty is presented as a shaded band below. The shaded area above COMPASS points corresponds to 100% uncertainty of SIDIS radiative corrections. The dashed line is a LEPTO MC prediction with fragmentation tuning from Ref. [11].

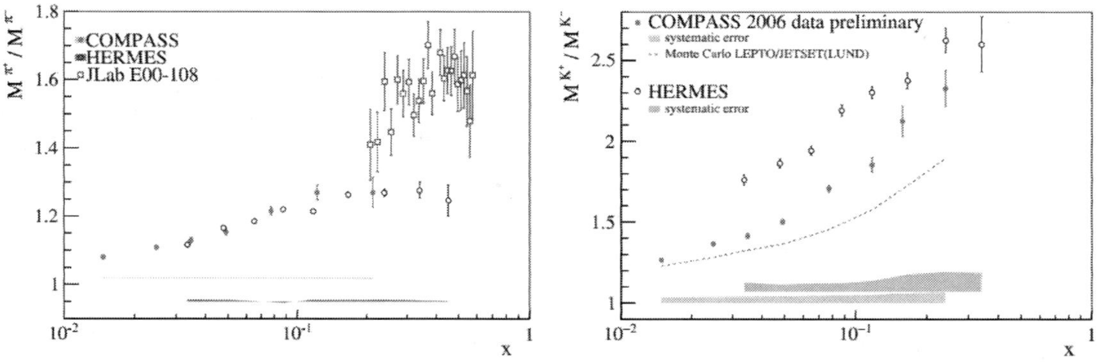

Figure 5. The π^+/π^- (left) and K^+/K^- (right) multiplicity ratio for COMPASS, HERMES [6] and in π case also JLab [12]. The data points are shown with statistical errors only, the systematic uncertainty is presented as a shaded band below.

the reference scale. The values of N_i, α_i, β_i are the fit parameters. In the case of pions a favoured and unfavoured fragmentation functions were considered in the fit ($i = \{fav, unf\}$), while in the kaon case in addition a strange fragmentation function was also included ($i = \{fav, unf, str\}$). In both cases the gluon fragmentation functions were also extracted. For the evolution of MSTW08L the LHAPDF library was employed [13], while for the evolution of the fragmentation function the code [14] was used. Results for the pion fragmentation functions (not shown here) agree well with the most recent NLO fits [15, 16]. For the kaon case the results for the D_{fav} and D_{unf} are presented in figure 6. The COMPASS fit suggests higher values of D^K_{fav} and D^K_{unf} than the DSS results [4]. This is not a surprise as the analysis of $\mathcal{M}^K$ points towards very different values of D_Q obtained from the COMPASS and DSS fits. The result for the strange quark fragmentation into kaons need further study.

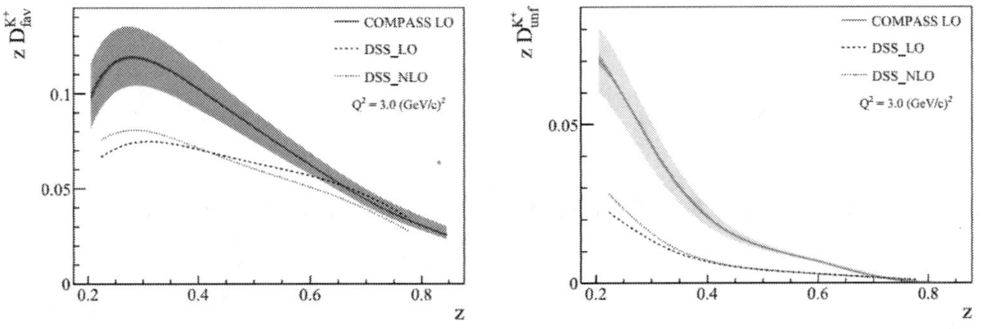

Figure 6. The results of COMPASS LO QCD fit to the kaon multiplicities: D^K_{fav} (left) and D^K_{unf} (right).

6 Summary

Preliminary results for charged hadrons, pion and kaon multiplicities from the COMPASS experiment were presented. They were compared to results of HERMES; observed differences were discussed and are subject of further studies. They also motivate importance of multiplicity measurements at the future electron-proton collider. The presented measurements give an important input to the global analyses of fragmentation functions.

References

[1] COMPASS Collaboration, P. Abbon *et al.*, Nucl. Instrum. and Meth. A **577**, 455 (2007).
[2] A. Sandacz and P. Sznajder, arXiv:1207.0333 .
[3] A. A. Akhundov, D.Yu. Bardin, L. Kalinovskaya and T. Riemann, Fortschr. Phys. **44**, 373 (1996).
[4] D. de Florian, R. Sassot and M. Stratmann, Phys Rev D **75**, 114010 (2007).
[5] EMC, J. Ashman, et al., Z. Phys. C **52**, 1 (1991).
[6] HERMES Collaboration, A. Airapetian, *et al.*, Phys. Rev. D **87**, 074029 (2013).
[7] HERMES Collaboration, A. Airapetian *et al.*, Phys Lett. B **666**, 446 (2008).
[8] A.D. Martin, W.J. Stirling, R.S. Thorne and G. Watt, Eur. Phys. J. C **63**, 653 (2009).
[9] E. Leader, A. V. Sidorov and D. B. Stamenov, Phys. Rev. D **84**, 014002 (2011).
[10] E. Leader, A. V. Sidorov and D. B. Stamenov, Phys. Rev. D **91**, 054017 (2015).
[11] COMPASS Collaboration, C. Adolph *et al.*, Phys. Lett. B **718**, 922 (2013).
[12] R. Asaturyan, *et al.*, Phys. Rev. C **85**, 015202 (2012).
[13] LHAPDF homepage, https://lhapdf.hepforge.org/.
[14] M. Hirai and S. Kumano, Comut. Phys. Commun. **183**, 1002 (2012).
[15] E. Leader, A.V. Sidorov and D. Stamenov, Proceedings of the 15th Workshop on High Energy Physics; A.V. Efremov and S.V. Goloskokov (editors), JINR Dubna, Russia (2013), arXiv:1312.5200
[16] D. de Florian *et al.*, Phys. Rev. D **91** 014045 (2015).

EPJ Web of Conferences **112**, 04001 (2016)
DOI: 10.1051/epjconf/201611204001

D^* and B^* mesons in strange hadronic medium at finite temperature

Rahul Chhabra[1,a] and Arvind Kumar[2,b]

[1] *National Institute of Technology, Jalandhar*
[2] *National Institute of Technology, Jalandhar*

Abstract. We calculate the effect of density and temperature of isospin symmetric strange medium on the shift in masses and decay constants of vector D and B mesons using chiral SU(3) model and QCD sum rule approach. In the present investigation the values of quark and gluon condensates are calculated from the chiral SU(3) model and these condensates are further used as input in the QCD Sum rule framework to calculate the in-medium masses and decay constants of vector D and B mesons. These in medium properties of vector D and B mesons may be helpful to understand the experimental observables of the experiments like CBM and PANDA under FAIR project at GSI, Germany. The results which are observed in present work are also compared with the previous predictions.

1 Introduction

Medium modification of heavy mesons is important to understand the QCD phase transition and the outcomes of different heavy ion collision experiments. In-medium properties of mesons are believed to be changed from the vacuum properties due to the interactions with the medium. These modifications can effect the various observed values, e.g, if the mass reduction of D(B) meson in the medium is very large then higher charmonium (bottomonium) states can decay to these D(B) mesons instead of $J/\psi(\Upsilon)$ and hence will suppress their production. On the other hand, if masses of these mesons increase to very high value then these mesons can facilitates the production of $J/\psi(\Upsilon)$. These statements are important because suppression of $J/\psi(\Upsilon)$ is considered as an important indication of production of Quark Gluon Plasma and therefore extra care has to be taken before to make definite conclusion. Negative mass shift of these mesons predicts the attractive potential in nuclear medium which shows that the bound states of $D^*N(B^*N)$ is possible. Positive mass shift predicts repulsive potential in the medium and shows that bound states of $D^*N(B^*N)$ mesons is not possible. These predictions can be confirmed from the data of future experiments like CBM and PANDA. In present work, we investigate the medium modification of D^* and B^* mesons in hot and dense strange medium using Chiral SU(3) model and QCD sum rule. We calculate the values of condensates through Chiral SU(3) model and these act as input to the QCD sum rules which we further use to calculate the medium modification of vector D and B mesons in hadronic medium.

[a] e-mail: rahulchhabra@ymail.com
[b] e-mail: kumara@nitj.ac.in

2 Chiral SU(3) Model

Chiral SU(3) model is an effective theory which we can use in a regime where basic theory of QCD interaction fails. Chiral SU(3) model is based on the chiral symmetry, broken scale invariance, non realization and broken scale invariance property of chiral symmetry. From the Lagrangian density of the model, using mean field approximation, we obtain the coupled equations of motion for the scalar fields σ, ζ and scalar dilaton field χ. These coupled equations are solved to investigate the density and temperature dependence of σ, ζ and χ in hadronic medium. By introducing explicit symmetry breaking term, we can calculate the values of quark condensates $< \bar{q}q >$ in terms of these mesons [1],

$$\sum_i m_i \bar{q}_i q_i = \left(\frac{\chi}{\chi_0}\right)^2 \left(m_\pi^2 f_\pi \sigma + (\sqrt{2} m_k^2 f_k - \frac{1}{\sqrt{2}} m_\pi^2 f_\pi)\zeta\right). \tag{1}$$

Similarly, broken scale invariance term can be used to evaluate the gluon condensates and we have,

$$\left\langle \frac{\alpha_s}{\pi} G^a{}_{\mu\nu} G^{a\mu\nu} \right\rangle = \frac{8}{9}\left[(1-d)\chi^4 + \left(\frac{\chi}{\chi_0}\right)^2 \left(m_\pi^2 f_\pi \sigma + (\sqrt{2} m_k^2 f_k - \frac{1}{\sqrt{2}} m_\pi^2 f_\pi)\zeta\right)\right]. \tag{2}$$

Using equations (1) and (2) we can calculate the medium modification of quark and gluon condensates through the medium modification of scalar fields σ, ζ and scalar dilaton field χ [2].

3 QCD SUM RULE FOR VECTOR D AND B MESONS

We start with the two point correlation function, $\Pi_{\mu\nu}(q)$ which can be divided into the vacuum part, nucleon part and pion bath terms[2]. We have,

$$\Pi_{\mu\nu}(q) = \Pi^0_{\mu\nu}(q) + \frac{\rho_B}{2m_N} T^N_{\mu\nu}(q) + \Pi^{P.B.}_{\mu\nu}(q), \tag{3}$$

In Eq. (3), third term on the right, i.e, pion bath term had been used in the literature to investigate the effect of temperature on the properties of mesons. In this investigation, we ignore pion bath term and calculate the effect of temperature on the mesons through the modification of scalar σ, ζ, and scalar dilaton χ fields with temperature. As discussed in [4], correlation function $T^N_{\mu\nu}(q)$ can be related to D^*N scattering T matrix through scattering lengths. Phenomenological density can be parametrized in terms of unknown parameters a, b and c [5]. Borel transformed equation is used to solve these unknown parameters and final mass shift can be defined as [4],

$$\delta m_{D^*} = \sqrt{m^2_{D^*} + \Delta m^2_{D^*}} - m_{D^*}, \tag{4}$$

where, shift of squared mass, $\Delta m^2_{D^*}$ of mesons can be written in terms of unknown parameter a and hence scattering length a_{D^*},

$$\Delta m^2_{D^*} = \frac{\rho_B}{2m_N} \frac{a}{f^2_{D^*} m^2_{D^*}} = -\frac{\rho_B}{2m_N} 8\pi(m_N + m_{D^*}) a_{D^*}. \tag{5}$$

In this problem, we use the centroid approximation and calculate the average mass shift of particle and antiparticle instead of their mass splitting. In [3], author investigated the mass splitting by taking the even and odd part of QCD sum rule. Shift in decay constant can be written as [6],

$$\delta f_{D^*} = \frac{1}{2 f_{D^*} m^2_{D^*}} \left(\frac{\rho_B}{2m_N} b - 2 f^2_{D^*} m_{D^*} \delta m_{D^*}\right). \tag{6}$$

Table 1. Numerical result to indicates the effect of density ρ_B, temperature, T(in MeV) and strangeness fractions, f_s of the medium, on the decay shift(in MeV) of vector mesons

	$\delta f_{D^*}(\delta f_{B^*})$				$\delta f_{D_s^*}(\delta f_{B_s^*})$			
	$\rho_B = \rho_0$		$\rho_B = 4\rho_0$		$\rho_B = \rho_0$		$\rho_B = 4\rho_0$	
fs	T=0	T=150	T=0	T=150	T=0	T=150	T=0	T=150
0	-18(-56)	-13(-43)	-28(-104)	-23(-91)	-11(-30)	-8(-23)	-18(-53)	-15(-46)
0.3	-22(-68)	-16(-54)	-31(-108)	-24(-95)	-16(-42)	-13(-34)	-28(-76)	-24(-65)

QCD sum rule equations for vector B meson can be found by simply using mass of bottom, b, quark instead of mass of charm, c, quark in sum rule equation for vector D meson [6].

4 Results and Discussions

The values of masses, decay constants and threshold parameter s_0 of $D^*(B^*)$ mesons are 2.01(5.325), 0.270(0.195) GeV and 6.5(35) GeV^2, respectively. For $D_s^*(B_s^*)$, we use the values of masses, decay constants and threshold parameter as 2.112(5.415), $1.16 f_{D^*}(1.16 f_{B^*})$ GeV, and 7.5(38) GeV^2, respectively. Nuclear matter saturation density chosen in the present case is $0.15 fm^{-3}$. We chose the value of coupling constant $g_{D^*N\Lambda_c} \approx g_{D^*N\Sigma_c} \approx \approx g_{B^*N\Lambda_b} \approx g_{B^*N\Sigma_b}$ as 3.86 [4]. The masses of quarks, u, d, c, b used in present investigation are 0.005, 0.007, 0.095, 1.35 and 4.7 GeV, respectively. In case of mass shift the Borel window chosen for D^*, B^*, D_s^* and B_s^* are (4.5-6.5), (30-33), (5.0-7.0) and (31-34) GeV^2, respectively. For decay shift, we chose Borel window as (3.3-4.9), (26-31), (3.8-5.3) and (27-31) GeV^2, respectively [2]. We chose proper Borel window so that there is almost no variation in the shift in masses and decay constant. In Fig. 1, we plot the shift in masses and decay constants of D^*, B^*, D_s^* and B_s^* mesons as a function of square of Borel mass parameter, i.e, M^2. We observe that in nuclear medium, at zero temperature, the value of mass shift for $D^*(B^*)$, $D_s^*(B_s^*)$ meson change from -76(-367), -46(-206) to -128(-707), -76(-377) MeV, when we move from $\rho_B = \rho_0$ to $\rho_B = 4\rho_0$. This clearly indicates the much large drop of masses of $D^*(B^*)$ vector mesons, with increase in the density of the medium and therefore, this may cause $J/\psi(\Upsilon)$ suppression in heavy ion collision experiments. The negative mass shift reveals the attractive interaction in the nuclear medium and hence it may be possible to form bound states $D^*N(B^*N)$. For constant value of density and temperature of the medium, increase in the value of mass shift is observed when we move from non strange medium, i.e, $f_s = 0$ to strange medium, i.e, $f_s = 0.3$. On the other hand, shift in mass for finite temperature of medium, $T = 150$ MeV is less as compared to the zero temperature values, $T = 0$ and therefore, temperature decrease the magnitude of the shift in masses of vector mesons. Table 1 shows the shift in decay constant of $D^*(B^*)$ and $D_s^*(B_s^*)$ mesons for baryonic density, ρ_0 and $4\rho_0$ and temperature, T = 0 and 150 MeV. The results are plotted for strangeness fractions, $f_s = 0$ and $f_s = 0.3$. The magnitude of the shift in decay constants of vector mesons increase with the increase in the density of the medium for constant value of strangeness fraction, f_s and temperature, T. Similarly, for constant value of density, ρ_B and temperature, T, with increase in the strangeness fraction, f_s, of the medium, the decay shift increase. The values of masses and decay constants depend on the values of quark and gluon condensates and these condensates are proportional to the scalar fields, σ and ζ. Therefore, the medium modification of masses and decay constants of vector D and B mesons are due to the medium modification of scalar fields, σ and ζ [2].

Medium modification of heavy vector mesons had also been studied in [1, 4, 6]. Using QCD sum rules the values of mass shift for D^* and B^* in [4] were -71 and -380 MeV, respectively. In [6] by including the leading order term in the QCD sum rules, author studied the shift in masses(decay con-

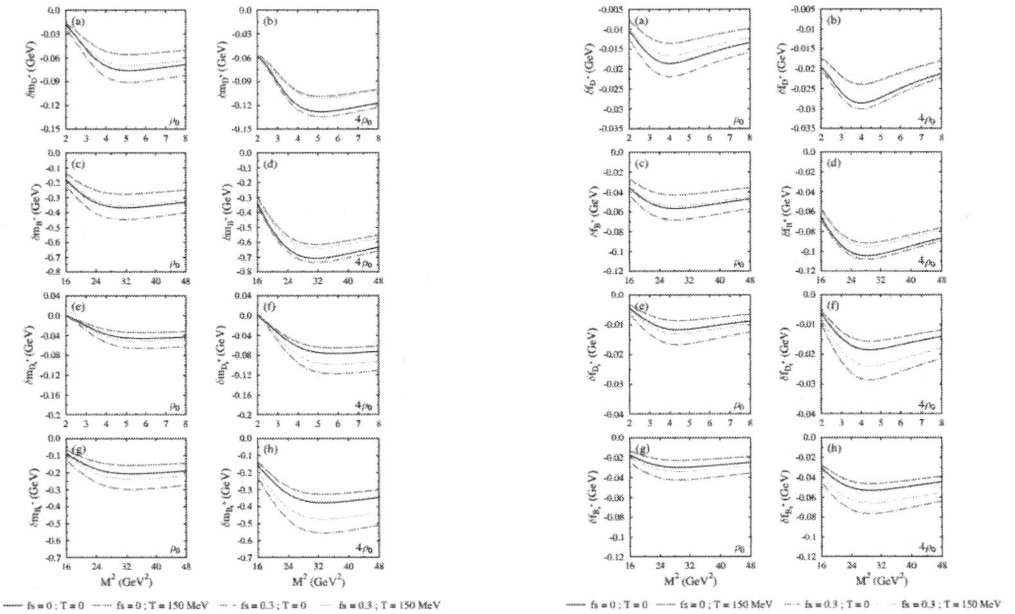

Figure 1. We plot the mass shift (left) and shift in decay constant (right) of D^*, B^*, D_s^* and B_s^* meson as a function of Borel mass parameter M^2. We show the result at temperature $T = 0$ and $T = 150$ MeV. For each value of temperature results are shown for strangeness fractions, $f_s = 0$ and 0.3.

stants) of vector D and B mesons as -70(-16) and -340(-55) MeV, whereas, including next to leading order term observation changed to -102(-26) and -687(-111) MeV, respectively. We can compare these values with our calculated values -76(-18) and -367(-56) MeV, respectively, at nuclear saturation density, zero temperature and in non-strange medium. Using self consistent coupled channel approach authors in [7] observed the positive values of mass shift, which indicates the repulsive interaction in the medium. In our present work, the negative values of shift indicates the attractive interaction in the nuclear medium and this indicates the possibility of bound states $D^*N(B^*N)$. Further, decrease in the masses of D and B meson can suppress J/ψ (Υ) production in heavy ion collision experiments.

References

[1] Arvind Kumar, Adv. in High Energy Physics **2014**, 549726 (2014).

[2] Arvind Kumar and Rahul Chhabra, Phys. Rev. C **92**, 035208 (2015).

[3] T. Hilger, R. Thomas and B. Kampfer, Phys. Rev. C **79**, 025202 (2009).

[4] Zhi-Gang Wang, Int. J. Mod. Phys. A **28**, 1350049 (2013)

[5] Yuji Koike and Arata Hayashigaki, Prog. Theo. Phys. **98**, 631 (1997)

[6] Zhi-Gang Wang, arXiv:1501.05093 [hep-ph].

[7] L. Tolos, C. Garcia - Recio, and J. Nieves, Phys. Rev. C **80**, 065202 (2009).

EPJ Web of Conferences **112**, 04002 (2016)
DOI: 10.1051/epjconf/201611204002

Centrality-dependent forward J/ψ production in high energy proton-nucleus collisions

B. Ducloué[1,2], T. Lappi[1,2], and H. Mäntysaari[3]

[1] *Department of Physics, University of Jyväskylä, P.O. Box 35, 40014 University of Jyväskylä, Finland*
[2] *Helsinki Institute of Physics, P.O. Box 64, 00014 University of Helsinki, Finland*
[3] *Physics Department, Brookhaven National Laboratory, Upton, NY 11973, USA*

Abstract. Forward J/ψ production and suppression in high energy proton-nucleus collisions can be an important probe of gluon saturation. In an earlier work we studied this process in the Color Glass Condensate framework and showed that using the Glauber approach to extrapolate the dipole cross section of a proton to a nucleus leads to results closer to experimental data than previous calculations in this framework. Here we investigate the centrality dependence of the nuclear suppression in this model and show a comparison of our results with recent LHC data.

1 Introduction

The study of the nuclear suppression of forward J/ψ production in high energy proton-nucleus collisions can be a valuable tool to better understand saturation dynamics. Indeed, it probes the target nucleus at very high densities, where saturation effects should be enhanced, and the charm quark mass is small enough to be sensitive to these dynamics while being large enough to provide a hard scale allowing a perturbative treatment. J/ψ mesons also have clean experimental signatures and their production and suppression have been the subject of many experimental studies. In a recent work [1] we re-evaluated the nuclear suppression of forward J/ψ production at high energy in the Color Glass Condensate (CGC) framework, showing that using the optical Glauber model to relate the dipole cross section of a nucleus to the one of a proton leads to a smaller suppression for minimum bias events than in previous works in the same formalism [2] and results closer to experimental data[1]. In this work we discuss the relation between the explicit impact parameter dependence of the optical Glauber model and centrality and we compare our results with recent data on the centrality dependence of J/ψ suppression at the LHC presented by the ALICE Collaboration [4].

2 Formalism

The formalism for gluon and quark pair production in the dilute-dense limit of the Color Glass Condensate has been studied in Refs. [5, 6] (see also Ref. [7]) and applied in several works, such as [2, 8–11]. This allows to compute the cross section for $c\bar{c}$ pair production, which is the central object needed

[1]The authors of [2] have recently presented updated results [3] similar to those obtained in [1].

to study J/ψ production. The expression for this cross section can be found in Ref. [2]. We use the simple color evaporation model to describe the hadronization of the produced $c\bar{c}$ pairs in J/ψ mesons. In this model, a fixed fraction of the $c\bar{c}$ pairs produced with an invariant mass between $2m_c$ and $2M_D$, where m_c is the charm quark mass and m_D is the D-meson mass, is assumed to hadronize into J/ψ mesons. Thus the cross section for J/ψ production with transverse momentum $\boldsymbol{P}_\perp$ and rapidity Y reads

$$\frac{\mathrm{d}\sigma_{J/\psi}}{\mathrm{d}^2\boldsymbol{P}_\perp\,\mathrm{d}Y} = F_{J/\psi}\int_{4m_c^2}^{4M_D^2}\mathrm{d}M^2\frac{\mathrm{d}\sigma_{c\bar{c}}}{\mathrm{d}^2\boldsymbol{P}_\perp\,\mathrm{d}Y\,\mathrm{d}M^2}, \tag{1}$$

where $\frac{\mathrm{d}\sigma_{c\bar{c}}}{\mathrm{d}^2\boldsymbol{P}_\perp\,\mathrm{d}Y\,\mathrm{d}M^2}$ is the cross section for the production of a $c\bar{c}$ pair with transverse momentum $\boldsymbol{P}_\perp$, rapidity Y and invariant mass M. In this expression $F_{J/\psi}$ is a non-perturbative constant which can be extracted from data. In the following we will focus on ratios of cross sections for which the value of this parameter plays no role.

In forward J/ψ production, the projectile proton is probed at relatively large x which justifies the use of the collinear approximation on this side. The gluon density in the projectile can thus be described using a collinear parton distribution function. Here we will use the MSTW 2008 [12] LO parametrization for this purpose. On the other hand, when working at forward rapidity, the target, which can be either a proton or a nucleus, is probed at very small x. The information about its gluon density is contained in the function

$$S_Y(\boldsymbol{x}_\perp - \boldsymbol{y}_\perp) = \frac{1}{N_c}\left\langle \mathrm{Tr}\, U^\dagger(\boldsymbol{x}_\perp)U(\boldsymbol{y}_\perp)\right\rangle, \tag{2}$$

where $U(\boldsymbol{x}_\perp)$ is a fundamental representation Wilson line in the target color field. The evolution of $S_Y(\boldsymbol{r}_\perp)$ is governed by the running coupling Balitsky-Kovchegov (rcBK) equation [13–15], which is solved numerically. We use as an initial condition the MV^e parametrization [16] which involves, as the AAMQS [17] one, parameters which are extracted from DIS measurements. In the MV^e parametrization the initial condition for a proton target reads

$$S^p_{Y=\ln\frac{1}{x_0}}(\boldsymbol{r}_\perp) = \exp\left[-\frac{\boldsymbol{r}_\perp{}^2Q_{s0}^2}{4}\ln\left(\frac{1}{|\boldsymbol{r}_\perp|\Lambda_{\mathrm{QCD}}}+e_c\cdot e\right)\right], \tag{3}$$

with $x_0 = 0.01$. The expression for the running coupling is

$$\alpha_s(r) = \frac{12\pi}{(33-2N_f)\log\left(\frac{4C^2}{r^2\Lambda_{\mathrm{QCD}}^2}\right)}. \tag{4}$$

In this case there is no explicit dependence on the impact parameter and the integration over $\boldsymbol{b}_\perp$ can be simply replaced by

$$\int \mathrm{d}^2\boldsymbol{b}_\perp \to \frac{\sigma_0}{2}, \tag{5}$$

where $\frac{\sigma_0}{2}$ corresponds to the effective proton transverse area measured in DIS experiments. The values of the parameters in these expressions obtained in Ref. [16] by a fit to HERA DIS data [18] are $Q_{s0}^2 = 0.060\ \mathrm{GeV}^2$, $C^2 = 7.2$, $e_c = 18.9$ and $\frac{\sigma_0}{2} = 16.36$ mb. To generalize the dipole cross section from a proton to a nucleus we use, as in Ref. [16], the optical Glauber model. In this approach the initial condition to the rcBK equation reads

$$S^A_{Y=\ln\frac{1}{x_0}}(\boldsymbol{r}_\perp, \boldsymbol{b}_\perp) = \exp\left[-AT_A(\boldsymbol{b}_\perp)\frac{\sigma_0}{2}\frac{\boldsymbol{r}_\perp{}^2Q_{s0}^2}{4}\ln\left(\frac{1}{|\boldsymbol{r}_\perp|\Lambda_{\mathrm{QCD}}}+e_c\cdot e\right)\right], \tag{6}$$

where the only additional quantity involved compared to the proton case (3) is the standard Woods-Saxon distribution $T_A(\boldsymbol{b}_\perp)$, where $\boldsymbol{b}_\perp$ is the impact parameter:

$$T_A(\boldsymbol{b}_\perp) = \int \mathrm{d}z \frac{n}{1 + \exp\left[\frac{\sqrt{{\boldsymbol{b}_\perp}^2 + z^2} - R_A}{d}\right]}, \tag{7}$$

with $d = 0.54\,\mathrm{fm}$, $R_A = (1.12A^{1/3} - 0.86A^{-1/3})\,\mathrm{fm}$ and n is a normalization constant defined such that $\int \mathrm{d}^2\boldsymbol{b}_\perp T_A(\boldsymbol{b}_\perp) = 1$. Besides T_A, all the parameters take the same value as in the proton case. The rcBK equation is then solved independently for each value of $\boldsymbol{b}_\perp$. In principle this explicit impact parameter dependence can be related to the centrality classes used by experiments in the following way: in the optical Glauber model a class $(c_1 - c_2)\%$ would be defined by limiting the integration over $\boldsymbol{b}_\perp$ between b_1 and b_2 defined such that

$$(c_1 - c_2)\% = \frac{1}{\sigma_{\mathrm{inel}}^{\mathrm{pA}}} \int_{b_1}^{b_2} \mathrm{d}^2\boldsymbol{b}_\perp p(\boldsymbol{b}_\perp). \tag{8}$$

Here $\sigma_{\mathrm{inel}}^{\mathrm{pA}}$ is the total inelastic proton-nucleus cross section, given by

$$\sigma_{\mathrm{inel}}^{\mathrm{pA}} = \int \mathrm{d}^2\boldsymbol{b}_\perp\, p(\boldsymbol{b}_\perp)\,, \tag{9}$$

with the scattering probability at impact parameter $\boldsymbol{b}_\perp$ being

$$p(\boldsymbol{b}_\perp) \approx 1 - e^{-AT_A(\boldsymbol{b}_\perp)\sigma_{\mathrm{inel}}}, \tag{10}$$

where σ_{inel} is the total inelastic nucleon-nucleon cross section. The particle yield in a given centrality class is then given by

$$\frac{\mathrm{d}N}{\mathrm{d}^2\boldsymbol{P}_\perp \mathrm{d}Y} = \frac{\int_{b_1}^{b_2} \mathrm{d}^2\boldsymbol{b}_\perp \frac{\mathrm{d}N(\boldsymbol{b}_\perp)}{\mathrm{d}^2\boldsymbol{P}_\perp \mathrm{d}Y}}{\int_{b_1}^{b_2} \mathrm{d}^2\boldsymbol{b}_\perp\, p(\boldsymbol{b}_\perp)}, \tag{11}$$

where the values of b_1 and b_2 are defined as in Eq. (8). However, if we use this procedure and compare the values of the average number of binary nucleon-nucleon collisions, given in this model by

$$\langle N_{\mathrm{coll}} \rangle_{\mathrm{opt.}} = \frac{\int_{b_1}^{b_2} \mathrm{d}^2\boldsymbol{b}_\perp N_{\mathrm{coll\,opt.}}(\boldsymbol{b}_\perp)}{\int_{b_1}^{b_2} \mathrm{d}^2\boldsymbol{b}_\perp p(\boldsymbol{b}_\perp)}, \tag{12}$$

where

$$N_{\mathrm{coll\,opt.}}(\boldsymbol{b}_\perp) = AT_A(\boldsymbol{b}_\perp)\sigma_{\mathrm{inel}}\,, \tag{13}$$

with the values of $\langle N_{\mathrm{coll}} \rangle$ estimated by ALICE [4] in each centrality class, we find a disagreement between the two as can be seen from Table 1. In the following, for a first comparison, we decide to perform our calculation in each centrality class at a fixed impact parameter $\boldsymbol{b}_\perp$ defined such that the number of binary collisions in the optical Glauber model corresponds to its average value estimated by ALICE, i.e. $N_{\mathrm{coll\,opt.}}(\boldsymbol{b}_\perp) = \langle N_{\mathrm{coll}} \rangle_{\mathrm{ALICE}}$. A more consistent comparison would require the use of distributions in impact parameter space but for this one would need to have access to the N_{coll} distributions at experiments and not only to $\langle N_{\mathrm{coll}} \rangle$.

Table 1. Average number of binary collisions in each centrality class as obtained in the optical Glauber model compared with the value estimated by ALICE [4].

Centrality class	$\langle N_{\text{coll}}\rangle_{\text{opt.}}$	$\langle N_{\text{coll}}\rangle_{\text{ALICE}}$
2–10%	14.7	11.7 ± 1.2 ± 0.9
10–20%	13.6	11.0 ± 0.4 ± 0.9
20–40%	11.4	9.6 ± 0.2 ± 0.8
40–60%	7.7	7.1 ± 0.3 ± 0.6
60–80%	3.7	4.3 ± 0.3 ± 0.3
80–100%	1.5	2.1 ± 0.1 ± 0.2

3 Results

The ALICE collaboration recently measured the nuclear suppression of J/ψ production in proton-lead collisions in different centrality classes at $\sqrt{s_{NN}} = 5$ TeV [4]. To compare our results with these data we use, as explained previously, a fixed impact parameter defined such that $N_{\text{coll. opt}}(\boldsymbol{b}_\perp) = \langle N_{\text{coll}}\rangle_{\text{ALICE}}$ in each centrality class considered by ALICE. This procedure would lead for the 80–100% class to an impact parameter for which the saturation scale of the nucleus would fall below the one of the proton. For this reason we will not consider this class in the following. In Fig. 1 we show the comparison of our results and ALICE data for the nuclear modification factor Q_{pPb}, defined as

$$Q_{\text{pPb}} = \frac{\frac{\mathrm{d}N^{\text{pPb}}}{\mathrm{d}^2\boldsymbol{P}_\perp\,\mathrm{d}Y}}{A\langle T_A\rangle \frac{\mathrm{d}\sigma^{\text{pp}}}{\mathrm{d}^2\boldsymbol{P}_\perp\,\mathrm{d}Y}}, \tag{14}$$

as a function of $P_\perp$ in the five most central classes considered by ALICE. We include in the uncertainty band the variation of m_c between 1.2 and 1.5 GeV and of the factorization scale between $M_\perp/2$ and $2M_\perp$ with $M_\perp = \sqrt{M^2 + P_\perp^2}$ where M is the $c\bar{c}$ pair's invariant mass. The description of the data is quite good in the first three bins but our calculation predicts values of Q_{pPb} which approach unity too quickly when N_{coll} decreases. This too strong dependence on centrality in our calculation can also be seen in Fig. 2, where we show the nuclear modification factor integrated over $P_\perp$ as well as the transverse momentum broadening, defined as the difference of $\langle P_\perp^2\rangle$ in proton-lead and in proton-proton collisions, as a function N_{coll}. However, one should keep in mind that the value of N_{coll} indicated for ALICE data is an average while in our calculation it is a fixed value. Taking into account the fluctuations in our calculation could have a significant impact.

4 Conclusions

In this work we have studied the centrality dependence of forward J/ψ nuclear suppression in the Color Glass Condensate. For this we used, as in Ref. [1], the optical Glauber model to generalize the dipole cross section of a proton to nuclei. In Ref. [1] this model was found to lead to values for the nuclear modification factor in minimum bias collisions closer to experimental results than previous calculations in a similar framework. However, when studying the centrality dependence in this model, we found here that this dependence appears to be much stronger than in recent ALICE data [4]. Nevertheless, we stress that the results shown here have been obtained using a fixed impact parameter for which, in the optical Glauber model, the number of binary collisions corresponds to the average value of this quantity estimated by ALICE. For a more consistent comparison with experimental data,

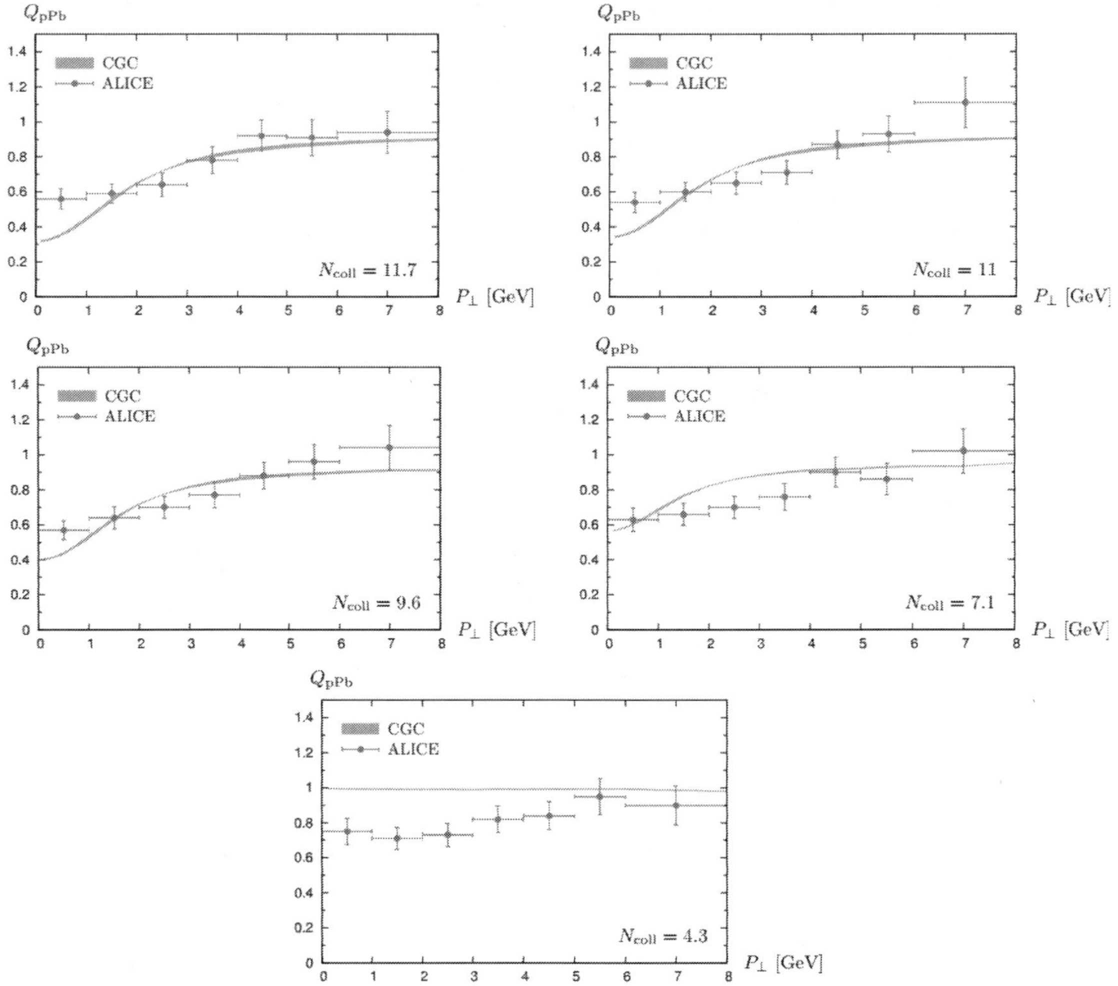

Figure 1. Nuclear modification factor $Q_{\rm pPb}$ as a function of $P_\perp$ in different centrality bins compared with ALICE data [4].

it would be necessary to use distributions in impact parameter space, but this would require access to the typical size of the fluctuations of the number of binary collisions in the experimental centrality classes.

Acknowledgments

T. L. and B. D. are supported by the Academy of Finland, projects 267321 and 273464. H. M. is supported under DOE Contract No. DE-SC0012704. This research used computing resources of CSC – IT Center for Science in Espoo, Finland. We would like to thank C. Hadjidakis and I. Lakomov for discussions on the ALICE data.

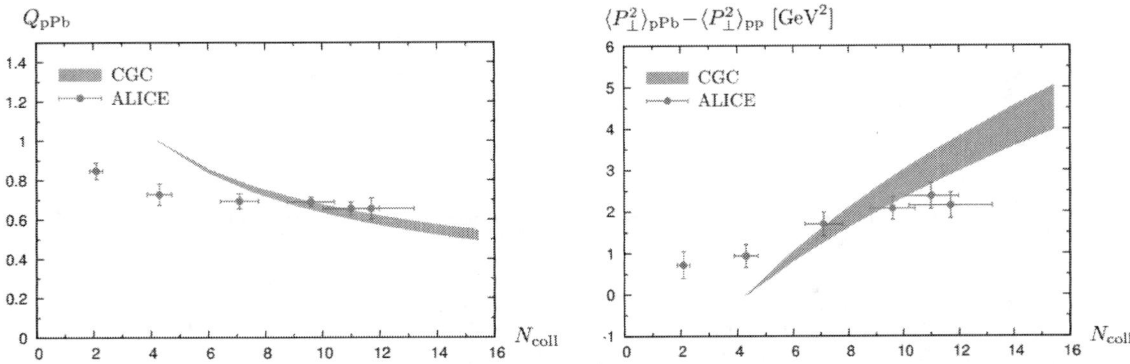

Figure 2. Nuclear modification factor $Q_{\rm pPb}$ (left) and nuclear transverse momentum broadening $\langle P_\perp^2\rangle_{\rm pPb} - \langle P_\perp^2\rangle_{\rm pp}$ (right) as a function of $N_{\rm coll}$ compared with ALICE data [4].

References

[1] B. Ducloué, T. Lappi, H. Mäntysaari, Phys. Rev. **D91**, 114005 (2015), 1503.02789
[2] H. Fujii, K. Watanabe, Nucl. Phys. **A915**, 1 (2013), 1304.2221
[3] H. Fujii, K. Watanabe (2015), 1511.07698
[4] J. Adam et al. (ALICE), JHEP **11**, 127 (2015), 1506.08808
[5] J.P. Blaizot, F. Gelis, R. Venugopalan, Nucl. Phys. **A743**, 13 (2004), hep-ph/0402256
[6] J.P. Blaizot, F. Gelis, R. Venugopalan, Nucl. Phys. **A743**, 57 (2004), hep-ph/0402257
[7] D. Kharzeev, E. Levin, K. Tuchin, Nucl .Phys. **A924**, 47 (2014), 1205.1554
[8] H. Fujii, F. Gelis, R. Venugopalan, Eur. Phys. J. **C43**, 139 (2005), hep-ph/0502204
[9] H. Fujii, F. Gelis, R. Venugopalan, Nucl. Phys. **A780**, 146 (2006), hep-ph/0603099
[10] H. Fujii, K. Watanabe, Nucl. Phys. **A920**, 78 (2013), 1308.1258
[11] Y.Q. Ma, R. Venugopalan, H.F. Zhang, Phys. Rev. **D92**, 071901 (2015), 1503.07772
[12] A. Martin, W. Stirling, R. Thorne, G. Watt, Eur. Phys. J. **C63**, 189 (2009), 0901.0002
[13] I. Balitsky, Nucl. Phys. **B463**, 99 (1996), hep-ph/9509348
[14] Y.V. Kovchegov, Phys. Rev. **D61**, 074018 (2000), hep-ph/9905214
[15] I. Balitsky, Phys. Rev. **D75**, 014001 (2007), hep-ph/0609105
[16] T. Lappi, H. Mäntysaari, Phys. Rev. **D88**, 114020 (2013), 1309.6963
[17] J.L. Albacete, N. Armesto, J.G. Milhano, P. Quiroga-Arias, C.A. Salgado, Eur. Phys. J. **C71**, 1705 (2011), 1012.4408
[18] F. Aaron et al. (H1 and ZEUS), JHEP **1001**, 109 (2010), 0911.0884

EPJ Web of Conferences **112**, 04003 (2016)
DOI: 10.1051/epjconf/201611204003

Kinetic theory of a longitudinally expanding system

Franҫis Gelis[1,a]

[1] *Institut de physique théorique, CEA, CNRS, Université Paris-Saclay*
F-91191 Gif-sur-Yvette, France

Abstract. We use kinetic theory in order to study the role of quantum fluctuations in the isotropization of the pressure tensor in a system subject to fast longitudinal expansion, such as the matter produced in the early stages of a heavy ion collision.

1 Introduction

Heavy ion collisions pose a number of interesting challenges to Quantum Chromodynamics (QCD). Despite the very high center of mass energy of these collisions, the transverse momentum of a typical final state particle is rather small compared to the hard scales typical in jet physics. Thus, a legitimate question is whether one can describe the bulk of this particle production using a weak coupling approach in QCD. The very high parton occupation numbers reached at these energies are the key to a positive answer to this question. At high energy, the gluon density in the hadron wavefunction increases exponentially with rapidity, and eventually the gluons reach an occupation such that their nonlinear interactions become important. These non-linearities tame the growth of the gluon occupation number, and introduce a dynamically generated dimensionful scale –the saturation momentum Q_s– in the problem.

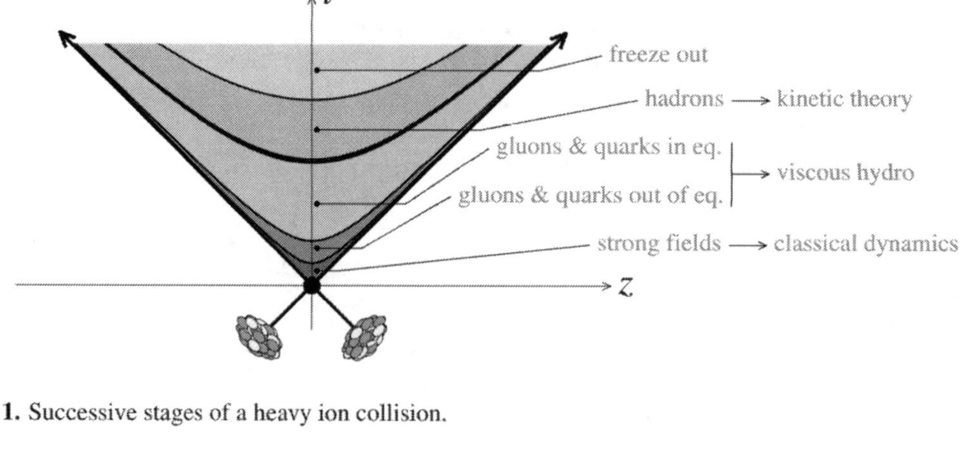

Figure 1. Successive stages of a heavy ion collision.

[a]e-mail: francois.gelis@cea.fr

An effective way of describing the physics of gluon saturation and calculating QCD processes in this regime came a bit later, firstly in the form of the McLerran-Venugopalan model [1] where the degrees of freedom were identified and the classical aspects of their dynamics recognized. In this model, the degrees of freedom with a large longitudinal momentum (in the observer's frame) contribute via the color flow they carry, and are treated as static variables thanks to Lorentz time dilation (the internal dynamics of a hadron appears frozen in the observer's frame, on the timescale of the collision). This simplification does not apply to gluons in the vicinity of the observer's rapidity, and they are treated as conventional gauge field operators. In the MV model, the distribution of color

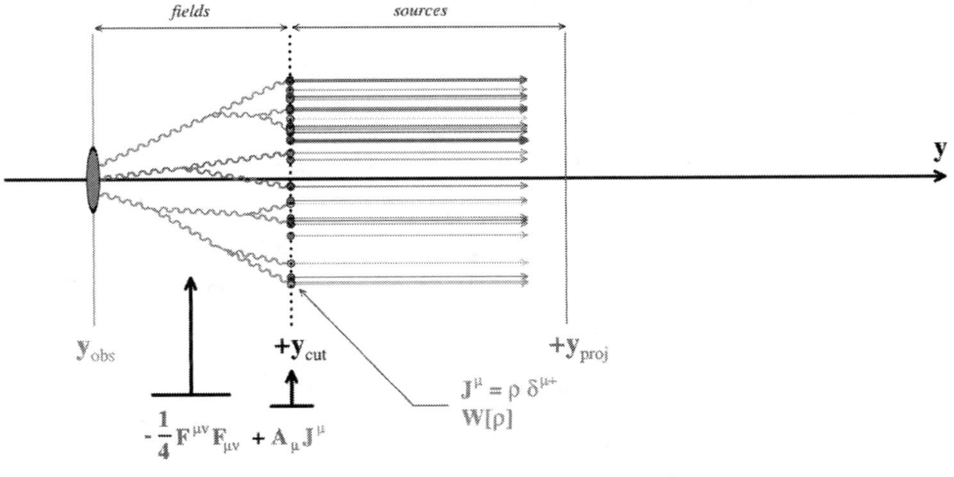

Figure 2. Degrees of freedom in the CGC and their interplay.

sources in the wavefunction of a high energy nucleus was argued to be Gaussian. A few years later, it was realized that this distribution must depend on a longitudinal momentum scale, due to large logarithms coming from loop corrections. This led to the Color Glass Condensate (CGC) effective theory, and to the JIMWLK evolution equation for the distribution of sources [2].

2 Evolution towards hydrodynamics in the CGC

The CGC provides a tool for systematically calculating observable quantities in heavy ion collisions, such as the gluon spectrum, the energy-momentum tensor, etc... One may think of using the energy-momentum tensor calculated in the CGC framework as an initial condition for hydrodynamical models of the expansion of the fireball produced in heavy ion collisions. Such a matching is justified theoretically provided that the two descriptions (CGC and hydrodynamics) agree over some non-empty time range.

At leading order, the CGC leads to a very simple answer, since all inclusive observables can be expressed in terms of a classical solution of the Yang-Mills equations. However, this classical approximation leads to an energy-momentum tensor whose behavior is difficult to reconcile with hydrodynamics. Just after the collision, the longitudinal pressure is negative. It increases and becomes mostly positive at a time $Q_s\tau \sim 1$. However, the ratio of longitudinal to transverse pressure decreases forever, while it increases in hydrodynamics.

In higher orders, the energy-momentum tensor receives quantum contributions, beyond that of the LO classical fields. Despite being suppressed by a power of α_s, these corrections may be large because

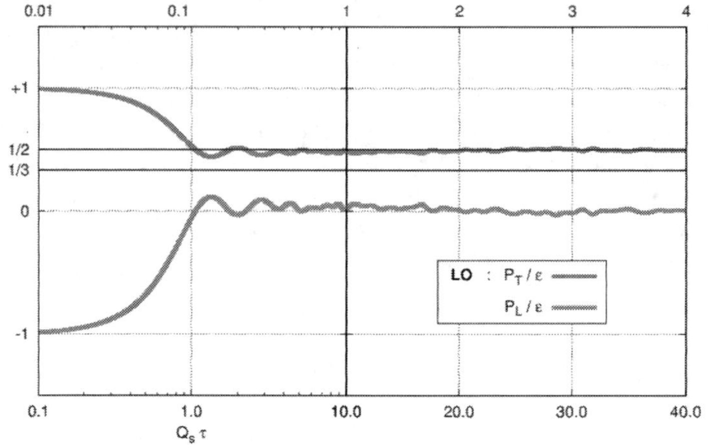

Figure 3. Ratio of the transverse and longitudinal pressures to the energy density in the CGC at leading order. The lower times axis is in units of $1/Q_s$. The upper axis gives absolute values for the specific choice $Q_s = 2$ GeV.

of instabilities that exist in classical solutions of Yang-Mills equations. Indeed, NLO corrections can be expressed in terms of perturbations around the LO classical solutions, that grow exponentially in the presence of unstable modes.

Thus, instead of a strict NLO calculation, one should seek for a resummation. One such framework is provided by the *classical statistical approximation* (CSA), in which one performs an average over classical solutions obtained from Gaussian fluctuating initial conditions. In the limit where the variance of the fluctuations goes to zero, one recovers a unique classical evolution. Several implementations of this scheme have been proposed in the literature, that differ in the distribution of fluctuating initial fields.

1. There exists an ensemble of initial fields such that the CSA contains the leading and next-to-leading orders exactly [3, 4], and a subset of all higher orders. In order to achieve this, the initial fields must be the sum of the leading order classical field $\mathcal{A}^{\mu a}(x)$ (non fluctuating) and a linear superposition of *mode functions* $a^{\mu a}_{\boldsymbol{k}\lambda c}(x)$:

$$A^{\mu a}(x) = \mathcal{A}^{\mu a}(x) + \sum_{\lambda,c} \int_{\boldsymbol{k}} \frac{d^3\boldsymbol{k}}{(2\pi)^3 2|\boldsymbol{k}|} \left[c_{\boldsymbol{k}\lambda c}\, a^{\mu a}_{\boldsymbol{k}\lambda c}(x) + \text{c.c.}\right],$$

$$\langle c_{\boldsymbol{k}\lambda c}\rangle = 0\,, \quad \langle c_{\boldsymbol{k}\lambda c} c^*_{\boldsymbol{k}'\lambda' c'}\rangle = \frac{1}{2}\, 2\big|\boldsymbol{k}\big|\, (2\pi)^3 \delta(\boldsymbol{k}-\boldsymbol{k}')\, \delta_{\lambda\lambda'}\, \delta_{cc'}\,. \tag{1}$$

In this formula, the functions $a^{\mu a}_{\boldsymbol{k}\lambda c}(x)$ are solutions of the Yang-Mills equations linearized about the field $\mathcal{A}^{\mu a}$; with plane wave initial conditions (of momentum $\boldsymbol{k}$, polarization λ and color c). The coefficients $c_{\boldsymbol{k}\lambda c}$ are random Gaussian distributed, with a null mean value and the variance given in the second line. The prefactor $1/2$ in this variance can be interpreted as the occupation number corresponding to the quantum fluctuations in each quantum mode. Since this setup reproduces the one loop result, it also leads to the usual one-loop ultraviolet divergences. However, it also contains extra ultraviolet divergences that do not exist in the underlying theory. These spurious divergences originate from the fact that the CSA is a non-renormalizable scheme: starting at two loops, it contains divergences that cannot be removed by a renormalization of the parameters of the Lagrangian [5].

This leads to a very strong dependence on the ultraviolet cutoff when it is chosen much larger than the physical scales [6].

2. Another form of the CSA uses as initial condition fluctuating fields that correspond to a gas of gluons with a distribution $f(\boldsymbol{k})$. These initial fields have no non-fluctuating part, and the variance of the random coefficients is proportional to $f(\boldsymbol{k})$ [7–9]:

$$A^{\mu a}(x) = \sum_{\lambda,c} \int_{\boldsymbol{k}} \frac{d^3\boldsymbol{k}}{(2\pi)^3 2|\boldsymbol{k}|} \left[c_{\boldsymbol{k}\lambda c}\, a^{\mu a}_{\boldsymbol{k}\lambda c}(x) + \text{c.c.} \right],$$

$$\langle c_{\boldsymbol{k}\lambda c} \rangle = 0\,, \quad \langle c_{\boldsymbol{k}\lambda c} c^*_{\boldsymbol{k}'\lambda' c'} \rangle = f(\boldsymbol{k})\, 2\big|\boldsymbol{k}\big|\, (2\pi)^3 \delta(\boldsymbol{k} - \boldsymbol{k}')\, \delta_{\lambda\lambda'}\, \delta_{cc'}\,. \tag{2}$$

(Note the fact that in eq. (2) the distribution $f(\boldsymbol{k})$ replaces the $1/2$ of eq. (1).) This implementation of the CSA leads to ultraviolet finite results provided that the distribution $f(\boldsymbol{k})$ decreases sufficiently fast, but it does not contain any quantum fluctuations.

Eqs. (1) and (2) describe rather different situations, despite their resemblance:

- The flat spectrum of Eq. (1) on top of a non-fluctuating classical field describes a quantum coherent state.
- The spectrum of Eq. (2), with compact support, describes an incoherent classical state.

When applied to simulations of the early stages of heavy ion collisions, these two types of fluctuating initial conditions lead to quite different behaviors of the pressure tensor. The setup **1** leads to a roughly constant ratio P_L/P_T on short timescales, of the order of a few Q_s^{-1}, while the setup **2** leads to a ratio P_L/P_T that decreases forever. From the time evolution of the particle distribution in this setup, it seems that this decrease of P_L/P_T should continue up to large times of order $\alpha_s^{-3/2} Q_s^{-1}$.

3 Kinetic theory approach

To clarify the role of quantum fluctuations in the process of isotropization, one should include quantum fluctuations while preserving renormalizability. A field theoretic framework that does this is the *2-particle irreducible* approximation [10–15] of the Kadanoff-Baym equations.

A computationally much cheaper alternative is to use kinetic theory [16] in order to investigate the role of quantum fluctuations. The Boltzmann equation with $2 \to 2$ scatterings reads:

$$\begin{aligned} \partial_t f_3 \quad \sim \quad & g^4 \int_{124} \cdots \left[f_1 f_2 (f_3 + f_4) - f_3 f_4 (f_1 + f_2) \right] \\ & + g^4 \int_{124} \cdots \left[f_1 f_2 - f_3 f_4 \right]. \end{aligned} \tag{3}$$

where the dots contain the cross-section and the delta functions for the conservation of energy and momentum. In Eq. (3), we the first line contains the terms that remain in the classical approximation of Eq. (2) (see refs. [17, 18]), and the terms on the second line come from quantum fluctuations.

The cubic terms of the first line may be dominant in the regime of large occupation number. But since this approximation is usually not uniform across all momentum space, this may lead to problems especially when the distribution is very anisotropic. In Fig. 4, we illustrate this for such a distribution, for which the incoming particles 1,2 have almost purely transverse momenta. Isotropization would require that one of the outgoing particles (3 or 4) has a nonzero p_z. With this kinematics, non vanishing contributions can only come from the second line, which is neglected in the classical approximation.

This has been seen in numerical studies of the Boltzmann equation for a longitudinally expanding system [19, 20]. Firstly, one can check that the Boltzmann equation with only the classical terms

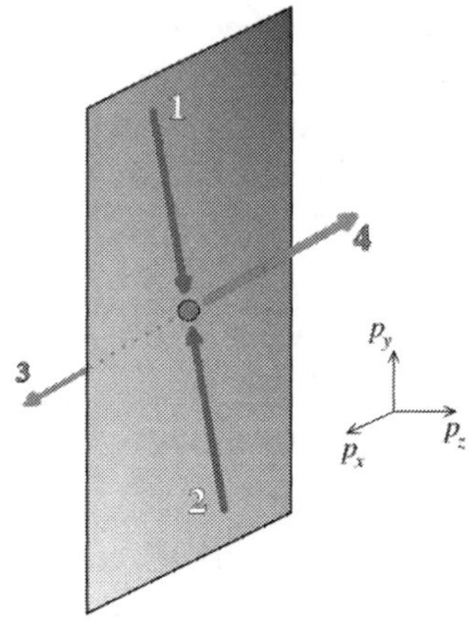

Figure 4. Kinematics of a $2 \to 2$ scattering process that would contribute to the isotropization of pressures when the particle distribution is very anisotropic (with $p_z \approx 0$). The momenta of the two particles in the initial state are purely transverse, and the outgoing particles have opposite p_z's.

leads to a forever decreasing P_L/P_T (see Fig. 5, and the curve $\lambda = 0$ in Fig. 7). Like in the setup **2** of the CSA, the classical approximation of the Boltzmann equation leads to a *classical attractor*: the asymptotic behavior of these classical evolutions is universal, regardless of the coupling and of the details of the initial condition.

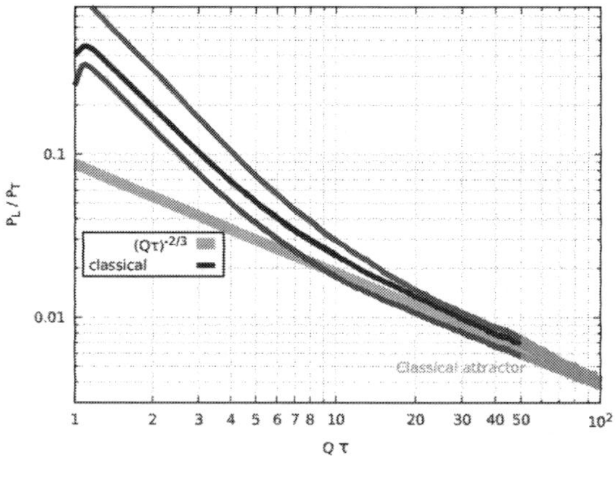

Figure 5. Time evolution of the ratio P_L/P_T in the classical approximation of the Boltzmann equation. When the initial condition is of the form $f_0(\boldsymbol{p}) = (n_0/g^2)\mathbf{f}(\boldsymbol{p})$, the classical evolution does not depend on the coupling and its value need not be specified. The three curves correspond to various CGC-like initial conditions that differ in the magnitude of the initial anisotropy. All these evolutions converge towards a universal classical attractor.

The figure 6 (for scalar fields) and the curves labeled $\lambda = 0.5$ to 10 (for gluons in Yang-Mills theory) in Fig. 4 shows results obtained with the full Boltzmann equation, including the quantum terms. In the computations, the initial conditions was a CGC-like initial condition, with a large occupation number $\sim g^{-2}$ up to a scale Q_s, at a time of order Q_s^{-1}. These results show that it takes very small couplings (or equivalently, very large values of the ratio η/s) and very large occupation numbers for the

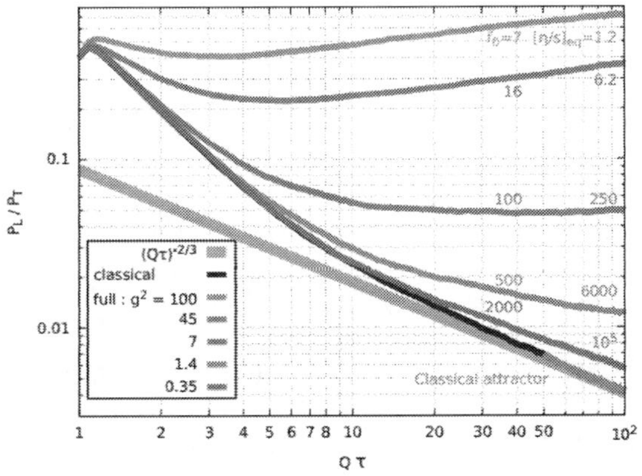

Figure 6. Time evolution of the ratio P_L/P_T obtained from the full Boltzmann equation, for various values of the coupling constant. The initial condition is of the form $f_0(\boldsymbol{p}) = (n_0/g^2)\mathbf{f}(\boldsymbol{p})$. The evolution in the classical approximation is shown for the same initial condition for comparison. The numbers f_0 and $[\eta/s]_{\rm eq}$ overlaid on the plot indicate the corresponding initial occupation number and the equilibrium value of the viscosity to entropy ratio (in equilibrium).

quantum evolution to get even close to the classical attractor regime. For more moderate couplings, the two evolutions depart from each other very early: for $\lambda = 0.5$ (i.e. $\alpha_s \approx 0.02$ for $N_c = 2$ colors), this happens around $Q_s\tau \approx 2$, well before the conjectured range of validity of the classical approximation, $Q_s\tau \approx \alpha_s^{-3/2} \approx 350$. This suggests that quantum fluctuations are essential for isotropization: purely classical approximations do not capture the relevant physics and largely overestimate their own range of validity.

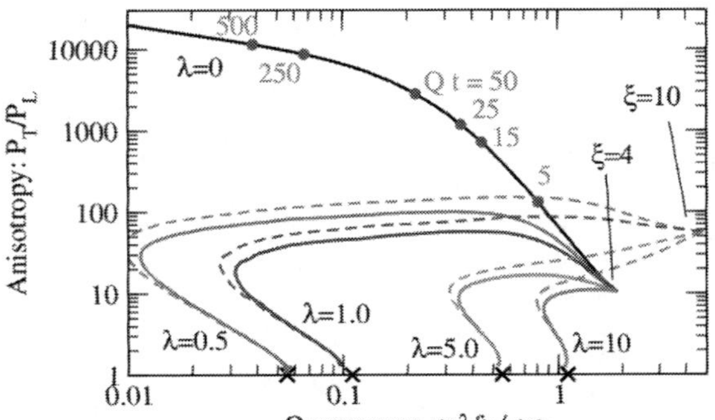

Figure 7. Evolution of the anisotropy and momentum weighted occupation number in a system of gluons described in kinetic theory, for initial conditions of the form $f_0(\boldsymbol{p}) = (n_0/\lambda)\mathbf{f}(\boldsymbol{p})$. The curve labeled $\lambda = 0$ is a classical approximation, while the curves for $\lambda = 0.5, 1, 5, 10$ are obtained with the full Boltzmann equation. Plot taken from [20], with time labels added by us for clarity.

References

[1] L.D. McLerran, R. Venugopalan, Phys. Rev. **D49**, 2233 (1994)

[2] F. Gelis, E. Iancu, J. Jalilian-Marian, R. Venugopalan, Ann. Rev. Nucl. Part. Sci. **60**, 463 (2010), 1002.0333

[3] T. Epelbaum, F. Gelis, Phys. Rev. **D88**, 085015 (2013), 1307.1765

[4] T. Epelbaum, F. Gelis, Phys. Rev. Lett. **111**, 232301 (2013), 1307.2214

[5] T. Epelbaum, F. Gelis, B. Wu, Phys. Rev. **D90**, 065029 (2014), 1402.0115

[6] J. Berges, K. Boguslavski, S. Schlichting, R. Venugopalan, JHEP **05**, 054 (2014), 1312.5216

[7] J. Berges, K. Boguslavski, S. Schlichting, R. Venugopalan, Phys. Rev. **D89**, 074011 (2014), 1303.5650

[8] J. Berges, K. Boguslavski, S. Schlichting, R. Venugopalan, Phys. Rev. **D89**, 114007 (2014), 1311.3005

[9] J. Berges, K. Boguslavski, S. Schlichting, R. Venugopalan, Phys. Rev. Lett. **114**, 061601 (2015), 1408.1670

[10] J.M. Luttinger, J.C. Ward, Phys. Rev. **118**, 1417 (1960)

[11] G. Aarts, D. Ahrensmeier, R. Baier, J. Berges, J. Serreau, Phys. Rev. **D66**, 045008 (2002), hep-ph/0201308

[12] E.A. Calzetta, B.L. Hu (2002), hep-ph/0205271

[13] J. Berges, Phys. Rev. **D70**, 105010 (2004), hep-ph/0401172

[14] Y. Hatta, A. Nishiyama, Nucl. Phys. **A873**, 47 (2012), 1108.0818

[15] Y. Hatta, A. Nishiyama, Phys. Rev. **D86**, 076002 (2012), 1206.4743

[16] T. Epelbaum, F. Gelis, N. Tanji, B. Wu, Phys. Rev. **D90**, 125032 (2014), 1409.0701

[17] A.H. Mueller, D.T. Son, Phys. Lett. **B582**, 279 (2004), hep-ph/0212198

[18] S. Jeon, Phys. Rev. **C72**, 014907 (2005), hep-ph/0412121

[19] T. Epelbaum, F. Gelis, S. Jeon, G. Moore, B. Wu, JHEP **09**, 117 (2015), 1506.05580

[20] A. Kurkela, Y. Zhu, Phys. Rev. Lett. **115**, 182301 (2015), 1506.06647

EPJ Web of Conferences **112**, 04004 (2016)
DOI: 10.1051/epjconf/201611204004

Multiplicity distributions and fluctuations in proton-proton and heavy-ion collisions

Maitreyee Mukherjee[a]

Variable Energy Cyclotron Centre, Kolkata, India

Abstract. The charged particle multiplicity is one of the simplest observable in collision of the hadrons, imposing constrains on the mechanisms of the particle production. Multiplicity distributions at lower energies have been studied earlier. In this work, it has been investigated how the multiplicity distributions and fluctuations have been explained in LHC energies. For this, a detailed study has been done from √s=7.7 GeV/A to √s= 2.76 TeV/A with event generators AMPT-Default and AMPT-String Melting. It will be also discussed how multiplicity distribution parameters are related to early stages of collision and how small-x helps to find information on initial states.

1 Introduction

Multiplicity, i.e, the number of charged particles produced per event is an important ingredient for the understanding of multi-particle production. Multiplicity of produced particles is an important quantity to characterize the evolving system. This is one of the very basic studies as many of the observables such as energy-density, transverse energy, momentum etc. are connected to the number of particles and its fluctuation event-by-event. Multiplicity is a global observable, and multiplicity distribution and fluctuation sheds light into mechanisms of particle production and provide constraints to the many models aiming at describing these collisions, it helps to reject or improve models [1,2].

Because of the production of large number of particles in each heavy-ion collision at RHIC as well as LHC, it has become possible to make precision event-by-event measurements of several quantities and study their event-by-event fluctuations. QCD phase transition can manifest itself by characteristic behavior of several observables which may vary dramatically from one event to the other. Fluctuations are usually characterized by scaled variances. Dynamical fluctuations (other than the statistical fluctuation and fluctuation in the number of participants) may provide insights on the intrinsic mechanisms of the particle production.

Event-by-event multiplicity fluctuations are basically connected to QCD-phase transition. Variance of the multiplicity distributions are directly related to the isothermal compressibility (k_T) of the system formed in heavy ion collision, i.e,

$$\sigma^2 = \frac{k_B T < N >^2}{V} k_T \tag{1}$$

where, k_B is Boltzmann's constant and <N> is the mean multiplicity. k_T is expected to increase by at least an order of magnitude close to the QCD critical point [3]. Additionally, determination of critical

[a] Corresponding author: maitreyee.mukherjee@gmail.com

exponent helps to determine the universality class in which QCD is grouped. Multilicity fluctuation expressed in terms of scaled variance, is basically the variance scaled over mean of the multiplicity distribution, which also is expected to increase by an order of magnitude near the QCD critical point.

2 Earlier measurements

Multiplicity distributions in e^+e^- collisions at √s = 29 GeV are described by Poisson distributions, indicates no correlation between the produced particles. p+p non-single diffractive events behave little differently, follow dispersion D∝√<n> upto max. ISR energy at √s = 62 GeV (where, <n> denotes mean multiplicity). In 1972, concept of KNO-scaling comes, which is an universal scaling followed at higher energies. Here, we get, <n>∝ln √s [2]. UA5 shows that KNO-scaling breaks at √s = 200 GeV. This is shown in Fig 1.

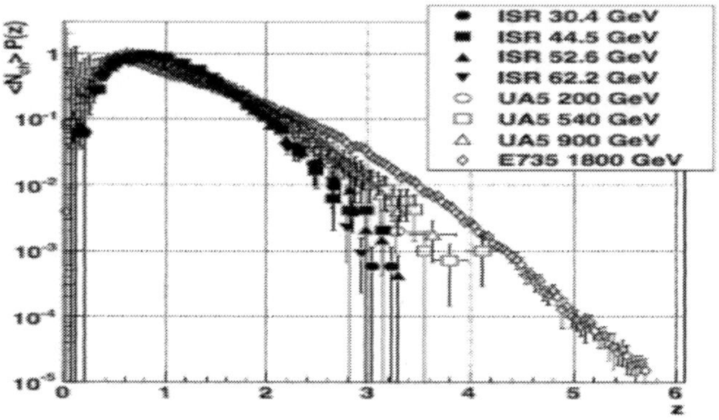

Figure 1. KNO-Scaling and its breaking

UA5 found that, upto √s = 540 GeV , multiplicity distributions follow NBD, which has two parameters <n> and k (related to width), and 1/k is approximately proportional to ln √s. Here, we can write the multiplicity distributions follow [3,4],

$$P(n) = \frac{\Gamma(n+k_{NBD})}{\Gamma(n+1)\Gamma(k_{NBD})} \frac{(\mu_{ch}/k_{NBD})^n}{(1+\mu_{ch}/k_{NBD})^{n+k_{NBD}}} \tag{2}$$

where,

$$\mu_{ch} = < N_{ch} > \tag{3}$$

and,

$$\sigma^2 = \mu + \mu^2 / k_{NBD} \tag{4}$$

and, the scaled variance (ω_{ch}) is given by,

$$\omega_{ch} = 1 + \frac{\mu_{ch}}{k_{NBD}} \tag{5}$$

Deviations from single NBD observed around √s = 900 GeV. A shoulder-like structure seen. Thus concept of double NBD comes, where the distributions are described as the combination of two NBD's, one having soft component and another having the semi-hard component. Another description was given from the point of view of multi-parton interaction, which becomes very much important at higher energies.

3 Multiplicity fluctuations : Results from AMPT

For this work, centrality has been selected using minimum-bias distribution of number of participants (N_{part}) as shown in Fig 2. N_{part} is directly proportional to some quantities that are experimentally observed, such as N_{ch}. So, centrality selection from N_{part} is quite justified. N_{ch}-minimum bias distribution is the convolution of N_{ch}-distributions in different centralities.

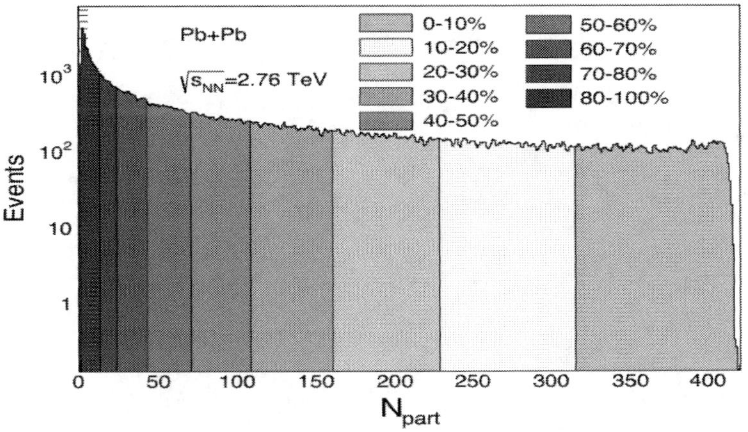

Figure 2. Centrality selection from N_{part}-distribution using AMPT-model

Finer bin in centrality needs to be selected for fluctuation studies. This will avoid inherent fluctuations in N_{part} within a given centrality. In this analysis, we select 0-1%, 1-2%, 2-3%, 3-4%, …… centrality bins and also check with centrality binwidth corrections. Now, centrality bin-width effect arises due to the impact parameter(or volume) variations due to the finite centrality bin. To correct for this effect, the best way is to divide one centrality bin into smaller bins and weight the moments [5]. Fig 3 illustrates the correction due to the centrality binwidth-effect.

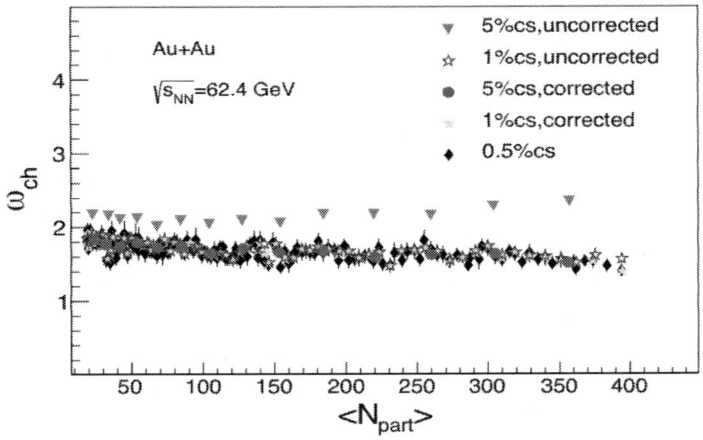

Figure 3. Centrality binwidth-effect

From Fig 3, it is evident that, the binwidth correction changes result for 5% cs by ~25% and for 1%cs by ~2-3% at the most central. So, doing analysis using narrower centrality bins(here, 0.5%cs) is quite

The centrality dependence of the multiplicity fluctuation has been investigated in this work. The whole work has been done using AMPT-default and AMPT-String Melting model. The transverse momentum and pseudorapidity range is selected as $0.2 < p_T < 2$ GeV/c and $|\eta|<0.5$, respectively. The result has been shown in Fig 4.

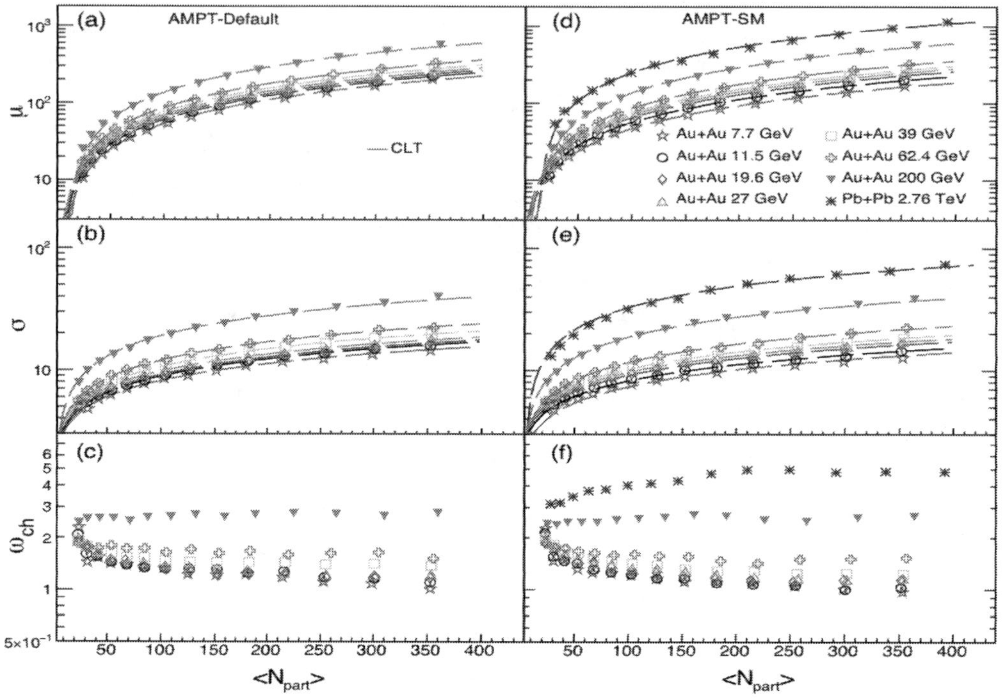

Figure 4. Centrality dependence of mean, variance and the scaled variance

We have observed that, Central Limit Theorem (CLT)-fits are better for μ and σ for lower energies. For energies from 7.7 GeV to 200 GeV, ω_{ch} shows increasing trend for central to peripheral collisions. For 2.76 TeV, the trend for scaled variance is somehow opposite. Result for AMPT-default has not been shown for 2.76 TeV, as we have observed that , results from String-Melting model are more reliable in such higher energies.

The beam-energy dependence of the scaled variance has also been investigated, which is shown here in Fig 5.

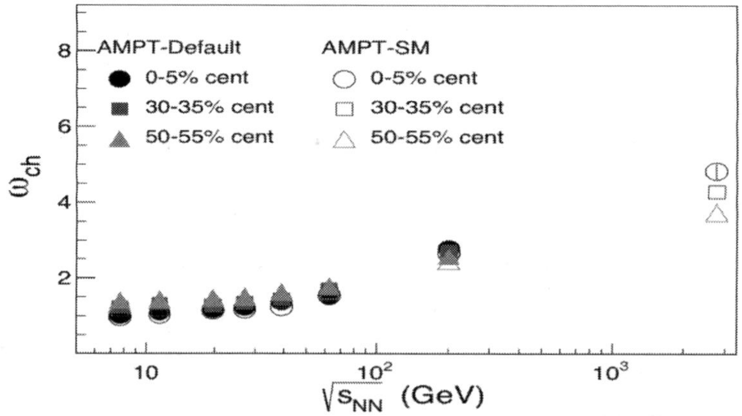

Figure 5. Beam-energy dependence of the scaled variance

We observe that the fluctuations show an increasing trend from lower to higher energies. Also, the fluctuations at lower energies are close to each other for all centralities.

4 Recent results from LHC

The ALICE Collaboration has carried out a detailed study of multiplicity distributions of primary charged particles produced in proton-proton collisions, at $\sqrt{s} = 0.9, 2.36, 2.76, 7$ and 8 TeV[6] (Result for 8 TeV is shown in Fig 6)

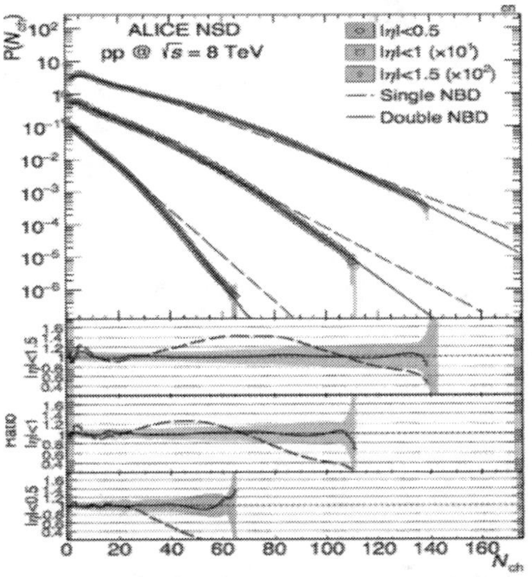

Figure 6. Multiplicity Distributions for NSD events

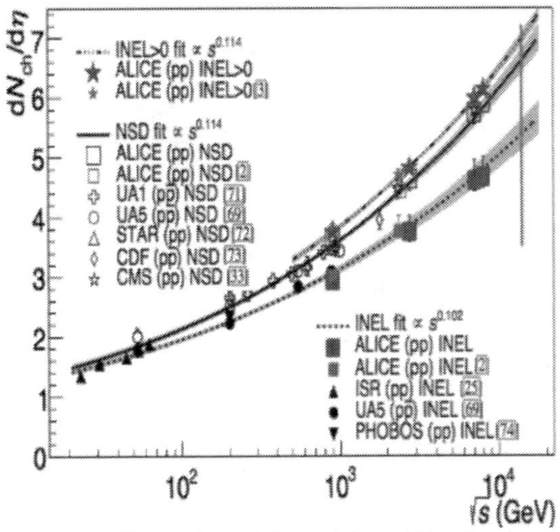

Figure 7. Charged-particle pseudorapidity density

It has been observed from Fig 6 that, at LHC energies, above 2 TeV, multiplicity distributions can no longer be represented by a single NBD, but a double NBD give a good representation of the data. Fig 7 shows that, the power law parameterization of $dN_{ch}/d\eta$ at $\eta = 0$, s^{δ}, provides a good description of the data from ISR to LHC energies: $\delta = 0.102 \pm 0.003, 0.114 \pm 0.003$ and 0.114 ± 0.001, for the INEL, NSD and INEL>0 event classes, respectively [6]. The charged particle pseudorapidity density distribution has been measured in Pb-Pb collisions at $\sqrt{s}$=2.76 TeV also in ALICE [7]. The measurement was performed in a wide pseudorapidity interval of $-5.0 < \eta < 5.5$ allowing the total number of charged particles produced at the LHC. Results are shown in Fig 8 [7].

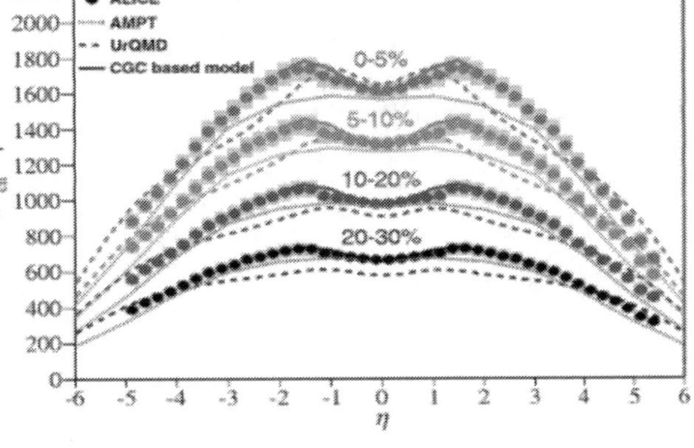

Figure 8. Comparison to models with data from ALICE

We observe that the theoretical predictions do not describe the data well, except the CGC based model within its limited pseudorapidity range.

5 Multiplicity distributions : connection to early stages of collision

At small enough Bjorken-x, we get a saturation region, where the non-linear effects due to high gluon density becomes very important [8]. Small Bjorken-x (expressed as $x = p_T/(\sqrt{s})\cdot e^{-\eta}$) values can be accessed by going to higher center-of-mass energy and higher rapidity, which is easier to achieve in LHC energies. So, this is an opportunity to study initial state effects. Yang Mills dynamics gives framework to study correlated multi-particle production. From Glasma flux-tube picture, NBD-k scaled over mean multiplicity can be expressed as [9,10],

$$\frac{k}{\bar{n}} = \frac{2N_c\pi}{c}(\kappa\alpha_s) \tag{6}$$

where, N_c are the number of gluon-colors, c is a constant and $\varkappa$ is a non-perturbative constant. So, we can say that for symmetric systems, k is proportional to mean multiplicity ($\tilde{n}$). For asymmetric systems (say, p-Pb), we have [11],

$$\bar{n} \sim k \ln^2 \frac{Q^2_{s,Pb}}{Q^2_{s,p}} \tag{7}$$

where, Q_s denotes the saturation-momentum. We see, this is not a proportional-dependence. Thus, we can differentiate between symmetric and asymmetric system. Thus, the multiplicity distribution parameters are related to early stages of collision.

References

1. M. Gazdzicki, *On the history of multiparticle production in high energy collisions*, Acta Phys.Polon.B **43** (2012) 791; arXiv:1201.0485 [hep-ph] (2012).
2. J. Fiete Grosse-Oetringhaus, K.Reygers, J.Phys. G **37** 083001 (2010).
3. PHENIX Collaboration, A. Adare *et al.*, Phys. Rev. C **78,** 044902 (2008).
4. WA98 Collaboration, M.M. Aggarwal *et al.*, Phys. Rev. C **65,** 054912 (2002).
5. N.R. Sahoo, S. De, T.K. Nayak, Phys.Rev. C **87,** 044906 (2013).
6. ALICE Collaboration, J. Adam *et al.*, arxiv : 1509.07541 [nucl-ex] (2015).
7. ALICE Collaboration, E. Abbas *et al.*, Physics Letters B **726** (2013) 610-622.
8. E. Iancu, arxiv : 1205.0579 [hep-ph] (2012).
9. B. Schenke, P. Tribedy, R. Venugopalan, Phys. Rev. C **86** (2012).
10. T. Lappi, arxiv : 1104.3725 [hep-ph] (2011).
11. A.H.Rezaeian, arxiv : 1308.4736 [hep-ph] (2013).

EPJ Web of Conferences **112**, 04005 (2016)
DOI: 10.1051/epjconf/201611204005

Nuclear suppression in p–A collisions from induced radiation

François Arleo[1], Rodion Kolevatov[2], Stéphane Peigné[3], and Taklit Sami[3]

[1] *Laboratoire Leprince-Ringuet (LLR), École polytechnique, CNRS/IN2P3 91128 Palaiseau, France*
[2] *Department of High Energy Physics, Saint-Petersburg State University Ulyanovskaya 1, 198504, Saint-Petersburg, Russia*
[3] *SUBATECH, UMR 6457, Université de Nantes, Ecole des Mines de Nantes, CNRS/IN2P3 4 rue Alfred Kastler, 44307 Nantes cedex 3, France*

Abstract. The current status of coherent energy loss is reviewed, both in theory and in its phenomenological applications to p–A collisions.

1 Introduction

The nuclear suppression of hadron production in proton-nucleus (p–A) with respect to proton-proton (p–p) collisions has been widely discussed, but is still an open question due to the presence of various competing effects, depending on the precise kinematics and collision energy. Until quite recently, most of the phenomenological approaches assumed hadron nuclear suppression to be due either to hadron nuclear absorption or to the shadowing of the gluon distribution in the target nucleus expected at small $x_2 \lesssim 10^{-2}$. However another fundamental phenomenon, namely coherent parton energy loss in cold nuclear matter, could play a decisive role in hadron nuclear suppression [1].

Medium-induced coherent gluon radiation is a QCD prediction which has now been obtained in various theoretical setups [1–7], and was successfully applied to the phenomenology of quarkonium nuclear suppression in p–A collisions [2, 8, 9]. Here we review the current status of coherent energy loss in theory (Sect. 2) and in its phenomenological applications to p–A collisions (Sect. 3).

2 Theory of induced energy loss: a brief history

It was first noted by Gavin and Milana [10] that the strong increase of J/ψ suppression with x_F observed in p–A collisions could be easily reproduced by *assuming* an induced parton energy loss ΔE scaling as the energy E of the energetic $c\bar{c}$ pair, namely $\Delta E \propto E$. However, the Gavin-Milana 'explanation' was soon put aside, because of a claim that any *induced* energy loss ΔE should be *bounded* in the $E \to \infty$ limit. It is now understood (see for instance Ref. [11]) that the latter claim is incorrect, more precisely the induced energy loss is not bounded in general, but only in the specific situation where the energetic parton is suddenly accelerated (as in deep inelastic scattering) in the nuclear medium. In the situation where the parton is 'asymptotic', *i.e.* 'prepared' at $t = -\infty$ and 'tagged' at $t = +\infty$ after crossing a nuclear medium of thickness L (a situation relevant to forward hadron production in p–A collisions), the behavior $\Delta E \propto E$ is correct [1]. Below we briefly review the features of induced radiation in these two different situations.

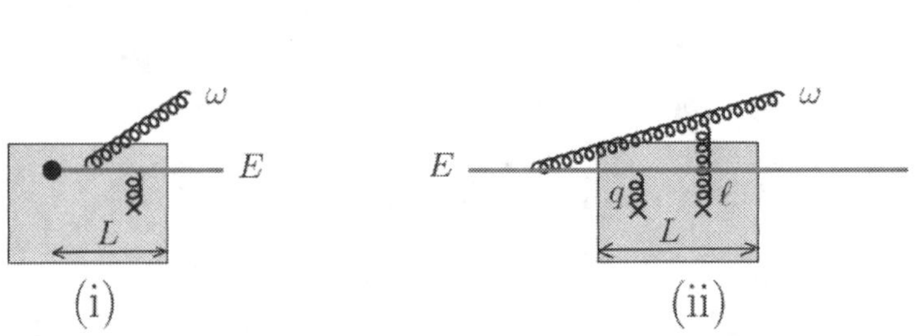

Figure 1. Two qualitatively different situations relevant to medium-induced parton energy loss: (i) parton suddenly accelerated in a medium and (ii) 'asymptotic parton' crossing a medium.

2.1 Parton suddenly produced in a medium

The case (i) of a parton suddenly created in a medium is sketched in Fig. 1 (left). The associated medium-induced (soft) gluon radiation spectrum $\omega dI/d\omega$ has been calculated before and is represented schematically in Fig. 2 (left).

It is useful to recall the different regimes identified in [12] depending on the value of the gluon formation time $t_f = \omega/k_\perp^2$ (for a more detailed discussion in QED and in QCD, see [11]). In the *incoherent* or *Bethe–Heitler* regime $t_f < \lambda_R$ (where λ_R is the parton mean free path in the medium), corresponding to very soft gluons $\omega < \hat{q}_R \lambda_R^2$ (where $\hat{q}_R \equiv \mu^2/\lambda_R$ with μ the typical transverse momentum exchange in a single scattering), each scattering center acts as an independent source of radiation. In this regime the energy spectrum $\omega dI/d\omega$ is independent of ω (up to logarithms). In the domain $\lambda_R < t_f < L$, corresponding to intermediate radiated energies $\hat{q}_R \lambda_R^2 < \omega < \hat{q}_R L^2$, a group of $\sim t_f/\lambda_R$ scattering centers acts as a single radiator, leading to a relative suppression $\propto 1/\sqrt{\omega}$ of the gluon radiation spectrum with respect to the Bethe-Heitler regime [12, 13]. This is the so-called Landau–Pomeranchuk–Migdal (LPM) suppression regime. Finally, in the domain of large formation time $t_f > L$ (or $\omega > \hat{q}_R L^2$), called the *fully coherent* domain in the following, all scattering centers in the medium act coherently as a source of radiation, and the induced radiation spectrum behaves as $\sim 1/\omega$ [14], *i.e.* is more strongly suppressed than in the LPM regime.

In the situation (i), the suppression of the radiation spectrum in the fully coherent domain can be simply understood as follows. For formation times $t_f \gg L$, the radiation spectrum must arise dominantly from time-ordered diagrams where the gluon emission vertex is far beyond the medium. The associated gluon emission amplitude is then $\propto (\vec{\theta} - \vec{\theta}_s)/(\vec{\theta} - \vec{\theta}_s)^2$ [1], where $\vec{\theta} \equiv \vec{k}_\perp/\omega$ and $\vec{\theta}_s \equiv \vec{p}_\perp/E$ are the final 'angles' of the radiated gluon and energetic parton, respectively. The double differential spectrum $\omega dI/d\omega d^2\vec{\theta} \propto 1/(\vec{\theta} - \vec{\theta}_s)^2$ depends on the medium only through $\vec{\theta}_s$, which is related to the medium size by the random walk estimate $\theta_s^2 \simeq (\mu^2/E^2) \cdot (L/\lambda_R)$ resulting from multiple scattering of the energetic parton across the medium. The spectrum *integrated over gluon angles*, $\omega dI/d\omega \sim \int d^2\vec{\theta}/(\vec{\theta} - \vec{\theta}_s)^2$, is independent of $\vec{\theta}_s$ (by a trivial change of variable in the $\vec{\theta}$-integral) and thus of L, leading to a strong suppression of the *induced* energy spectrum. The above heuristic discussion illustrates why large formation times $t_f \gg L$ are suppressed in situation (i).

Due to this suppression, the spectrum decreases as $1/\omega$ when $\omega > \hat{q}_R L^2$ [14], leading to an *average* energy loss in situation (i) effectively dominated by $\omega \sim \hat{q}_R L^2$ and given by (up to logarithms)

$$\Delta E \equiv \int d\omega\, \omega \frac{dI}{d\omega} \sim \alpha_s N_c\, \hat{q}_R L^2 = \alpha_s C_R\, \hat{q} L^2 , \tag{1}$$

where C_R is the energetic parton color charge, and $\hat{q} \equiv \hat{q}_g = \mu^2/\lambda_g$ the *gluon* transport coefficient. For a rigorous derivation of the induced radiation spectrum and associated average energy loss in situation (i), see Refs. [12, 13].

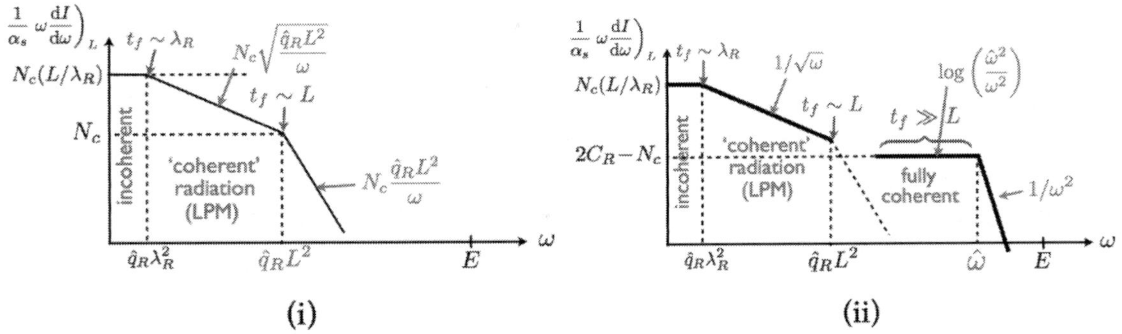

Figure 2. Log-log schematic plot showing the parametric dependence of the induced radiation spectrum $\omega dI/d\omega$ (divided by α_s), in case (i) of a parton produced in a medium (left) and in case (ii) of an 'asymptotic parton' (right). See text for details.

2.2 Fully coherent energy loss

We now turn to the case (ii) of an asymptotic parton (see Fig. 1, right), which is relevant to forward hadron production in p–A collisions. The fundamental difference with the case (i) reviewed in the previous section is that large formation times $t_f \gg L$ are not suppressed any longer, but on the contrary *dominant* in $\omega dI/d\omega$. This can be simply understood as follows. Supposing $t_f \gg L$, three types of contributions to the radiation spectrum can emerge: purely final state radiation, where the gluon emission vertex is associated with a large *positive* time $t \gg L$ in both the amplitude and its conjugate, purely initial state radiation corresponding to large *negative* emission times in the amplitude and its conjugate, and an interference contribution where the gluon is emitted with $t \gg L$ in the amplitude and $-t \gg L$ in the conjugate amplitude (or vice versa). The two former contributions are suppressed in the *induced* spectrum, for the same reason as in the case (i), see Sect. 2.1. On the contrary, the latter interference contribution to $\omega dI/d\omega$ is of the form $\sim \int d^2\vec{\theta}\,(\vec{\theta}/\theta^2)\cdot[(\vec{\theta}-\vec{\theta}_s)/(\vec{\theta}-\vec{\theta}_s)^2]$, and its dependence on $\vec{\theta}_s$ (and thus on L) cannot be removed by a simple shift in the $\vec{\theta}$-integral. It thus survives in the induced spectrum, and actually gives the dominant contribution in case (ii) [1]. Moreover, the dependence of the interference on the parton scattering 'angle' $\vec{\theta}_s$ leads to an explicit dependence of the induced spectrum on the hard exchange $\vec{q}_\perp = E\vec{\theta}_s$ suffered by the parton.

As can be seen in Fig. 2 (right), the resulting induced radiation spectrum in the region $t_f \gg L$ is logarithmic, up to the scale $\hat{\omega} \equiv (\sqrt{\hat{q}L}/M_\perp)\,E$ (where $M_\perp \equiv (M^2 + q_\perp^2)^{1/2}$ with M the parton mass), above which it is suppressed as $1/\omega^2$. The resulting average energy loss is thus dominated by $\omega \sim \hat{\omega}$ and reads [1]

$$\Delta E \sim \alpha_s\,\hat{\omega} = \alpha_s\frac{\sqrt{\hat{q}L}}{M_\perp}\,E\,. \tag{2}$$

In summary, the interference contribution, which is absent in the situation (i), leads in the situation (ii) to an average parton energy loss proportional to E and arising from the *fully coherent* domain of formation times $t_f \gg L$.

Let us remark that for the induced radiation associated with $t_f \lesssim L$, the precise way the parton is produced (in the medium or long before) does not matter. Thus, the induced spectra in the corresponding ω-domain, $\omega < \hat{q}_R L^2$, are parametrically the same in both cases (i) and (ii), see Fig. 2. Note that in case (ii), this domain gives a contribution to ΔE of the LPM type (1), but at large E this contribution is negligible compared to the coherent energy loss (2).

The rigorous calculation of medium-induced coherent radiation has been addressed by several groups, using different formalisms and considering different particular cases. Ref. [1] (see also [2]) studied the induced coherent radiation associated with the hard forward $g \to Q\bar{Q}$ process (mediated by a single hard gluon exchange in the t-channel), the final $Q\bar{Q}$ pair being a massive *pointlike* color octet, in a Feynman diagram calculation and at first order in the opacity expansion formalism [15]. Induced coherent radiation was also studied using a semi-classical method in Refs. [3, 4], at first order [3] and all orders [4] in opacity, and in a similar kinematical setup as that of Ref. [1], however for the $q \to q$ case of a massless quark experiencing a hard scattering mediated by a *color singlet* exchange in the t-channel. In Ref. [5], induced coherent radiation was revisited and derived to all orders in the opacity expansion, for any $1 \to 1$ hard forward process. A general expression encompassing the particular cases studied before was found for the induced (soft) coherent spectrum [5],

$$\omega \frac{dI}{d\omega} = (C_R + C_{R'} - C_t)\, \frac{\alpha_s}{\pi} \log\left(1 + \frac{\hat{\omega}^2}{\omega^2}\right); \qquad \hat{\omega} \equiv \frac{\sqrt{\hat{q}L}}{M_\perp}\, E\,, \tag{3}$$

with C_R, $C_{R'}$ the incoming and outgoing color charges, and C_t the color charge of the hard t-channel exchange.[1] Note that the transport coefficient $\hat{q}$ to be used in the expression of $\hat{\omega}$ is the *gluon* transport coefficient $\hat{q} \equiv \hat{q}_g$, independently of the incoming and outgoing parton color states [5]. The limiting behaviors of (3) in the fully coherent domain, namely, the (logarithmic) 'plateau' at $\omega \ll \hat{\omega}$, and the $1/\omega^2$ suppression at $\omega \gg \hat{\omega}$, are displayed in Fig. 2 (right), where for the sake of illustration the incoming and outgoing partons are chosen of the same type, $C_R = C_{R'}$. With a color octet t-channel exchange, $C_t = N_c$, the height of the plateau is thus $\propto (2C_R - N_c)$, see Fig. 2 (right).[2]

The induced coherent radiation associated with $1 \to 2$ forward processes was also addressed (in the leading logarithm and large N_c limits), for $g \to q\bar{q}$ and $q \to qg$ in the dipole formalism [6] and for $q \to qg$ and $g \to gg$ using the opacity expansion [7]. In the leading logarithm approximation, the soft induced coherent radiation does not probe the size of the final two-parton system, and thus only depends on its *total* color charge. Hence the conjecture, proposed and explicitly checked for $q \to qg$ and $g \to gg$ in Ref. [7], that the spectrum associated with $1 \to n$ hard forward processes is given by an incoherent sum of spectra associated with $1 \to 1$ processes, weighted by the probabilities $P_{R'}$ for the n-parton state to be produced in color representation R' in the hard process. This conjecture is expected to hold for any finite N_c (but only in the leading logarithm approximation) [7].

3 Coherent energy loss in p–A collisions

Induced coherent radiation arises in the production of a single forward particle [1–5] and in forward dijet production [6, 7], suggesting the broad relevance of coherent energy loss in forward processes.

3.1 Quarkonium suppression from fixed-target to collider energies

Coherent energy loss is expected in quarkonium production in p–A collisions, where typically (at sufficiently large $\sqrt{s}$) a high-energy gluon from the projectile proton is scattered to a compact color octet heavy $Q\bar{Q}$ pair [1].

[1] The color factor in (3) has a very simple explanation [5]. The interference term is of the form $\sim 2T^a_R T^a_{R'}$ (where T^a_R and $T^a_{R'}$ are the color generators of the incoming and outgoing parton color representations R and R'), which can be written as $2T^a_R T^a_{R'} = (T^a_R)^2 + (T^a_{R'})^2 - (T^a_R - T^a_{R'})^2 = C_R + C_{R'} - C_t$.

[2] In the $q \to q$ case, $C_R = C_F = (N_c^2 - 1)/(2N_c)$, we have $2C_F - N_c < 0$ and the *induced* coherent energy loss is negative, see Ref. [5] for a discussion and simple interpretation of this result.

In Refs. [2, 8], coherent energy loss is implemented as follows. The single differential p–A production cross section as a function of the quarkonium (labelled ψ) energy reads

$$\frac{1}{A}\frac{d\sigma^{\psi}_{pA}}{dE}(E) = \int_0^{\varepsilon^{max}} d\varepsilon\, \mathcal{P}(\varepsilon, E, \hat{q}L)\, \frac{d\sigma^{\psi}_{pp}}{dE}(E+\varepsilon)\ , \tag{4}$$

where E (respectively, ε) is the energy (respectively, energy loss) of the compact $Q\bar{Q}$ pair in the rest frame of the nucleus A. The upper limit on the energy loss is $\varepsilon^{max} = \min\left(E, E_p - E\right)$, where E_p is the beam energy in that frame. The energy loss probability distribution, or *quenching weight*, $\mathcal{P}$, is simply related [2] to the induced coherent radiation spectrum for the $g \to Q\bar{Q}$ process, which spectrum is given by Eq. (3) in the particular case $C_R = C_{R'} = C_t = N_c$. The quenching weight depends on the *gluon* nuclear broadening $\sqrt{\hat{q}L}$, with $\hat{q}$ parametrized as [2]

$$\hat{q} \equiv \hat{q}_0 \left[\frac{10^{-2}}{\min(x_0, x_2)}\right]^{0.3}; \quad x_0 \equiv \frac{1}{2m_p L}\ , \tag{5}$$

where $\hat{q}_0$ is the only free parameter of the model, and $x_2 = M_\perp\, e^{-y} / \sqrt{s}$ (with y the quarkonium rapidity in the center-of-mass frame of an elementary proton–nucleon collision, and $M = 3\,\text{GeV}$ ($M = 9\,\text{GeV}$) for the $c\bar{c}$ ($b\bar{b}$) mass in the expression of $M_\perp$). The parameter $\hat{q}_0$ is determined by fitting the J/ψ suppression measured by E866 [16] in p–W over p–Be collisions ($\sqrt{s}$ = 38.7 GeV), see [2]. The obtained value is $\hat{q}_0 = 0.075^{+0.015}_{-0.005}\ \text{GeV}^2/\text{fm}$. The p–p production cross section appearing in Eq. (4) is obtained from a fit to p–p measurements.

Quite remarkably, all available J/ψ and Υ suppression measurements from fixed-target experiments (SPS, HERA, FNAL) to RHIC could be described within the above simple model on a broad kinematical range in rapidity y (or x_F) [2].[3] The predictions for J/ψ and Υ suppression in p–Pb collisions at $\sqrt{s}$ = 5.02 TeV [2] are shown in Fig. 3. As can be seen, the model predicts a rather strong J/ψ suppression at forward rapidity (say, $y \gtrsim 3$) and a slight enhancement in the most backward rapidity bins ($y < -4$). The suppression predicted in the Υ channel shares the same features, however the suppression is less pronounced than that of J/ψ, since the (average) coherent energy loss scales as $M_\perp^{-1}$, see Eq. (2). The latter predictions proved to be in excellent agreement with ALICE [17] and LHCb [18] data.

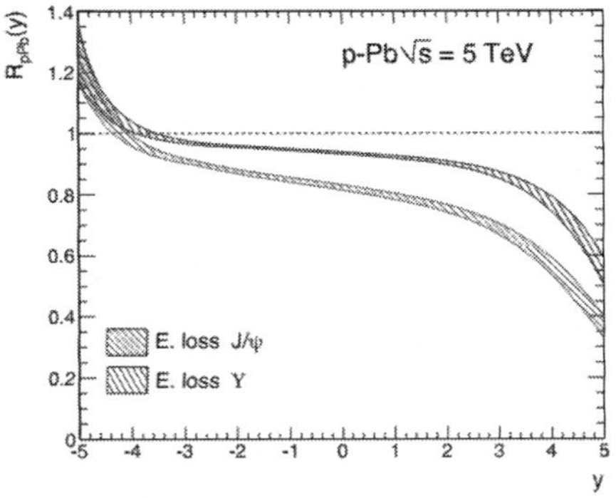

Figure 3. J/ψ and Υ suppression in p–Pb collisions at $\sqrt{s}$ = 5.02 TeV from induced coherent energy loss [2].

[3]A simple extension of the model (4) to double differential cross sections $d\sigma^{\psi}_{pA}/dEd^2\vec{p}_\perp$ similarly yields a good description of the quarkonium suppression data as a function of transverse momentum [9].

3.2 Light-hadron suppression at the LHC

Coherent energy loss can also be applied to light-hadron production in p–A collisions. The picture of light-hadron production we use is as follows [19]. In the target rest frame, a projectile parton i of high energy E_i scatters on the target and produces a dijet made of two partons, j and k, with approximately back-to-back transverse momenta, $|\vec{K}_{j\perp}| \simeq |\vec{K}_{k\perp}| \equiv K_\perp$, and energy fractions $E_j/E_i = x_{\rm h}$ and $E_k/E_i = 1-x_{\rm h}$. The label j is chosen for the parton which then fragments (on a long timescale) into the detected hadron carrying transverse momentum $p_\perp = z\,K_\perp$ and energy $E \simeq z\,x_{\rm h}\,E_i$, where z is the fragmentation variable. For simplicity we shall consider only the partonic process $g \to gg$ (with single gluon exchange in the t-channel), which dominates at the LHC at not too large $p_\perp$.

The dijet may be produced in different color states R. Since in p–A collisions the amount of induced coherent radiation depends on R (via the global charge C_R), it is convenient to first express the p–p production cross section of a hadron with energy E as an incoherent sum over color states,

$$\frac{d\sigma^h_{\rm pp}(E)}{dE} = \sum_R \int dx_{\rm h}\, P_R(x_{\rm h})\, \frac{d\sigma^h_{\rm pp}(E, x_{\rm h})}{dE\, dx_{\rm h}}\,, \tag{6}$$

where we introduced the probability P_R for the dijet to be in color state R (which in general depends on the energy fraction $x_{\rm h}$) and used $\sum_R P_R(x_{\rm h}) = 1$. In Eq. (6), the quantity $d\sigma^h_{\rm pp}(E, x_{\rm h})/dE\, dx_{\rm h}$ is formally the cross section to find a hadron with energy E in a gluon dijet with energy fractions $x_{\rm h}$ and $1 - x_{\rm h}$. For $N_c = 3$, the probability $P_R(x_{\rm h})$ for the $g \to gg$ channel vanishes in the decuplet, $P_{\mathbf{10}\oplus\overline{\mathbf{10}}}(x_{\rm h}) = 0$, and the gg dijet can thus be color singlet, octet or vigintiseptuplet ($R = \mathbf{1}, \mathbf{8} \equiv \mathbf{8_a} \oplus \mathbf{8_s}, \mathbf{27}$) [19].

In p–A collisions, the medium-induced coherent radiation *does not probe* the dijet [7], and thus leaves unchanged the dijet internal structure, implying that the energies of the dijet and its constituents are simply scaled down by a common factor $1 + \hat{\varepsilon}$, where $\hat{\varepsilon} \equiv \varepsilon/E_i$ is the dijet fractional energy loss. Thus, for each dijet color state R, the p–A and p–p hadron production cross sections are related similarly to Eq. (4) in the quarkonium case. Using (6) we get

$$\frac{1}{A}\,\frac{d\sigma^h_{\rm pA}(E)}{dE} = \sum_R \int d\hat{\varepsilon}\, \hat{\mathcal{P}}_R(\hat{\varepsilon}) \int dx_{\rm h}\, P_R(x_{\rm h})\, \frac{d\sigma^h_{\rm pp}(E(1+\hat{\varepsilon}), x_{\rm h})}{dE\, dx_{\rm h}}\,, \tag{7}$$

where the quenching weight $\hat{\mathcal{P}}_R(\hat{\varepsilon})$ is related to the induced coherent radiation spectrum corresponding to the $g \to gg$ process with the final *compact* gluon pair in color state R. The latter spectrum is given by the expression (3), with $C_R \to N_c$, $C_t \to N_c$, $C_{R'} \to C_R$ (where $R = \mathbf{1}, \mathbf{8}$, or $\mathbf{27}$), $M_\perp \to K_\perp$ [7], and $\hat{q}$ still given by (5) but with now $x_2 = p_\perp e^{-y_g}/(x_{\rm h}\sqrt{s})$ the momentum fraction of the target gluon in the $gg \to gg$ subprocess (viewed in the proton–nucleon center-of-mass frame, y_g being the rapidity in this frame of the gluon fragmenting into the hadron).

Dividing (7) by (6) gives the nuclear attenuation factor $R^h_{\rm pA}$. Since $P_R(x_{\rm h})$ is a rather flat function of $x_{\rm h}$ [7, 19], we approximate it as a constant in the integrand of (7), and will eventually vary $x_{\rm h}$ within a realistic range in order to evaluate the associated theoretical uncertainty. Thus,

$$R^h_{\rm pA}(E) \equiv \frac{1}{A}\,\frac{d\sigma^h_{\rm pA}(E)/dE}{d\sigma^h_{\rm pp}(E)/dE} \simeq \sum_R P_R(x_{\rm h})\, R^R(E)\,; \quad R^R(E) \equiv \int d\hat{\varepsilon}\, \hat{\mathcal{P}}_R(\hat{\varepsilon})\, \frac{d\sigma^h_{\rm pp}(E(1+\hat{\varepsilon}))/dE}{d\sigma^h_{\rm pp}(E)/dE}\,, \tag{8}$$

where $R^R(E)$ is the nuclear suppression factor of a hadron produced from a parent dijet in color state R. As anticipated in [7], the hadron suppression factor $R^h(E)$ is the average of the R^R's over the accessible dijet color representations, weighted by the corresponding probabilities $P_R(x_{\rm h})$.

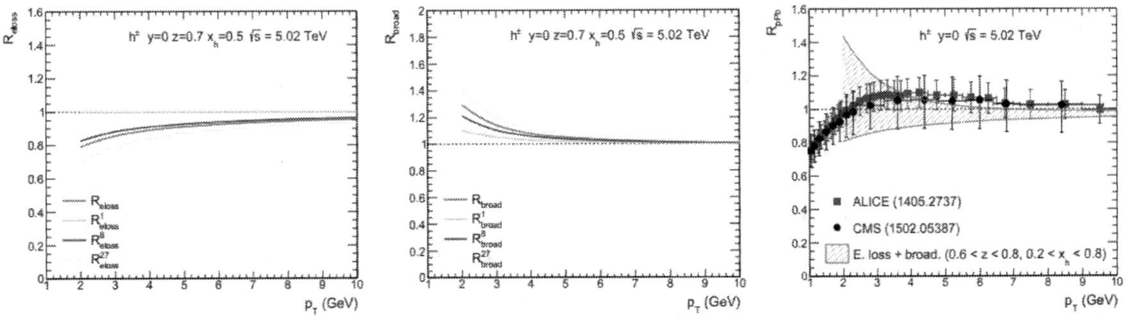

Figure 4. Nuclear attenuation factor of charged hadrons in p–Pb collisions at $\sqrt{s}$ = 5 TeV including the effect of coherent energy loss only (left) and nuclear broadening only (middle) for different dijet color representations, and including both effects (right) as predicted in the model (band) and measured by ALICE [20] and CMS [21].

Instead of the variables E and $\hat{\varepsilon}$, we can equivalently choose the hadron rapidity y (in the center-of-mass frame of the proton–nucleon collision) and the rapidity shift $\delta \equiv \ln(1+\hat{\varepsilon})$. Making the dependence on the hadron transverse momentum $p_\perp$ explicit, we can rewrite (8) as

$$R^h_{\mathrm{pA}}(y,p_\perp) \simeq \sum_R P_R(x_{\mathrm{h}})\, R^R(y,p_\perp)\,; \quad R^R(y,p_\perp) = \int_0^{\delta_{\max}(y)} \mathrm{d}\delta\, \hat{\mathcal{P}}_R(x)\, \frac{\mathrm{d}\sigma^h_{\mathrm{pp}}(y+\delta,p_\perp)/\mathrm{d}y\mathrm{d}p_\perp}{\mathrm{d}\sigma^h_{\mathrm{pp}}(y,p_\perp)/\mathrm{d}y\mathrm{d}p_\perp}\,. \tag{9}$$

Here, $x \equiv e^\delta - 1$ and $\delta_{\max}(y) = \min(\ln 2, y_{\max} - y)$ where $y_{\max} = \ln(\sqrt{s}/p_\perp)$.

The sole effect of parton energy loss is encoded in Eq. (9). In addition to energy loss, $p_\perp$-broadening can be simply implemented by shifting $\vec{p}_\perp$ by the nuclear broadening $\Delta\vec{p}_\perp$ when expressing the double differential p–A cross section in terms of that in p–p, as was done in Ref. [9]. When $|\Delta\vec{p}_\perp| \ll K_\perp$, nuclear broadening cannot resolve the dijet, and is thus transferred to the dijet as a whole. Consequently, all dijet constituents undergo the same *rotation* of 'angle' $\Delta\vec{p}_\perp/E_i$. The fragmenting gluon and the tagged hadron respectively acquire a transverse momentum shift $x_{\mathrm{h}}\,\Delta\vec{p}_\perp$ and $z\,x_{\mathrm{h}}\,\Delta\vec{p}_\perp$. Taking into account the effect of $p_\perp$-broadening modifies (9) to

$$R^h_{\mathrm{pA}}(y,p_\perp) \simeq \sum_R P_R(x_{\mathrm{h}})\, R^R(y,p_\perp)\,; \quad R^R(y,p_\perp) = \int_0^{\delta_{\max}(y)} \mathrm{d}\delta\, \hat{\mathcal{P}}_R(x) \int \frac{\mathrm{d}\varphi}{2\pi}\, \frac{\frac{\mathrm{d}\sigma^h_{\mathrm{pp}}(y+\delta+\delta',\,|\vec{p}_\perp - z\,x_{\mathrm{h}}\,\Delta\vec{p}^{\,R}_\perp|)}{\mathrm{d}y\,\mathrm{d}p_\perp}}{\frac{\mathrm{d}\sigma^h_{\mathrm{pp}}(y,p_\perp)}{\mathrm{d}y\,\mathrm{d}p_\perp}}\,, \tag{10}$$

where $\Delta\vec{p}^{\,R}_\perp$ (of azimutal angle φ) is the broadening of the dijet in color state R, and the rapidity shift δ' induced by the broadening reads $\delta' \equiv \ln(p_\perp/|\vec{p}_\perp - z\,x_{\mathrm{h}}\,\Delta\vec{p}^{\,R}_\perp|)$.[4]

In order to compute R^h_{pPb} for charged hadron production at the LHC, we determine $\mathrm{d}\sigma^h_{\mathrm{pp}}/\mathrm{d}y\,\mathrm{d}p_\perp$ from a fit to 7 TeV p–p data. The fragmentation variable z and the dijet momentum fraction x_{h} are chosen as $z = 0.7$ and $x_{\mathrm{h}} = 0.5$ in average (based on estimates from perturbative calculations) but allowed to vary in the range $0.6 < z < 0.8$ and $0.2 < x_{\mathrm{h}} < 0.8$.

We define the attenuation factors R_{eloss} and R_{broad} obtained from (10) and corresponding respectively to the effect of energy loss only, and of nuclear broadening only. In Fig. 4 (left) we compute R^R_{eloss} in p–Pb collisions at $\sqrt{s}$ = 5 TeV, for the accessible representations R of the parent dijet, as well

[4]The broadening $\Delta\vec{p}^{\,R}_\perp$ is given by, in average, $(\Delta\vec{p}^{\,R}_\perp)^2 = \frac{N_c+C_R}{2N_c}(\hat{q}L - \hat{q}_{\mathrm{p}}L_{\mathrm{p}})$. The color factor arises from the average dijet color charge in the nucleus $(N_c + C_R)/2$ (since the dijet enters the nucleus with charge N_c and exits with charge C_R) and after normalization by the gluon charge N_c (recall that $\hat{q}$ is the *gluon* transport coefficient).

as for the color average $R_{\text{eloss}} = \sum_R P_R(x_h)\, R^R_{\text{eloss}}$, for $z = 0.7$ and $x_h = 0.5$. As can be seen, energy loss leads to a moderate suppression at low $p_\perp$ for $R = \mathbf{8}$ and slightly more pronounced for $R = \mathbf{27}$ due to the larger Casimir factor. (There is no coherent energy loss in the singlet channel $R = \mathbf{1}$.) An opposite trend is reported in Fig. 4 (middle) regarding R^R_{broad}: the higher the color representation R, the stronger the enhancement at small $p_\perp$. Energy loss and nuclear broadening effects thus tend to balance each other. In this respect, the *rapidity dependence* of R^h_{pPb} for charged hadrons (inclusive in $p_\perp \geq p_\perp^{\text{cut}} \sim 1$ GeV) would be interesting, since it is expected to be mostly driven by energy loss. The attenuation factor R^h_{pPb} including both coherent energy loss and nuclear broadening is shown in Fig. 4 (right). Although the theoretical uncertainty, resulting from the variation of $\hat{q}_0$, x_h and z on the one hand and from the shape of the p–p cross section on the other hand, is rather large, the model prediction is compatible with ALICE [20] and CMS [21] data.

More details on the present approach will follow in a forthcoming publication.

4 Outlook

Investigating further coherent energy loss in p–A collisions, in addition to its theoretical interest, should also help disentangling hot from cold nuclear effects in nucleus-nucleus (A–A) collisions. Indeed, the sole effect of coherent energy loss in cold nuclear matter, when extrapolated to A–A collisions, has been shown to yield a significant J/ψ nuclear suppression [22]. This stresses the need to fully understand quarkonium/hadron nuclear suppression in p–A collisions before extracting the additional *hot* effects expected in A–A collisions.

References

[1] F. Arleo, S. Peigné and T. Sami, Phys. Rev. D **83**, 114036 (2011).
[2] F. Arleo and S. Peigné, JHEP **1303**, 122 (2013).
[3] N. Armesto *et al.*, Phys. Lett. B **717**, 280 (2012).
[4] N. Armesto *et al.*, JHEP **1312**, 052 (2013).
[5] F. Arleo, R. Kolevatov and S. Peigné, Phys. Rev. D **93**, 014006 (2016).
[6] T. Liou and A. H. Mueller, Phys. Rev. D **89**, no. 7, 074026 (2014).
[7] S. Peigné and R. Kolevatov, JHEP **1501**, 141 (2015).
[8] F. Arleo and S. Peigné, Phys. Rev. Lett. **109**, 122301 (2012).
[9] F. Arleo, R. Kolevatov, S. Peigné and M. Rustamova, JHEP **1305**, 155 (2013).
[10] S. Gavin and J. Milana, Phys. Rev. Lett. **68**, 1834 (1992).
[11] S. Peigné and A. V. Smilga, Phys. Usp. **52**, 659 (2009) [Usp. Fiz. Nauk **179**, 697 (2009)].
[12] R. Baier *et al.*, Nucl. Phys. B **483**, 291 (1997).
[13] B. G. Zakharov, JETP Lett. **65**, 615 (1997).
[14] B. G. Zakharov, JETP Lett. **73**, 49 (2001) [Pisma Zh. Eksp. Teor. Fiz. **73**, 55 (2001)].
[15] M. Gyulassy, P. Levai and I. Vitev, Nucl. Phys. B **594**, 371 (2001).
[16] M. J. Leitch *et al.* [NuSea Collaboration], Phys. Rev. Lett. **84**, 3256 (2000).
[17] B. B. Abelev *et al.* [ALICE Collaboration], JHEP **1402**, 073 (2014).
[18] R. Aaij *et al.* [LHCb Collaboration], JHEP **1402**, 072 (2014).
[19] F. Arleo, R. Kolevatov and S. Peigné, work in progress.
[20] B. B. Abelev *et al.* [ALICE Collaboration], Eur. Phys. J. C **74**, no. 9, 3054 (2014).
[21] V. Khachatryan *et al.* [CMS Collaboration], Eur. Phys. J. C **75**, no. 5, 237 (2015).
[22] F. Arleo and S. Peigné, JHEP **1410**, 73 (2014).

EPJ Web of Conferences **112**, 04006 (2016)
DOI: 10.1051/epjconf/201611204006

Forward di-jet production in dilute-dense collisions

E. Petreska[1,2,a], P. Kotko[3], K. Kutak[4], C. Marquet[1], S. Sapeta[4,5], and A. van Hameren[4]

[1] *Centre de Physique Théorique, École Polytechnique, CNRS, Université Paris-Saclay, F-91128 Palaiseau, France*
[2] *Departamento de Física de Partículas and IGFAE, Universidade de Santiago de Compostela, 15782 Santiago de Compostela, Spain*
[3] *Department of Physics, Penn State University, University Park, 16803 PA, USA*
[4] *The H. Niewodniczański Institute of Nuclear Physics PAN, Radzikowskiego 152, 31-342 Kraków, Poland*
[5] *CERN PH-TH, CH-1211, Geneva 23, Switzerland*

Abstract. We derive a factorization formula for forward production of two jets in dilute-dense collisions that is valid for an arbitrary value of the momentum imbalance of the jets, k_t. This generalizes the transverse momentum dependent (TMD) factorization formula that has been derived before by Dominguez et al. Their formula is valid only for small values of the transverse momentum of the small-x gluon from the target; it has k_t dependent TMD gluon distributions, but on-shell hard matrix elements. We extend the TMD formula to all ranges of k_t by including off-shell matrix elements. We also add finite N_c corrections. The new formula encompasses both, the TMD factorization for small k_t on the order of the saturation scale, and the High Energy Factorization (HEF) for large k_t on the order of the momentum of the jets. The TMD and HEF factorizations can be derived from the Color Glass Condensate (CGC) formula for forward di-jet production in the appropriate limits. We show explicitly the equivalence of HEF and CGC in the dilute target approximation.

1 Introduction

The production of two jets in the forward rapidity direction of ultra-relativistic proton-proton and proton-nucleus collisions is sensitive to the small-x gluon distributions in the target, and the large-x partons in the projectile. The longitudinal momentum fraction x of the partons is defined with respect to the momentum of the parent hadron. The wave function of the projectile is obtained from perturbative quantum chromodynamics (pQCD), while the gluon distributions in the target are in the non-linear saturation regime. Di-jet production in asymmetric collisions has been studied in different theoretical frameworks, with different ranges of values for the momentum imbalance.

The transverse momentum of the produced jets, P_t, is one of the hardest scales in the problem, while the saturation momentum, Q_s, is one of the softest scales. The third scale in this problem is the momentum imbalance of the jets (or equivalently the transverse momentum of the interacting gluon from the target), k_t. The color glass condensate (CGC) approach [1] does not assume any particular ordering of these scales, and in the most general case there is no k_t factorization formula for the

[a] e-mail: elena.petreska@usc.es

cross section. The transverse momentum dependent (TMD) factorization [2], [3] is valid when the momentum imbalance is close to Q_s, and both are much smaller than the hard momentum of the jets, $k_t \sim Q_s \ll P_t$. The high-energy factorization (HEF) [4] [5] describes the region of large k_t values on the order of P_t, $Q_s \ll k_t \sim P_t$. We unify the different approaches by deriving a formula valid for the whole range of k_t values between Q_s and P_t [6].

In section 2 we derive the HEF formula from a CGC calculation in the limit of a dilute target, by restricting the scattering to two gluon exchanges. In section 3 we revise the TMD factorization formula that was derived in Ref. [3] in the large-N_c limit. We include all finite N_c correction, introduce three new gluon distributions, and reduce the number of independent distributions to two per channel. We improve the finite-N_c TMD factorization in section 4 by restoring the k_t dependence in the matrix elements.

2 Derivation of the high-energy factorization from the color glass condensate cross section in the dilute target limit

The dilute projectile in the HEF formula for di-jet production is represented with a parton distribution function of collinear factorization, $f_{a/p}(x_1, \mu^2)$, the dense target with one k_t-dependent gluon distribution, $\mathcal{F}_{g/A}(x_2, k_t)$, and the hard part of the scattering with off-shell matrix elements, $\overline{\mathcal{M}_{ag^*\to cd}}$ [5], [7]:

$$\frac{d\sigma^{pA\to \text{dijets}+X}}{dy_1 dy_2 d^2p_{1t} d^2p_{2t}} = \frac{1}{16\pi^3 (x_1 x_2 s)^2} \sum_{a,c,d} x_1 f_{a/p}(x_1,\mu^2)\, \overline{|\mathcal{M}_{ag^*\to cd}|^2} \mathcal{F}_{g/A}(x_2,k_t) \frac{1}{1+\delta_{cd}} . \tag{1}$$

In the above expression s is the center of mass energy squared, p_{1t} and p_{2t} are the transverse momenta of the outgoing particles, and y_1 and y_2 are their rapidities. The longitudinal momentum fractions of the parton from the projectile and the gluon from the target are x_1 and x_2, respectively. The matrix elements have been calculated in Refs. [5], [8] and [9]. The HEF formula is an ansatz formula which turns out to be valid for large k_t values, with $\mathcal{F}_{g/A}(x_2, k_t)$ equal to the unintegrated gluon distribution entering in the cross section for deep inelastic scattering (DIS). This distribution is associated with the S-matrix for a quark-anitquark dipole scattering:

$$\mathcal{F}_{g/A}(x_2,k_t) = \frac{N_c}{\alpha_s (2\pi)^3} \int d^2\mathbf{v} d^2\mathbf{v}' e^{-ik_t\cdot(\mathbf{v}-\mathbf{v}')} \nabla^2_{\mathbf{v}-\mathbf{v}'} \left[1 - S^{(2)}_{q\bar{q}}(\mathbf{v},\mathbf{v}')\right] , \tag{2}$$

where $S^{(2)}_{q\bar{q}}(\mathbf{v},\mathbf{v}')$ is a correlator of two fundamental Wilson lines,

$$S^{(2)}_{q\bar{q}}(\mathbf{v},\mathbf{v}') = \frac{1}{N_c} \left\langle \mathrm{Tr}\, U(\mathbf{v}) U^\dagger(\mathbf{v}') \right\rangle . \tag{3}$$

The Wilson lines are path ordered exponentials of the gauge field of the target, A_μ, and here, we define them in the light-cone gauge $A^+ = 0$ of the projectile,

$$U(\mathbf{x}) = \mathcal{P} \exp\left[ig \int dx^+ A_a^-(x^+,\mathbf{x}) t^a \right] . \tag{4}$$

The multi-gluon scatterings of the projectile parton with the dense field of the target, in the CGC cross section, are described with more complicated objects than the dipole $S^{(2)}_{q\bar{q}}$. The cross section for the $qg^* \to qg$ channel, for example, was calculated in Ref. [10], and it was shown that it involves four-point and three-point correlators of fundamental and/or adjoint Wilson lines as well, describing the multiple scatterings of a quark and a gluon in the amplitude and its complex conjugate. The four-point correlator is an expectation value of two fundamental Wilson lines, U, and two adjoint Wilson lines, V, (all at different transverse coordinates):

$$S^{(4)}_{qg\bar{q}g}(\mathbf{b},\mathbf{x},\mathbf{b}',\mathbf{x}) = 1/N_c \left\langle \mathrm{Tr}\left(U(\mathbf{b})U^\dagger(\mathbf{b}')t^d t^c\right)\left[V(\mathbf{x})V^\dagger(\mathbf{x}')\right]^{cd}\right\rangle_{x_2} . \tag{5}$$

The three-point correlator involves two fundamental and one adjoint Wilson line:

$$S^{(3)}_{qg\bar{q}}(\mathbf{b},\mathbf{x},\mathbf{z}') = 1/N_c \left\langle \mathrm{Tr}\left(U^\dagger(\mathbf{z}')t^c U(\mathbf{b})t^d\right) V^{cd}(\mathbf{x})\right\rangle_{x_2} . \tag{6}$$

In the dilute target approximation (a target with two scattering centres), the $S^{(4)}_{qg\bar{q}g}$ and $S^{(3)}_{qg\bar{q}}$ correlators can be written in terms of the dipole $S^{(2)}_{q\bar{q}}$, and they can be related to the gluon distribution $\mathcal{F}_{g/A}(x_2,k_t)$ from the HEF formula. This approximation amounts to expanding the Wilson lines to second order in the background field A_μ. In this approximation $S^{(4)}_{qg\bar{q}g}$ and $S^{(3)}_{qg\bar{q}}$ are [6]:

$$\begin{aligned}
S^{(4)}_{qg\bar{q}g}(\mathbf{b},\mathbf{x},\mathbf{b}',\mathbf{x}') &\simeq S^{(2)}(\mathbf{b},\mathbf{b}') - \frac{C_A}{C_F}\left[1 - S^{(2)}(\mathbf{x},\mathbf{x}')\right] \\
&- \frac{C_A}{2C_F}\left[S^{(2)}(\mathbf{x}',\mathbf{b}) + S^{(2)}(\mathbf{x},\mathbf{b}') - S^{(2)}(\mathbf{x},\mathbf{b}) - S^{(2)}(\mathbf{x}',\mathbf{b}')\right] , \\
S^{(3)}_{qg\bar{q}}(\mathbf{b},\mathbf{x},\mathbf{v}') &\simeq \frac{C_A}{2C_F}\left[S^{(2)}(\mathbf{b},\mathbf{x}) + S^{(2)}(\mathbf{x},\mathbf{v}') - \frac{1}{C_A^2}S^{(2)}(\mathbf{b},\mathbf{v}')\right] - \frac{C_A}{C_F} .
\end{aligned} \tag{7}$$

With the above results we express the CGC di-jet cross section only in terms of the DIS distribution $\mathcal{F}_{g/A}(x_2,k_t)$, and we reproduce the matrix elements $\overline{\mathcal{M}_{ag^*\to cd}}$ from the HEF formula exactly [6]. We derive the HEF formula from the CGC cross section for a dilute target for all three channels: $qg^* \to qg$, $gg^* \to q\bar{q}$ and $gg^* \to gg$.

3 Simplified TMD factorization for finite N_c

A TMD factorization formula for forward di-jet production for small k_t values, $Q_s \ll k_t \sim P_t$, was derived in Ref. [3], in the large-N_c limit. This is the scenario of nearly back-to-back di-jets. The target is described with five k_t dependent gluon distributions, instead of one, while the k_t dependence is not present in the matrix elements (the incoming gluon from the target is put on shell). The different TMD distributions in the cross section represent the resummation of collinear gluons from the target that couple to the hard part. The resummation depends on the color flow in the $2 \to 2$ sub-process, and brings different gauge link structure in the gluon densities for different $2 \to 2$ Feynman diagrams. The gauge links, products of such resummation, turn a generic correlator of gluon field strength tensors

$$\mathcal{F}(x_2,k_t) \overset{\text{naive}}{=} 2\int \frac{d\xi^+ d^2\xi}{(2\pi)^3 p_A^-} e^{ix_2 p_A^- \xi^+ - ik_t\cdot\xi} \left\langle A\middle|\mathrm{Tr}\left[F^{i-}(\xi^+,\xi)\,F^{i-}(0)\right]\middle|A\right\rangle \tag{8}$$

into several (different) gauge invariant TMD gluon distributions that will emerge in the factorized cross section. The TMD's for all types of $2 \to 2$ diagrams have been calculated in Ref. [2].

We revise the derivation of the TMD formula for this process, keeping N_c finite. For each hard factor from Ref. [3] we identify the corresponding gluon distributions from Ref. [2], for all three channels. The finite N_c corrections bring three new TMD distributions compared to the large-N_c case, as well as corrections to the hard parts that were previously omitted. The TMD gluon distributions are:

$$\mathcal{F}_{qg}^{(1)} = \int \frac{2d\xi^+ d^2\xi}{(2\pi)^3 p_A^-} e^{ix_2 p_A^- \xi^+ - ik_t\cdot\xi} \left\langle \mathrm{Tr}\left[F(\xi)\, \mathcal{U}^{[-]\dagger} F(0)\, \mathcal{U}^{[+]}\right]\right\rangle ,$$

$$\mathcal{F}_{qg}^{(2)} = \int \frac{2d\xi^+ d^2\xi}{(2\pi)^3 p_A^-} e^{ix_2 p_A^- \xi^+ - ik_t\cdot\xi} \left\langle \mathrm{Tr}\left[F(\xi)\, \frac{\mathrm{Tr}\left[\mathcal{U}^{[\square]}\right]}{N_c} \mathcal{U}^{[+]\dagger} F(0)\, \mathcal{U}^{[+]}\right]\right\rangle ,$$

$$\mathcal{F}_{gg}^{(1)} = \int \frac{2d\xi^+ d^2\xi}{(2\pi)^3 p_A^-} e^{ix_2 p_A^- \xi^+ - ik_t\cdot\xi} \left\langle \mathrm{Tr}\left[F(\xi)\, \frac{\mathrm{Tr}\left[\mathcal{U}^{[\square]}\right]}{N_c} \mathcal{U}^{[-]\dagger} F(0)\, \mathcal{U}^{[+]}\right]\right\rangle ,$$

$$\mathcal{F}_{gg}^{(2)} = \int \frac{2d\xi^+ d^2\xi}{(2\pi)^3 p_A^-} e^{ix_2 p_A^- \xi^+ - ik_t\cdot\xi} \frac{1}{N_c} \left\langle \mathrm{Tr}\left[F(\xi)\, \mathcal{U}^{[\square]\dagger}\right] \mathrm{Tr}\left[F(0)\, \mathcal{U}^{[\square]}\right]\right\rangle ,$$

$$\mathcal{F}_{gg}^{(3)} = \int \frac{2d\xi^+ d^2\xi}{(2\pi)^3 p_A^-} e^{ix_2 p_A^- \xi^+ - ik_t\cdot\xi} \left\langle \mathrm{Tr}\left[F(\xi)\, \mathcal{U}^{[+]\dagger} F(0)\, \mathcal{U}^{[+]}\right]\right\rangle ,$$

$$\mathcal{F}_{gg}^{(4)} = \int \frac{2d\xi^+ d^2\xi}{(2\pi)^3 p_A^-} e^{ix_2 p_A^- \xi^+ - ik_t\cdot\xi} \left\langle \mathrm{Tr}\left[F(\xi)\, \mathcal{U}^{[-]\dagger} F(0)\, \mathcal{U}^{[-]}\right]\right\rangle ,$$

$$\mathcal{F}_{gg}^{(5)} = \int \frac{2d\xi^+ d^2\xi}{(2\pi)^3 p_A^-} e^{ix_2 p_A^- \xi^+ - ik_t\cdot\xi} \left\langle \mathrm{Tr}\left[F(\xi)\, \mathcal{U}^{[\square]\dagger} \mathcal{U}^{[+]\dagger} F(0)\, \mathcal{U}^{[\square]} \mathcal{U}^{[+]}\right]\right\rangle ,$$

$$\mathcal{F}_{gg}^{(6)} = \int \frac{2d\xi^+ d^2\xi}{(2\pi)^3 p_A^-} e^{ix_2 p_A^- \xi^+ - ik_t\cdot\xi} \left\langle \mathrm{Tr}\left[F(\xi)\, \mathcal{U}^{[+]\dagger} F(0)\, \mathcal{U}^{[+]}\right] \left(\frac{\mathrm{Tr}\left[\mathcal{U}^{[\square]}\right]}{N_c}\right)^2 \right\rangle . \quad (9)$$

The new distributions are $\mathcal{F}_{gg}^{(3-5)}$. The gauge links are defined as $\mathcal{U}^{[\pm]} = U(0, \pm\infty; \mathbf{0}) U(\pm\infty, \xi^+; \xi)$, and $\mathcal{U}^{[\square]} = \mathcal{U}^{[+]}\mathcal{U}^{[-]\dagger} = \mathcal{U}^{[-]}\mathcal{U}^{[+]\dagger}$, where $U(a, b; \mathbf{x}) = \mathcal{P}\exp[ig \int_a^b dx^+ A_a^-(x^+, \mathbf{x}) t^a]$. The gluon distribution $\mathcal{F}_{qg}^{(1)}$ is the *dipole distribution*, and $\mathcal{F}_{gg}^{(3)}$ is the *Weizsäcker-Williams gluon distribution*.

Not all of the matrix elements that correspond to the above distributions are independent. We reduce the number of independent matrix elements and their corresponding distributions to two per channel, and we write a more compact TMD factorization formula for forward di-jet production at finite N_c [6]:

$$\frac{d\sigma^{pA\to \mathrm{dijets}+X}}{d^2P_t d^2k_t dy_1 dy_2} = \frac{\alpha_s^2}{(x_1 x_2 s)^2} \sum_{a,c,d} x_1 f_{a/p}(x_1, \mu^2) \sum_{i=1}^{2} K_{ag\to cd}^{(i)} \Phi_{ag\to cd}^{(i)}(k_t) \frac{1}{1+\delta_{cd}} . \quad (10)$$

The new hard factors, $K_{ag\to cd}^{(i)}$, are given in Table 1. The respective TMD's are linear combinations of $\mathcal{F}_{ag}^{(i)}$. For the $qg \to qg$ channel:

$$\Phi_{qg\to qg}^{(1)} = \mathcal{F}_{qg}^{(1)} \quad \text{and} \quad \Phi_{qg\to qg}^{(2)} = \frac{1}{(N_c^2-1)}\left(-\mathcal{F}_{qg}^{(1)} + N_c^2 \mathcal{F}_{qg}^{(2)}\right). \quad (11)$$

For the $gg \to q\bar{q}$ channel:

$$\Phi_{gg\to q\bar{q}}^{(1)} = \frac{1}{(N_c^2-1)}\left(N_c^2 \mathcal{F}_{gg}^{(1)} - \mathcal{F}_{gg}^{(3)}\right) \quad \text{and} \quad \Phi_{gg\to q\bar{q}}^{(2)} = -N_c^2 \mathcal{F}_{gg}^{(2)} + \mathcal{F}_{gg}^{(3)}. \quad (12)$$

For the $gg \to gg$ channel:

$$\Phi_{gg\to gg}^{(1)} = \frac{1}{2N_c^2}\left(N_c^2 \mathcal{F}_{gg}^{(1)} - 2\mathcal{F}_{gg}^{(3)} + \mathcal{F}_{gg}^{(4)} + \mathcal{F}_{gg}^{(5)} + N_c^2 \mathcal{F}_{gg}^{(6)}\right) \quad \text{and}$$

$$\Phi_{gg\to gg}^{(2)} = \frac{1}{2N_c^2}\left(N_c^2 \mathcal{F}_{gg}^{(2)} - 2\mathcal{F}_{gg}^{(3)} + \mathcal{F}_{gg}^{(4)} + \mathcal{F}_{gg}^{(5)} + N_c^2 \mathcal{F}_{gg}^{(6)}\right). \quad (13)$$

Table 1. The hard factors $K^{(i)}_{ag\to cd}$ for an on-shell gluon accompanying the gluon TMDs $\Phi^{(i)}_{ag\to cd}$. The Mandelstam variables are defined as $\hat{s} = (p_1 + p_2)^2$, $\hat{t} = (p_1 - k)^2$ and $\hat{u} = (p_2 - k)^2$.

	$K^{(1)}_{ag\to cd}$	$K^{(2)}_{ag\to cd}$
1	$-\frac{C_F}{N_c}\frac{\hat{s}(\hat{s}^2+\hat{u}^2)}{\hat{t}^2\hat{u}}$	$-\frac{\hat{s}^2+\hat{u}^2}{2\hat{t}^2\hat{s}\hat{u}}\left[\hat{u}^2+\frac{\hat{s}^2-\hat{t}^2}{N_c^2}\right]$
2	$\frac{1}{2N_c}\frac{(\hat{t}^2+\hat{u}^2)^2}{\hat{s}^2\hat{t}\hat{u}}$	$-\frac{1}{2C_FN_c^2}\frac{\hat{t}^2+\hat{u}^2}{\hat{s}^2}$
3	$\frac{2N_c}{C_F}\frac{(\hat{s}^2-\hat{t}\hat{u})^2(\hat{t}^2+\hat{u}^2)}{\hat{t}^2\hat{u}^2\hat{s}^2}$	$\frac{2N_c}{C_F}\frac{(\hat{s}^2-\hat{t}\hat{u})^2}{\hat{t}\hat{u}\hat{s}^2}$

Table 2. The hard factors $K^{(i)}_{ag^*\to cd}$ for an off-shell gluon accompanying the gluon TMDs $\Phi^{(i)}_{ag\to cd}$. The bared versions of the Mandelstam variables are defined as $\bar{s} = (x_2p_A + p)^2$, $\bar{t} = (x_2p_A - p_1)^2$ and $\bar{u} = (x_2p_A - p_2)^2$. We have also denoted $v = \bar{s}^4 + \bar{t}^4 + \bar{u}^4$ and $w = \bar{u}\hat{u} + \bar{t}\hat{t} - \bar{s}\hat{s}$.

	$K^{(1)}_{ag^*\to cd}$	$K^{(1)}_{ag^*\to cd}$
1	$-\frac{\bar{u}\left(\bar{s}^2+\bar{u}^2\right)}{2\bar{t}\hat{t}\hat{s}}\left(1+\frac{\bar{s}\hat{s}-\bar{t}\hat{t}}{N_c^2\,\bar{u}\hat{u}}\right)$	$-\frac{C_F}{N_c}\frac{\bar{s}\left(\bar{s}^2+\bar{u}^2\right)}{\bar{t}\hat{t}\hat{u}}$
2	$\frac{1}{2N_c}\frac{\left(\bar{t}^2+\bar{u}^2\right)\left(\bar{u}\hat{u}+\bar{t}\hat{t}\right)}{\bar{s}\hat{s}\hat{t}\hat{u}}$	$\frac{\left(\bar{t}^2+\bar{u}^2\right)w}{4N_c^2C_F\bar{s}\hat{s}\hat{t}\hat{u}}$
3	$\frac{N_c}{C_F}\frac{v\left(\bar{u}\hat{u}+\bar{t}\hat{t}\right)}{\bar{t}\hat{t}\bar{u}\hat{u}\bar{s}\hat{s}}$	$-\frac{N_c}{2C_F}\frac{v\,w}{\bar{t}\hat{t}\bar{u}\hat{u}\bar{s}\hat{s}}$

4 Unifying TMD factorization formula

As we mention in the previous section, the matrix elements in the TMD factorization were derived for an on-shell incoming gluon from the target, which limits the applicability of the formula to small values of k_t. We propose a solution to this limitation by including off-shell matrix elements in the factorization formula and by restoring the k_t dependence in the hard part. We calculate the k_t dependent matrix elements with two methods. The high energy factorization approach is a Feynman diagram based calculation, with light-cone gauge for the on-shell gluons, with the gauge vector n set to be equal to the four momentum of the target, $n = p_A$, and with prescribing a longitudinal polarization vector to the off-shell gluon from the target of the form $\epsilon^0_\mu = i\sqrt{2}\,x_2 p_{A\mu}/|k_t|$ [4]. The advantage of this particular prescription is that it generates gauge invariant matrix elements, but the drawback is that the gauge invariance is not clearly manifested for the separate Feynman diagrams. The second method, the helicity method for TMD amplitudes [11] [12], reveals the gauge invariance already on the level of the amplitudes. These are the so-called *color ordered amplitudes* that represent the coefficients of a color decomposition of a generic amplitude (involving an arbitrary number of gluons and/or quarks) into a color part and a kinematic part. The color ordered amplitudes are functions of kinematic arguments only, and are gauge invariant. They give the hard factors, while the color part of the decomposition, after squaring, indicates the corresponding gluon TMDs. In addition, this procedure takes advantage of the relations between the coefficients, removes the dependent hard factors from the beginning, and results in a compact factorization formula without redundancy.

We derive a factorization formula, similar to the one derived in the previous section, Eq. (10), but now with k_t dependent matrix elements $K^{(i)}_{ag^*\to cd}(k_t)$ [6]. We apply both of the methods described above independently. The new formula is:

$$\frac{d\sigma^{pA\to\mathrm{dijets}+X}}{d^2P_t d^2k_t dy_1 dy_2} = \frac{\alpha_s^2}{(x_1x_2s)^2}\sum_{a,c,d} x_1 f_{a/p}(x_1,\mu^2)\sum_{i=1}^{2} K^{(i)}_{ag^*\to cd}(k_t)\Phi^{(i)}_{ag\to cd}(k_t)\,\frac{1}{1+\delta_{cd}}\,, \tag{14}$$

with the matrix elements given in Table 2. The above formula unifies the HEF approach to forward dijet production with the TMD factorization by establishing a framework applicable for hard jets, $P_t \gg Q_s$, but arbitrary k_t, and it is the main result of this work.

For phenomenological applications to di-jet azimuthal correlations in high-energy collisions, we study the new framework at large N_c, and with analytical models for the gluon distributions. We calculate the gluon densities that survive the large N_c limit in the Golec-Biernat-Wusthoff model [13]:

$$\begin{aligned}
\mathcal{F}^{(1)}_{qg}(x_2,k_t) &= \frac{N_cS_\perp}{2\pi^3\alpha_s}\frac{S_\perp}{Q_s^2(x_2)}k_t^2\exp\left[-\frac{k_t^2}{Q_s^2(x_2)}\right],\\
\mathcal{F}^{(2)}_{qg}(x_2,k_t) &= \frac{N_cS_\perp}{2\pi^3\alpha_s}\exp\left[-\frac{k_t^2}{Q_s^2(x_2)}\right]\int_1^\infty\frac{dt}{t(t+2)}\exp\left[\frac{2k_t^2}{(t+2)Q_s^2(x_2)}\right],\\
\mathcal{F}^{(1)}_{gg}(x_2,k_t) &= \frac{N_cS_\perp}{16\pi^3\alpha_s}\exp\left[-\frac{k_t^2}{2Q_s^2(x_2)}\right]\left(2+\frac{k_t^2}{Q_s^2(x_2)}\right),\\
\mathcal{F}^{(2)}_{gg}(x_2,k_t) &= \frac{N_cS_\perp}{16\pi^3\alpha_s}\exp\left[-\frac{k_t^2}{2Q_s^2(x_2)}\right]\left(2-\frac{k_t^2}{Q_s^2(x_2)}\right),\\
\mathcal{F}^{(6)}_{gg}(x_2,k_t) &= \frac{N_cS_\perp}{4\pi^3\alpha_s}\exp\left[-\frac{k_t^2}{2Q_s^2(x_2)}\right]\int_1^\infty\frac{dt}{t(t+1)}\exp\left[\frac{k_t^2}{2(t+1)Q_s^2(x_2)}\right].
\end{aligned} \tag{15}$$

In the above expressions $S_\perp$ is the transverse area of the target. We also obtain their perturbative behaviour at large k_t in the McLerran-Venugopalan model [14]. We derive the leading order term in Q_s^2/k_t^2, and we find that all of them scale as

$$\simeq \frac{N_cS_\perp Q_s^2(x_2)}{4\pi^3\alpha_s k_t^2} + O\left(\frac{Q_s^4(x_2)}{k_t^4}\log\frac{k_t^2}{\Lambda^2}\right)\,, \tag{16}$$

except $\mathcal{F}^{(2)}_{gg}$ that goes to zero. The above expressions for the densities, as well as numerical results that implement small-x evolution, will be used for phenomenological application of the unifying formula in a forthcoming publication.

Acknowledgments

EP thanks the organizers of the Physics Opportunities at an Electron-Ion Collider 2015 conference for the invitation for this talk.

References

[1] F. Gelis, E. Iancu, J. Jalilian-Marian and R. Venugopalan, Ann. Rev. Nucl. Part. Sci. **60** (2010) 463.

[2] C. J. Bomhof, P. J. Mulders and F. Pijlman, Eur. Phys. J. C **47** (2006) 147.
[3] F. Dominguez, C. Marquet, B. -W. Xiao and F. Yuan, Phys. Rev. D **83** (2011) 105005.
[4] S. Catani, M. Ciafaloni and F. Hautmann, Nucl. Phys. B **366** (1991) 135.
[5] M. Deak, F. Hautmann, H. Jung and K. Kutak, JHEP **0909** (2009) 121.
[6] P. Kotko, K. Kutak, C. Marquet, E. Petreska, S. Sapeta and A. van Hameren, JHEP **1509** (2015) 106
[7] K. Kutak and S. Sapeta, Phys. Rev. D **86** (2012) 094043.
[8] A. van Hameren, P. Kotko and K. Kutak, JHEP **1212** (2012) 029.
[9] A. van Hameren, P. Kotko and K. Kutak, JHEP **1301** (2013) 078.
[10] C. Marquet, Nucl. Phys. A **796** (2007) 41.
[11] M. L. Mangano and S. J. Parke, Phys. Rept. **200** (1991) 301.
[12] L. J. Dixon, arXiv:1310.5353 [hep-ph].
[13] K. J. Golec-Biernat and M. Wusthoff, Phys. Rev. D **59**, (1998) 014017
[14] L. D. McLerran and R. Venugopalan, Phys. Rev. D **49**, 2233 (1994), Phys. Rev. D **49**, 3352 (1994);

EPJ Web of Conferences **112**, 04007 (2016)
DOI: 10.1051/epjconf/201611204007

Thermalization of mini-jets in a quark-gluon plasma

Edmond Iancu[1,a] and Bin Wu[1,2,b]

[1] *Institut de Physique Théorique, CEA Saclay, CNRS UMR 3681, F-91191 Gif-sur-Yvette, France*
[2] *Department of Physics, The Ohio State University, Columbus , OH 43210, USA*

Abstract. We present the complete physical picture for the evolution of a high-energy jet propagating through a weakly-coupled quark-gluon plasma (QGP) by analytical and numerical investigation of thermalization of the soft components of the jet. Our results support the following physical picture: the leading particle emits a significant number of mini-jets which promptly evolve via multiple branching and thus degrade into a myriad of soft gluons, with energies of the order of the medium temperature T. Via elastic collisions with the medium constituents, these soft gluons relax to local thermal equilibrium with the plasma over a time scale which is considerably shorter than the typical lifetime of the mini-jet. The thermalized gluons form a tail which lags behind the hard components of the jet. Together with the background QGP, they behave hydrodynamically.

1 Introduction

A striking imbalance in dijet transverse momentum has been observed in central lead-lead collisions at the LHC[1, 2]. Detailed data analysis indicates that the momentum imbalance is accompanied by a softening of the fragmentation pattern of the second most energetic away-side jet[2]. One natural theoretical interpretation of such a so-called jet quenching phenomenon is that the away-side jet has enough time to degrade into soft gluons due to medium-induced multiple branching. The emergent new picture for the in-medium jet evolution is as follows: the energy of the hard components of the jet is efficiently transmitted, via multiple, quasi-democratic, branchings, to a multitude of comparatively soft gluons[3–8], which relax to local thermal equilibrium with the plasma over a time scale considerably shorter than the branching time of the jet[9]. In particular, Ref. [9] presented for the first time the picture of this evolution in longitudinal phase-space, with the longitudinal axis defined as the direction of propagation of the leading particle. This picture will be succinctly summarized in what follows.

2 Kinetic theory for jet evolution

We study the parton distribution produced by a high-energy jet propagating through a weakly-coupled quark-gluon plasma in thermal equilibrium at temperature T. The leading particle (LP) which initiates

[a] e-mail: edmond.iancu@cea.fr
[b] e-mail: bin.wu.phys@gmail.com

the jet has a high energy $E \gg T$ and a comparatively small virtuality. We concentrate on the *medium-induced* evolution, as triggered by the collisions between the partons from the jet and the constituents of the medium. One can distinguish between two types of collisions:

(i) elastic, 2→2, collisions, which entail energy and momentum transfer between the jet and the medium;

(ii) inelastic collisions, like 2 → 3 or, more generally, 1+(many)→ 2+(many), in which a parton from the jet undergoes a 1 → 2 branching.

At weak coupling, such processes can be described by a kinetic equation for the gluon distribution [3, 10],

$$\left(\frac{\partial}{\partial t} + \boldsymbol{v} \cdot \nabla_{\boldsymbol{x}}\right) f(t, \boldsymbol{x}, \boldsymbol{p}) = C_{\rm el}[f] + C_{\rm br}[f]. \tag{1}$$

Here, $f(t, \boldsymbol{x}, \boldsymbol{p})$ is the gluon occupation number, $\boldsymbol{v} = \boldsymbol{p}/p$ with $p \equiv |\boldsymbol{p}|$ is the gluon velocity. In diffusion approximation (see [11, 12] and references therein), the elastic collision term takes the form:

$$C_{\rm el}[f] \simeq \frac{1}{4}\hat{q}\, \nabla_{\boldsymbol{p}} \cdot \left[\left(\nabla_{\boldsymbol{p}} + \frac{\boldsymbol{v}}{T}\right) f\right], \tag{2}$$

with $\hat{q} \sim \alpha_s^2 T^3 \ln(1/\alpha_s)$ the jet quenching parameter. One can easily check that Eq. (2) admits the Maxwell-Boltzmann thermal distribution $f_{\rm eq} \propto {\rm e}^{-p/T}$ as a fixed point, which shall be reached for a gluon with $p \sim T$ within a time interval of order

$$t_{\rm rel} \equiv \frac{4T^2}{\hat{q}} \sim \frac{1}{\alpha_s^2 T \ln(1/\alpha_s)}. \tag{3}$$

The inelastic collision integral $C_{\rm br}[f]$ including only medium-induced gluon branching in the LPM regime is of the form[3]

$$C_{\rm br}[f] \simeq \frac{1}{t_{\rm br}(p)} \int_0^1 {\rm d}x\, \mathcal{K}(x) \left[\frac{1}{x^{\frac{5}{2}}} f\left(t, \boldsymbol{x}, \frac{\boldsymbol{p}}{x}\right) - \frac{1}{2} f(t, \boldsymbol{x}, \boldsymbol{p})\right] \tag{4}$$

with $t_{\rm br}(p) = \frac{1}{\alpha_s}\sqrt{\frac{p}{\hat{q}}}$ and the splitting kernel[13, 14]

$$\mathcal{K}(x) \equiv \frac{[1 - x(1-x)]^{\frac{5}{2}}}{[x(1-x)]^{\frac{3}{2}}} \qquad \text{with} \qquad 0 \le x \le 1. \tag{5}$$

During most stages of the branching process, the cascade is built with relatively hard gluons, which are nearly collinear with the LP: $p_z \gg p_\perp$. It therefore makes sense to focus on the *longitudinal* dynamics, as obtained after integrating out the transverse phase-space. This motivates the following, relatively simple, kinetic equation, for the *longitudinal gluon distribution* $f_\ell(t, z, p_z) \equiv \int {\rm d}^2\boldsymbol{x}_\perp {\rm d}^2\boldsymbol{p}_\perp\, f(t, \boldsymbol{x}, \boldsymbol{p})$ [9] :

$$\begin{aligned}(\partial_t + v\partial_z) f_\ell(t, z, p) &= \frac{\hat{q}}{4}\partial_p \left[\left(\partial_p + \frac{v}{T}\right) f_\ell(t, z, p)\right] \\ &+ \frac{1}{t_{\rm br}(p)} \int_r {\rm d}x\, \mathcal{K}(x) \left[\frac{1}{\sqrt{x}} f_\ell\left(t, z, \frac{p}{x}\right) - \frac{1}{2} f_\ell(t, z, p)\right].\end{aligned} \tag{6}$$

Here, $p \equiv p_z$, $v \equiv p/|p|$ and the subscript r on the integral over x means that the branching process is cut off at the soft scale $p = T$. The two terms within the inelastic collision integral are recognized as the *gain* term and *loss* term, respectively. The initial condition reads

$$f_\ell(t = 0, z, p) = \delta(p - E)\delta(z), \tag{7}$$

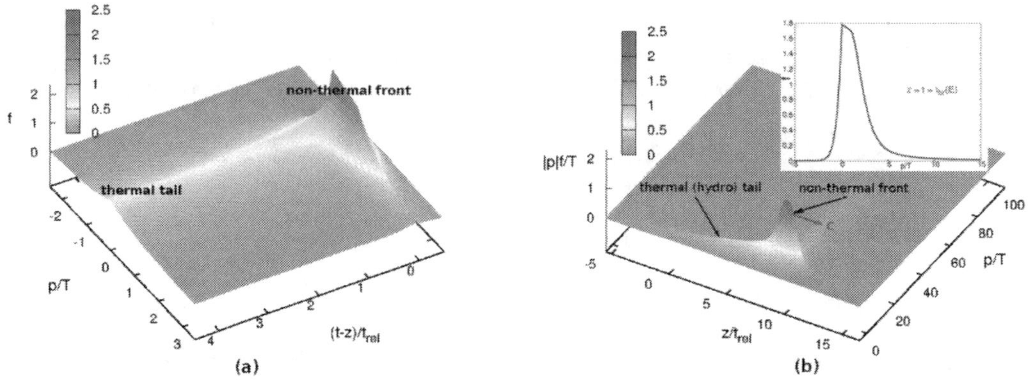

Figure 1. Charateristic "front+tail" struction of the gluon distribution: (a) the analytical solution with the steady source (8) and $p_0 = T$; (b) the numerical solution to Eq. (6) with $E = 90\,T$ and $t = t_{\rm br} = 9.5\,t_{\rm rel}$.

corresponding to a LP with longitudinal momentum $p = E$ which enters the medium at $t = 0$ and $z = 0$. The subscript l of f_l shall be omitted in the following discussions.

3 The complete physical picture of jet evolution

A high-energy jet typically has the branching time $t_{\rm br} \gg t_{\rm rel}$. And the energy flux carried away by the radiated gluons is a slowly varying function of time within a time interval $\Delta t \sim t_{\rm rel}$. Motivated by such an observation, we first consider a simpler problem, where the inelastic collision term in the r.h.s. of (6) is replaced by a steady source for particles with longitudinal momentum p_0 which propagate at the speed of light:

$$C_{\rm br}[f_\ell] \;\rightarrow\; T\delta(p-p_0)\delta(z-t) \qquad \text{with} \qquad p_0 \sim T. \tag{8}$$

We find that analytical solutions exist in this case[9], which is illustrated in Fig. 1 (a). In this figure, one sees a two component structure, with a front and a tail. The *front* is made with gluons travelling at the speed of light. These gluons are injected by the source at $z \gtrsim t - t_{\rm rel}$ before t and do not have enough time to establish local thermalization. The *tail* lies behind the front. At $z \lesssim t - t_{\rm rel}$, it is made of gluons which have relaxed to local thermal equilibrium with the QGP. This picture can be justified by the Green function for this problem, which takes the form[9]

$$f_G \simeq \frac{e^{-|p|/T}}{2}\,\frac{e^{-\frac{(z-p_0t_{\rm rel}/T)^2}{4t}}}{2\sqrt{\pi t/t_{\rm rel}}} \qquad \text{at } t \gtrsim \frac{t_{rel}p_0}{T} \text{ and } t_{\rm rel}. \tag{9}$$

It also tells us that these gluons, together with the background QGP, behave hydrodynamically with the (longitudinal) size of disturbance $\sqrt{\langle (z - \frac{t_{rel}p_0}{T})^2\rangle} \propto \sqrt{t}$. Numerical solutions to (6) also show the "front+tail" structure (See Fig. 1 (b) for example). Therefore, the analytical solution in this simplified problem captures the generic feature of jet evolution before $t \sim t_{\rm br}$. Our calculations support that the motion of these gluons in the tail can be described by hydrodynamics[15].

We now turn to the solutions to the general equation (6) with the initial condition (7). The time scales inherent in this equation are the branching time $t_{\rm br}(E)$ for the LP and the relaxation time $t_{\rm rel}$. We shall take $E = 90\,T$ for example to discuss the features of jet evolution. At $t \gtrsim 1.5$, the jet is found to be fully quenched. The LP disappears completely and almost all the gluons resulted from multiple branching of the jet establish local thermal equilibrium with the plasma. At the LHC, one has $L < t_{\rm br}(E)$, as already mentioned, hence the LP is expected to survive in the final state. This is indeed visible in the numerical results displayed in Fig. 2, as numerically obtained for $E = 90\,T$ and $t_{\rm br}(E) = 9.5\,t_{\rm rel}$ [9].

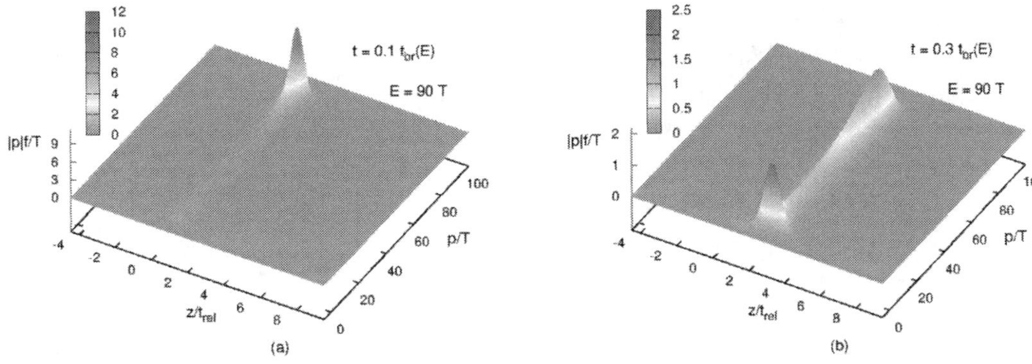

Figure 2. The phase-space energy distribution $|p|f_\ell/T$ produced by an energetic jet with $E = 90\,T$ at two successive times: (a) an early time $t = 0.1t_{\rm br}(E)$, when the jet is almost unquenched; (b) a larger time $t = 0.3t_{\rm br}(E)$, when the jet is partially quenched.

Fig. 2 (a) shows the distribution at the very early time $t = 0.1t_{\rm br}(E)$, when most of the energy is still carried by the LP. Hence the energy distribution $|p|f_\ell/T$ shows a pronounced peak at $p/T = 90$ and at $z = t$. Yet, this peak shows some spreading in p, as a consequence of early emissions, which are necessarily soft: the typical quanta emitted up to time t have $p \lesssim \omega_{\rm br}(t) = \alpha_s^2\hat{q}t^2$.

At the larger time $t = 0.3t_{\rm br}(E)$, cf. Fig. 2 (b), the softening of the distribution in p is clearly visible, albeit a pronounced LP peak still exists. One can now distinguish the characteristic "front+tail" structure anticipated in Fig. 1. The front at $z = t$ involves relatively hard gluons, whose momentum distribution (within the range $T < p \ll E$) is given by the scaling spectrum $f_\ell \propto 1/p^{3/2}$, as expected for the LPM spectrum [3, 5, 13, 14]. Similar results have been obtained in a kinetic theory study of the thermalization of the background quark-gluon plasma [16].

The front in Fig. 2 (b) also shows a secondary peak at $p = T$, due to the accumulation of gluons at the lower end of the cascade. Such gluons are abundantly produced via branchings and they cannot thermalize instantaneously — rather, they need a time $\sim t_{\rm rel}$ to that aim. Yet, since $t = 0.3t_{\rm br}(E) \simeq 3t_{\rm rel}$ is relatively large compared to $t_{\rm rel}$, a thermalized tail at $z \lesssim t - t_{\rm rel}$ develops indeed, as visible too in Fig. 2.b. This tail carries the energy lost by the jet towards the medium.

These results allow for a qualitative comparison with the phenomenology of di-jet asymmetry at the LHC [1, 2]. (For a more quantitative comparison, one could take $T = 0.5\,$GeV and $t_{\rm rel} = 1\,$fm.) The early situation in Fig. 2 (a), where the jet is essentially unquenched, is illustrative for the leading jet, which crosses at most a very narrow slab of matter. The situation in Fig. 2 (b), where the jet looks partially quenched, is representative for the subleading jet in a di-jet event characterized by a large asymmetry. For even larger times, $t \gtrsim t_{br}(E)$, both the LP and the front would disappear and the whole energy would be found in the thermalized tail [9].

References

[1] G. Aad, et al., Phys. Rev. Lett. 105 (2010) 252303.
[2] S. Chatrchyan, et al., Phys. Rev. C84 (2011) 024906.
[3] R. Baier, A. H. Mueller, D. Schiff, D. Son, Phys.Lett. B502 (2001) 51.
[4] J.-P. Blaizot, F. Dominguez, E. Iancu, Y. Mehtar-Tani, JHEP 1301 (2013) 143; JHEP 1406 (2014) 075.
[5] J.-P. Blaizot, E. Iancu, Y. Mehtar-Tani, Phys.Rev.Lett. 111 (2013) 052001.
[6] L. Fister, E. Iancu, JHEP 03 (2015) 082.
[7] A. Kurkela, U. A. Wiedemann, Phys.Lett. B740 (2015) 172.
[8] L. Apolinário, N. Armesto, J. G. Milhano, C. A. Salgado, JHEP 02 (2015) 119.
[9] E. Iancu, B. Wu, JHEP **1510**, 155 (2015).
[10] P. B. Arnold, G. D. Moore, L. G. Yaffe, JHEP 0301 (2003) 030.
[11] J. Hong and D. Teaney, Phys. Rev. C **82**, 044908 (2010).
[12] J. P. Blaizot, B. Wu and L. Yan, Nucl. Phys. A **930**, 139 (2014).
[13] R. Baier, Y. L. Dokshitzer, A. H. Mueller, S. Peigne, D. Schiff, Nucl. Phys. B483 (1997) 291–320.
[14] B. G. Zakharov, JETP Lett. 63 (1996) 952.
[15] Y. He, T. Luo, X. N. Wang and Y. Zhu, Phys. Rev. C **91**, 054908 (2015)
[16] A. Kurkela and E. Lu, Phys.Rev.Lett. 113 (2014) 182301.

AUTHOR INDEX

AUTHOR INDEX